The New Geographies of Energy

The New Geographies of Energy: Assessment and Analysis of Critical Landscapes is a pioneering collection of new geographic scholarship. It examines such vitally important research topics as energy dilemmas of the United States, large trends and patterns of energy consumption including China's role, "peak oil", energy poverty, and ethanol and other renewable energy sourcing.

The book offers advances in key emerging areas of energy research, each distinguished in the following sections: (i) geographic approaches to energy modeling and assessment; (ii) fossil fuel landscapes; (iii) the landscapes of renewable energy; (iv) landscapes of energy consumption; and (v) an overview of the new geographies of energy (Karl Zimmerer, *Annals* Nature-Society and Energy issue editor) and an essay on America's oil dependency (Vaclav Smil, renowned energy geographer). In addition there is a specially commissioned book review.

This book was published as a special issue of the *Annals of the Association of American Geographers*.

Karl Zimmerer is professor and head of the department of geography at Pennsylvania State University and a member of the Earth and Environmental Systems Institute. Currently Dr. Zimmerer's research is focused on three areas: landscape-based social-ecological models of energy, water resources, and food production (*Land Use Policy*, 2012; *Global Environmental Change-Human and Policy Dimensions*, 2011; *Annals of the Association of American Geographers*, 2010, 2011; *Roots of Conflict*, 2010; *Knowing Nature, Transforming Ecologies*, 2010); agrobiodiversity and global change (*Biodiversity in Agriculture*, 2012; *Annual Review of Environment and Resources*, 2010; *Professional Geographer*, 2010); and the conservation-agriculture interface (*Latin American Research Review*, 2011; *Mapping Latin America*, 2011).

The New Geographies of Energy

Assessment and Analysis of Critical Landscapes

Edited by
Karl Zimmerer

Routledge
Taylor & Francis Group
LONDON AND NEW YORK

First published 2013
by Routledge
2 Park Square, Milton Park, Abingdon, Oxfordshire OX14 4RN

Simultaneously published in the USA and Canada
by Routledge
711 Third Avenue, New York, NY 10017

First issued in paperback 2014

Routledge is an imprint of the Taylor & Francis Group, an informa business

This book is a reproduction of *Annals of the Association of American Geographers*, vol. 101, issue 4. The Publisher requests to those authors who may be citing this book to state, also, the bibliographical details of the special issue on which the book was based.

British Library Cataloguing in Publication Data
A catalogue record for this book is available from the British Library

ISBN 978-0-415-62387-2 (hbk)
ISBN 978-1-138-81037-2 (pbk)

Typeset in Goudy
by Taylor & Francis Books

Publisher's Note
The publisher would like to make readers aware that the chapters in this book may be referred to as articles as they are identical to the articles published in the special issue. The publisher accepts responsibility for any inconsistencies that may have arisen in the course of preparing this volume for print.

Contents

Citation Information vii
Dedication xi

1. Approaching the New Geographies of Global Energy: Analytics and Assessment of Current Energy Landscapes and Alternatives
Karl Zimmerer 1

Invited Essay

2. America's Oil Imports: A Self-Inflicted Burden
Vaclav Smil 11

Energy Modeling and Assessment

3. Modeling and Assessment of Wind and Insolation Resources with a Focus on Their Complementary Nature: A Case Study of Oklahoma
Weiping Li, Steve Stadler, and Rama Ramakumar 16

4. "Papering" Over Space and Place: Product Carbon Footprint Modeling in the Global Paper Industry
Joshua P. Newell and Robert O. Vos 29

5. Phenology-Based Assessment of Perennial Energy Crops in North American Tallgrass Prairie
Cuizhen Wang, Felix B. Fritschi, Gary Stacey, and ZhengWei Yang 41

6. A Geographic Approach to Sectoral Carbon Inventory: Examining the Balance Between Consumption-Based Emissions and Land-Use Carbon Sequestration in Florida
Tingting Zhao, Mark W. Horner, and John Sulik 51

7. Toward an Integrated GIScience and Energy Research Agenda
Mark W. Horner, Tingting Zhao, and Timothy S. Chapin 65

8. The Role of Climate Change Litigation in Establishing the Scale of Energy Regulation
Hari M. Osofsky 74

Fossil Fuel Landscapes

9. Energy and Identity: Imagining Russia as a Hydrocarbon Superpower
Stefan Bouzarovski and Mark Bassin 82

10. The Changing Structure of Energy Supply, Demand, and CO2 Emissions in China
Michael Kuby, Canfei He, Barbara Trapido-Lurie, and Nicholas Moore 94

11. Mountaintop Removal and Job Creation: Exploring the Relationship Using Spatial Regression
Brad R. Woods and Jason S. Gordon 105

12. Enforcing Scarcity: Oil, Violence, and the Making of the Market
Matthew T. Huber 115

Landscapes of Renewable Energy

13. Constructing Sustainable Biofuels: Governance of the Emerging Biofuel Economy
Robert Bailis and Jennifer Baka 126

14. Social Perspectives on Wind-Power Development in West Texas
Christian Brannstrom, Wendy Jepson, and Nicole Persons 138

15. Farmer Attitudes Toward Production of Perennial Energy Grasses in East Central Illinois: Implications for Community-Based Decision Making
Miriam A. Cope, Sara McLafferty, and Bruce L. Rhoads 151

16. Renewable Energy and Human Rights Violations: Illustrative Cases from Indigenous Territories in Panama
Mary Finley-Brook and Curtis Thomas 162

17. Downstream Effects of a Hybrid Forum: The Case of the Site C Hydroelectric Dam in British Columbia, Canada
Nichole Dusyk 172

18. A Study of the Emerging Renewable Energy Sector Within Iowa
Peter Kedron and Sharmistha Bagchi-Sen 181

19. A Regional Evaluation of Potential Bioenergy Production Pathways in Eastern Ontario, Canada
Warren E. Mabee and Jaconette Mirck 196

20. Opposing Wind Energy Landscapes: A Search for Common Cause
Martin J. Pasqualetti 206

21. Burning for Sustainability: Biomass Energy, International Migration, and the Move to Cleaner Fuels and Cookstoves in Guatemala
Matthew J. Taylor, Michelle J. Moran-Taylor, Edwin J. Castellanos, and Silvel Elías 217

22. The Impact of Brazilian Biofuel Production on Amazônia
Robert Walker 228

Landscapes of Energy Consumption

23. Shifting Networks of Power in Nicaragua: Relational Materialisms in the Consumption of Privatized Electricity
Julie Cupples 238

24. "Because You Got to Have Heat": The Networked Assemblage of Energy Poverty in Eastern North Carolina
Conor Harrison and Jeff Popke 248

25. Powering "Progress": Regulation and the Development of Michigan's Electricity Landscape
Jordan P. Howell 261

Book Review Essay

26. The Geography of Energy and the Wealth of the World
Martin J. Pasqualetti 270

Index 280

Citation Information

The following chapters were originally published in the *Annals of the Association of American Geographers*, volume 101, issue 4 (July 2011). When citing this material, please use the original page numbering for each article, as follows:

Chapter 2
America's Oil Imports: A Self-Inflicted Burden
Vaclav Smil
Annals of the Association of American Geographers, volume 101, issue 4 (July 2011) pp. 712-716

Chapter 3
Modeling and Assessment of Wind and Insolation Resources with a Focus on Their Complementary Nature: A Case Study of Oklahoma
Weiping Li, *Steve Stadler*, and *Rama Ramakumar*
Annals of the Association of American Geographers, volume 101, issue 4 (July 2011) pp. 717-729

Chapter 4
"Papering" Over Space and Place: Product Carbon Footprint Modeling in the Global Paper Industry
Joshua P. Newell and *Robert O. Vos*
Annals of the Association of American Geographers, volume 101, issue 4 (July 2011) pp. 730-741

Chapter 5
Phenology-Based Assessment of Perennial Energy Crops in North American Tallgrass Prairie
Cuizhen Wang, *Felix B. Fritschi*, *Gary Stacey*, and *ZhengWei Yang*
Annals of the Association of American Geographers, volume 101, issue 4 (July 2011) pp. 742-751

Chapter 6
A Geographic Approach to Sectoral Carbon Inventory: Examining the Balance Between Consumption-Based Emissions and Land-Use Carbon Sequestration in Florida
Tingting Zhao, *Mark W. Horner*, and *John Sulik*
Annals of the Association of American Geographers, volume 101, issue 4 (July 2011) pp. 752-763

Chapter 7
Toward an Integrated GIScience and Energy Research Agenda
Mark W. Horner, *Tingting Zhao*, and *Timothy S. Chapin*
Annals of the Association of American Geographers, volume 101, issue 4 (July 2011) pp. 764-774

Chapter 8
The Role of Climate Change Litigation in Establishing the Scale of Energy Regulation
Hari M. Osofsky
Annals of the Association of American Geographers, volume 101, issue 4 (July 2011) pp. 775-782

Chapter 9
Energy and Identity: Imagining Russia as a Hydrocarbon Superpower

Stefan Bouzarovski and *Mark Bassin*
Annals of the Association of American Geographers, volume 101, issue 4 (July 2011) pp. 783-794

Chapter 10
The Changing Structure of Energy Supply, Demand, and CO2 Emissions in China
Michael Kuby, *Canfei He*, *Barbara Trapido-Lurie*, and *Nicholas Moore*
Annals of the Association of American Geographers, volume 101, issue 4 (July 2011) pp. 795-805

Chapter 11
Mountaintop Removal and Job Creation: Exploring the Relationship Using Spatial Regression
Brad R. Woods and *Jason S. Gordon*
Annals of the Association of American Geographers, volume 101, issue 4 (July 2011) pp. 806-815

Chapter 12
Enforcing Scarcity: Oil, Violence, and the Making of the Market
Matthew T. Huber
Annals of the Association of American Geographers, volume 101, issue 4 (July 2011) pp. 816-826

Chapter 13
Constructing Sustainable Biofuels: Governance of the Emerging Biofuel Economy
Robert Bailis and *Jennifer Baka*
Annals of the Association of American Geographers, volume 101, issue 4 (July 2011) pp. 827-838

Chapter 14
Social Perspectives on Wind-Power Development in West Texas
Christian Brannstrom, *Wendy Jepson*, and *Nicole Persons*
Annals of the Association of American Geographers, volume 101, issue 4 (July 2011) pp. 839-851

Chapter 15
Farmer Attitudes Toward Production of Perennial Energy Grasses in East Central Illinois: Implications for Community-Based Decision Making
Miriam A. Cope, *Sara McLafferty*, and *Bruce L. Rhoads*
Annals of the Association of American Geographers, volume 101, issue 4 (July 2011) pp. 852-862

Chapter 16
Renewable Energy and Human Rights Violations: Illustrative Cases from Indigenous Territories in Panama
Mary Finley-Brook and *Curtis Thomas*
Annals of the Association of American Geographers, volume 101, issue 4 (July 2011) pp. 863-872

Chapter 17
Downstream Effects of a Hybrid Forum: The Case of the Site C Hydroelectric Dam in British Columbia, Canada
Nichole Dusyk
Annals of the Association of American Geographers, volume 101, issue 4 (July 2011) pp. 873-881

Chapter 18
A Study of the Emerging Renewable Energy Sector Within Iowa
Peter Kedron and *Sharmistha Bagchi-Sen*
Annals of the Association of American Geographers, volume 101, issue 4 (July 2011) pp. 882-896

Chapter 19
A Regional Evaluation of Potential Bioenergy Production Pathways in Eastern Ontario, Canada
Warren E. Mabee and *Jaconette Mirck*
Annals of the Association of American Geographers, volume 101, issue 4 (July 2011) pp. 897-906

Chapter 20
Opposing Wind Energy Landscapes: A Search for Common Cause
Martin J. Pasqualetti
Annals of the Association of American Geographers, volume 101, issue 4 (July 2011) pp. 907-917

Chapter 21
Burning for Sustainability: Biomass Energy, International Migration, and the Move to Cleaner Fuels and Cookstoves in Guatemala
Matthew J. Taylor, *Michelle J. Moran-Taylor*, *Edwin J. Castellanos*, and *Silvel Elías*
Annals of the Association of American Geographers, volume 101, issue 4 (July 2011) pp. 918-928

Chapter 22
The Impact of Brazilian Biofuel Production on Amazonia
Robert Walker
Annals of the Association of American Geographers, volume 101, issue 4 (July 2011) pp. 929-938

Chapter 23
Shifting Networks of Power in Nicaragua: Relational Materialisms in the Consumption of Privatized Electricity
Julie Cupples
Annals of the Association of American Geographers, volume 101, issue 4 (July 2011) pp. 939-948

Chapter 24
"Because You Got to Have Heat": The Networked Assemblage of Energy Poverty in Eastern North Carolina
Conor Harrison and *Jeff Popke*
Annals of the Association of American Geographers, volume 101, issue 4 (July 2011) pp. 949-961

Chapter 25
Powering "Progress": Regulation and the Development of Michigan's Electricity Landscape
Jordan P. Howell
Annals of the Association of American Geographers, volume 101, issue 4 (July 2011) pp. 962-970

Chapter 26
The Geography of Energy and the Wealth of the World
Martin J. Pasqualetti
Annals of the Association of American Geographers, volume 101, issue 4 (July 2011) pp. 971-980

Dedicated to Wes Jackson and the late Marty Bender of the Land Institute, teachers, researchers, and proponents extraordinaire of sustainable energy and agriculture.

Approaching the New Geographies of Global Energy: Analytics and Assessment of Current Energy Landscapes and Alternatives

Karl Zimmerer

Professor and Head, Department of Geography, Earth and Environmental Systems Institute, The Pennsylvania State University

Overview: Evolving Geographic Approaches to Rapidly Changing Energy Issues

Energy issues worldwide are leading to major concern and fueling a diverse spectrum of social-ecological impacts and potential transitions. Complex geographical and historical forces drive the current energy dilemmas, the profusion of responses, and the urgent searching for sustainable alternatives. Societies around the world face a firestorm of volatile energy markets and financial risk; scrambles for energy-yielding territories and technological innovation; palpable uncertainties in the provisioning of basic economic goods and services; escalating social justice impacts and vulnerabilities; declining supplies of easy-to-get oil; rising impacts of climate change along with mitigation and adaptation imperatives; issues of geopolitical security and international relations related to energy; political and regulatory failures and limitations, and energy-fueled impacts on varied resource systems (water, ecosystems, agriculture) and their management. Importance and influence of these issues is cascading across local and global scales together with many in-between levels. Indeed energy, in general, is increasingly recognized as central—perhaps the resource-based center per se—within global environmental change and prospects of sustainability (Wilbanks 1994; Zimmerer 2010, 2011).

Energy likewise has become the defining focus of a majority of the largest companies, parastatal firms, broad private-sector consortia and concerns, and national enterprises, as well as governments, non-state institutions, and concerned citizenries worldwide. Public and private initiatives in energy have strengthened, albeit inadequately (e.g., among governments see Smil 2011), in support of multi-prong energy efficiency in potential low-carbon transitions along with conversion to much-needed renewable sources and hydrogen-based futures. Oil, still the world's principal energy source, reigns supreme as an unparalleled global commodity, mythic in scope and substance (Huber 2011a; Watts 2001). Embedded in the current energy crisis is also a surge of advances in technical knowledge and technological innovation, as well as scholarly and scientific approaches to understanding energy.

Geography is crucial to understanding and addressing the multiple, interconnected dimensions of the current potpourri of global energy dilemmas and opportunities (Solomon et al. 2003; Wilbanks 1983, 2011; Zimmerer 2011). The geographic systems of energy production, technology, distribution, and consumption are thoroughly entwined as social-environmental interactions occurring as networks and in territorial configurations across multiple scales. In the ongoing organization and contestation of most energy systems, global and national levels are principal circuits and often the most visible manifestations of power, both that of energy and human societies. At the same time, many influential policies and much of the political action on energy issues, as well as related climate and environmental concerns, are being framed at levels of sub-national units (the equivalent of state-level governments in the U.S.), cities, and locales. This volume seeks to offer a timely contribution and a general approach referred to here as the "new geographies of global energy." It examines changing energy landscapes by combining the perspective of globalization processes operating among diverse networks at multiple scales–including many national and sub-national–with a focus on environmental change and resource systems.

The multi-scale and combined social-environmental perspective of this volume thus follows generally similar approaches that are being advanced with respect to the "new geographies" of related global environmental issues such as global climate change (Liverman 1999; O'Brien and Leichenko 2000; Leichenko and O'Brien 2008); the globalization of environmental conservation and the sustainability of agriculture and food (Zimmerer 2006a, 2006b); and globalization entwined with the rapid growth of urban environmental issues (Seto et al. 2010). It also builds upon the geographic idea of overarching "landscapes of power" within the study of energy geography (Solomon et al. 2003: 309). Analytically, new geographies of energy emerge through core topics of research such as resources systems, technological change, and broad socio- and political economic processes (Bridge and Wood 2005). The goal of this volume is to advance geography's contributions to these core areas, both existing and newly expanding.

The new geographies of energy cover and connect a range of dynamic landscapes that, in turn, intermingle across a variety of scales, territories, and networked spatialities. Accordingly the chapters of this volume are organized into a series of heuristically designated sections: (i) this introduction and the following essay (2 chapters); (ii) geographic approaches to energy modeling and assessment (5 chapters); (iii) fossil fuel landscapes (4 chapters); (iv) landscapes of renewable energy (10 chapters); and (v) landscapes of energy consumption (3 chapters). As expected, a number of chapters provide bridge-like connections among multiple sections, rather than belonging solely to a single type of energy landscape. Furthermore, the energy landscapes, as distinguished here, intermingle in dynamic, place-based processes. For example, landscapes of energy production most commonly co-occur together with landscapes of energy consumption. As a consequence, the analysis of this introductory chapter is designed to distinguish the *principal* area of focus of each of these geographic contributions, as well as key areas of overlap and potential synergies suited to geographic analysis and understandings. Distinguishing the principal themes of the volume's five s ections has thus emerged through systematic analysis and the opportunity to contribute productively to the advance of themes and concept-building within energy geography—an undertaking made possible through the original editorial process of the volume and, in particular, it's open submission and ample peer review (rather than pre-selected themes or authors).

New Geographic Approaches to Energy Modeling and Assessment

Geographic approaches to energy modeling and assessment are focused on a range of topics involving production and consumption landscapes. Much of the development and application of energy modeling is designed to address the potential of renewable energy sources. In the volume's first chapter, Li, Stadler, and Ramakumar demonstrate that geographic modeling of the geophysical inputs (solar, wind) for renewable energy generating capacity can contribute to rigorous specification of the regional co-occurrences of these sustainable energy sources (Li, Stadler, and Ramakumar 2011), and thus toward the analytics needed for energy policies and planning aimed at the spatial density of renewable energy sourcing and the sorts of capacity increase expected to be crucial. In a related chapter, also found in this same section, Wang et al. use remote sensing analysis to offer new insights into the phenology-based geospatial categorization of switchgrass (*Panicum virgatum* L.) (Wang et al. 2011), which is an ecologically versatile perennial grass seen to offer the prospect of a significantly more sustainable scenario of biofuels production than corn-based ethanol (see also van der Horst and Evans 2010).

Energy accounting and spatial inventory assessments, including the use of index and footprint models and techniques, are vital to low-carbon energy transitions (Kammen and Pacca 2004; Brown and Sovacool 2007). In this volume, Zhao, Horner, and Sulik pioneer a geographically framed approach to estimating overall carbon inventories in order to account for both emissions and sequestration in a net balance model of the state of Florida (Zhao, Horner, and Sulik 2011). In the same section Newell and Vos develop and apply GIS-based transportation and energy models to account for globally dispersed networks of carbon (e.g., carbon footprints for the paper industry; Newell and Vos 2011). More generally, GIScience-based approaches are shown to enable advances in carbon estimation and inventorying, infrastructure placement amid the transition to a low-carbon energy economy, and household-level energy consumption (Horner, Zhao, and Chapin 2011; see also Horner 2004). Finally, multi-scalar assessment of energy regulation will need to account for the geographic influence of mounting litigation on cases involving climate change since it is via this avenue, particularly in the U.S., whereby the legal system is expanding its treatment of the emissions from transportation and power plants (Osofsky 2011).

Raging Debate Over Fossil Fuel Landscapes

Critical geopolitics is increasingly central to understanding the changing landscapes of hydrocarbons in particular. In the first paper of this section, Bouzarovski and Bassin elucidate how nationalist discourses of Russia as a "global energy superpower" are politically powerful and implicated as a widespread symbol in the processes of identity formation of the citizens within that country (Bouzarovski and Bassin 2011). Contesting these predominant identities must be seen in the context of the massive political-economic forces currently at play. Substantial geographic work discusses the role of modern-day imperialism, including that of the U.S. and China, which is aimed at oil extraction across extensive regions ranging from the Middle East, Latin America, Africa, and Asia (Watts 2001, 2008a, 2008b; Harvey 2003). Clashes and contestation are commonplace in the escalating emergence of conflictive energy landscapes in these diverse locales. "Resource curse" impacts, the limits and failure of oil-based development, and dispossession of land and other resources are extensive and intensifying in various places; also common are oil-fueled conflicts, referred to as "petro-violence," that includes the rise of new forms of armed insurgencies in such countries as Nigeria (Watts 2001, 2008a: 38, 2008b; Harvey 2003: 67, 137; Bradshaw 2010; Orta-Martínez and Finer 2010).

"Energy security" and "energy scarcity," as well as prospective "energy sustainability," are shown to denote contested concepts that arise as consequences of both geophysical factors as well as broadly defined social forces and those of multi-scale political economies continually emerging through entwined national contexts and international relations (Goldthau and Witte 2010; Heiman 2002; Sovacool and Brown 2010). For example, energy security is comprised of the interacting conditions influencing availability, affordability, technology development, sustainability, and regulation (Sovacool et al. 2011), while the concept of "energy scarcity" is used to show the roles of many non-market influences, ranging from the broad political economy to violence in and near regions of hydrocarbon deposits (Huber 2009). This volatile mix of factors, which has energized a spate of geographic analysis, contributes to the actual pricing and availability of oil including the powerful construct of "peak oil" (Huber 2009, 2011a, 2011b; see also Kaufman and Cleveland 2001; Cleveland and Kaufmann 2003; Bridge 2010; Bridge and Wood 2010; Harvey 2010: 73, 79-80; Kaufmann 2011; Labban 2010).

At the same time, the changing landscapes of natural gas extraction are undergoing vast transformations. In some circles, gas is increasingly seen as integral to energy-sourcing scenarios needed to propel national economies to potential low-carbon transitions (e.g., the U.S. Natural Gas and Sustainable Energy Initiative of the Worldwatch Institute in Washington, D.C., see Flavin and Kitasei 2010). This interest is intensifying despite concerns over growing shale gas extraction, liquefied natural gas facilities (Harrison 2008), and the social conflicts, territorial disputes, human rights abuses, and environmental degradation in geographic areas of gas production, transport, and processing (e.g., Ecuador and Bolivia; Bebbington 2009; Bridge 2004; Hindery 2004; Perreault 2006; Perreault and Valdivia 2010; Schroeder 2007; Valdivia 2007). In the latter countries (Ecuador and Bolivia) the booming energy economy and the still growing exploitation of fossil fuels are deeply embedded, and in some ironic ways, as principal revenue supplies undergirding the distinctive "postneoliberal" social and political agendas of the national administrations of Rafaél Correa and Evo Morales, respectively.

Coals, which vary widely in chemical composition and quality (both energetic and environmental; on coal characteristics including pollution see Smil 2006: 90-1), were overtaken by hydrocarbons as a source of global energy production only in the mid-1960s. Today coals are still especially important to various countries, including China and the U.S., where dozens of new coal-powered generating plants are under consideration. As Kuby et al. demonstrate, the decline of coal use in residential heating and cooking, agriculture, and transportation in China has been more than offset by the large increase of use for electricity and industrial purposes (Kuby et al. 2011). The extractive techniques utilized for coals involve some of the most dramatic reorganization of energy landscapes. One of these, mountaintop removal (MTR) mining, involves removing the vegetation, soils, and geologic material of surface layers that cover coal deposits. In the U.S., mountaintop removal mining has become increasingly common in Appalachia and has led to major adverse environmental consequences ranging from the destruction of rivers to biodiversity loss. While often justified by claims of contributing a social good in the form of local jobs, Woods and Gordon show through West Virginia case studies that MTR mining has not delivered on the promise of employment benefits to local communities (Woods and Gordon 2011).

Institutional and technological innovation have become central to economic-, governance-, and infrastructure-based approaches designed to guide a shift to more sustainable use of fossil fuels, their landscapes, and a general bridging to sustainable energy. The accounting of carbon, as a molecular-level component, is a building block of these approaches. In particular, the construction of carbon markets is being undertaken to control greenhouse gas emissions and thus help enable the transition to more environmentally sustainable energy (Fawcett 2010; Knox-Hayes 2009, 2010; Solomon and Heiman 2010). These new carbon markets, which involve the complex incorporation of social and economic processes, represent a significant stage—arguably a "neo-modernity"—in the further extension of commodification and the disembodied representations of the materiality of energy in the current and future trajectories of profoundly altered human-environment interactions and nature-society relations (Knox-Hayes 2011).

Technological innovation, along with human learning, occurs through an ongoing evolution of the geographic dimension of fossil-fuel landscapes and the hoped-for transitions to sustainable energy. For example, the industrial sequestration of carbon through the processing of fossil-fuel combustion by-products (CCS, or CO_2 Capture and Storage) is a high-risk environmental solution whose viability is contingent on "learning curves" involving spatial components at the firm- and intra-firm level with potential industry-wide policy initiatives (Bielicki 2012). At the same time, and ironically, the technology-intensive emphasis on geographic systems of carbon-centered accounting, in which abstract metrics referred to generally as "carbon counting" are granted primacy, may in fact undermine landscape-level understandings of energy transitions (van der Horst and Evans 2010).

Critical Landscapes of Renewable Energy

Wind, bioenergy, nuclear, hydropower, solar, geothermal, and tidal energy sources possess renewable qualities, albeit with highly varying characteristics and social-environmental impacts (Smil 2003). Significant reorganization of energy production landscapes is associated with each of these renewable sources. Critical geographic perspectives of growing research have been central to our capacity to understand the panoply of new renewable energy landscapes. Bioenergy (the extraction of energy from recently living plants through biomass) as well as biofuels (principally ethanol and biodiesel) are a particular focus of recent geographic research. These sources currently account for approximately one half of renewable energy production in the United States (Kedron and Bagchi-Sen 2011). Kedron and Bagchi-Sen focus on the ethanol production value-chain in Iowa, analyzing the capacity for innovation as well as energy policy impacts and opportunities (see also the geographic approaches and analysis of Gillon 2010; Hollander 2010). While also using the tools of economic geography and planning, Mabee and Mirck provide a region-level assessment of potential forest-derived bioenergy stemming from Ontario's 2009 Green Energy and Green Economy Act, and particularly its Feed-In Tariff (FIT) designed to stabilize producer prices (Mabee and Mirck 2011).

Bailis and Baka explore the political geographies of prominent governance regimes that have arisen in the case of biofuels, focusing especially on environmental impacts, and emphasizing the role of non-nation-state-actors (NNSAs) (Bailis and Baka 2011; see also Bailis, Ezzati, and Kammen 2005). Cope, McLafferty, and Rhoads employ public participation GIS (PPGIS) to determine perceived barriers to the adoption of switchgrass production in Illinois; their findings include an important multi-factorial definition, rather than strictly environmental delimitations, of conditions that account for marginal growing environments (Cope, McLafferty, and Rhoads 2011). Such information will be crucial to much-needed policy innovations involving the certification of biofuels that avoid problems of soil degradation, such as fertility depletion, and that also minimize the reduction of food crops. In Guatemala, Taylor et al. show how existing household-level cooking and heating practices demonstrate the prospective sustainability of biomass energy in the form of woodfuels in conjunction with the economic and cultural roles of international migration (Taylor et al. 2011; see also Babanyara and Saleh 2010). In Brazil, and the extensive Amazon region in particular, Walker provides an incisive analysis of existing and possible future landscapes of biofuel production, ranging from a purported green panacea to the downsides of a "resource curse" akin to previous Latin American energy booms built on fossil fuel extraction that resulted in negative impacts on overall socioeconomic development (Walker 2011; see also Walker et al. 2009).

Globally installed capacity of wind and solar are increasing 30-50 percent per year (currently the global growth of wind capacity is 38 percent annually; Pasqualetti 2011), with wind energy in the United States expanding at 24 percent annually since 2000 (Brannstrom, Jepson, and Persons 2011). Wind, like other renewable energies, is relatively less dense in terms of the geographic concentration of sourcing and thus presents challenges of scaling in relation to the spatial distribution of energy demand (Smil 2003: 242-3; Smil 2006: 165). While general support is high, local opposition is also especially noteworthy in many wind projects (Pasqualetti 2001; Pasqualetti, Gipe, and Righter 2002; Phadke 2010). Creating higher density sources of wind energy tends to fuel greater concerns over impacts on landscape processes. Even then, while the spatial externalities of many wind projects are actually small or medium in scope, they often appear quite large. Pasqualetti provides a comprehensive overview and a series of skillfully designed case studies of oppositional sentiments, ranging from identities, visual qualities of landscape, and sense of place, to economic considerations, which have become integral social and cultural processes within most landscapes of wind energy (Pasqualetti 2011). In a case study of West Texas, Brannstrom, Jepson, and Persons demonstrate the important role of specific forms of ownership and participation, including perceptions of tax incentives, in determining the opinion of individuals on wind-power development and, more generally, the discourses of support/opposition with regard to wind energy (Brannstrom, Jepson, and Persons 2011). Environmental and aesthetic barriers, whether real or imagined, continue to exert an active force and sometimes surprising impacts on planning, policy, and projects related to wind development.

Hydropower is often regarded as constrained in scope for future development due to social and environmental limitations of this resource use. (Bioenergy prospects may be seen as possessing generally similar concerns.) Still hydropower is being actively expanded through such policies as the Clean Development Mechanism (CDM) of the World Bank and other agencies at global, international, and local levels (Finley-Brook and Thomas 2011). Finley-Brook and Thomas find that new hydropower projects in Panama are in fact violations of the "indigenous rights" of ethnic minorities (Finley-Brook and Thomas 2011). In British Columbia, Dusyk shows how the intent to incorporate participatory processes into the planning of a hydroelectric dam contributed to conflict and opposition; her analysis advances a group of concepts associated with social-technological networks, hybrid forums, and their socionatural spatialities (Dusyk 2011). The findings of these papers resonate with more general concerns about large dams and hydropower (Smil 2006: 167-8). Such results are a powerful reminder that the "natural resources" of renewable energy are cultural, economic, and technological appraisals of landscapes that take shape through individual activities and institutional agendas as well as through industrial forms of organization.

Contested Landscapes of Energy Consumption

Landscapes of energy consumption, ranging from distribution networks and industrial use to the household level, are extremely important in understanding existing impacts and the often complex prospect of sustainability transitions. Since household-level energy use depends on personal decisions, choice-constraining technological innovations, for example, may be limiting among many consumers, at least those accustomed to a high level of buying options. At the same time national-level expansion of global energy demand, including the roles of China and India (Kuby et al. 2011; see also Bradshaw 2010), is an epochal force in current energy landscapes worldwide. Landscapes of consumption also offer striking examples of technologic momentum in energy use–such as electricity grids and transportation networks–whereby factors such as cultural dispositions and sunk investments exert powerful influences (on the cultural and social shaping of such technological momentum in broad historical perspective see Nye 1998). Energy poverty and electricity networks, an emphasis of this volume, offer prime examples of such "consumption landscapes."

Harrison and Popke draw on current theoretical approaches in nature-society geography to represent the concept of energy poverty as a "geographical assemblage of networked materialities and socio-economic relations" in order to examine causes and patterns of household-level consumption in Eastern North Carolina (Harrison and Popke 2011). Their study extends energy policy to a broadening template of related social welfare and care (on other new geographies of energy poverty see also Buzar 2007). In Nicaragua, Cupples reveals how the material effects of neoliberal-guided policies of energy privatization, namely the flows of electricity generation and distribution, have contributed to social responses and mobilization as well as to everyday hardship that are incurred especially among the urban poor (Cupples 2011). These case studies underscore how electricity is projected to be the fastest-growing energy sector globally. Howell skillfully conducts a case study of the Michigan's Public Service Commission in historic hearings on the Midland Nuclear Facility, this turning our attention in this volume toward the timely topic of regulation/deregulation (Howell 2011; on geographies of electricity restructuring and regulation see also Heiman and Solomon 2004). Howell's analysis shows the involvement and influence of U.S. utility companies in the powerful shaping of attitudes and policies that contribute to excess electricity capacity and that deter energy conservation and adoption of renewable fuels. These chapters are also generally relevant to the interests of energy policy analysts who advocate for the decentralization of electrical resources as a key step toward distributed generation and increased efficiency. Their focus on electricity also underscores the role of its generation as the single largest source of greenhouse gas emissions, and thus a topic of significant social-environmental import.

Further Geographic Dimensions of Energy

The chapters in this volume also add to advances and identify important opportunities in other dimensions of geographic research on energy. In particular it is vital to highlight the urgency of energy concerns and the insights of active geographic research in the sectors of industry, transportation, and building (e.g., Cidell 2009a, 2009b, Cidell and Beata 2009; Heiman and Solomon 2007). While covered to some extent in the above sections, each of these topical areas is well suited to the development of significant new avenues of geographic research (see also Horner 2004). Of equal importance, each of these major areas—industry, transportation, and buildings—holds immediate large-scale opportunities for policy interventions and marked efficiency improvements ripe for the expanded contributions of geography.

Similarly the all-around intensified relevance of energy issues puts into high relief a multitude of close-knit interactions with such topics as the geographies of peace, conflict, and political mobilization, as well as those of climate change (Kobayashi 2009; Aspinall 2010; see also Johnson 2010; North 2010; Rice 2010). Energy topics traverse broad social-environmental spectra of concerns and challenges; within contemporary geographic research these topics offer the promise of strengthening a core focus on the human-environment interactions and nature-society relations of industry (Huber 2010). Energy-fueled competition over resources and impacts on their quantity and quality—water in particular—is escalating over vast swaths of ecosystems and landscapes worldwide. Other growing interactions and dimensions of energy that are particularly well-suited topics of geographic research include the relations or linkages to overall socioeconomic development, urbanization, health and disease, and food and agriculture (to name only a few). The probing consideration of these issues is subject to a productive legacy of analysis and ideas over the past several decades (e.g., Harte and Jassby 1978; Knight 1989; Solomon and Karthik 2011).

Conclusions

The chapters in this volume cover an ample range of geographic research topics, approaches, and contributions related to the new geographies and landscapes of global energy. These studies illustrate many of the principal geographic dimensions of energy emerging from multi-scale social-environmental influences and impacts embedded in dynamic landscapes of fossil fuels, renewable energy, and energy consumption. Specific contributions are concentrated in the following chief areas: (i) GIS modeling and assessment of energy flows—including an emphasis on carbon stocks and flows—with special interest and pronounced promise in footprint approaches, infrastructure (e.g., transportation, buildings), and industry- and household-level use; (ii) integrated social, political economic, and environmental analysis of oil and coal landscapes, with special emphasis on perspectives and concepts such as geopolitics, imperialism, peak oil, social justice, and resource-curse effects on development; (iii) original analysis of the renewable energy "alternatives" of biofuels, bioenergy, hydropower, and wind power, including detailed place-based examinations, that demonstrate the intricate interplay of potential productivity as well as limitations (typically the first three of these sources are held to have less scope for increase because of concerns over environmental resource impacts, while a combination of concerns also occur in the development of the nascent wind industry, as introduced above, as well as solar industry); and (iv) new geographies of energy networks (e.g., electricity regulation, generation, distribution, and use) and the thoroughly entwined socioeconomic-material nodes of energy consumption (e.g., energy poverty). A proliferating array of prime opportunities exist for continued and expanded geographic contributions in each of these principal areas, the several related and similarly compelling topics that are referred to only briefly above, and the numerous other dimensions of energy—all in urgent need of timely research and initiatives in geography and through geography's myriad connections to interdisciplinary approaches.

Acknowledgments

Editing of the special issue devoted to energy (in the journal *Annals of the Association of American Geographers*, 2011) that led to this book volume relied from start to finish on the editorial assistance, guidance, and expertise of Martha G. Bell who worked skillfully and independently from Britain and Peru. Pre- and post-publication ideas and insights benefitted from to opportunities to hold conversations and hear presentations involving such leading figures in energy geography and related fields as Janelle Knox-Hayes, Jeff Bielicke, Rob Bailis, Benjamin Sovacool, Julie Cidell, Jennifer Rice, Matt Huber, Gail Hollander, Tom Perreault, Gabriela Valdivia, Christian Brannstrom, Sean Gillon, and Gavin Bridge. My editorial capacity benefited also from input and knowledge of fellow faculty and graduate students in the College of Earth and Mineral Sciences and the Institutes for Energy and the Environment, both at Penn State. I note especially my interactions with colleagues working on topics related to the gas extraction from the nearby Marcellus Shale, currently a major "energy play." Personally the engagement with energy issues is a welcome continuation of my activities begun in the U.S. energy crisis of the 1970s when, as an undergraduate, I interned in the solar energy division of the National Center for Appropriate Technology in Butte, Montana (1978) and, in a subsequent internship, in research and applied activities on wind and biomass energy, all in the context of perennial polycultures, at The Land Institute in Salina, Kansas (1979). In the course of having prepared the special Energy issue it was a pleasure to work with the large pool of talented authors. The AAG and its specialty groups offered unending use of their listserves and other means of opening up the opportunities for interested researchers to contribute. Routledge Publishing was a productive partner in the creation of this book volume. Finally, this volume is dedicated to Wes Jackson and the late Marty Bender of the Land Institute, my teachers, fellow researchers, and proponents extraordinaire of sustainable energy and agriculture.

References

Babanyara, Y. Y., and U. F. Saleh. 2010. Urbanisation and the choice of fuel wood as a source of energy in Nigeria. *Journal of Human Ecology* 31(1): 19-26.

Bailis, R., and Baka, J. 2011. Constructing sustainable biofuels: Governance of the emerging biofuel economy. *Annals of the Association of American Geographers* 101(4): 827-838.

Bailis, R., M. Ezzati, and D. M. Kammen. 2005. Mortality and greenhouse gas impacts of biomass and petroleum energy futures in Africa. *Science* 308(5718): 98-103.

Bebbington, A. 2009. Latin America: Contesting extraction, producing geographies. *Singapore Journal of Tropical Geography* 30(1): 7–12.

Bielicki, J. M. 2012. Technological change, learning, and spillovers in geologic CO_2 injection for enhanced oil recovery and the implications for CO_2 storage. Presentation, Department of Geography, The Pennsylvania State University (10 February 2012).

Bouzarovski, S., and M. Bassin. 2011. Energy and identity: Imagining Russia as a hydrocarbon superpower. *Annals of the Association of American Geographers* 101(4): 783-794.

Bradshaw, M. J. 2010. Global energy dilemmas: A geographical perspective. *The Geographical Journal* 176(4): 275-90.

Brannstrom, C., Jepson, W., and N. Persons. 2011. Social perspectives on wind-power development in West Texas. *Annals of the Association of American Geographers* 101(4): 839-851.

Bridge, G. 2004. Gas, and how to get it. *Geoforum* 35: 395-7.

———. 2010. Geographies of peak oil: The other carbon problem. *Geoforum* 41: 523-30.

Bridge, G., and A. Wood. 2005. Geographies of knowledge, practices of globalization: learning from the oil exploration and production industry. *Area* 37(2): 199-208.

———. 2010. Less is more: Spectres of scarcity and the politics of resource access in the upstream oil sector. *Geoforum* 41: 565-76.

Brown, M. A., and B. K. Sovacool. 2007. Developing an 'energy sustainability index' to evaluate energy policy. *Interdisciplinary Science Reviews* 32(4): 335-49.

Buzar, S. 2007. The 'hidden' geographies of energy poverty in post-socialism: Between institutions and households. *Geoforum* 38: 224-9.

Cidell, J. 2009a. Building Green: The emerging geography of LEED-certified buildings and professionals. *The Professional Geographer* 61(2): 200-215.

———. 2009b. A political ecology of the built environment: LEED certification for green buildings. *Local Environment: The International Journal of Justice and Sustainability* 14(7): 621-633.

Cidell, J., and A. Beata. 2009. Spatial variation among greenbuilding certification categories: Does place matter? *Landscape and Urban Planning* 91(3): 142–151.

Cleveland, C. J., and R. K. Kaufmann. 2003. Oil supply and oil politics. Déjà vu all over again. *Energy Policy* 31(6): 485-489.

Cope, M., McLafferty, S. and Rhoads, B. 2011. Farmer attitudes toward production of perennial energy grasses in East Central Illinois: Implications for community-based decision making. *Annals of the Association of American Geographers* 101(4): 852-862.

Cupples, J. 2011. Shifting networks of power in Nicaragua: Relational materialisms in the consumption of privatized electricity. *Annals of the Association of American Geographers* 101(4): 939-948.

Dusyk, N. 2011. Downstream effects of a hybrid forum: The case of the Site C hydroelectric dam in British Columbia, Canada. *Annals of the Association of American Geographers* 101(4): 873-881.

Fawcett, T. 2010. Personal carbon trading: A policy ahead of its time? *Energy Policy* 38(11): 6868-6876.

Finley-Brook, M., and C. Thomas. 2011. Renewable energy and human rights violations: Illustrative cases from indigenous territories in Panama. *Annals of the Association of American Geographers* 101(4): 863-872.

Flavin, C., and S. Kitasei. 2010. *The role of natural gas in a low-carbon energy economy*. Briefing Paper. Washington, D. C.: Worldwatch Institute.

Gillon, S. 2010. Fields of dreams: Negotiating an ethanol agenda in the Midwest United States. *The Journal of Peasant Studies: Critical Perspectives on Rural Politics and Development* 37(4): 723-48.

Goldthau, A., and J. M. Witte, eds. 2010. *Global energy governance: The new rules of the game*. Berlin: Global Public Policy Institute and Washington, D. C.: Brookings Institution Press.

Harrison, B. 2008. Offshore threats: Liquified natural gas, terrorism, and environmental debate in Connecticut. *Annals of the Association of American Geographers* 98(1): 135-59.

Harrison, C. and J. Popke. 2011. "Because you got to have heat": The networked assemblage of energy poverty in eastern North Carolina. *Annals of the Association of American Geographers* 101(4): 949-961.

Harte, J., and A. Jassby. 1978. Energy technologies and natural environments – search for compatibility. *Annual Review of Energy* 3: 101-146.

Harvey, D. 2003. *The new imperialism*. Oxford: Oxford University Press.

———. 2010. *The enigma of capital and the crises of capitalism*. Oxford: Oxford University Press.

Heiman, M. K. 2002. Power for the people: A comparison of the US and German commitments to renewable energy. In *Renewable Energy Trends and Prospects*, eds. S. K. Majumdar, E. W. Miller, and A. I. Panah, 403-411. Easton, PA: Pennsylvania Academy of Science.

Heiman, M. K. and B. D. Solomon. 2004. Power to the people: Electric utility restructuring and the commitment to renewable energy. *Annals of the Association of American Geographers* 94(1): 94-116.

———. 2007. Fueling US transportation: the hydrogen economy and its alternatives. *Environment* 49(8): 10-25.

Hindery, D. 2004. Social and environmental impacts of World Bank/IMF-funded economic restructuring in Bolivia: An analysis of Enron and Shell's hydrocarbon projects. *Singapore Journal of Tropical Geography* 25(3): 281–303.

Hollander, G. 2010. Power is sweet: Sugarcane in the global ethanol assemblage. *The Journal of Peasant Studies* 37(4): 699–721.

Horner, M. W. 2004. Spatial dimensions of urban commuting: A review of major issues and their implications for future geographic research. *The Professional Geographer* 56(2): 160-74.

Horner, M. W., Zhao, T., and T. S. Chapin. 2011. Towards an integrated GIScience and energy research agenda. *Annals of the Association of American Geographers* 101(4): 764-774.

Howell, J. P. 2011. Powering "progress": Regulation and the development of Michigan's electricity landscape. *Annals of the Association of American Geographers* 101(4): 962-970.

Huber, M. 2009. Energizing historical materialism: Fossil fuels, space and the capitalist mode of production. *Geoforum* 40(1): 105-15.

———. 2010. Hyphenated geographies: The deindustrialization of nature-society geography. *Geographical Review* 100 (1): 74–89.

———. 2011a. Oil, life, and the fetishism of geopolitics. *Capitalism Nature Socialism* 22(3): 32-48.

———. 2011b. Enforcing scarcity: Oil, violence and the making of the market. *Annals of the Association of American Geographers* 101(4): 816-826.

Johnson, L. 2010. The fearful symmetry of Arctic climate change: Accumulation by degradation. *Environment and Planning D: Society and Space* 28(5) 828-47.

Kammen, D. M., and S. Pacca. 2004. Assessing the costs of energy. *Annual Review of Environment and Resources* 29: 301-44.

Kaufmann, R. K. 2011. The role of market fundamentals and speculation in recent price changes for crude oil. *Energy Policy* 39(1): 105-115.

Kaufmann, R. K., and C. J. Cleveland. 2001. Oil production in the lower 48 states: Economic, geologic, and institutional determinants. Energy Journal 22(1): 27-49.

Kedron, P. and S. Bagchi-Sen. 2011. A study of the emerging renewable energy sector within Iowa. *Annals of the Association of American Geographers* 101(4): 882-896.

Knight, C. G. 1989, Environmental economic models for energy resource development. *Socio-Economic Planning Sciences* 23(1-2): 55-66.

Knox-Hayes, J. 2009. The developing carbon financial service industry: Expertise, adaptation and complementarity in London and New York. *Journal of Economic Geography* 9: 749-777.

———. 2010. Constructing carbon market spacetime: Climate change and the onset of neomodernity. *Annals of the Association of American Geographers* 100(4): 953-962.

Kuby, M., He, C., Trapido-Lurie, B., and N. Moore. 2011. The changing structure of energy supply, demand, and CO_2 emissions in China. *Annals of the Association of American Geographers* 101(4): 793-805.

Labban, M. 2010. Oil in parallax: Scarcity, markets, and the financialization of accumulation. Geoforum 41(4): 541-52.

Leichenko, R. M., and K. L. O'Brien. 2008. *Environmental change and globalization: Double exposures.* Oxford: Oxford University Press.

Li, W, Stadler, S. J., and R. Ramakumar. 2011. Modeling and assessment of wind and insolation resources with a focus on their complementary nature: A case study of Oklahoma. *Annals of the Association of American Geographers* 101(4): 717-729.

Liverman, D. M. 1999. Geography and the global environment. *Annals of the Association of American Geographers* 89 (1): 107-20.

Mabee, W. E., and J. Mirck. 2011. A regional evaluation of potential bioenergy production pathways in Eastern Ontario, Canada. *Annals of the Association of American Geographers* 101(4): 897-906.

Newell, J., and R. O. Vos. 2011. "Papering" over space and place: Product carbon footprint modeling in the global paper industry. *Annals of the Association of American Geographers* 101(4): 730-741.

North, P. 2010. Eco-localisation as a progressive response to peak oil and climate change – A sympathetic critique. Geoforum 41(4): 585-94.

Nye, D. E. 1998. *Consuming power: A social history of American energies.* Cambridge, Massachusetts: MIT Press.

O'Brien, K. L., and R. M. Leichenko. 2000. Double exposure: Assessing the impacts of climate change within the context of economic globalization. *Annals of the Association of American Geographers* 93(1): 221-32.

Orta-Martíneza, M., and M. Finer. 2010. Oil frontiers and indigenous resistance in the Peruvian Amazon. *Ecological Economics* 70(2): 207-18.

Osofsky, H. 2011. Rescaling energy regulation through litigation: Issues of fixity and fluidity. *Annals of the Association of American Geographers* 101(4): 775-782.

Pasqualetti, M. J. 2001. Wind energy landscapes: Society and technology in the California desert. *Society and Natural Resources* 14(8): 689-99.

———. 2011. Opposing wind energy landscapes: A search for common cause. *Annals of the Association of American Geographers* 101(4): 907-917.

Pasqualetti, M. J., P. Gipe, and R. Righter. 2002. *Wind power in view: Energy landscapes in a crowded world.* San Diego: Academic Press.

Perreault, T. 2006. From the Guerra del Agua to the Guerra del Gas: Resource governance, neoliberalism and popular protest in Bolivia. *Antipode* 38(1): 150-172.

Perreault, T., and G. Valdivia. 2010. Hydrocarbons, popular protest and national imaginaries: Ecuador and Bolivia in comparative context. *Geoforum* 41(5): 689-99.

Phadke, R. 2010. Steel forests and smoke stacks: The politics of visualization in the Cape Wind controversy. *Environmental Politics* 19(1): 1-20.

Rice, J. L. 2010. Climate, carbon, and territory: Greenhouse gas mitigation in Seattle, Washington. *Annals of the Association of American Geographers* 100(4): 929-37.

Schroeder, K. 2007. Economic globalization and Bolivia's regional divide. *Journal of Latin American Geography* 6(2): 99-120

Seto, K. C., R. Sánchez-Rodríguez, and M. Fragkias. 2010. The new geography of contemporary urbanization and the environment. *Annual Review of Resources and the Environment* 35: 167-94.

Smil, V. 2003. *Energy at the crossroads: Global perspectives and uncertainties.* Cambridge, Massachusetts: MIT Press.

———. 2006. *Energy.* Oxford: Oneworld Publications.

———. 2011. America's oil imports: A self-inflicted burden. *Annals of the Association of American Geographers* 101(4): 712-716.

Solomon, B. D., and Heiman, M. K. 2010. Integrity of the emerging global markets in greenhouse gases. *Annals of the Association of American Geographers* 100(4): 973-982.

Solomon, B. D., and Karthik, K. 2011. The coming sustainable energy transition: History, strategies, and outlook. Energy Policy 39(11): 7422-7431.

Solomon, B. D., Pasqualetti, M. J., et al. 2003. Energy geography. In *Geography in America*, eds. G. L. Gaile and C. J. Wilmott, 302-313. Oxford: Oxford University Press.

Sovacool, B. K., and K. E. Sovacool. 2009. Identifying future electricity-water tradeoffs in the United States. *Energy Policy* 37: 2763-2773.

Sovacool, B. K., and M. A. Brown. 2010. Competing dimensions of energy security: An international perspective. *Annual Review of Resources and the Environment* 35: 77-108.

Sovacool, B. K., Mukherjeee, I., Drupady, I. M., and A. L. D'Agostino. 2011. Evaluating energy security performance from 1990 to 2010 for eighteen countries. *Energy* 36: 5846-5853.

Taylor, M. J., Moran-Taylor, M. J., Castellanos, E. J., and S. Elías Gramajo. 2011. Burning for sustainability: Biomass energy, international migration, and the move to cleaner fuels and cookstoves in Guatemala. *Annals of the Association of American Geographers* 101(4): 918-928.

Valdivia, G. 2007. The 'Amazonian Trial of the Century': Indigenous identities, transnational networks, and petroleum in Ecuador. *Alternatives: Global, Local, Political* 32(1): 41-72.

van der Horst, D., and J. Evans. 2010. Carbon claims and energy landscapes: Exploring the political ecology of biomass. *Landscape Research* 35: 173-193.

Walker. R. 2011. The impact of Brazilian biofuel production on Amazônia. *Annals of the Association of American Geographers* 101(4): 929-938.

Walker, R., R. DeFries, M. de Carmen Vera-Diaz, Y. Shimabukuro, and A. Venturieri. 2009. The expansion of intensive agriculture and ranching in Brazilian Amazonia. In *Amazonia and global change*, Geophysical Monograph Series Volume 186, eds. M. Keller, M. Bustamente, J. Gash, and P. Dias, 63-81. Washington, D.C.: American Geophysical Union.

Wang, C., Fritschi, F. B., Stacey, G., and Z. W. Yang. 2011. Phenology-based assessment of energy crops in North American tallgrass prairie. *Annals of the Association of American Geographers* 101(4): 742-751.

Watts, M. 2001. Petro-violence: Community, extraction, and political ecology of a mythic commodity. In *Violent environments*, eds. N. L. Peluso and M. Watts, 189-212. Ithaca: Cornell University Press.

———. 2008a. *Imperial Oil: The anatomy of a Nigerian oil insurgency*. Erdkunde 62(1): 27-39.

Watts, M., ed. 2008b. *Curse of the black gold: 50 years of oil in the Niger Delta*. Brooklyn, New York: PowerHouse Books.

Wilbanks, T. J. 1983. Geography and our energy heritage. *Materials and Society* 7(3-4): 437-452.

———. 1994. Sustainable development in geographic perspective. *Annals of the Association of American Geographers* 84(4): 541-556.

———. 2011. Inducing transformational energy technological change. *Energy Economics* 33(4): 699-708

Woods, B. R., and J. Gordon. 2011. Mountaintop removal and job creation: Exploring the relationship using spatial regression. *Annals of the Association of American Geographers* 101(4): 806-815.

Zhao, T., Horner, M. W., and J. Sulik. 2011. A geographic approach to sectoral carbon inventory: Examining the balance between consumption-based emissions and land-use carbon sequestration in Florida. *Annals of the Association of American Geographers* 101(4): 752-763.

Zimmerer, K. S., ed. 2006a. *Globalization and new geographies of conservation*. Chicago: University of Chicago Press.

———. 2006b. Cultural ecology: at the interface with political ecology—the new geographies of environmental conservation and globalization. *Progress in Human Geography* 30(1): 63-78.

———. 2010. Retrospective on nature-society geography: Tracing trajectories (1911-2010) and reflecting on translations. *Annals of the Association of American Geographers* 100(5): 1076-1094.

———. 2011. New geographies of energy: Introduction to the special issue. *Annals of the Association of American Geographers* 101(4): 705-711.

America's Oil Imports: A Self-Inflicted Burden

Vaclav Smil

Faculty of Environment, University of Manitoba

Burdensome dependence on crude oil imports is a key challenge for America's energy policy; its principal cause is excessive consumption of refined oil products, which is mainly the result of an inefficient automotive fleet, the virtual absence of diesel-powered cars, and the complete absence of modern high-velocity trains. Addressing these challenges will require major infrastructural investment, a reality that precludes any early attainment of energy independence. *Key Words: crude oil imports, energy consumption, energy independence, high-velocity trains, U.S. energy policy.*

对原油进口的累赘依赖是美国能源政策的一个关键挑战，其主要原因是成品油产品的过度消费，这主要是由低效的汽车车队，柴油动力车事实上的缺失，和现代高速列车的完全缺失所造成的。应对这些挑战将需要大量的基建投资，这一现实妨碍了能源独立的早日实现。*关键词：原油进口，能源消耗，能源独立，高速列车，美国的能源政策。*

La onerosa dependencia de importaciones de petróleo en crudo es un reto clave para la política energética americana; su causa principal es el excesivo consumo de productos refinados del petróleo, lo cual resulta principalmente de tener una flota automotora ineficiente, la virtual ausencia de carros movidos por diesel y la completa ausencia de trenes modernos de alta velocidad. Enfrentar estos retos demandará mayores inversiones infraestructurales, realidad que impide lograr una independencia energética en el corto plazo. *Palabras clave: importaciones de crudo, consumo de energía, independencia energética, trenes de alta velocidad, política energética de EE.UU.*

American energy debates have suffered from a surfeit of shallow generalities, from uncritical proffers of naive solutions, and from the persistence of many seemingly ineradicable myths. I have spent a lifetime trying to infuse them with a modicum of common sense and requisite doses of scientific and engineering realities, an endeavor that is summarized in three of my most recent energy books (Smil 2008, 2010a, 2010b). I use the limited space of this contribution to focus on a key problem, the continuing burden of which is overwhelmingly self-inflicted, and the negative impacts of which could have been minimized by actions that called for no extraordinary technical advances and were successfully accomplished by less affluent countries during a single generation.

America's crude oil imports burden the country with enormous strategic, financial, political, social, and environmental costs. A single number illustrates the strategic implications: In 2005, the year of record domestic oil consumption, the country imported 67 percent of its oil supply, and the share remained above 60 percent in 2009 (U.S. Energy Information Administration [USEIA] 2010). Financial considerations have become particularly important because of the new, post-2008 economic realities. During the first decade of the twenty-first century, the United States paid nearly $1.7 trillion for foreign oil, a total equal to 30 percent of its cumulative trade deficit between 2000 and 2009 (U.S. Census Bureau 2010). In 2008, high oil prices pushed the total to the annual record of nearly $350 billion and similarly wounding outlays will return once oil prices begin, yet again, their climb toward $150 per barrel. Higher costs are made more likely even if U.S. imports were to fall: Because international oil trade is denominated in dollars, any substantial weakening of the U.S. currency will be reflected by commensurately rising oil prices, and strong demand by large Asian economies (particularly China) will prevent any large supply surpluses by easily claiming any oil released by declining U.S. demand.

These large imports are not a consequence of America's low domestic oil extraction: The country was the world's largest oil producer until 1975 when it was surpassed by the USSR; two years later it was also surpassed by Saudi Arabia, but as Saudi output fell, the United States regained second place in 1982 and kept it until 1991 (BP 2010). By the year 2000, U.S. extraction was nearly 10 percent ahead of Russia's and equivalent to 77 percent of Saudi Arabia's rising output. In 2009, the United States held a firm third place, more than 70 percent ahead of Iran. The U.S. imports are not caused by a meager hydrocarbon patrimony but by the country's extraordinarily high per capita demand: If the United States consumed its oil at the French rate (1.35 rather than 2.70 metric tons [t] per capita), its 2009 imports would be only about 22 percent of its consumption.

Americans (ubiquitous media commentators and energy experts alike) consider any comparison with France insulting and view it with derision: Is it not obvious that America's huge territory, its climate, its superpower economic structure, and its military might combine to dictate significantly higher levels of per-capita energy consumption in general and crude oil in particular? Actually, they do not. The direct energy cost of America's military is surprisingly low: In 2009 the Department of Defense claimed less than 2 percent of the country's total crude oil consumption (USEIA 2010). Decades of American deindustrialization mean that the shares of gross economic value added by industry are now virtually identical in the two countries: In 2008 they were 21.8 percent in the United States and 20.5 percent in France (Organization for Economic Co-operation and Development [OECD 2010]), and the French industrial sector includes some of the world's largest companies specializing in such energy-intensive products, such as aluminum and steel (Pechiney and Arcelor-Mittal), aircraft (Airbus), and nuclear power plants (Areva).

Climate differences are also largely irrelevant when comparing the consumption of refined oil products: Natural gas and electricity are the dominant energizers of heating and cooling in both countries, and in the United States less than 6 percent of all residential energy used in the year 2009 came from petroleum and only about 1 percent of all electricity is generated from fuel oil (USEIA 2010). Differences are thus overwhelmingly due to the demand for transportation, above all to road transport, which accounts for more than 85 percent of the sector's petroleum demand in the United States and more than 95 percent in France. Average annual U.S. consumption of gasoline and diesel fuel was 1.73 t per capita in 2007, compared to just 0.65 t per capita in France, a nearly 2.7-fold difference, and the disparity for aviation fuel is even greater, with the United States consuming ten times as much per capita as France (OECD 2009).

America's size (8.08 million km^2 for the forty-eight contiguous states compared to 552,000 km^2 for mainland France, a nearly fifteen-fold difference) and intercity distances (almost 4,000 km from Los Angeles to New York compared to just over 900 km from Calais to Perpignan) are repeatedly cited as the most obvious reason for these differences. It is true that American drivers logged almost exactly twice the distance their French counterparts did in 2008 (European Union 2010), but distances dictate neither the typical performance of road vehicles nor the availability of desirable alternatives, and it is these realities that account for most of America's inferior performance. This failure has three major components: indefensibly low average efficiency of the U.S. vehicular fleet, a virtual absence of diesel vehicles, and an unpardonable absence of modern intercity trains.

Few Americans are aware of this inexcusable case of a gross technical retrogression: Average fuel efficiency of America's passenger cars actually declined during the two generations when impressive innovations and performance gains were sweeping entire industrial sectors (think of aviation, chemical syntheses, and electronics): Between 1936 (the first year for which the nationwide mean can be calculated) and 1973, it fell from more than 15 to just 13.4 miles per gallon (mpg; Sivak and Tsimhoni 2009). Introduction of Corporate Average Fuel Economy (CAFE) standards in 1975 ushered in a period of rapid gains: By 1985 the standard had reached 27.5 mpg, but in the subsequent twenty-five years, it has been kept at the same, now inexcusably low, level. Because of a massive adoption of inefficient light trucks and sport utility vehicles (SUVs) during the late 1980s and throughout the 1990s (their worst models had efficiency well below 20 mpg), the overall performance of the U.S. vehicle fleet was no better in 2006 (25.8 mpg) than it was in 1986 when it stood at 25.9 mpg (U.S. Department of Transportation [USDoT] 2010). These are truly stunning inefficiencies.

Post-2006 gains pushed the fleet average to 29.2 mpg by 2010, well below the average performance of the best fleets currently on the market and far below the technical possibilities attainable even without any electric or fuel-cell vehicles. In 2010 Honda's U.S.-made passenger car fleet averaged 34.7 mpg and imported models rated at 40.9 mpg, and the analogical figures for Toyota were, respectively, 36.4 and 44.4 mpg (USDoT 2010). If the CAFE standards were not frozen at 27.5 mpg in 1985 and continued to improve at just half the annual rate they had sustained between 1975 and 1985 (i.e., about 0.7 mpg/year), America's vehicular fleet would deliver at least 45 mpg in 2010, 54 percent better than its actual performance.

Achieving that efficiency would not have been contingent on any heroic measures or unprecedented technical advances: It could have been done by a dedicated pursuit of three lines of action. First, simply widely emulating today's best car manufacturers whose fleets are, without any enforcement, already close to or above 40 mpg would have helped. GM behaving as Honda was an incomprehensible idea to pre-2008 (prebankruptcy) U.S. car executives but one that would have repaid in more efficient cars and in tens of billions of dollars of profit. The second element is vigorous promotion of steady technical improvements in the dominant automotive prime mover that will continue to power most American cars for many years to come: Otto-cycle gasoline engines can be made more efficient while concurrent adoption of lighter but stronger car bodies could further boost the overall performance. The third ingredient should have been a progressively expanding adoption of modern clean passenger diesel vehicles.

Diesel engines are inherently more efficient prime movers than Otto-cycle gasoline-fueled engines (Smil 2010c): The gain is at least 20 percent and a quick comparison of performance data for German vehicles now sold in the United States in both gasoline and diesel versions shows differences mostly between 25 percent and 30 percent and the highway rating for the VW Diesel Golf is 40 percent higher (42 vs. 30 mpg) than for its gasoline version (U.S. Department of Energy 2010). Moreover, new ultra-low-sulfur diesels conform to emission standards for gasoline-fueled cars and minimize the emission of nitrogen oxides (Mercedes-Benz 2010). Not surprisingly, in 2009 diesel vehicles accounted for more than 56 percent of the French car fleet (Institute National de la Statistique et des Études Économiques 2010), but they were only about 2 percent of all U.S. cars.

If the United States had followed this triple-pronged approach since the mid-1970s, its car fleet's performance would now average close to 50 mpg. Unfortunately, this has not been the only unrealized transportation opportunity. Missing trains have been the second most important ingredient of America's transportation neglect. Japan, with its Tokyo-Osaka *shinkansen* introduced in 1964, pioneered the era of modern rapid trains (Shima 1994), and France followed and surpassed that achievement. Its era of *trains à grande vitesse* (TGV) began in 1981 with the Paris–Lyon link, and by 2010 the TGV network operated on routes totaling 1,540 km and connecting Paris with the country's southeast (*Sud-Est*), west (*Atlantique*), north (*Nord-Europe*), the Alps region (*Rhône-Alpes*), the Mediterranean (*Méditerranée*), and the southwest (*Aquitaine*), as well as with its neighbors, traveling at speeds of between 250 and 300 km/h with a perfect safety record (Soulié and Tricoire 2002; TGVweb 2010).

But (the spatial objection rising once again), do not America's enormous territory and low population density make it a quite unsuitable place for emulating the French fast train model? This perennial argument is valid only if somebody would advocate TGV-like links between Miami and Seattle or between Los Angeles and Boston. But consider this: France, a nation of 65 million people, has a population density of less than 120 people/km^2 and nationwide connections centered on the capital require a radial rail network with its longest links close to 1,000 km. In contrast, the Northeastern U.S. megalopolis, an area with more than 50 million inhabitants, has population density three times as high (360 people/km^2), mostly in a fairly narrow linear arrangement along the Atlantic coast, with large cities regularly strung along a route of less than 700 km. Which one of these territories is more suited for rapid downtown-to-downtown train links?

Whereas the French zip at speeds close to 300 km/h more than 900 km from Paris to Nice, Americans unwilling or unable to drive have the less attractive option of taking shuttle flights from Washington to New York, Philadelphia, or Boston (still only 600+ km away) with all the attendant joys of taxi rides to and from the airports, check-in lines, security searches, and flight delays. Acela is an unacceptably poor substitute for the real thing: It is not a real fast train, as it averages only about 120 km, no better than the best steam engine-powered trains of the 1930s. The Northeastern megalopolis is not the only region that should have had rapid trains for decades. Thalys trains from Paris (11.2 million) to Amsterdam (2.5 million) run more than 500 km twelve times a day (Thalys 2010): Why could not their counterparts run between Dallas/Fort Worth (6.5 million) and Houston (5.9 million), a distance of 360 km that could take just a bit more than one hour? Or between Los Angeles and San Francisco or New York and Montreal?

America has lost its global technical leadership, its exemplary innovative drive, and its political will to invest in transformative infrastructures with inevitably substantial (but by no means crippling) initial capital costs that are repaid by decades of efficient and convenient service. Even its existing infrastructures are falling apart: In its latest report card, the American Society of Civil Engineers (2009) awarded D to the country's aviation and D– to its roads. Four decades of regressing automotive performance followed by a decade of improvements and then by two more decades of stagnation, virtual absence of diesel-powered cars, and an inexplicable refusal to participate in the greatest land-transportation innovation of the past two generations, in building convenient and highly energy-efficient rapid trains, is a combination that has left the country with an extraordinarily inefficient transportation system.

America now looks inept even when compared with China, a country whose per capita gross domestic product (even when expressed by a liberally calculated purchasing power parity) is one seventh of the U.S. rate. By the summer of 2010 China had completed 6,920 km of high-speed rails, including nearly 2,000 km of links capable of speeds up to 350 km/h ("High-speed rail operations in China" 2010). Is the United States permanently incapable of investing in its essential transportation infrastructure as much as France has done since the 1980s or as much as China has done since the year 2000? No technical problems and no excessive capital costs could have prevented the United States from lowering its demand for transportation hydrocarbons and reducing its dependence on oil imports: Two dozen affluent countries now have considerably better automotive efficiency than does the United States and (besides the pioneering Japan and France and determined China) rapid train links have been built in such less affluent countries as South Korea and Spain.

In his State of the Union address in January 1974, Richard M. Nixon set a new national goal: "At the end of this decade, in the year 1980, the United States will not be dependent on any other country for the energy we need to provide our jobs, to heat our homes, and to keep our transportation moving" (Nixon 1974). That was a patently unrealistic goal, but during the subsequent decades the United States could have gradually reduced its crude oil demand at least to the level that would have required only imports from Canada and Mexico, its two North American Free Trade Agreement neighbors (in 2009 they amounted to about 22 percent of total U.S. crude oil consumption). Inexplicably, obvious opportunities are still ignored: CAFE standards were finally raised (to reach 35 mpg by 2020), diesels are slowly gaining a slightly larger market share, hybrid drives have become somewhat popular, and electrics are touted as a new solution, but all of this is still too little to make a real difference. Moreover, all plans for high-speed trains have been shelved indefinitely ("High-speed trains" 2010). America still waits to join the late twentieth century and get its first really fast train, but it is instead content to keep transferring trillions of dollars to the Middle Eastern theocrats and autocrats and pretending that it is possible to run permanently deep trade deficits.

References

American Society of Civil Engineers. 2009. *Report card for America's infrastructure*. http://www.infrastructurereportcard.org/report-cards (last accessed 28 March 2011).

British Petroleum (BP). 2010. BP statistical review of world energy. http://www.bp.com/liveassets/bp_internet/globalbp/globalbp_uk_english/reports_and_publications/statistical_energy_review_2008/STAGING/local_assets/2010_downloads/statistical_review_of_world_energy_full_report_2010.pdf (last accessed 28 March 2011).

European Union. 2010. Performance of passenger transport expressed in passenger-kilometers. In *Energy and transport in figures*. Luxembourg: European Union. http://ec.europa.eu/energy/publications/statistics/doc/2010_energy_transport_figures.pdf (last accessed 28 March 2011).

High-speed rail operations in China have reached 6,920 km. 2010. China.com. http://news.china.com.cn/rollnews/2010–07/02/content_2986085.htm (last accessed 28 March 2011).

High-speed trains: Running out of steam. 2010. *The Economist* 11 December 2010:40.

Institute National de la Statistique et des Études Économiques. 2010. *Annuaire statistique de la France* [Statistical yearbook of France]. Paris: INSEE. http://www.insee.fr/fr/publications-et-services/sommaire.asp?codesage=ASF08 (last accessed 28 March 2011).

Mercedes-Benz. 2010. *BlueTEC clean diesel*. Stuttgart: Mercedes-Benz. http://www.mbusa.com/mercedes/innovation/thinking_green/bluetec (last accessed 28 March 2011).

Nixon, R. M. 1974. *State of the union address 1974*. http://stateoftheunionaddress.org/category/richard-nixon (last accessed 28 March 2011).

Organization for Economic Co-operation and Development (OECD). 2009. *Energy balances of OECD countries*. Paris: OECD.

———. 2010. *StatExtracts*. Paris: OECD. http://stats.oecd.org/index.aspx (last accessed 28 March 2011).

Shima, H. 1994. Birth of the Shinkansen—A memoir. *Japan Railway & Transport Review* 3:45–48.

Sivak, M., and O. Tsimhoni. 2009. Fuel efficiency of vehicles on US roads: 1923–2006. *Energy Policy* 37:3168–70.

Smil, V. 2008. *Energy in nature and society: General energetics of complex systems*. Cambridge, MA: MIT Press.

Smil, V. 2010a. *Energy myths and realities: Bringing science to the energy policy debate*. Washington, DC: American Enterprise Institute.

———. 2010b. *Energy transitions: History, requirements, prospects*. Santa Barbara, CA: Praeger.

———. 2010c. *Prime movers of globalization: The history and impact of diesel engines and gas turbines*. Cambridge, MA: MIT Press.

Soulié, C., and J. Tricoire. 2002. *Le grand livre du TGV* [Great book of TGV]. Paris: Vie du rail.

TGVweb. 2010. *TGVweb*. http://www.trainweb.org/tgvpages/tgvindex.html (last accessed 28 March 2011).

Thalys. 2010. Thalys timetables. Paris: Thalys. http://www.thalys.com/fr/en/timetables/Paris/Amsterdam (last accessed 28 March 2011).

U.S. Census Bureau. 2010. *Foreign trade*. Washington, DC: U.S. Census Bureau. http://www.census.gov/foreign-trade/statistics/historical/ (last accessed 28 March 2011).

U.S. Department of Energy. 2010. Find and compare cars. Washington, DC: U.S. Department of Energy. http://www.fueleconomy.gov/ (last accessed 28 March 2011).

U.S. Department of Transport. 2010. *Summary of fuel economy performance*. Washington, DC: U.S. Department of Transportation. http://www.nhtsa.gov/staticfiles/rulemaking/pdf/cafe/Oct2010_Summary_Report.pdf (last accessed 28 March 2011).

U.S. Energy Information Administration (USEIA). 2010. *Annual energy review*. Washington, DC: USEIA. http://www.eia.doe.gov/aer/ (last accessed 28 March 2011).

Modeling and Assessment of Wind and Insolation Resources with a Focus on Their Complementary Nature: A Case Study of Oklahoma

Weiping Li,* Steve Stadler,* and Rama Ramakumar†

*Geography Department, Oklahoma State University
†School of Electrical and Compter Engineering, Oklahoma State University

Energy in wind and incoming solar radiation has been observed to be complementary over time. That is, as one of these resources slackens, the other tends to increase. Considering the intermittency of wind and solar radiation, complementarity provides the potential to better exploit the two resources in combination. How well a hybrid system using both resources can improve the reliability of energy generation is partially a function of their complementarity. This study developed an approach to calculate the Complementarity Index of Wind and Solar Radiation (CIWS). Geographic analyses including principal component analysis and geographically weighted regression modeling were applied to investigate the impacts of diverse geographic factors on complementarity. Oklahoma was used as the case study area because of the availability of quality-controlled five-minute data from 127 Mesonet sites. The results indicate average CIWS (unitless) is 10.99, about 46 percent of the theoretical maximum value of 24; the standard deviation is 1.5. Approximately 57 percent of the sites have above-average CIWS values. Also, complementarity is spatially skewed, with the highest CIWS values falling in the east, where both wind and solar energy are less abundant. Geographic analysis shows three groups of geographic factors including moisture, temperature, and landscape explain 86.6 percent of variations in complementarity. *Key Words: complementarity, geographically weighted regression, Oklahoma, solar energy, wind energy.*

本研究观察到风力和太阳辐射能量随着时间的互补性。也就是说，当这些资源中的一种放缓时，其他则呈上升趋势。考虑到风力和太阳辐射的间歇性，此互补性提供了更好地一起利用两种资源的可能。混合动力系统多么恰当地使用两种资源，以提高发电的可靠性，部份地由这两种资源的互补性决定。本研究开发了一种方法来计算风和太阳辐射 (CIWS) 互补性指数。采用包括主成分分析和地理加权回归模型的地理分析，以探讨不同地理因素对互补性的影响。俄克拉荷马州被用作研究区域是由于其质量的可及性-来自 127 个气象网 Mesonet 站点的五分钟受控数据。结果表明，平均风和太阳辐射互补性指数 CIWS (无单位) 为 10.99，约最大理论值 24 的百分之46，标准差为 1.5。大约百分之 57 的网站都高于平均 CIWS 值。此外，互补性在空间上不对称，最高的 CIWS 值分布于风能和太阳能资源相对少的东部地区。地理分析显示三组地理因素，包括湿度，温度，以及景观，解释了互补性中百分之 86.6 的变化。*关键词：互补性，地理加权回归，俄克拉荷马州，太阳能，风能。*

La energía del viento y de la insolación ha sido considerada complementaria a través del tiempo. Es decir, a medida que uno de estos recursos mengua, el otro tiende a incrementarse. Teniendo en cuenta la intermitencia del viento y de la radiación solar, la complementariedad suministra el potencial para explotar los dos recursos en combinación. Qué tan bien puede un sistema híbrido que utilice ambos recursos mejorar la confiabilidad en la generación de energía es parcialmente una función de su complementariedad. En este estudio se desarrolló un enfoque para calcular el Índice de Complementariedad del Viento y la Radiación Solar (CIWS, sigla en inglés). Se aplicaron análisis geográficos que incluyen el análisis de componentes principales y modelos de regresión geográficamente ponderada para investigar los impactos de diversos factores geográficos sobre la complementariedad. Se adoptó a Oklahoma como el área del estudio de caso debido a la disponibilidad de datos sobre el control de la calidad con frecuencia de cada cinco minutos en 127 sitios de la Mesonet. Los resultados indican que el promedio del CIWS (sin unidades) es de 10.99, cerca del 46 por ciento del valor teórico máximo de 24; la desviación estándar es de 1.5. Aproximadamente el 57 por ciento de los sitios registran valores de CIWS por encima del promedio. La complementariedad también está sesgada espacialmente, presentándose los más altos valores de CIWS hacia el este, donde tanto el viento como la radiación solar son menos intensos. El análisis geográfico muestra tres grupos de factores geográficos, incluyendo humedad, temperatura y paisaje, que explican el 86.6 por ciento de las variaciones en complementariedad. *Palabras clave: complementariedad, regresión geográficamente ponderada, Oklahoma, energía solar, energía eólica.*

Environmental degradation and energy exhaustion have become prominent concerns in relation to increased consumption of nonrenewable fossil fuels (Energy Information Administration 2010). Projections for the middle of the twenty-first century indicate that the supply of fossil fuels as primary energy sources will not keep up with accelerating demands (Lincoln 2005).

Renewable energy, abundant and with minimal pollution, presents attractive opportunities for sustained development. Renewable energy is "energy derived from resources that are regenerative or for all practical purposes cannot be depleted" (U.S. Department of Energy 2010). Importantly, renewable energy "comes with the territory" and is "free." However, costs of conversion of renewable energy sources into electrical energy can be significant and vary in a complex manner by geography and scale of generation.

Wind and incoming solar radiation (insolation) are the most abundant and ubiquitous renewable resources (Sawin 2004). It has been estimated that solar energy delivered to Earth's surface is about 5,500 times of the current world energy consumption (Meyer 2008). The global wind power potential is at least five times current global power demand (Archer and Jacobson 2005). These two renewable resources represent significant opportunities for worldwide electrical generation with minor impacts on Earth's natural systems.

Despite availability of wind and solar resources, the implementation of widespread conversion of these resources to electricity is not straightforward. The essential argument is not whether renewable hybrid electric generation is wise in the long run but how to achieve a sustainable energy regime in which renewable hybrid generation plays a significant role in world electrical generation. In more developed countries this means diversified energy generation portfolios and, hopefully, increased power reliability. In lesser developed countries, hybrid generation can offer cost-effective electricity at off-grid locations. Standing in the way are nontrivial hurdles of reliability, conversion efficiency, land use conflicts, and variations in the economic development of world regions. Space is limited here, but the reader is referred to Owen (1996); Abbasi and Abbasi (2000); Kaltschmitt, Streicher, and Weise (2009); and Simon (2009) for the flavor of ongoing debates.

Hybrid systems combining wind and solar generation at single sites have started to appear around the world. To date, hybrid systems are of modest output, capable of powering individual homes and villages. Wind and solar resources vary appreciably over short time periods such as hours and days and their dependability as hybrid fuel sources requires complex engineering. In particular, it is expensive to store electrical output so that it is desirable to mix and match renewable sources to meet immediate electrical demand (Zhou et al. 2010; Zimmerer 2011).

As solar and wind technologies become more efficient and significant battery storage becomes cheaper, hybrid generation should flourish. Yet, the geographic nature of the joint occurrence of wind and solar resources has not been fully explored. We argue that increased knowledge of geographic patterns of complementarity will aid the efficient tapping of these resources. Looking to the future, this article presents a measure by which the temporal variability of wind and solar resources can be jointly assessed and a method by which to model complementarity over area.

Complementarity

The notion of complementarity was first used in economics by Edgeworth (1925). According to Edgeworth, "activities are Edgeworth complements if doing (more of) any one of them increases the returns to doing (more of) the others" (Milgrom and Roberts 1995, 181). Nobel Laureate Niels Bohr's wide-ranging work introduced complementarity as a basis for quantum physics and strongly asserted that complementarity is important in disciplines beyond physics (Bohr 1999). Diverse disciplines such as molecular biology (International Human Genome Sequencing Consortium 2004), mathematics (Kuyk 1997), and social psychology (Scott and Marshall 2010) have explored various forms of complementarity.

Complementarity is an attractive paradigm that has received some attention in power generation applications. Our expectation is that colocating wind and solar generation of electricity can assure higher total output of the two when either one of the two sources increases and can improve the stability and reliability of the power provided by such systems.

The Importance of the Complementarity of Wind and Solar Resources

Over 1.5 billion persons on Earth have no access to electricity (International Energy Authority 2010). The shortage of transmission line capacities and environmental concerns have increased interest in the use of distributed power generation, also called *on-site generation* and *dispersed generation* (Wang and Singh 2008). Such generation represents scattered power from many small energy sources usually located near the end demand for the electricity. In developed nations with sophisticated transmission grids, distributed generation can serve as capacitors to help stabilize the entire grid (Macken, Bollen, and Belmans 2004).

The ubiquity of wind and solar resources make them especially intriguing for use in distributed, hybrid systems in which wind and solar generation units are integrated. Any complementarity between their outputs would further serve to increase their reliability in terms of total power generated.

Currently, capital costs of energy favors wind farms if a site is grid-connected, whereas electric load matching usually favors a hybrid wind–solar plant (Reichling and Kulacki 2008). An assessment of the complementarity of wind and solar energy should be a precursor to deciding the configuration of a wind–solar hybrid unit, but previous research has not developed a quantitative method.

A renewable portfolio standard (RPS) is a set of regulations mandating electricity suppliers to use renewable sources for a percentage of total generation or absolute amount of generation (Wiser et al. 2007). In the United States, over half of the states have adopted RPS and several others have adopted nonbinding standards. Achieving RPS goals requires greater penetration of multiple renewable resources. This milieu secures a place for the importance of knowledge of complementarity, which will help to reduce cost from inherent fluctuations in wind and solar resources.

Defining Complementarity of Wind and Solar Resources

When jointly considering these wind and solar resources in many middle latitude locations, it can be informally observed that wind speeds are lowest in the summer when the sun shines brightest and longest, whereas the wind is strongest in the winter when cloudiness and shorter day lengths limit solar energy. In this context, the complementarity of wind and solar resources is the degree to which one energy source can be depended on to take up the slack for the other. We wish to gauge the consistency of combined wind–solar resources. Locations with "high" complementarities would have major advantages if a design purpose was to situate wind–solar hybrid systems to produce some minimum amount of electricity from the two sources.

The complementarity of wind and solar resources is herein defined as the extent to which the energy fluxes of the two sources tend to be inversely related. The more inverse the tendency, the more they complement each other. Figure 1 illustrates hypothetical constructs. The vertical axes represent energy amount and the horizontal axes the months of the year. Figure 1A shows perfect complementarity, which would result in the greatest of consistency incident energy from wind and solar radiation combined. Figure 1B illustrates the condition in which wind power and solar energy follow the same curve and would produce the greatest annual fluctuation in the total energy resource. Figure 1C is an illustration of partial complementarity in which the solar source somewhat makes up for times of lesser wind power availability and vice versa; it is this situation that we would expect to be representative of most real-world conditions.

This line of thought begs some questions: What method can be used to quantify the complementarity between the two site-specific energy forms? Importantly, are there significant geographic differences? Do geographic factors affect the nature of the complementarity?

Related Research

The complementary nature of wind and solar radiation has been qualitatively noted, but there have been few studies targeting its quantitative aspects. Tackle and Shaw (1977) studied the complementary nature of daily wind and solar energy in Des Moines, Iowa. In March and April, solar energy typically reached its minimum when wind energy was at annual peak; in July and August there was a tendency for both energy sources to rise and fall in concert instead of being complementary.

Kimura, Onai, and Ushiyama (1996) installed a small wind and solar hybrid system for nine months in Ashikaga, Japan and found complementarity between wind and solar energy. Barley et al. (1997) found "combinations of wind and PV [photovoltaic] are more cost-effective than using either one alone" in their study of Inner Mongolia.

Sahin (2000) characterized the complementarity of one-minute wind and solar power over a year at an Arabian Gulf shoreline site in northeastern Saudi Arabia. The author calculated the Pearson correlation coefficient between the solar and wind power densities.

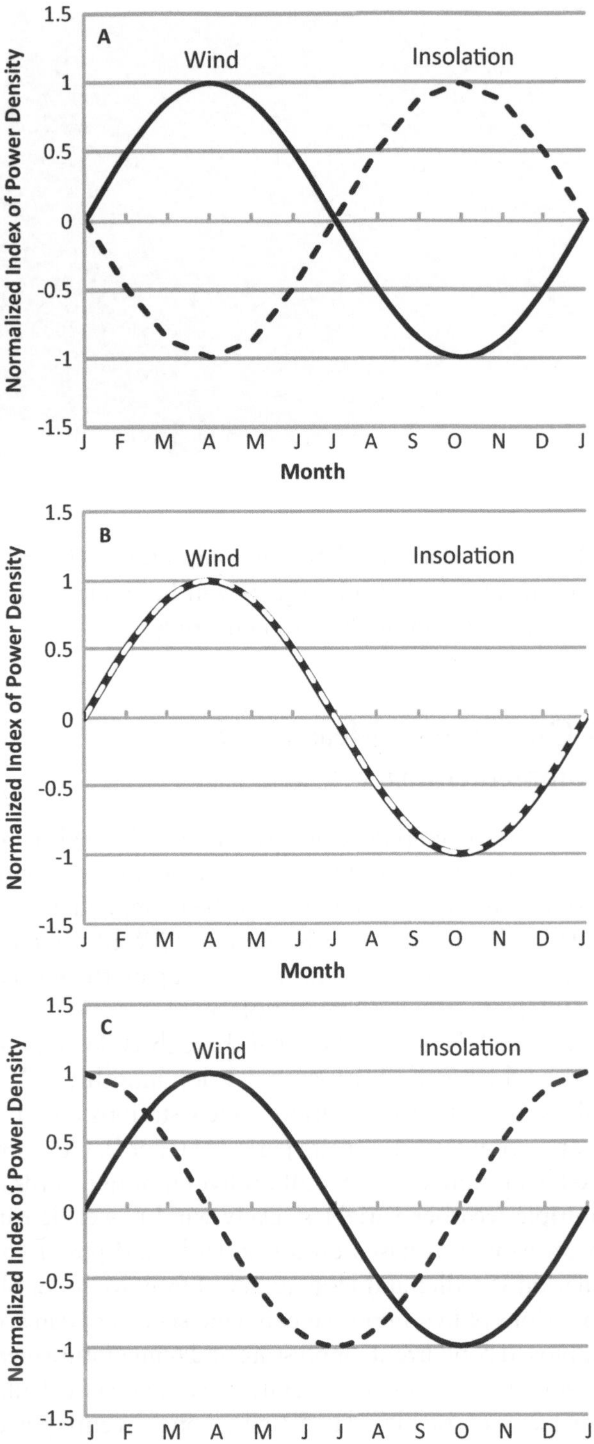

Figure 1. Hypothetical example of annual wind and solar resources. (A) Perfectly complementary, (B) Perfectly concordant, (C) Partially complementary.

The daily average solar power showed greater consistency than wind power and monthly averages of combined solar and wind power potential indicated some complementarity.

Ai et al. (2003) developed a set of calculations for optimum sizing of photovoltaic–wind hybrid systems for given electric demands. Hourly wind, solar radiation, and electrical load data for Wagland Island southeast of Hong Kong indicated seasonal complementary, but they did not employ a formal analysis.

Felder (2004) advanced an alternative strategy to match regional end-use energy demand with local renewable energy resources. Over the annual cycle at two locations in central Pennsylvania, a large portion of end-use energy demand could be satisfied through a combination of local wind, solar, and hydropower, with superinsulated housing. Her graphs show evidence of complementarity between wind and solar power.

Jacobson (2009) presented a range of future energy solutions by examining combinations of energy sources and their environmental consequences. Using combined modeled energy output from five California locations, he showed that resource intermittency can be managed so that renewables might provide 80 percent of the total demand. The results were agglomerated and not focused on analysis of spatial variation of complementarity.

Stoutenberg, Jenkins, and Jacobson (2010) have estimated theoretical combined power capacity from colocated offshore wind turbines and wave energy converters situated on the California coastline. Their results indicate the hybridization of the two energy sources would provide less hours of zero generation and outperform wind or wave generation alone.

The results from these studies at disparate locations hint that the complementarity between wind and solar energy is a global feature. However, none of the studies used spatial modeling. Greater knowledge of complementarity's geographic characteristics should be of value.

The Case of Oklahoma

Oklahoma is a suitable test bed for the study of complementarity because of the physical diversity over its substantial extent (181,000 km^2). At its northern boundary the state is 769 km across, incorporating humid subtropical and middle latitude steppe climates as well as middle latitude forest and middle latitude grassland biomes. The state is monitored by over one hundred Oklahoma Mesonetwork (Mesonet) stations. Its sites are shown in Figure 2, and its characteristics are given by McPherson et al. (2007). These data have the advantage of having undergone thorough temporal

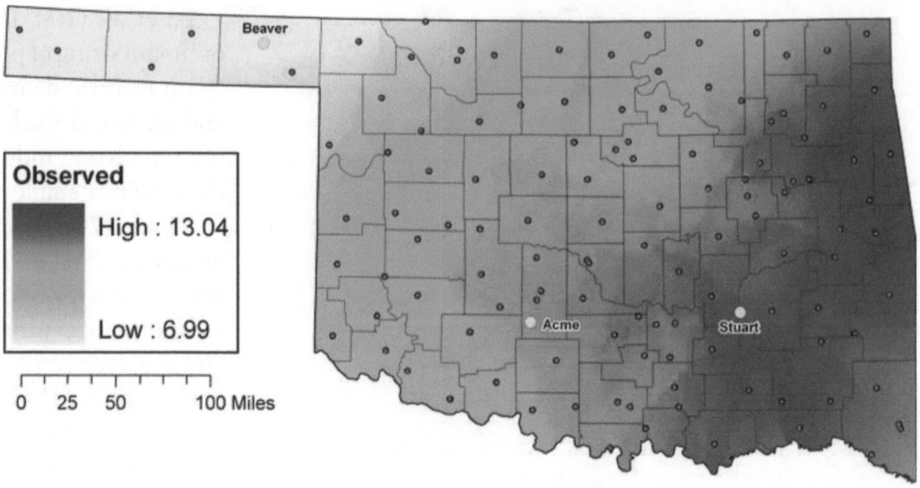

Figure 2. Locations of the Oklahoma Mesonetwork sites and annual Complementary Index of Wind and Solar Radiation values.

and spatial quality control. We used Mesonet temperature, rainfall, pressure, humidity, wind speed, and solar radiation. To calculate wind power we used five-minute averages at the standard observation heights of 10 m for wind and 1.5 m for solar radiation. In our subsequent geographically weighted regression (GWR) analysis of geographic factors, we used a number of weather and landscape attributes of each Mesonet site.

Our analysis was rurally based. The Oklahoma Mesonet stations serving as our data source were originally sited to be representative of rural surroundings within a few kilometers. This avoided the complicated urban temperature and wind variations so well established over the micro- and mesoscales (Geiger, Aron, and Todhunter 2005). We surmise that statewide complementarity maps such as those presented in this article would not change with the inclusion of urban data and that data from urban areas would need to be explored for individual siting decisions within cities. Moreover, wind–solar hybrid systems are more readily sited in rural areas because of land use conflicts.

We excluded economic and social factors that might be vital to successful installation of hybrid generation. Likewise, determination of optimal scales of generation and equipment specifications were beyond the scope of this study. Our analysis was dictated by the natural resources of Oklahoma rather than extant wind and solar technologies.

In Oklahoma, hybrid projects using wind and solar energy might or might not be attached to the electrical grid, depending on economics and the wishes of the developer. However, large expanses of the lesser developed world lacking electrical transmission will likely adopt off-grid generation as the only viable manner to achieve electrification. These are important differences in terms of how hybrid projects will be implemented, but have not been included in our study.

A Quantitative Measure of Complementarity

To aid geographic comparisons, we devised a numeric index to quantify the complementarity of wind and solar resources and named it the Complementarity Index of Wind and Solar Radiation (CIWS). Calculation of CIWS provides a means to compare the level of complementarity for different places.

We used data from the 1994 through 2006 archives of the Oklahoma Mesonet to develop monthly CIWS values for each of 127 stations (a few stations were not in operation over the entire period of record). Our work used approximately 158 million five-minute records of multiple weather variables. Daily wind power density in watts/m^2 was based on a standard wind power calculation (Stadler and Hughes 2005) from Mesonet observations of five-minute mean wind speed at 10 m and employed concurrent air pressure and temperature measurements at 1.5 m. The resulting wind power calculations were extrapolated to 50 m by a one-seventh power law commonly applied to such data (Stadler and Hughes 2005); this is more realistic than using 10 m wind measurements because large wind turbines have hub heights of 50 m or more to reach greater wind power. Long-term average monthly incoming solar radiation (insolation) density (watts/m^2) used five-minute average insolation (watt/m^2) summed over each calendar day, averaged by month, and averaged over the thirteen years in the period of record. Figure 3A shows the long-term monthly

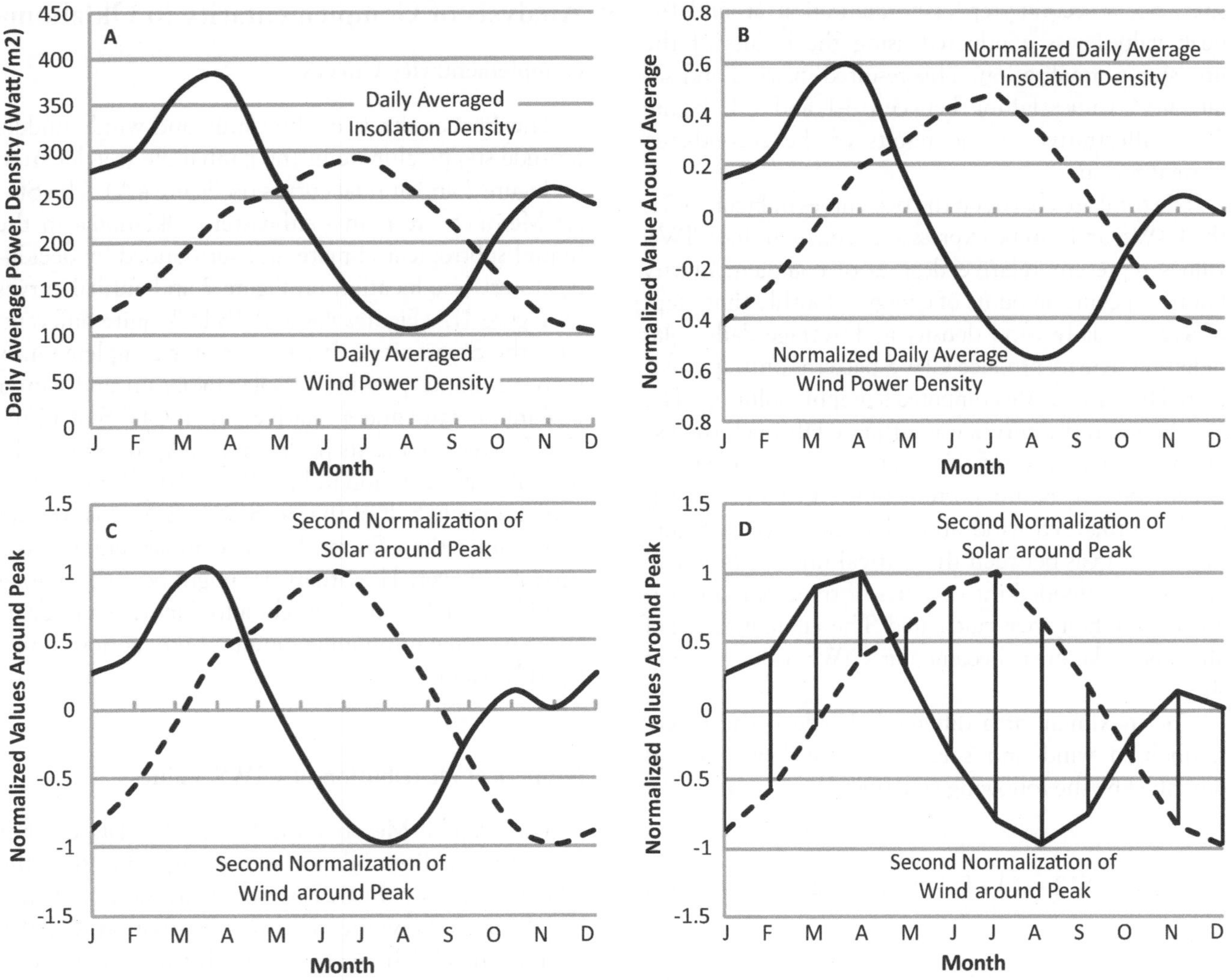

Figure 3. Data from the Acme, Oklahoma, Mesonet station. See Figure 2 for its location. (A) Long-term average daily incoming solar radiation density and average daily wind power density by month. (B) Normalization of Figure 3A data around the yearly mean. (C) Second normalization of Figure 3B data around peak values. (D) Example of areas between normalized wind and insolation values of Figure 3C data. Complementary Index of Wind and Solar Radiation is the summed area of all trapezoids.

averages of wind and solar densities at the Mesonet Acme station in southwestern Oklahoma. The Acme site is relatively dry and surrounded by grassland with interspersed trees. Qualitatively, there seems to be substantial annual complementarity.

Normalization

The complementarity between wind power and solar radiation is relative to a specific period of time at each place on Earth's surface. It can be derived from the raw data of the actual temporal distribution profiles for wind and solar power output but is not definable for a single instant of time. Moreover, over any time period these profiles would vary depending on the location and the height of wind towers. To facilitate exploration of the spatial features of complementarity, a two-step normalization procedure was employed.

The first-step normalization was based around average daily wind density and average daily solar radiation density by month for thirteen years at each site. The results from this step were unitless values ranging around 0 (the long-term annual mean). For instance, in Figure 3B, +0.20 on the vertical axis represents a monthly value 20 percent above the annual average value for either wind or solar radiation. The normalized values had differing extremes by resource (wind or solar) and by Mesonet station.

Peak amounts (watts/m^2) of daily incident solar and wind energy varied widely at single sites and between sites. So, a second step of normalization around the peak values was conducted using the results of the first-step normalization. This resulted in wind and solar curve values falling between +1 and −1. Figure 3C is illustrative of the results of the second-step normalization.

The total area between the two curves in Figure 3C is the CIWS and can be expressed in square units. CIWS values represent relative degrees of complementarity but not specific amounts of energy. Monthly time steps of average daily wind density and average daily solar radiation density produce CIWS patterns only approximated by curves. We computed a graphic solution. The area between the curves was estimated through the use of an automated script. Vertical lines were drawn between the two points corresponding to monthly values of normalized wind and solar energy (e.g., Figure 3D). The areas between the vertical lines and the two curves were divided into a series of trapezoids and triangles and their areas calculated. The summarization of all trapezoidal areas became our CIWS value for that site.

The maximum area difference between the twice-normalized wind and solar curves in Figure 1A is described by the following equation:

$$(12 - 0) * [1 - (-1)] = 24 \tag{1}$$

The theoretical upper limit for CIWS is 24 and the theoretical lower limit is 0.

Analysis of Complementarity in Oklahoma

Complementarity Curves

The Beaver site is in the sunny and windy middle latitude steppe climate in the panhandle (see location on Figure 2 and annual curves on Figure 4A). The Stuart Mesonet site is in southeastern Oklahoma in the humid subtropical climate and surrounded by deciduous forest (see location on Figure 2 and annual curves on Figure 4B). Figures 4A and 4B look quite different, with the Stuart site exhibiting greater complementary with a summer peaking of solar radiation and winter peaking of wind and a relatively high CIWS of 12.67. The Beaver site has its peak wind in April followed by its peak solar radiation in July. Beaver's CIWS of 6.42 is approximately half of that at Stuart. Beaver is windier and sunnier than Stuart, but its complementarity is a good deal lower. The Stuart site might be superior for a hybrid system because it might afford more hours, guaranteeing some minimum amount of production from a hybrid system.

Numerical Distribution of CIWS Values

The original Mesonet wind and solar data did not possess normality and it was not surprising that the CIWS values for the 127 Mesonet sites were not normal. Figure 5 shows the frequency distribution of CIWS values. The Oklahoma frequency distribution is decidedly nonnormal and statistics assuming normality must be applied with caution.

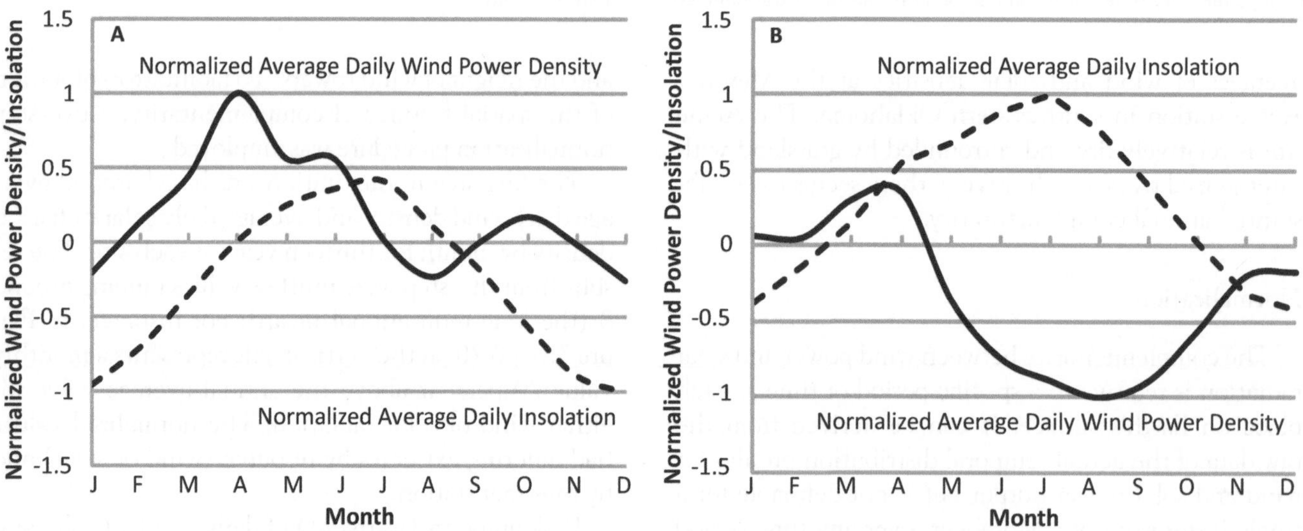

Figure 4. Twice-normalized wind and solar radiation data. (A) Stuart Mesonet site. (B) Beaver Mesonet site.

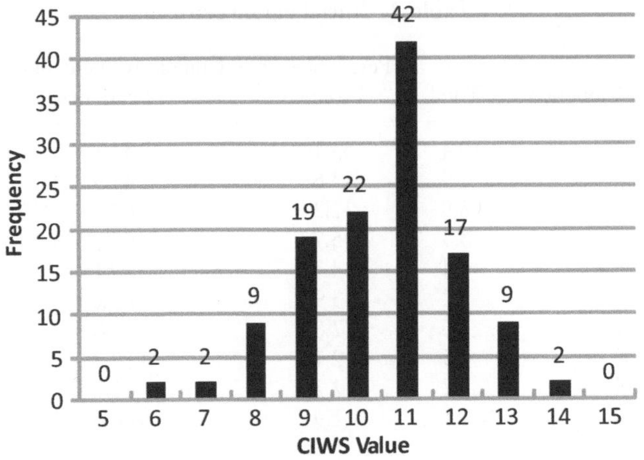

Figure 5. Frequency distribution of Complementary Index of Wind and Solar Radiation (CIWS) by integer values for all Mesonet sites. Mean is 10.99, standard deviation is 1.50, the maximum is 14.79, minimum is 6.99, skewness is –0.48, and kurtosis is +50.

Spatial Distribution of CIWS Values

In themselves, CIWS values do not represent amounts of power and assume meaning only when placed in a spatial context. The index values of Mesonet sites were converted into a CIWS surface using kriging (Oliver and Webster 1990). We experimented with several other interpolation methods and they did not produce smooth patterns as did kriging. Figure 2 shows that the locations having the highest CIWS values are in eastern Oklahoma, and the western portion of the state has lower CIWS values. The absolute annual totals of wind and solar energy together are greater at western sites (like Beaver), whereas complementarity is best at eastern sites (like Stuart). What do the spatial differences between CIWS values mean? Our index can only qualitatively approach this, but the definite west-to-east differences in Figure 2 lead us to believe that the essence of Oklahoma's climates, topography, and biomes are embedded in the CIWS values.

Geographic Analysis

Both wind and solar radiation vary from place to place and from moment to moment. Physical factors might well act in combination and at various spatial scales. Landscape variations—such as topography, vegetation, or buildings—can create complex flows of wind (Geiger, Aron, and Todhunter 2005). Furthermore, solar radiation is affected by latitude and content of the air. Because these absolute and relative geographic factors are known to affect wind and insolation, they should affect complementarity as well.

We attempted to identify significant geographic factors affecting complementarity over Oklahoma. We used land cover, terrain, relative locations, and Mesonet data within a geographic information system (GIS). In all, we identified more than four dozen candidate variables, listed in Table 1. We subjected the values of these variables at Mesonet sites to principal component analysis (Giordani and Kiers 2006). Ten initial eigenvectors were significant at the 0.01 level (see Table 2), explaining 86.6 percent of the variance in the CIWS. An examination of the components showed the first three (comprising 60.1 percent of the variance) to be explainable, and we termed them the *temperature*, *moisture*, and *landscape dimensions*. The variables included in each of the dimensions are given in Table 3. Of note is the logical nature of signs of most of the simple correlations between each variable and CIWS.

The eigenvectors in Table 2 were placed in a GWR model, also known as *local geographic modeling*. This technique provides an alternative to spatially unweighted standard multiple regression. Fotheringham, Brunsdon, and Charlton (2002) developed GWR to reduce spatial autocorrelation by limiting independent variables to smaller moving windows rather than the areal domain of a study area as a whole. Certainly, this makes sense for study of Oklahoma with its substantial natural diversity. GWR emphasizes spatially varying relationships, which we believe to be the nature of complementarity in Oklahoma. Our GWR model used CIWS as the dependent variable and the eigenvectors from Table 2 as independent variables.

Of the ten original eigenvectors, seven entered into the model as statistically significant. The adjusted R^2 was 0.65 (significance < 0.001). This result was taken to mean that we had some success in interpolating the CIWS across distance in Oklahoma. Also, it represented an improvement on the results of our parallel non-geographically weighted multiple regression, which produced an adjusted R^2 of 0.53 (significance < 0.001). The GWR produced an Akaike information criterion (Akaike 1974) value of 365.5. The information criterion is not a formal test of significance because the values are specific to the particular model, but rather a measure of the trade-off between bias and variance when the model is constructed. The regular multiple regression produced an information criterion value of 377.6, meaning that the GWR model had outperformed it. (In this measure, larger numbers are less desirable.)

Table 1. Variables used in the principal components analysis

Variable	Variable
1 Latitude	29 Average rain in fall
2 Longitude	30 Average precipitation in winter
3 Elevation	31 Average pressure in spring
4 Distance to Gulf of Mexico	32 Average pressure in fall
5 Aspect 1 km range	33 Average pressure in summer
6 Aspect 5 km range	34 Average pressure in winter
7 Aspect 10 km range	35 Relative humidity in spring
8 Slope 1 km range	36 Relative humidity in summer
9 Slope 5 km range	37 Relative humidity in winter
10 Slope 10 km range	38 Relative humidity in fall
11 Curvature 1 km range	39 Average temperature in spring
12 Curvature 5 km range	40 Average temperature in summer
13 Curvature 10 km range	41 Average temperature in winter
14 Hill shade within 10 km	42 Maximum temperature in spring
15 Relative elevation 5 km east	43 Minimum temperature in spring
16 Relative elevation 5 km south	44 Temperature range in spring
17 Relative elevation 5 km north	45 Maximum temperature in fall
18 Relative elevation 10 km east	46 Minimum temperature in fall
19 Relative elevation 10 km south	47 Temperature range in fall
20 Maximum temperature in summer	48 Average cloudiness
21 Minimum temperature in summer	49 Temperature range in winter
22 Temperature range in summer	50 Percentage pasture within 10 km
23 Maximum temperature in winter	51 Percentage barren within 10 km
24 Minimum temperature in winter	52 Percentage shrub within 10 km
25 Relative elevation 10 km west	53 Water surface within 10 km
26 Relative elevation 10 km north	54 Percentage urban within 10 km
27 Average rain in spring	55 Percentage forest within 10 km
28 Average rain in summer	56 Relative elevation 5 km west
	57 Average temperature in fall

We created interpolated surfaces by kriging GWR model results. Figure 6 maps the intercepts of the models. Intercepts represent the average effect on the dependent variable of all the independent variables excluded from the model. The intercept values increased in a southeasterly direction and can be interpreted as the model performing modestly differently over the state.

Table 2. Initial eigenvectors

Component	Total	Percentage of variance	Cumulative percentage of variance
1	19.4	34.1	34.1
2	7.9	13.8	47.9
3	6.9	12.2	60.1
4	3.3	5.7	65.8
5	2.4	4.3	70.1
6	2.2	3.9	74 .0
7	1.6	2.8	76.8
8	1.5	2.7	79.5
9	1.2	2.2	81.7
10	1.1	1.9	86.6

Note: All ten eigenvectors are significant at the 0.01 level.

Figure 7 maps the moisture dimension. The lowest regression coefficients are in the southeastern mountains, which are the part of the state most often in the train of low-level moisture from the Gulf of Mexico. Figure 8 places the temperature dimension coefficients with lowest values associated with the forested mountains in the southeast and highest values over broadly rolling terrain to the north and west; in all it is very similar to Figure 7 and hints that moisture and temperature act together on complementarity. Figure 9 shows the distribution of the landscape dimension coefficients; they are highest in the southeast (greatest topographic relief) and the panhandle (highest elevations). Figure 10 presents the geography of the GWR model residuals.

Table 3. Explanation of first three principal components

Component number	Dimension name	Sign of components' simple correlations with Complementary Index of Wind and Solar Radiation
1	Moisture	Relative humidity + Longitude – Elevation – Cloud Index +
2	Temperature	Latitude – Distance from Gulf of Mexico – Elevation – Temperature +
3	Landscape	Relative elevation 5 km south + Curvature[a] –

Note: All correlations in this table are statistically significant at the 0.01 level.

[a]Curvature derived by smoothing of a raster surface by a function within a geographic information system.

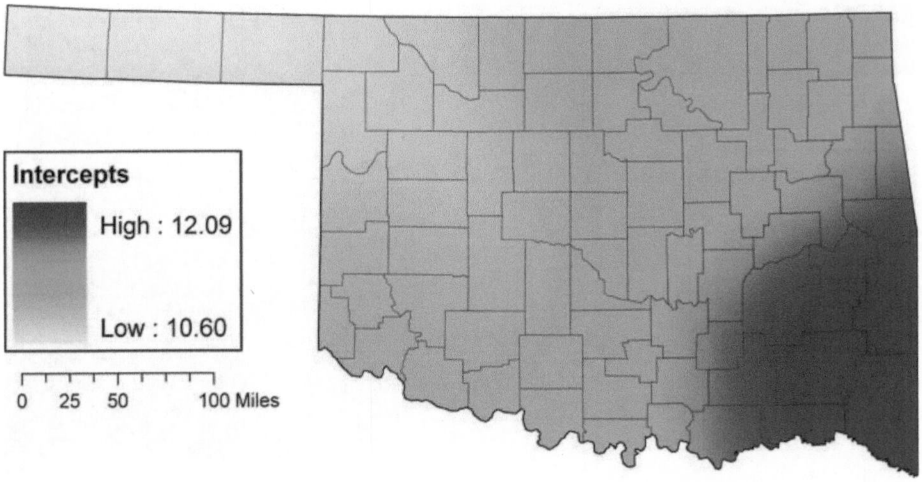

Figure 6. Kriged surface of geographically weighted regression model intercepts.

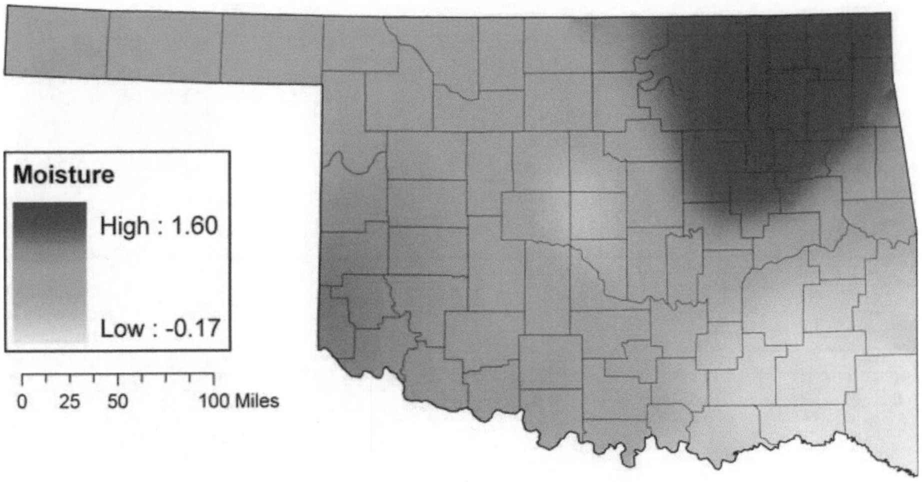

Figure 7. Kriged surface of geographically weighted regression model moisture dimension coefficients.

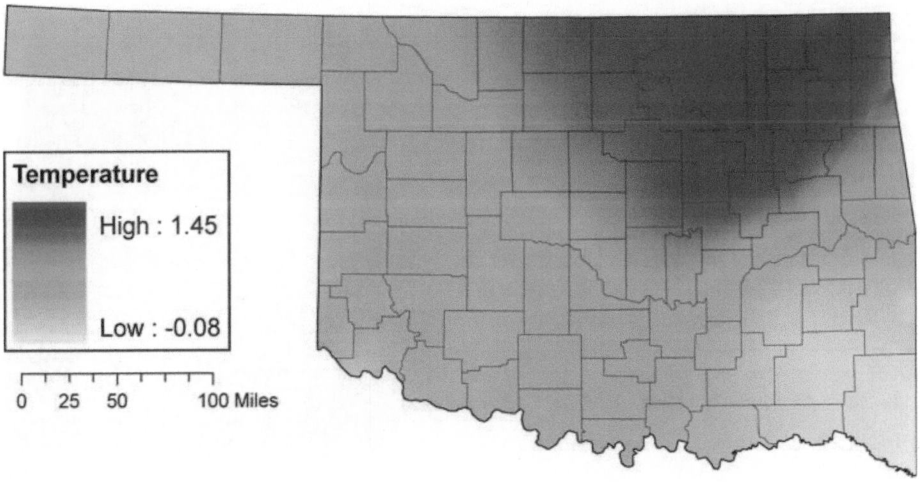

Figure 8. Kriged surface of geographically weighted regression model temperature dimension coefficients.

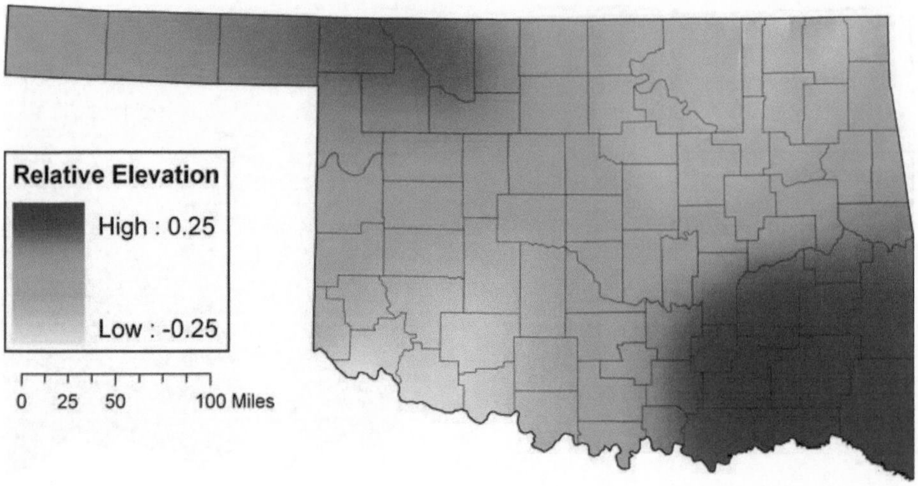

Figure 9. Kriged surface of geographically weighted regression model landscape dimension coefficients.

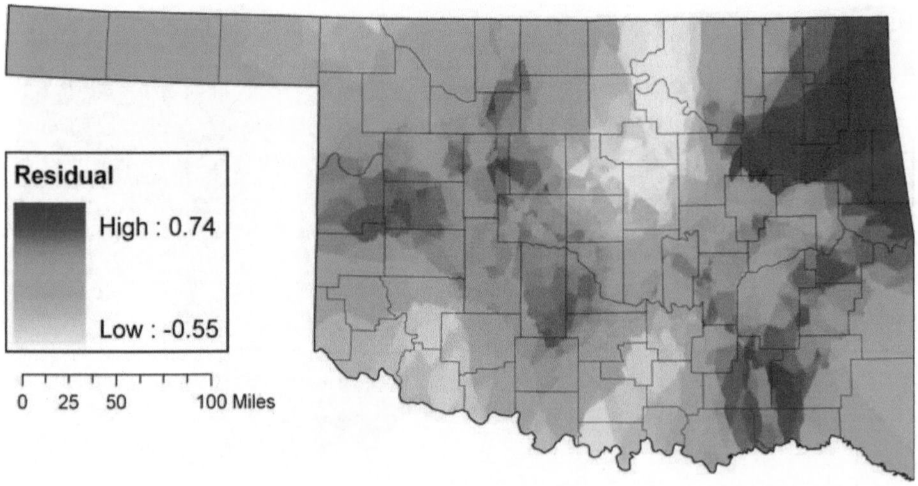

Figure 10. Kriged surface of geographically weighted regression model residuals.

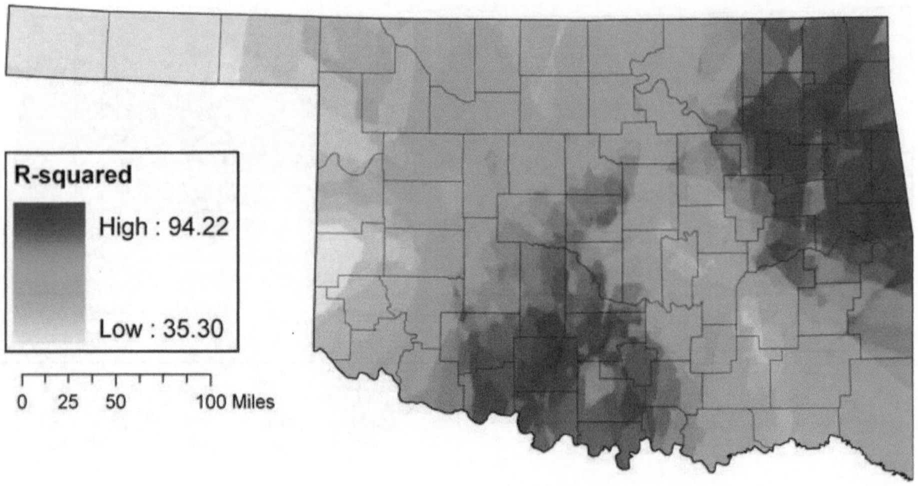

Figure 11. Kriged surface of geographically weighted regression model local R^2.

Model residuals are modest and have no patterns related to climate or vegetation. Figure 11 is the kriged map of local R^2 between the Mesonet-based calculations and the GWR-modeled CIWS values The R^2 is best on the Ozark Plateau of the northeast and Oklahoma's south central plains; this is of note because neither of these are areas where the regression coefficients of the independent variables are highest or lowest. Perhaps this indicates the existence of other factors we have not captured well in our GWR.

Conclusions

Several conclusions can be derived from this study. First, the complementary nature of wind and insolation is a feature that can be quantified. Quantification can be either based on historical wind–solar data or predicted based on reliably known local geographic factors. Next, complementarity varies from place to place. Perhaps surprisingly, the highest complementarities are found in locations with the least total incident wind and solar power. The average level of complementarity between wind and insolation for all of Oklahoma is 46 percent of the theoretical maximum CIWS value. The frequency distribution of CIWS among 127 Mesonet sites of Oklahoma is skewed to the effect that about 57 percent of the sites have above-average CIWS values and the occurrence of CIWS is spatially skewed in that the highest values are in eastern Oklahoma. Finally, annual cycles of CIWS noticeably vary over the state (e.g., the differences between the Stuart and Beaver sites as shown in Figure 4) because of pronounced climatic and biome diversity. Adjacent Mesonet sites have similar CWIS values, so we have some confidence that the CIWS is based in reality.

The CIWS can be used in GIS or other explorative spatial data analysis tools to illuminate its spatial variations. It also can be used in quantitative models relating CIWS to sets of geographic factors. In Oklahoma, three groups of spatial factors that we identified as moisture, temperature, and landscape dimensions explained 86.6 percent of the variation in complementarity. Would the percentage explained be similar in other regions or are other combinations of factors important? The methods applied in this study should be applied to other regions so that the GWR approach can be tested and verified. We intentionally positioned our study in a data-rich state. Other index methods could be developed given varying quantity and quality of wind, solar, and geographic data in other regions of the world.

The CIWS values and their annual complementarity profiles of wind and insolation are of direct use to guide the planning and design of the hybrid systems that hold much promise for the dual exploitation of wind and solar energy in many places around the world. Extending this reasoning, relatively rich wind and solar resources with good complementarity could be identified to facilitate identification of generation sites with favorable costs and potential power production.

We are not suggesting that the method we employed to calculate CIWS is the penultimate approach for the quantification of complementarity. The normalization procedure could be further simplified and improved. Larger-area variations of the geographic factors related to the index invite additional study. We have conducted some preliminary quantification of diurnal CIWS using subdaily data for selected days; this appears promising but is complicated and not included here. Further study of the complementary nature of wind and solar radiation is necessary.

Acknowledgments

We gratefully acknowledge the support of the Oklahoma Department of Commerce, the Oklahoma Mesonetwork, and the U.S. Department of Energy. Our cartographer, Mike Larson, produced the final versions of our maps. Rebecca Sheehan provided some invaluable manuscript suggestions. We appreciate the issue editor and the three anonymous reviewers who provided constructive comments that improved our work.

References

Abbasi, S. A., and N. Abbasi. 2000. The likely adverse environmental impacts of renewable energy sources. *Applied Energy* 65:121–44.

Ai, B., H. Yang, H. Shen, and X. Liao. 2003. Computer-aided design of PV/wind hybrid systems. *Renewable Energy* 28:1491–1512.

Akaike, H. 1974. A new look at the statistical model identification. *IEEE Transactions on Automatic Control* 19 (6): 716–23.

Archer, C., and M. Jacobson. 2005. Evaluation of global wind power. *Journal of Geophysical Research—Atmospheres* 110:D12110.

Barley, C., L. Dennis, D. J. Lew, and L. T. Flowers. 1997. Sizing wind/photovoltaic hybrids for households in Inner Mongolia. Golden, CO: National Renewable Energy Laboratory. http://www.nrel.gov/docs/legosti/fy97/23116.pdf (last accessed 10 January 2011).

Bohr, N. 1999. *Neils Bohr collected works*. Vol. 10. *Complementarity beyond physics (1928–1962)*, ed. D. Flavholdt. Amsterdam: Elsevier.

Edgeworth, F. Y. 1925. The pure theory of monopoly. In *Papers relating to political economy*, 111–42. London: Macmillan.

Energy Information Administration. 2010. Table of primary energy review 1949–2009. http://www.eia.doe.gov/emeu/aer/txt/ptb0101.html (last accessed 10 January 2011).

Felder, D. 2004. A regionally based energy end-use strategy: Case studies from Centre Country, Pennsylvania. *The Professional Geographer* 56:185–200.

Fotheringham, A., C. Brunsdon, and M. Charlton. 2002. *Geographically weighted regression: The analysis of spatially varying relationships*. Chichester, UK: Wiley.

Geiger, R., R. Aron, and P. Todhunter. 2005. *The climate near the ground*. Lanham, MD: Rowman and Littlefield.

Giordani, P., and H. Kiers. 2006. A comparison of three methods for principal component analysis of fuzzy interval data. *Computational Statistics & Data Analysis* 51 (1): 379–97.

International Energy Authority. 2010. *Outlook, executive summary*. Paris: International Energy Authority.

International Human Genome Sequencing Consortium. 2004. Finishing the euchromatic sequence of the human genome. *Nature* 431 (7011): 931–45.

Jacobson, M. Z. 2009. Review of solutions to global warming, air pollution, and energy security. *Energy and Environmental Science* 2:148–73.

Kaltschmitt, M., W. Streicher, and A. Weise, eds. 2009. *Renewable energy, technology, economics, and environment*. Berlin: Springer.

Kimura, Y., Y. Onai, and I. Ushiyama. 1996. A demonstrative study for the wind and solar hybrid power system. *Renewable Energy* 9:895–98.

Kuyk, W. 1997. *Complementarity in mathematics: A first introduction to the foundations of mathematics and its history*. Dordrecht, The Netherlands: D. Reidel.

Lincoln, S. 2005. Fossil fuels in the 21st century. *Ambio* 34:621–27.

Macken, K. J. P., M. H. J. Bollen, and R. J. M. Belmans. 2004. Mitigation of voltage dips through distributed generation systems. *IEEE Transactions on Industry Applications* 40:1686–92.

McPherson, R. A., C. Fiebrich, K. C. Crawford, R. L. Elliott, J. R. Kilby, D. L. Grimsley, J. E. Martinez, et al. 2007. Statewide monitoring of the mesoscale environment: A technical update on the Oklahoma Mesonet. *Journal of Atmospheric and Oceanic Technology* 24: 301–21.

Meyer, R. 2008. The potential of solar energy for replacing fossil fuels. Paper presented at The VII ASPO conference, Barcelona, Spain. http://www.aspo-spain.org/asp07/presentations/Meyer-CSP-ASP07.pdf (last accessed 15 August 2010).

Milgrom, P., and J. Roberts. 1995. Complementarities and fit strategy, structure, and organizational change in manufacturing. *Journal of Accounting and Economics* 19(2–3): 179–208.

Oliver, M. A., and R. Webster. 1990. Kriging: A method of interpolation for geographical information systems. *International Journal of Geographical Information Science* 4 (3): 313–32.

Owen, A. D. 1996. Renewable energy: Externality costs as market barriers. *Energy Policy*. 34:632–41.

Reichling, J., and F. Kulacki. 2008. Utility scale hybrid wind–solar thermal electrical generation: A case study for Minnesota. *Energy* 33 (8): 626–38.

Sahin, A. Z. 2000. Applicability of wind–solar thermal hybrid power systems in the northeastern part of the Arabian Peninsula. *Energy Sources, Part A: Recovery, Utilization, and Environmental Effects* 22:845–50.

Sawin, J. 2004. *Mainstreaming renewable energy in the 21st century*. Washington, DC: Worldwatch Institute.

Scott, J., and G. Marshall, eds. 2005. *Oxford dictionary of sociology*. Oxford, UK: Oxford University Press.

Simon, C. A. 2009. Cultural constraints on wind and solar energy in the U.S. context. *Comparative Technology Transfer and Society* 7 (3): 252–69.

Stadler, S., and T. Hughes. 2005. Wind power climatology. In *Encyclopedia of world climatology*, ed. J. Oliver, 807–13. Dordrecht, The Netherlands: Springer.

Stoutenberg, E. D., N. Jenkins, and M. Z. Jacobson. 2010. Power output variations of co-located offshore wind turbines and wave energy in California. *Renewable Energy* 35 (12): 2781–91.

Tackle, E. S., and R. H. Shaw. 1977. Complimentary nature of wind and solar energy at a continental mid-latitude station. *International Journal of Energy Research* 3: 103–12.

U.S. Department of Energy. 2010. Glossary of energy-related terms. http://www1.eere.energy.gov/site_administration/glossary.html#R (last accessed 10 January 2011).

Wang, L., and C. Singh. 2008. Hybrid design of electric power generation including renewable sources of energy. *Bulletin of Science, Technology & Society* 28 (3): 192–99.

Wiser, R., C. Namovicz, M. Gielecki, and R. Smith. 2010. The experience with electrical portfolio standards in the United States. *The Electricity Journal* 20 (4): 8–20.

Zhou, W., C. Lou, Z. Li, C. Lu, and H. Yang. 2007. Current status of research on optimum sizing of stand-alone hybrid solar–wind power generation systems. *Applied Energy* 87:380–89.

Zimmerer, K. S. 2011. New geographies of energy: Introduction to special issue. *Annals of the Association of American Geographers* 101 (4): 705–11.

"Papering" Over Space and Place: Product Carbon Footprint Modeling in the Global Paper Industry

Joshua P. Newell* and Robert O. Vos†

*School of Natural Resources & Environment, University of Michigan
†Spatial Sciences Institute, University of Southern California

We are witnessing an explosion in carbon calculators for estimating the greenhouse gas (GHG) emissions (i.e., carbon footprint) of households, buildings, cities, and processes. Seeking to capitalize on the emergent "green" consumer, corporations are leading the next iteration in carbon footprinting: consumer products. This potentially lucrative low-carbon frontier, however, faces steep challenges due to complexities of *scale*, largely a function of the number of actors and geographies involved in globalized commodity and energy networks, and *scope*, which increasingly demands inclusion of emissions due to land use change (e.g., biofuel production, timber harvest, livestock grazing, mining). Life cycle assessment (LCA)—the principal method behind product-level GHG emissions footprint protocols—frequently avoids these challenges by narrowly delineating system boundaries, thereby excluding the "messiness" of space and place. Through a comparative model of energy sources and emissions in the globalized paper industry, this article reveals how complexities associated with geographic variation and land use change create indeterminacy in footprints based on these protocols. Using industry and trade data, the authors develop geographic information system transportation and energy models to map the globally dispersed pulp supply networks and to rescale Intergovernmental Panel on Climate Change GHG inventory guidelines to include carbon loss associated with land use change in the carbon footprint of coated paper. Given their integrative abilities to conceptualize and model coupled human–ecological systems, sophisticated understanding of time–space dynamics and critical theoretical insights, geographers have much to contribute to the LCA and product carbon footprinting enterprise, which to date has been largely the intellectual domain of engineers. *Key Words: carbon labels, emissions, land use change, life cycle assessment, paper industry.*

我们正在目睹一个估算住户，建筑，城市和过程温室气体（GHG）排放的（如碳足迹）碳计算器的激增。那些试图利用这一新兴的"绿色"消费的公司正带领下一次碳足迹的叠代：消费产品。然而，这种潜在的有利可图的低碳前沿，由于规模，主要是全球化商品和能源网络中的行为者和地理的数量，以及范围，即日益俱增的要求罗列土地利用变化产生的排放量（例如，生物燃料生产，木材采伐，放牧，采矿），而面临着极端的挑战。生命周期评估（LCA），这一产品级别 GHG 背后的排放足迹协议的主要方法，通过狭义地划定系统边界，从而排除了空间和地点的"混乱"，因此频繁地避免了这些挑战。通过在全球化造纸工业中能源和排放的比较模型，本文揭示了与地理和土地利用变化有关的复杂性是如何创建基于这些协议上的足迹的不确定性。利用工业和贸易数据，本文作者开发了地理信息系统运输和能源模型，以测绘分散在全球的纸浆供应网络，并重新调整政府间专门小组在气候变化 GHG 的清单指南以包括有关土地利用在涂碳纸的碳足迹变化的损失。鉴于他们概念化和模型耦合人类生态系统的综合能力，对时空动态和重要理论观点的复杂的理解，地理学家对迄今主要是工程师智力领域的 LCA 和产品的碳足迹事业，都作出了重大贡献。*关键词：碳标签，排放，土地利用的变化，生命周期评价，造纸工业。*

Estamos presenciando una explosión de calculadoras de carbono para calcular las emisiones de gases de invernadero (GHG, sigla en inglés), o polución atmosférica (es decir, la huella del carbono) originadas en viviendas, edificios, ciudades y procesos. Buscando capitalizar a expensas del emergente consumidor "verde", las corporaciones están apuntándole a la siguiente edición de la huella de carbono: productos de consumo. Esta frontera de bajo carbono potencialmente lucrativa, enfrenta, sin embargo, retos escabrosos debido a complejidades de *escala*, lo que en gran medida es una función del número de actores y geografías involucradas en las cadenas energéticas y de mercaderías globalizadas, y de *ámbito*, que cada vez más clama por la inclusión de las emisiones debido al cambio de usos del suelo (e.g., producción de biocombustibles, explotación maderera, pastoreo ganadero, minería). La evaluación del ciclo de vida (ECV)—principal método en el que se basan los protocolos de huella del carbono para emisiones de GHG a nivel de producto—con frecuencia le sacan el bulto a estos retos delineando con estrechez los sistemas de límites, para así descartar los "caos" de espacio y lugar. A través de un modelo comparativo de fuentes de energía y emisiones en la industria papelera globalizada, este artículo revela cómo las complejidades asociadas con variación geográfica y cambio de uso del suelo crean indeterminación en la huella de carbono con base en estos protocolos. Utilizando datos de industria y comercio, los autores desarrollaron modelos de sistemas de información geográfica para transporte y energía para cartografiar las cadenas de suministro de pulpa de papel dispersas globalmente y para cambiar la escala de las guías del inventario de los GHG del Panel Intergubernamental de Cambio Climático, a fin de incluir la pérdida de carbono asociada con los cambios de uso del suelo en la huella del carbono del papel esmaltado. Dadas sus habilidades integradoras para conceptualizar y modelar los sistemas humano-ecológicos en acoplamiento, su comprensión sofisticada de la dinámica tiempo-espacio y su perspicacia teórica crítica, los geógrafos pueden contribuir mucho a la tarea de la ECV y de generar la huella del carbono, que hasta hoy ha sido el dominio intelectual de los ingenieros. *Palabras clave: etiquetas de carbono, emisiones, cambio de uso del suelo, evaluación del ciclo de vida, industria papelera.*

Carbon labels now appear on potato chips, milk, breakfast cereal, sugar, bread, Japanese beer, and a wide range of other products. Facing increased pressure to reduce greenhouse gas (GHG) emissions in their operations and enticed by the lucrative prospects of an emergent green consumer, corporations have been active proponents of these labeling systems. Retail giants Tesco (UK) and Wal-Mart (U.S.) are engaged in major efforts to fund research and shape international protocols for product labeling (Brenton, Edwards-Jones, and Jensen 2009). Embracing the neoliberal faith that "buyer-driven" global commodity chains (Gereffi 1995) can use quality conventions and standards (e.g., certification, corporate social responsibility policies) to rescale governance, nongovernmental organizations (NGOs) and governments view carbon labels as a means to harness the power of the "green" markets to forge a more sustainable world. The governments of Japan, South Korea, Germany, and the European Union (EU) are developing carbon label standards (Bolwig and Gibbon 2009), and legislators in California have called for debate about a Carbon Labeling Act (AB 19), designed to facilitate carbon labels for products sold in the state. Meanwhile, recent research reveals the significance of consumer products in overall GHG emissions, sparking trade debates about who should take responsibility, producer countries (e.g., China) or consumer countries (e.g., the United States), and providing further impetus for product footprint accounting systems (Kejun, Cosbey, and Murray 2008).[1]

The race to establish an industry-standard international protocol for product carbon footprinting has narrowed to three hybrid public–private efforts. Publicly Available Standard (PAS) 2050 is the most specific and rigorous protocol to date (British Standards Institution, Carbon Trust, and Department for Environment 2008; Sinden 2009). The NGO in this partnership, Carbon Trust, which secured funding from Tesco and PepsiCo to implement specifications of the protocol on selected products, has created the Carbon Labeling Company, a private firm expanding corporate labeling efforts across the globe. But PAS 2050 faces stiff competition from the Product Life Cycle Accounting and Reporting Standard, which was developed by World Resources Institute and the World Business Council for Sustainable Development, with financial support from Wal-Mart. Finally, the International Organization for Standardization (ISO) has issued draft standard ISO 14067 (Carbon footprint of products).

The methodology underlying these protocols and labels is life cycle assessment (LCA), a central tool within the field of industrial ecology (Graedel and Allenby 2003; Matthews and Lifset 2007). Starting in the late 1970s in Europe, LCA methods developed rapidly, culminating in international standardization under the ISO 14040 protocol (SETAC Europe LCA Steering Committee 2008). LCA quantifies environmental impacts of products and processes for major phases of the life cycle, from material extraction to disposal (see top of Figure 1). The standard LCA method consists of sequential steps: definition of goal and functional unit, delimitation of scope or system boundary, life cycle inventory (LCI) and life cycle impact assessment (LCIA). LCI refers to the accounting of pollution and resource extraction in each life cycle phase, and LCIA is a decision-support model built on LCI to measure impacts (e.g., on human health or ecosystem quality). The field of LCA has expanded rapidly in recent years, with studies on a diverse set of products and processes, including buildings, fuels, renewable energy sources, nanotechnologies, and water.

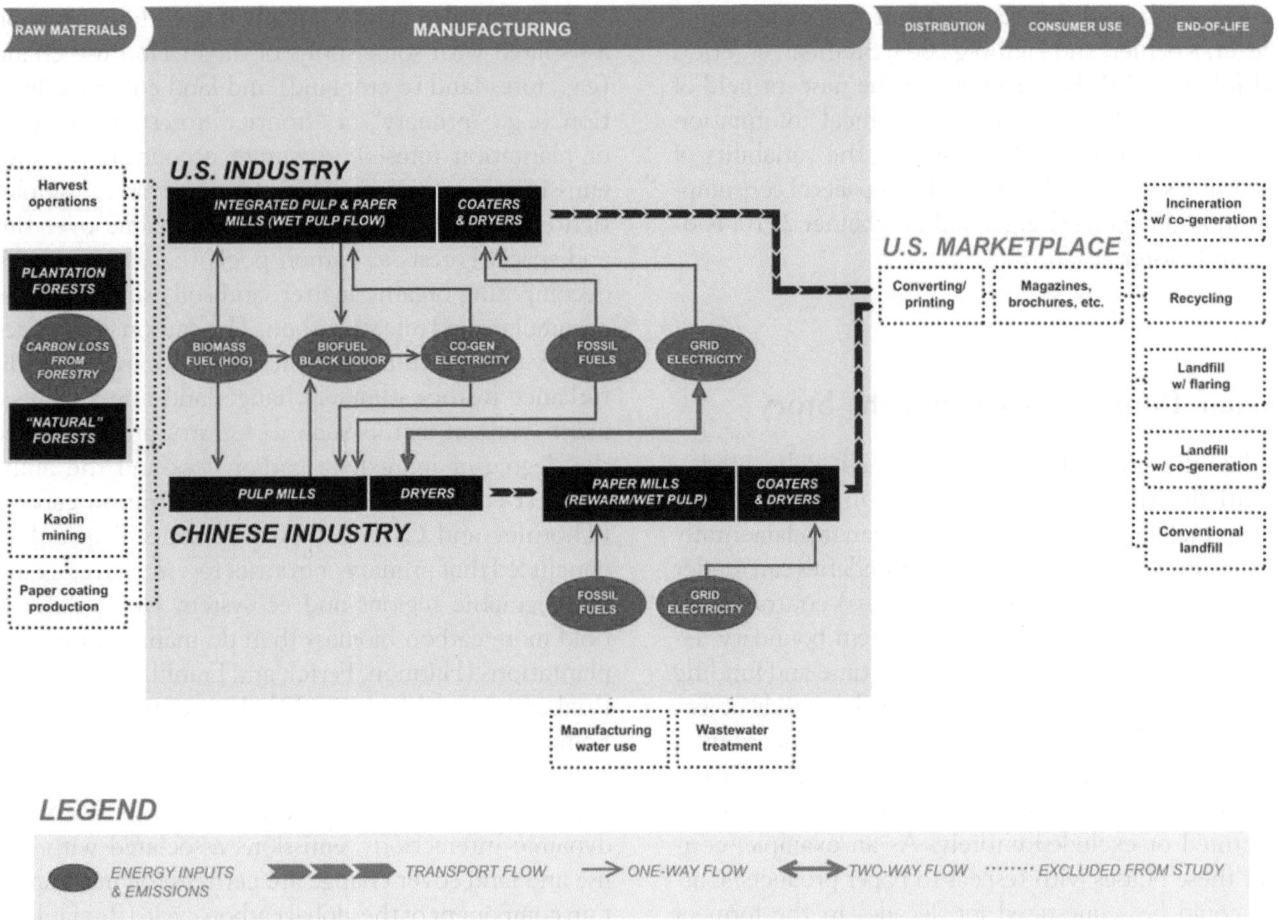

Figure 1. Comparative life cycle inventory of U.S. and Chinese coated paper: System flow and boundary scope diagram.

Flattened Geographies in Life Cycle Assessment

Basic building blocks of LCA include activity data and emission factors. To reduce uncertainty in LCA studies, company-specific rather than industry average data are preferable. Similarly, data for emission factors (i.e., GHGs emitted per unit of energy) should be as site-specific as possible. In globalized post-Fordist supply chains, however, the production and distribution of a single product can involve dozens of actors, including material suppliers, manufacturers, shippers, wholesalers, and retailers scattered across multiple countries. Shifts in price, resource availability, production method, and demand can abruptly shift the input–output structure and territoriality of these complex supply networks. In stark contrast, a carbon label offers the consumer a static and fixed accounting of the GHGs embedded in a particular product.

In theory, LCIs and impact assessments require data collection within these spatially and temporally contingent supply networks—following complex chains of interaction in production systems wherever and however production occurs. But hindered by an inability to situate production in space and to obtain regionally specific data, LCIs as practiced often draw on activity data and emission factors that are essentially global averages and therefore decidedly aspatial. Or they privilege regions where LCIs are well-developed (Curran 2006), such as Western Europe, by applying emission factors from these LCIs for products manufactured in other regions.

In essence, by minimizing areal differentiation for the purpose of expediency, these practices "flatten" geography. LCA software, which draws heavily on inventory databases generated from European case studies, has facilitated these tendencies, although practitioners do sometimes build out regionalized grid averages for electricity use. Some LCA scholars recognize the implications of this lack of geographic variability and nuance, and are developing procedures to make LCIs more spatially explicit (Weidema 2004) and to better incorporate spatial differentiation in impact modeling (Pfister, Koehler, and Hellweg 2009; Steinberger, Friot, and Jolliet 2009). For example, in the nascent field of product water footprinting, geographical information system (GIS)-based models reflecting the variability of freshwater supplies help specify the impacts of consumption in arid regions (Berger and Finkbeiner 2010; Ridoutt and Pfister 2010).

System Boundaries Change the Story

How the system boundary or scope is delineated—essentially which life cycle phases, inputs, and outputs to include or exclude—in LCA can fundamentally change the result. Varying system boundaries can render studies of the same product (or process) contradictory or incompatible for comparison. System boundary delineation hinges on the level of data, time and funding constraints, as well as other factors such as study objectives, geographic complexity, and levels of uncertainty.

To reduce complexity, temporal boundaries for use and disposal phases in particular are often narrowly structured or excluded entirely. As an example, consider these phases with respect to paper products. Carbon could be sequestered for decades in the form of books or it might be disposed of after one use in the case of copy paper. In the disposal phase, paper can release carbon as CO_2 following waste incineration, flaring or cogeneration from landfill gas, or as untreated methane gas from a landfill (Figure 1). The emission profile of disposal phases varies temporally and spatially, contingent on the type of local solid waste system in place.

Furthermore, most product LCAs exclude emissions associated with land use. Biomass fuel generally has been treated as "carbon neutral," based on the rationale that biomass stock (e.g., forest) will grow back and, over time, cancel out the global warming potential of the GHG emissions from the initial combustion. Recent research on biofuels and indirect land use change calls this into question. Searchinger et al. (2008) demonstrated that biofuel crops might displace food crops, leading to widespread conversion of forests or grasslands for agricultural production. Other research on indirect land use changes suggests more moderate emission scenarios (Hertel et al. 2010). Debate over the emissions associated with indirect land use change remains contentious, as evidenced by uncertainty about how to account for it in California's Low Carbon Fuel Standard, a state regulatory mandate.

There are also significant, albeit uncertain, emissions associated with some forms of direct land use change (e.g., forestland to cropland) and land cover modification (e.g., "primary" or "frontier" forests to managed or plantation forests). Accurate accounting of these emissions requires modeling land use as a complex, tightly coupled human–ecological dynamic over time and space. Forests as carbon pools (e.g., live biomass, decomposing organic matter, and soil) simultaneously accumulate and release carbon. The carbon flux of these pools varies depending on forest type, location, age, disturbance history, climate change, and forest management. Human factors such as forestry practices affect the degree of immediate carbon loss and the ability of forest ecosystems to recover sequestration capacity (Thornley and Cannel 2000). Forest ecologists have concluded that primary or frontier forests, across a range of geographic regions and ecosystem types, generally hold more carbon biomass than do managed forests or plantations (Harmon, Ferrel, and Franklin 1990; Dean, Roxburgh, and Mackey 2003; Luyssaert et al. 2008). Some NGOs advocate accounting for the opportunity cost of such logging because it could take decades to regain carbon sink capacity (Ford 2009). With such dynamic interactions, emissions associated with land use and land cover change are perhaps the most uncertain component of the global carbon cycle (Ramankutty et al. 2007).

Product footprint protocols make limited attempts to incorporate simplified elements of land use. PAS 2050 stipulates inclusion of GHG emissions resulting from direct land use change but excludes indirect land use change and land cover modification. Direct land use emissions are assessed in accordance with 2006 Intergovernmental Panel on Climate Control (IPCC) *Guidelines for National Greenhouse Gas Inventories*, using a twenty-year time scale. However, the *Guidelines* are for country-level reporting and provide no specific guidance on how to rescale these methodologies for products.

To explore how spatiality, land use issues, supply chain complexity, and system boundary decisions are negotiated in LCA modeling, this article models and compares LCI phases for coated freesheet paper (i.e., the sort of paper that is glossy in feel and commonly used in magazines, etc.) produced in China and in the United States. Specifically, we critique LCA practice by deploying the most rigorous product carbon footprint protocol available: PAS 2050. We focus on CO_2 inventories for three life cycle phases: carbon loss from timber harvest, transportation, and pulp and paper production.

Following the protocol's general guidance but supplied with no specific method to actually incorporate direct land use change, we develop an initial methodology to rescale IPCC guidelines for specific products. More broadly, our spatially explicit approach illustrates how geographic variation in the fiber supply structure for the U.S. and Chinese industries shapes the CO_2 emissions profiles for these phases and creates indeterminacy in terms of calculating an accurate product-level carbon footprint.

Modeling a Spatially Explicit Product Footprint for Coated Paper

Approximately 40 percent of the world's industrial wood harvest is used to produce paper with a rise to 50 percent predicted by 2050 (Abramovitz and Mattoon 1999). The paper industry is the third-largest consumer of industrial energy in the United States and about one third of the municipal solid waste stream consists of paper (U.S. Department of Energy 2005). China and the United States are the world's two largest producers of paper and paper products (Haley 2010). Biomass fuel including burning of timber harvest residue ("hog" in industry parlance) is a significant source of energy, especially for coated freesheet paper because it relies almost entirely on "virgin" rather than recycled wood fiber. A tightly networked North American industry, long the global leader in coated paper production, is gradually being supplanted by Chinese producers, who manage an increasingly complex, global web of fiber sourcing, pulping, paper production, and converting operations.

To elucidate how these contrasting production network structures shape and problematize product carbon footprint modeling, we compare CO_2 emissions from coated paper produced by the Chinese and U.S. industries. Six mills making coated paper for the largest North American manufacturer represent the U.S. supply chain. Given that the largest Chinese manufacturer has just one mill producing coated paper, we broaden the comparative analysis to include the eight largest mills in China for various manufacturers. Smaller producers are virtually impossible to track using industry data, so by including a number of mills as well as the largest producer in each country, the underlying models provide comparable representations for the two supply chains.

Our functional unit is the delivery of one metric ton of finished paper to Los Angeles (2007). We look solely at the paper's wood fiber, excluding other primary materials such as clay and treatment chemicals, additives, and processes such as wastewater treatment, harvesting equipment emissions, and emissions specific to cultivating plantations (Figure 1). We only quantify CO_2 emissions, excluding other GHGs, and our study scopes out the use and disposal phases, including landfill emissions, elements researched in previous paper LCAs (Leach and Givnish 1996). In short, this article probes underlying geographic variation in carbon footprints for coated paper using a partial comparative life cycle inventory approach for the timber harvest, transportation, and pulp and production phases of the production process.

To develop a spatially explicit product footprint, we first map the fiber supply structure and production processes for both supply chains. We then develop weighted averages for timber harvest yield at the national scale and for pulp and paper production at the facility scale. This model of spatial variation underlies calculations for emissions from transportation and energy used in manufacturing, as well as carbon loss from timber harvest.

Although we use the same industry data set (Resource Information Systems [RISI] 2007) to capture activity data for production processes at the mills, due to data availability, our methodology for the two supply chains differs slightly for fiber supplies. Through personal interviews, we acquired firm-specific fiber supply data for the six U.S. mills. Like virtually all major North American manufacturers, this U.S. manufacturer has integrated, facility-level pulp and paper production and sources logs within a 100-mile radius. About 10 percent of the wood fiber is imported, essentially softwood pulp from Canada, which we were able to track to individual Canadian mills. In contrast, China imports more than 90 percent of its pulp from mills across the globe (Wood Resources International and Seneca Creek Associates 2007). Unable to obtain firm-specific data for China's industry, we model the fiber supply structure for the eight mills using industry (RISI 2007) and trade data (Global Trade Information Services 2008).

The virgin fibers required for coated paper production are pulped in two basic commodity types: bleached hard kraft pulp (BHKP) and bleached soft kraft pulp (BSKP). In 2007, over 75 percent of the BHKP China imported came from Indonesia, Brazil, and Chile, and about 71 percent of the BSKP came from Canada, Chile, Russia, and the United States. A large number of countries (including New Zealand, Finland, and Thailand) supply the remainder. To identify pulp mills, the model estimates locations based on the countries (including China) where pulp is produced, followed by the size of each mill's production as a share of total BHKP and BSKP produced within each country.[2]

Transport

Following standard LCI procedures and PAS 2050 specifications, the model applies transport mode (U.S. Department of Energy 2008) and fuel type (U.S. Energy Information Administration 2008) emission factors based on the mass, distance, and modes of pulp and paper (i.e., ton mile by ship, rail, and truck). We model distances by data mining for exact locations of 116 global pulp mills, the Chinese paper mills, and U.S. integrated mills and use ArcGIS (Environmental Systems Research Institute 2009) to estimate the total distance that pulp is traveling from mills to papermaking facilities, and onto the consumer market (i.e., Port of Los Angeles). Calculations for route variations and distances rely on a suite of tools, including NetPas Distance (http://www.netpas.net), which allows the user to identify origin and destination ports to calculate shipping routes and distances. For mills less than 250 miles from a marine port or final destination mill, we model truck (rather than rail) as the transport mode of choice.

Pulp and Paper Production

To quantify the facility-level CO_2 emissions from electricity and fossil fuel use, the model assigns fuel use profiles for each facility on the basis of a weighted average for the overall supply chain, based on the mass of total production given in RISI (2007). From these mill use profiles, we then apply the weighted average to generate a GHG emissions factor for embedded CO_2 for BHKP, BSKP, and finished paper for both supply chains. The general formula is

$$P * E_f * E_{mf} = CO_2 \text{ per finished metric ton} \qquad (1)$$

where P = percentage of pulp supply (BHKP or BSKP), or percentage of finished paper supply (i.e., weighted average); E_f = energy factor for each fuel type, terajoules (TJ)/air dried metric ton (ADMT); and E_{mf} = emissions factor for each fuel type, metric tons of CO_2/TJ. We then sum values for each facility to obtain overall estimates of BHKP, BSKP, and finished coated paper.

Different Fuel Types

For specific distillate fuel oil and natural gas types listed by facility (RISI 2007), the model uses energy content factors from the International Energy Agency (IEA 2008) and emission factors from the IPCC (2006). *Coal* is nomenclature for a continuous range of solid organic fuels with varying energy content values and GHG emission factors, so using IEA (2008) consumption data we develop a soft coal (subbituminous and lignite)–to–hard coal (anthracite and bituminous) ratio for each country and then assign it to each facility. For electricity grid emissions, the model uses IEA (2008) data on the energy source mix (e.g., coal, natural gas, nuclear, oil, hydropower, renewables) for each country. Pulp mills and integrated mills both run on biomass sources, including residuals from production. Residual waste from the delignification process, known as *black liquor*, is almost always burned in recovery boilers (see Figure 1). Previous studies treat these biomass fuels as carbon neutral, but emissions from these sources are inherent to our model because we include carbon loss from timber harvest as detailed next.

Carbon Loss from Timber Harvest

The model includes two scenarios for carbon loss from timber harvest. Scenario 1 assumes that, in the supply chains, the country of pulp production is known but the specific land use changes from timber harvest in that country are unknown. As instructed by PAS 2050 for products with this unknown origin, Scenario 1 assumes forestland-to-annual-cropland change following logging in each producer country. Scenario 2 adds geographic nuance by using estimates from Wood Resources International and Seneca Creek Associates (2007) to develop "natural" versus plantation forest ratios for BHKP and BSKP from each country. Scenario 2 assumes all direct land use change associated with conversion to plantations to be pre-January 1990, which—based on PAS 2050 and IPCC guidance—renders plantations carbon neutral. We recognize they are not neutral due to uncertain levels of emissions specific to plantation cultivation (e.g., seeding, thinning, and fertilizing) and indirect land use change, but to be consistent in both scenarios we excluded emissions associated with forest management and the actual harvesting. Furthermore, the model only considers the above-ground biomass carbon pool, with changes to other pools (e.g., dead organic matter, below-ground biomass, and soil carbon) excluded. Essentially, our model accounts for the release of biogenic carbon from fuel combustion during pulp and paper production, from wood residue left at logging sites, and from solid residuals due to production.

The model uses Food and Agriculture Organization (FAO 2001) data for average timber harvest yields per hectare and to identify major forest ecosystem types (e.g., tropical, temperate, boreal) for each country. It then averages IPCC Tier I estimated biomass values for each forest type to develop a per-hectare average biomass profile. The customized formula is

$$(B_{before} - B_{after}) * (C_{fd}) = C_{removal} \quad (2)$$

$$C_{removal} \,/\, Y_{yield} = C_{loss} \quad (3)$$

where B_{before} = biomass stocks before conversion in tons dry matter/hectare (average for each country), B_{after} = biomass stock for annual cropland (IPCC factor of 5 tons dry matter/hectare), C_{fd} = carbon fraction of dry matter (IPCC factor of 0.5), $C_{removal}$ = carbon removal in kg of carbon/hectare, Y_{yield} = yield given in m^3/hectare (FAO factor for each country), and C_{loss} = kg of carbon per m^3 of wood (for each country).

Then based on the model of weighted averages, we obtain carbon loss estimates per m^3 for BHKP and BSKP in the China and U.S. supply chains. For finished paper, the formula is

$$P * C_{loss} * W_{ef} * 44 / 12 = \text{kg of } CO_2 \text{ per finished metric ton} \quad (4)$$

where P = percent of pulp or wood supply for each country (supply chain weighted average), W_{ef} = wood efficiency factor of 3.65 m^3/finished metric ton, and 44/12 = conversion of elemental carbon to CO_2.

Biogenic carbon is embedded in the coated paper until it is released during disposal or remains embedded in the form recycled paper products. As our study excludes the end-of-life phase, we estimate and subtract embedded carbon in the product from the direct land use change CO_2 emissions using the following formula:

$$D_{average} * C_{fd} = C_{density} \quad (5)$$

$$C_{density} * W_{ef} * 44 / 12 = \text{kg of } CO_2 \text{ embedded per finished metric ton} \quad (6)$$

where $D_{average}$ = average carbon density of wood species (IPCC 2006), in oven dry tons/m^3; and $C_{density}$ = carbon density in kg/m^3 of wood.

Space and Place Change the Story: Indeterminacy in the Product Footprint

Our comparative analysis of U.S. and Chinese coated paper illustrates how geographic variation and system boundary exclusions fundamentally shape the carbon footprint of products. Including direct land use change in the raw material phases of the system boundary changes the story dramatically, as our model reveals the potential magnitude of CO_2 emissions due to changes in above-ground forest biomass (Figure 2). In contrast, studies of the paper life cycle that exclude land

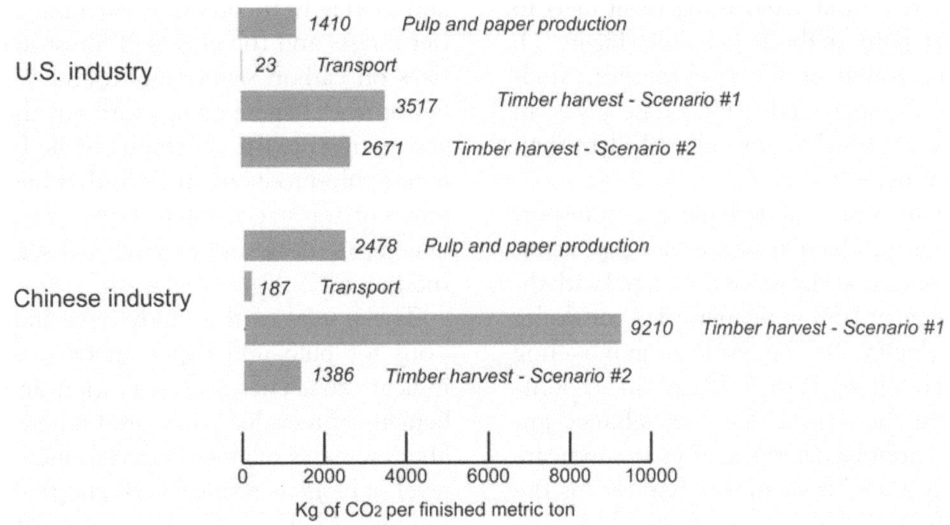

Figure 2. Comparative coated-paper carbon footprints for three life cycle phases, U.S. and Chinese industry.

use emissions conclude the pulp and paper production phase is the most carbon intensive (Gower 2006; Miner and Perez-Garcia 2007).

Consistent with earlier studies of paper (Subak and Craighill 1999; Gower 2006) and those for other land-based products, such as meat (Basset-Mens and van der Werf 2005), transportation emissions are a comparatively small portion of overall emissions in both supply chains. For China's industry, transport-related emissions are higher due to greater wood fiber to paper mill (BHKP average of 8,800 km) and consumer market delivery distances (Figure 2). The U.S. industry sources most fiber locally for integrated mills, with the imported Canadian BSKP averaging 2,400 km. A sensitivity analysis—in a typical uncertainty range (±20%–±40%) for transport emissions factors (Kioutsioukis et al. 2004)—changes neither the significance of the transport phase nor the overall result, because the transport footprint is small and the comparative difference between the U.S. and Chinese industries is large. From a product carbon footprint perspective, these findings counteract the "buy local" cliché in popular energy sustainability discourse—which, by overemphasizing transport as an emission source, conflates "greenness" with local sourcing.

The difference in the pulp and paper production phase (China: 2,478 kg/metric ton; U.S. 1,410/kg metric ton) stems from the Chinese industry's lack of integrated pulp mills and its greater dependence on carbon-intensive coal both for process heat and in the electricity grid. Integrated production in the U.S. supply chain is made possible by colocation of the industry near the forests. The integrated mills use cogenerated electricity to run both pulp and papermaking machinery (including coaters) and avoid using fossil fuels to rewarm and rewet pulp at the paper mill (Figure 1). Ashby's (2009) meta-analysis of carbon footprint studies of the pulp and paper production phase gives an uncertainty range of ±11.5%, considerably less than the difference seen here.

Geographical differences underlying the transport and pulp and paper production phases are significant but, as noted, the potential emissions associated with direct land use change, at least in Scenario 1, overwhelm these two phases (Figure 2). The results from modeling Scenario 1—which follows PAS 2050 carbon accounting guidance when the specific "land use change impact of an input cannot be determined" (10)—indicate much higher carbon loss from timber harvest for the Chinese supply chain (9,210 kg) than for the U.S. supply chain (3,517 kg). This model essentially assumes that all forests are "natural." As such, the Chinese industry imports more pulp from tropical forests, which (based on IPCC methodology) carry a larger CO_2 emissions conversion burden than do temperate and boreal forests. But if we go beyond PAS 2050 default guidance in Scenario 2, by incorporating natural forest-to-plantation ratios for each pulp-producing country, the results shift so much as to invert the comparison. The Chinese industry generates less CO_2 during the timber harvest phase (1,368 kg) than does the U.S. industry (2,671 kg) due to greater sourcing from countries that rely on plantations to produce pulp (e.g., Brazil and Chile).

Although these results provide insight into the relative importance of key life cycle phases, the more significant outcome of the study is the revelation that the overall footprint of coated paper is essentially indeterminate. Land use assumptions dictate the result. Scenario 1 illustrates a much larger overall footprint for Chinese paper than for U.S. paper. In Scenario 2, however, the two footprints are nearly even. These scenarios demonstrate the inadequacy of PAS 2050 to model essential system boundary inclusions when confronted with the spatial variation of globalized pulp supply chains.

As a way forward, the model we have presented offers an innovative attempt to incorporate land use change in product carbon footprinting by rescaling and customizing IPCC (2006) guidelines. As the results show, however, indeterminacy in the footprint persists. To refine this method, we would need to address geographic and temporal scales in forest growth cycles, include carbon pools currently excluded in PAS 2050 (below-ground biomass, soil carbon, and dead organic matter), and gradually incorporate emerging science on frontier forests and the effects of forest management practices on carbon sequestration and regrowth rates. To better model these complexities at the outset, we suggest narrowing the geographic scale by comparing two major pulp-producing regions that harvest timber from different forest ecosystem types (e.g., frontier forests, managed forests, and plantations) such as Canada and Indonesia.

The results of this study have important ramifications for pulp and paper energy use models, which indicate that energy conservation and environmental benefits of recycled paper are limited because recycled fiber increases reliance on fossils fuels due to the lower level of biomass residuals (i.e., hog) available as fuel at the mills (Ruth and Harrington 1998; Villanueva and Wenzel 2007; Gaudreault, Samson, and Stuart 2010).

If CO_2 emissions from biomass are included in these models, rather than being treated as carbon neutral, using more recycled paper will have obvious benefits in terms of reducing demand for fiber from forests. How can the models developed here be extended to other paper types that, unlike coated freesheet paper, incorporate recycled content? Future modeling would need to investigate the carbon intensity trade-offs between higher levels of fossil fuels and reduced use of biomass fuels, hinging primarily on spatial variation in fuel and land cover types.

Although this study is limited to the pulp and paper sector, LCA modelers have recently called for similarly spatially explicit LCIs that include land use to fully capture the emissions associated with emerging technologies in renewable energy sectors like biofuels, wind, and solar energy (Kim, Kim, and Dale 2009; Seager, Miller, and Kohn 2009). Studies indicate the impacts from these sectors vary based on scale, location, and production practice (Potting and Hauschild 2006; Canals et al. 2007). Case studies of product systems that do not depend on materials requiring significant land use change (or modification) are also needed to indicate the degree to which spatially mismatched inventory models and fundamental land use exclusions are pervasive problems in carbon footprinting.

Inserting Spatiality into LCA and Product Carbon Footprints

To more accurately model the energy and carbon footprints of coated paper, in this article, we have argued that LCA must be spatially explicit and account for emissions associated with land use change. So how might geographers insert more spatiality into the LCA and product carbon footprint enterprise? A spatially robust LCA requires the ability to conceptualize and model complex natural–human systems, a particular strength of geographers working within the broad traditions of land cover science and political ecology (Turner and Robbins 2008). Advances in remote sensing might make sophisticated accounting of land use tractable. Since the 1990s, industrial ecologists have recognized how GIS might build areal differentiation into LCA (Bengtsson et al. 1998). But in part because LCA has remained ensconced in engineering, efforts to couple GIS and LCA have been limited, although Geyer et al.'s (2010) efforts to do so with geographers—by modeling the impacts of land use change on biodiversity—introduce exciting possibilities.

But neglect of areal differentiation extends beyond modeling complexity and disciplinary turf. Those who deploy LCA for specific projects, such as developing footprint protocols and carbon labels, have economic and political motivations that inevitably shape how geographic complexity and system boundary inclusions are negotiated. PAS 2050 and the attendant carbon labels can be seen as an emergent form of market-based carbon governance (Bailey, Gouldson, and Newell 2010), since they have emerged due to the collective effort of the private sector, NGOs, and government agencies operating within a horizontal network structure (Bulkeley 2005). These actors might be woven together by a faith in the power of markets to rescale governance, but their individual underlying (often contradictory) motivations provide specific insight into *why* spatial complexity is avoided. In principle, the transnational corporation might support a carbon label because of its potential to forestall government standards and regulation or to differentiate products in the marketplace. In practice, however, implementing a spatially robust protocol might be incompatible if it necessitates a hardening and spatial concentration of the complex, fluid supply networks on which the corporation depends to efficiently accumulate capital. NGOs, meanwhile, might feel pressured to develop a "practical" protocol, enticed by the potential financial rewards of pioneering an industry-wide international standard.

One ramification of the papering over space and place is that rather than clarifying, the carbon label obscures. It becomes a warped, arguably more dangerous form of the commodity fetish than the one it intended to replace. The label appears to provide transparency for the consumers, yet the actual spaces of production (and the processes within them) remain obscure. In the consumer's imagination, purchasing products with such labels becomes a form of ethical consumption (Barnett et al. 2005; Clarke et al. 2007). This provides the illusory power of the green consumer to make a difference, further obfuscating the need to reduce overall levels of consumption (Lovell, Bulkeley, and Liverman 2009). As such, carbon labels and product footprint protocols might be theorized as a corporate strategy of "accumulation by [apparent] decarbonization," similar to carbon offsets (Bumpus and Liverman 2008, 127), and they could be readily situated within broader critical literatures on ethical consumption, commodity fetishism (Castree 2001), global commodity networks, and "climate capitalism" (Newell and Paterson 2010).

The few human geographers to write about LCA have generally framed it as aspatial and technocratic, nesting it within the broader ecological modernization movement (Keil and Desfor 2004; Robbins 2004).

Although we clearly see these tendencies in terms of how LCA traditionally has been developed and applied, we do not view the methodology as inherently so. LCA is fundamentally a process useful for thinking through and mapping out the complex assemblages associated with the production, consumption, and disposal of products. A relatively young methodology, LCA can still be as readily shaped and deployed by geographers as it can by engineers. We see the possibility, for example, of using LCA as a form of progressive praxis because of its potential to ground globalization by reconnecting spaces of production and consumption (Hartwick 2000). But left alone and poorly applied, as we have demonstrated with the PAS 2050 protocol, the practice of LCA threatens to paper over geographic variation and complexity and exclude fundamental inputs and processes. We will be left with protocols shaped by self-interested corporate actors and a confusing array of carbon labels that are impoverished and misleading representations of the carbon footprints of the products that line our supermarket shelves.

Acknowledgments

We are grateful to two anonymous reviewers and to Jim Ford, Mark Harmon, and Mansour Rahimi for their insightful comments and suggestions. We also thank Jingfen Sheng and Christine Lam of USC's Spatial Sciences Institute for help with GIS modeling, Jennifer Wolch for intellectual guidance and administrative support, and Robin Maier for her patience throughout the revision process. Finally, we acknowledge the sustainability-consulting firm Clean Agency, Inc., which provided funding for the project and provided access to paper industry data.

Notes

1. For example, after subtracting for imports, China's surplus embodied CO_2 emissions in exports represent approximately 18 percent of that country's total production-based emissions. In contrast, the United States had an export deficit of –7.3 percent (Peters and Hertwich 2007).
2. This study draws on data from *Cornerstone* (RISI 2007), which tracks production inputs and outputs for major paper facilities throughout the world in a materials balance framework.

References

Abramovitz, J., and A. T. Mattoon. 1999. *Paper cuts: Recovering the paper landscape*. Washington, DC: Worldwatch Institute.

Ashby, M. F. 2009. *Materials and the environment: Eco-informed material choice*. New York: Elsevier.

Bailey, I., A. Gouldson, and P. Newell. 2010. Ecological modernization and the governance of carbon: A critical analysis. Working Paper No. 26, Center for Climate Change Economics and Policy, Norwich, UK.

Barnett, C., P. Cloke, N. Clarke, and A. Malpace. 2005. Consuming ethics: Articulating the subjects and spaces of ethical consumption. *Antipode* 37 (3): 23–45.

Basset-Mens, C., and H. M. G. van der Werf. 2005. Scenario-based environmental assessment of farming systems: The case of pig production in France. *Agriculture, Ecosystems & Environment* 105:127–44.

Bengtsson, M., R. Carlson, S. Molander, and B. Steen. 1998. An approach for handling geographical information in life cycle assessment using a relational database. *Journal of Hazardous Materials* 61:67–75.

Berger, M., and M. Finkbeiner. 2010. Water footprinting: How to address water use in life cycle assessment? *Sustainability* 2:914–44.

Bolwig, S., and P. Gibbon. 2009. *Overview of product carbon footprinting schemes and standards*. Paris: OECD.

Brenton, P., G. Edwards-Jones, and M. F. Jensen. 2009. Carbon labeling and low-income country exports: A review of the development issues. *Development Policy Review* 27 (3): 243–67.

British Standards Institution, Carbon Trust, and Department for Environment, Food, and Rural Affairs. 2008. *PAS 2050: 2008 specification for the assessment of the life cycle greenhouse gas emissions of goods and services*. London: British Standards Institution.

Bulkeley, H. 2005. Reconfiguring environmental governance: Towards a politics of scales and networks. *Political Geography* 24 (8): 875–902.

Bumpus, A., and D. Liverman. 2008. Accumulation by decarbonization and the governance of carbon offsets. *Economic Geography* 84:127–56.

Canals, L. M., C. Bauer, J. Depestele, A. Dubreuil, R. F. Knuchel, G. Gaillard, O. Michelsen, R. Muller-Wenk, and B. Rydgren. 2007. Key elements in a framework for land use impact assessment within LCA. *International Journal of Life-Cycle Assessment* 12 (1): 5–15.

Castree, N. 2001. Commodity fetishism, geographical imaginations and imaginative geographies. *Environment and Planning A* 33:1519–29.

Clarke, N., C. Barnett, P. Cloke, and A. Malpas. 2007. Globalizing the consumption: Doing politics in an ethical register. *Political Geography* 26 (3): 231–49.

Curran, M. A. 2006. Report on activity of task force 1: Data registry—Global life cycle inventory data resources. *The International Journal of Life Cycle Assessment* 11 (4): 284–89.

Dean, C., S. Roxburgh, and B. Mackey. 2003. Growth modeling of *Eucalyptus regnans* for carbon accounting at the landscape scale. In *Modeling forest systems*, ed. A. Amaro, D. Reed, and P. Soares, 27–39. Cambridge, MA: CABI Publishing.

Environmental Systems Research Institute. 2007. ArcMap 9.2. Redlands, CA: ESRI.

Food and Agriculture Organization of the United Nations. 2001. *Global forest resources assessment 2000*. Rome: FAO.

Ford, J. 2009. *Carbon neutral paper: Fact or fiction?* Asheville, NC: Environmental Paper Network.

Gaudreault, C., R. Samson, and P. R. Stuart. 2010. Energy decision making in a pulp and paper mill: Selection of LCA system boundary. *International Journal of Life Cycle Assessment* 15 (2): 198–211.

Gereffi, G. 1995. The organization of buyer-driven global commodity chains: How US retailers shape overseas production networks. In *Commodity chains and global capitalism*, ed. G. Gereffi and M. Korzeniewicz, 95–122. Westport, CT: Praeger.

Geyer, R., D. M. Stoms, J. P. Lindner, F. W. Davis, and B. Wittstock. 2010. Coupling GIS and LCA for biodiversity assessments of land use. Part 1: Inventory modeling. *International Journal of Life Cycle Assessment* 15 (5): 454–67.

Global Trade Information Services. 2008. *World trade atlas*. Columbia, SC: Global Trade Information Services. http://www.gtis.com/english/GTIS_WTA.html (last accessed 10 September 2010).

Gower, S. T. 2006. *Following the paper trail: The impact of magazine and dimensional lumber on greenhouse gas emissions*. Washington, DC: Heinz Center.

Graedel, T. E., and B. Allenby. 2003. *Industrial ecology*. Saddle River, NJ: Prentice Hall.

Haley, U. 2010. No paper tiger. Briefing Paper No. 264, Economic Policy Institute, Washington, DC. http://www.epi.org/publications/entry/no_paper_tiger/ (last accessed 10 December 2010).

Harmon, M. E., W. K. Ferrel, and J. F. Franklin. 1990. Effects on carbon storage of conversion of old-growth forests to young forests. *Science* 247:699–702.

Hartwick, E. 2000. Towards a geographical politics of consumption. *Environment & Planning A* 32:1177–92.

Hertel, T. E., A. A. Golub, A. D. Jones, M. O'Hare, R. J. Plevin, and D. M. Kammen. 2010. Effects of U.S. maize ethanol on global land use and greenhouse gas emissions: Estimating market-mediated responses. *BioScience* 60 (3): 223–31.

Intergovernmental Panel on Climate Change. 2006. *2006 IPCC guidelines for national greenhouse gas inventories*. Kanagawa, Japan: Institute for Global Environmental Strategies.

International Energy Agency (IEA). 2008. *Coal information (with 2007 data)*. Paris: IEA Statistics Head of Communication and Information Office.

Keil, R., and G. Desfor. 2004. *Nature and the city: Making environmental policy in Toronto and Los Angeles*. Tucson: University of Arizona Press.

Kejun, J., A. Cosbey, and D. Murray. 2008. *Embedded carbon in traded goods*. Winnipeg, MB, Canada: International Institute for Sustainable Development.

Kim, H., S. Kim, and B. E. Dale. 2009. Biofuels, land use change, and greenhouse gas emissions: Some unexplored variables. *Environmental Science & Technology* 43:961–67.

Kioutsioukis, I., S. Tarantola, A. Saltelli, and D. Gatteli. 2004. Uncertainty and global sensitivity analysis of road transportation emission estimates. *Atmospheric Environment* 38 (38): 6609–20.

Leach, M. K., and T. J. Givnish. 1996. Ecological determinants of species loss in remnant prairies. *Science* 273:1555–58.

Lovell, H., H. Bulkeley, and D. Liverman. 2009. Carbon offsetting: Sustaining consumption? *Environment and Planning A* 41:2357–79.

Luyssaert, S., E. D. Schulze, A. Borner, A. Knohl, D. Hessenmoller, B. E. Law, P. Ciais, and J. Grace. 2008. Old-growth forest as global carbon sinks. *Nature* 455:213–15.

Matthews, S. H., and R. Lifset. 2007. The life-cycle assessment and industrial ecology communities. *Journal of Industrial Ecology* 11 (4): 1–4.

Miner, R., and J. Perez-Garcia. 2007. *The greenhouse gas and carbon profile of the global forest products industry*. Research Triangle Park, NC: National Council for Air and Stream Improvement.

Newell, P., and M. Paterson. 2010. The politics of the carbon economy. In *The politics of climate change*, ed. M. T. Boykoff, 80–99. London and New York: Routledge.

Peters, G. P., and E. G. Hertwich. 2007. CO_2 embodied in international trade with implications for global climate policy. *Environmental Science & Technology* 43:6416–20.

Pfister, S., A. Koehler, and S. Hellweg. 2009. Assessing the environmental impact of freshwater consumption in life cycle assessment. *Environmental Science & Technology* 43 (11): 4098–104.

Potting, J., and M. Z. Hauschild. 2006. Spatial differentiation in life cycle impact assessment. *International Journal of Life-Cycle Assessment* 11 (1): 11–13.

Ramankutty, N., H. K. Gibbes, F. Achard, R. Defries, J. A. Foley, and R. A. Houghton. 2007. Challenges to estimating carbon emissions from tropical deforestation. *Global Change Biology* 13:51–66.

Resource Information Systems (RISI). 2007. *Analytical cornerstone: Q3 update*. Bedford, MA: RISI.

Ridoutt, B. G., and S. Pfister. 2010. A revised approach to water footprinting to make transparent the impacts of consumption and production on global freshwater scarcity. *Global Environmental Change* 20:113–20.

Robbins, P. 2004. *Political ecology: A critical introduction*. Oxford, UK: Blackwell.

Ruth, M., and T. Harrington. 1998. Dynamics of material and energy use in U.S. pulp and paper manufacturing. *Journal of Industrial Ecology* 1 (3): 147–68.

Seager, T. P., S. A. Miller, and J. Kohn. 2009. Land use and geospatial aspects in life cycle assessment of renewable energy. Paper presented at the Proceedings of the 2009 IEEE International Symposium on Sustainable Systems and Technology, Phoenix, AZ.

Searchinger, T., R. Heimlich, R. A. Houghton, F. Dong, A. Elobeid, J. Fabiosa, S. Tokgoz, D. Hayes, and T. H. Yu. 2008. Use of U.S. croplands for biofuels increases greenhouse gases through emissions from land-use change. *Science* 319:1238–40.

SETAC Europe LCA Steering Committee. 2008. Standardization efforts to measure greenhouse gases and "carbon footprinting" for products. *International Journal of Life Cycle Assessment* 13 (2): 87–88.

Sinden, G. 2009. The contribution of PAS 2050 to the evolution of international greenhouse gas emission standards. *International Journal of Life Cycle Assessment* 14:195–203.

Steinberger, J. K., D. Friot, and O. Jolliet. 2009. A spatially explicit life cycle inventory of the global textile chain. *International Journal of Life Cycle Assessment* 14:443–55.

Subak, S., and A. Craighill. 1999. The contribution of the paper cycle to global warming. *Mitigation and Adaptation Strategies for Global Change* 4:113–35.

Thornley, J. H., and M. G. Cannel. 2000. Managing forests for wood yield and carbon storage: A theoretical study. *Tree Physiology* 20:477–88.

Turner, B. L., and P. Robbins. 2008. Land-change science and political ecology: Similarities, differences, and implications for sustainability science. *Annual Review of Environmental Resources* 33:295–318.

U.S. Department of Energy. 2005. *Forest products: Industry of the future*. Washington, DC: U.S. Department of Energy.

———. 2008. Domestic consumption of transportation energy by mode and fuel type, 2006. In *Transportation energy data book (edition 27)*, ed. S.C. Davis, S. W. Diegel, and R. G. Boundy, 2-8–2-10.

U.S. Energy Information Administration. 2008. Fuel and energy source codes and emissions coefficients. In *Voluntary reporting of greenhouse gases program*. Washington, DC: U.S. Energy Information Administration. http://www.eia.doe.gov.oiaf.1605/emission_factors.html (last accessed 10 September 2010).

Villanueva, A., and H. Wenzel. 2007. Paper waste—Recycling, incineration or landfilling? A review of existing life cycle assessments. *Waste Management* 27:29–46.

Weidema, B. P. 2004. Geographical, technological, and temporal delimitation in LCA. *Environmental News, Danish Ministry of the Environment* 74:1–69.

Wood Resources International and Seneca Creek Associates. 2007. *Wood for paper: Fiber sourcing in the global pulp and paper industry*. Washington, DC: American Forest & Paper Association.

Phenology-Based Assessment of Perennial Energy Crops in North American Tallgrass Prairie

Cuizhen Wang,* Felix B. Fritschi,† Gary Stacey,‡ and ZhengWei Yang§

*Department of Geography, University of Missouri
†Division of Plant Sciences, University of Missouri
‡Center for Sustainable Energy, Divisions of Plant Sciences and Biochemistry, University of Missouri
§Research and Development Division, National Agricultural Statistics Service, United States Department of Agriculture

Biomass is the largest source of renewable energy in the United States, and corn ethanol currently constitutes the vast majority of the country's biofuel. Extended plantation of annual crops for biofuel production, however, has raised concerns about long-term environmental, ecological, and socioeconomic consequences. Switchgrass (*Panicum virgatum* L.), along with other warm-season grasses, is native to the precolonial tallgrass prairie in North America and is identified as an alternative energy crop for cellulosic feedstocks. This article describes a phenology-based geospatial approach to mapping the geographic distribution of this perennial energy crop in the tallgrass prairie. Time series of Moderate Resolution Imaging Spectroradiometer (MODIS) satellite imagery (500-m resolution, eight-day interval) in 2007 were processed to extract five phenology metrics: end of season, season length, peak season, summer dry-down, and cumulative growth. A multitier decision tree was developed to map major crops, especially native prairie grasses in the region. The geographic context of the 20 million ha of perennial native grasses extracted in this study could be combined with economic and environmental considerations in a geographic information system to assist decision making for energy crop development in the prairie region. *Key Words: bioenergy, crop phenology, MODIS imagery, time-series analysis.*

生物量是美国可再生能源的最大来源，玉米乙醇目前占美国生物燃料的绝大部分。然而，用于生物燃料生产的一年生作物的延长种植现象，已经引起了人们对其长期的环境，生态，和社会经济后果的关注。柳枝（柳枝稷属），以及其它暖季型草，原产于北美前殖民地高草草原，被确定为纤维素原料的替代能源作物。本文介绍了一种能够映射此高草草原多年生能源作物之地理分布的，基于物候的地理空间方法。本研究对 2007 年的中分辨率成像光谱仪（MODIS）卫星影像（500 米分辨率，八天间隔）时间系列进行处理，提取了五个物候指标：季节结束期，季节长度，旺季，夏季干萎，和累计增长。并开发了一个多层的决策树以绘制该地区主要作物，特别是原生草原草的地图。在本研究中提取的两千万公顷的天然多年生草本植物的地理背景可结合地理信息系统中的经济和环境的考虑，以协助决策者在草原地区能源作物发展的决策。*关键词：生物能源，作物物候，MODIS 遥感影像，时间序列分析。*

La biomasa es la mayor fuente de energía renovable de los Estados Unidos, y actualmente el etanol de maíz constituye una vasta línea de biocombustibles del país. La ampliación del área de plantación anual para la producción de biocombustible, sin embargo, ha despertado preocupaciones sobre consecuencias ambientales, ecológicas y socioeconómicas a largo plazo. El pasto varilla o "switchgrass" (*Panicum virgatum* L.), junto con otros pastos de estación cálida, es nativo de la pradera de pastos altos en América del Norte y se le identifica como un cultivo energético alternativo para producir concentrados para animales a base de celulosa. Este artículo describe un enfoque geoespacial de base fenológica para cartografiar la distribución geográfica de este pasto energético perenne de la pradera de pastos altos. Se procesaron series de tiempo de imágenes satelitales del Moderate Resolution Imaging Spectroradiometer, o MODIS (resolución de 500-m, a intervalos de ocho días) de 2007, para extraer cinco métricas de fenología: final de la estación, longitud de la estación, estación pico, aridez del verano, y crecimiento acumulativo. Se desarrolló un árbol de decisiones multifila para cartografiar los principales cultivos de la región, especialmente los de pastos nativos de pradera. El contexto geográfico de los 20 millones de hectáreas de pastos nativos perennes considerados en este estudio podría combinarse con consideraciones económicas y ambientales en un sistema de información geográfica, para ayudar en la toma de decisiones sobre desarrollo de cultivos energéticos en la región de las praderas. *Palabras clave: bioenergía, fenología de cosechas, imágenes MODIS, análisis de series de tiempo.*

Bioenergy is of increasing interest in agriculture as biomass becomes the largest source of renewable energy in the United States. The recent Billion-Ton Annual Supply study (Perlack et al. 2005) estimated that biomass feedstocks (e.g., corn stover, native grasses, and short-rotation woody crops) could replace 30 percent of domestic petroleum consumption by 2030. Corn ethanol is currently the primary source of domestic biofuel (Farrell et al. 2006). The U.S. Department of Agriculture (USDA) Economic Research Service (2010) estimated that U.S. biofuel refiners will buy 4.2 billion bushels of corn in the 2009–2010 marketing year, which accounts for one third of total domestic corn production. Extended acreage of annual crops, however, has raised concerns about long-term environmental, ecological, and socioeconomic consequences (Paine et al. 1996). Increased use of fertilizers and pesticides causes water quality deterioration, and removal of large quantities of residues from croplands promotes nutrient runoff and leads to soil carbon loss, which in turn lowers crop productivity and profitability (Perlack et al. 2005).

Since 1978, the U.S. Department of Energy (DOE) has sponsored research to evaluate a wide variety of bioenergy alternatives (Wright 1994). In the early 1990s, the DOE identified switchgrass (*Panicum virgatum* L.) as a model cellulosic energy crop (McLaughlin and Kszos 2005; Wright 2007). Switchgrass and other warm-season grasses such as big bluestem (*Andropogon gerardi* Vitman), little bluestem (*Andropogon scoparius*), and indiangrass (*Sorghastrum nutans* (L.) Nash) are native to the precolonial tallgrass prairie in North America (Hitchcock 1935). Although lands in the prairie have been largely cultivated, the adaptability of these native grasses to poor soil conditions (e.g., low pH, low fertility) leads researchers to examine their competitive potentials for cellulosic feedstocks on low-productivity croplands (Kort, Collins, and Ditsch 1998; McLaughlin and Kszos 2005). Highly erodible and other environmentally sensitive croplands can be enrolled in the Conservation Reserve Program (CRP) and can be planted to native prairie grasses (Paine et al. 1996). The CRP lands, and amendments to the management of CRP lands permitted in the 2002 Farm Security and Rural Investment Act, could play important roles in determining the acreage planted with native prairie grasses in the future.

Efficient integration of crops into energy supply systems requires their geographic context to assess regional potentials and costs (Gehrung and Scholz 2009). Current predictions of biomass supplies of energy crops, however, are mostly derived from statistical crop production scenarios at county or state levels. The lack of spatially explicit information about energy crops limits our understanding of current and future bioenergy supplies in major U.S. agricultural regions.

Remote sensing (i.e., satellite imagery) techniques could be applied to extract detailed spatial information on biomass supplies in agricultural regions. Such information is essential in moving beyond currently available coarser geographic representations (e.g., county level) to local scales (e.g., fields) to better estimate biomass potential of a region (Gehrung and Scholz 2009). Assisted with satellite images, spatially resolved crop land use and production in the United States are well documented from various analyses and surveys such as the USDA Large Area Crop Inventory (Boatwright and Whitehead 1986) in early years and the Crop Explorer by the USDA Foreign Agricultural Service (FAS 2009; Cropland Explorer 2010). Satellite images acquired in critical stages of a growing season were found especially useful to differentiate major crops such as corn, soybean, and perennial grasses (Chang et al. 2007; Wang, and Spicci 2010). With monthly composites of satellite images acquired in March through October, the U.S. Geological Survey (USGS) created the 1-km land-cover database and identified 159 seasonal land-cover regions in the conterminous United States; each was characterized with internally homogeneous crops or natural land cover types (Loveland et al. 1995). Since the 1970s, the USDA National Agricultural Statistics Service (NASS) has developed annual products of the Cropland Data Layers (CDL) and acreage estimations from fine-resolution (30–56 m), multitemporal satellite images in major U.S. agricultural regions (Allen, Hanuschak, and Craig 2002; NASS 2010b). In crop year 2009 the CDL product covers all forty-eight states in the conterminous United States (Boryan et al. forthcoming). However, warm-season native prairie grasslands have not been specifically mapped in any of the U.S. agricultural databases.

With the availability of coarse-resolution daily observations such as imagery from the National Oceanic and Atmospheric Administration Advanced Very High Resolution Radiometer (AVHRR) and Terra/Aqua Moderate Resolution Imaging Spectroradiometer (MODIS), phenological variability in cropping systems became well recognized in assisting cropland monitoring in regional scales (Schwartz 1999). Reed et al. (1994) was among the earliest efforts to extract phenological metrics to support vegetation monitoring. Using the AVHRR-extracted normalized difference vegetation index (NDVI) time series, they measured a set of phenological metrics including onset of greenness, time of peak NDVI, rate of greenup/senescence, and integrated NDVI and explored their differences in row crops, grasslands, and deciduous and coniferous forests. Zhang et al. (2003) developed an approach to fitting the NDVI time-series curves into piecewise logistic functions so that phenological metrics can be more accurately extracted. Recently this approach has been well adopted in global phenology (Zhang, Friedl, and Schaaf 2006) and phenology-assisted crop mapping at regional scales (Wardlow, Egbert, and Kastens 2007; Wardlow and Egbert 2008).

Phenology-based time-series analysis could also be performed to identify energy crops and to assess their regional biomass production in a spatially resolved manner. This study aims to apply time-series satellite images to explore current land use patterns of energy crops in the tallgrass prairie. With this information, geospatial and economic tools could be applied to generate spatially explicit data sets to support bioenergy policies in the region.

Study Area and Data Sets

The tallgrass prairie is a triangular region covering ten states in the Midwest and central United States and is extended to Canada in the north (Figure 1).

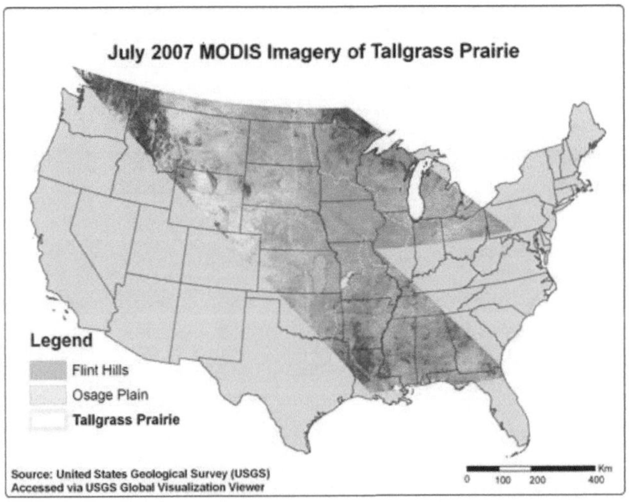

Figure 1. The North American tallgrass prairie and three example Moderate Resolution Imaging Spectroradiometer (MODIS) scenes acquired in July 2007. The color-infrared display of the image shows vegetation in red. Also marked in the figure are the training site (Cherokee Plain, Missouri) and the validation site (Flint Hills, Kansas).

Since European settlement, more than two thirds of the prairie has been converted to annual crops (Schroeder 1983; Risser 1988), primarily corn and soybean in the Corn Belt. Less productive areas tend to be converted to hayfields and pastures of cool-season grass species that were introduced to increase forage production. Native grasses often remain in prairie remnants of the region. Replanted acreages of native grasses also become common in environmentally sensitive croplands designated in federal or state conservation programs.

The tallgrass prairie can be almost fully covered in three MODIS scenes (as shown in Figure 1). In the 500-m MODIS Terra Surface Reflectance products (MOD09A1), each pixel represents a $500 \times 500\ m^2$ area on the ground. Each image is an eight-day composite by selecting the best pixel value from eight daily images with a maximal value compositing technique (Lovell and Graetz 2001). For each scene, a total of forty-six MOD09A1 composites in 2007 were downloaded from the Land Processes Distributed Active Archive Center (http://lpdaac.usgs.gov). The NDVI time series were then extracted from these image products. The sixteen-day MODIS imagery had been most commonly applied in past studies because it reached better reduction of cloud contamination. It was not used in this study, as subtle phenology variations of perennial crops in critical stages tend to be lost at such intervals.

Crop training data were collected in the Cherokee Plain in southwest Missouri (marked in Figure 1). For annual crops (corn, soybean, and winter wheat), training data were extracted from the CDL map in the Plain that were developed by USDA NASS using 56-m satellite images acquired in 2007 (Cropland Explorer 2010). In previous research funded by the Missouri Department of Conservation (MDC), warm-season native prairie grasses (WSG) and cool-season forage grasses (CSG) were collected during field surveys in 2007 and from the Grassland Coalition Focus Areas managed by MDC (Wang, Jamison, and Spicci 2010). To reduce mixed fields in coarse-resolution MODIS pixels, each data point represented a polygon of at least 1 km^2 (four MODIS pixels). A total of seventeen points for corn, fifteen for soybean, twelve for wheat, twenty-three for WSG, and nineteen for CSG were collected as training data in this study.

The Flint Hills grassland (marked in Figure 1) covers an area of 2.5 million ha in Kansas and is the largest unplowed tallgrass prairie remnant in North America. More than 80 percent of the area is covered with native prairie grasses. Inside the grassland, the 4,500 ha Tallgrass Prairie National Preserve (TPNP) is managed by the National Park Service with a single conservation strategy (Gu et al. 2007). According to the geographic information system (GIS) data downloaded from the Kansas Geological Survey (http://www.kgs.ku.edu), native prairie grass covers 90.95 percent of the TPNP. Moreover, a total of nineteen preserved prairies managed by MDC in southwest Missouri were collected from Wang, Jamison, and Spicci (2010). These prairies served as validation sites of native prairie grasses in this study. The CDL maps of the ten prairie states were also downloaded. As reported in CDL metadata files, the CDL products reached an overall accuracy around 90 percent for major crops (NASS 2010a). Comparing with the 500-m resolution of MODIS imagery, the 56-m CDL map provided fine-scale details and was selected as validation source of annual crops.

Methods

Because of the nature of herbaceous land covers and the spectral similarity of WSG and CSG, the CDL products group them into a single category—pasture/grass. Wang, Jamison, and Spicci (2010), in previous research, found that (1) CSG reached peak NDVI in spring (May), whereas WSG had peak values in early summer (July); and (2) the NDVI of WSG decreased gradually from summer to fall, whereas a second NDVI peak was observed for CSG. Therefore, it was feasible to separate these two grass types based on their phenological features. Phenology of annual crops (corn, soybean, and winter wheat) was also examined in this study. Non-crop land covers (e.g., forest, wetland, water, urban) were extracted from the CDL map and were masked out of the process.

Crop Phenology Metrics

As illustrated by the jagged NDVI curves in Figure 2, raw NDVI was influenced by cloud residuals as well as atmospheric and sun-sensor illumination conditions. A set of smoothing algorithms have been adopted to reduce these noises. For example, Reed et al. (1994) used a nonlinear median smoother to remove cloud-contaminated NDVI values, although it also reduced peak NDVI values that were assumed noise-free. Gu et al. (2006) applied an upper-envelope three-point smoothing process to provide the local "best estimate" of leaf area index from time-series MODIS products.

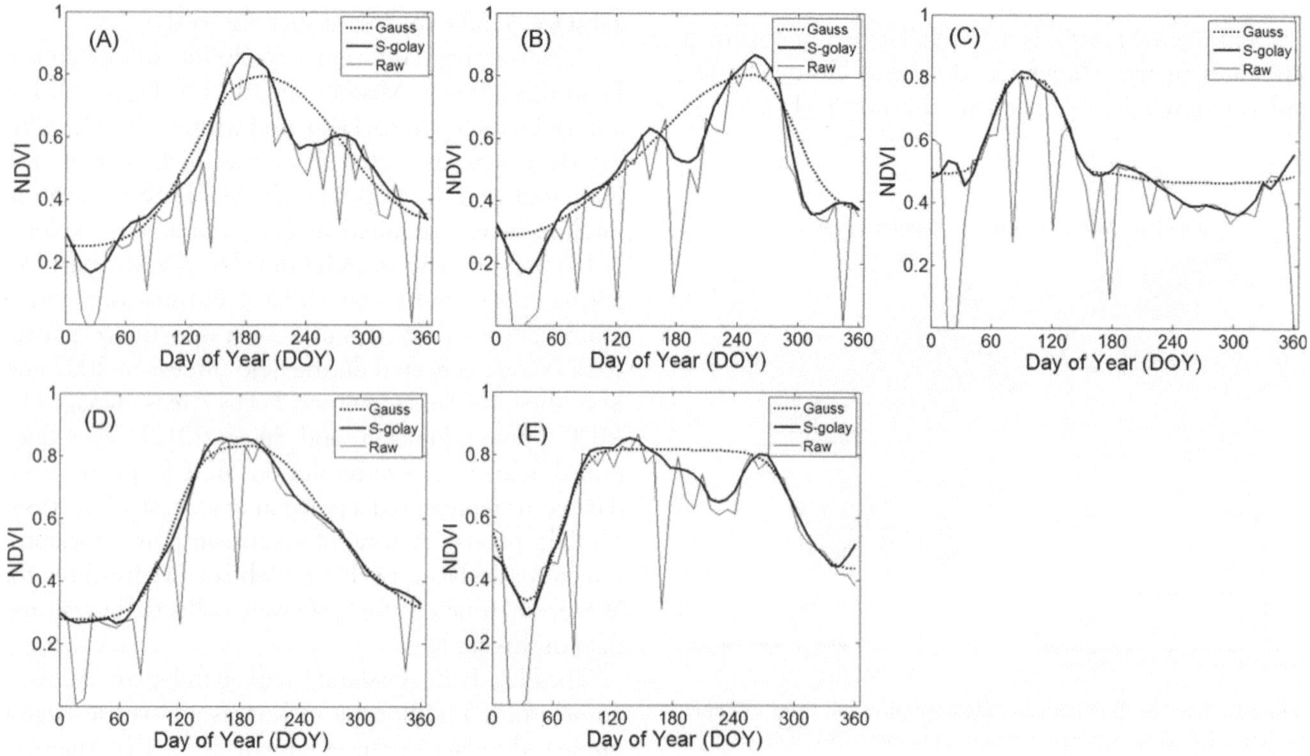

Figure 2. Normalized difference vegetation index (NDVI) variations of major crops in 2007 in the tallgrass prairie: (A) corn, (B) soybean, (C) winter wheat, (D) warm-season native prairie grass, and (E) cool-season forage grass. The three curves are raw NDVI, Savitzky–Golay smoothed NDVI, and asymmetric Gaussian fitted NDVI.

To optimize smoothing process of NDVI curves in this study, a five-point median filter was first applied to remove spikes of heavily contaminated NDVI values. If the filtered NDVI was lower than the original value, it was replaced with the original one to remain peak NDVI points in the curve. Then a commonly applied Savitzky–Golay filtering method (Savitzky and Golay 1964) was used to smooth the time-series curves (Figure 2).

The smoothed data were simulated to extract crop phenology metrics. de Beurs and Henebry (2010) compared a group of thresholding and simulating approaches in phenology studies. Among these, the widely applied ones in crop phenology include piecewise logistic functions (Zhang et al. 2003) and quadratic fitting (De Beurs and Henebry 2004). In this study, however, these approaches turned out to be ill-functioned because grass phenology is often characterized with prolonged growth duration and asymmetric growing trends in early and late seasons. Here we adopted the TIMESAT program (Jonsson and Eklundh 2004) to fit the smoothed MODIS NDVI curves in an asymmetric Gaussian function (Figure 2). The program applied different Gaussian simulation functions to smooth curves before and after growing peak, which was especially useful for crops with asymmetric growing patterns such as perennial grasses and winter wheat.

The asymmetric Gaussian fitted curve at each pixel was analyzed in the TIMESAT program to extract a set of phenology metrics. With metrics of all training data, statistical properties revealed that three metrics had significant differences for the five crops (corn, soybean, winter wheat, CSG, and WSG): (1) *end of season*, which represents the date when NDVI has decreased to 20 percent of the amplitude after peak NDVI; (2) *season length*, which represents the dates from start of season (when NDVI has increased to 20 percent of the amplitude) to end of season; and (3) *cumulative growth* ($\Sigma NDVI$), which is calculated as the integral of NDVI from start of season to end of season. Other metrics, such as start of season and derivatives of green-up and brown-down, tend to be overlaid for different crops in their feature spaces and box-plots. These metrics were not used in the study.

The Savitzky–Golay smoothed curves in Figure 2 revealed that annual crops displayed apparent temporal differences in peak NDVI over the course of a growing season, reflecting their variations in growth development. To quantify these differences, three growing stages were defined: early (Day of Year [DOY] 1–161), middle (DOY 145–193), and late (DOY 161–313). Two

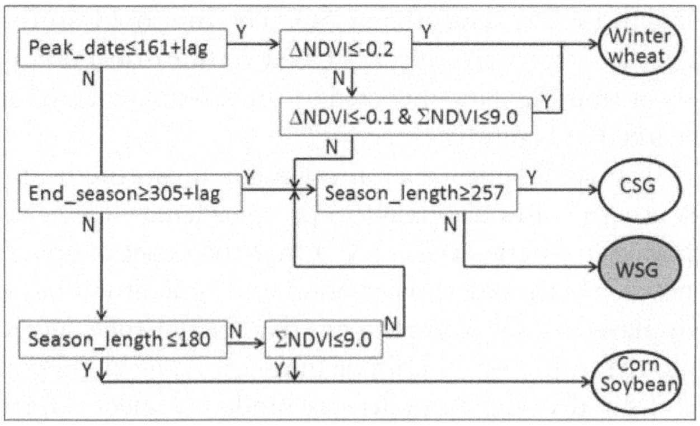

Figure 3. Flowchart of the phenology-based decision tree. The perennial native prairie grass (WSG) is highlighted in the outputs. The numbers 161 and 305 represent Day of Year (DOY). The numbers 180 and 257 represent growth duration (days). NDVI = normalized difference vegetation index; CSG = cool-season grasses; WSG = warm-season grasses.

additional phenology matrices were thus identified: (1) *peak season*, which indicates the date when peak NDVI falls; and (2) *summer dry-down* ($\Delta NDVI$), which is the maximal decrease of NDVI in early-middle stages if peak NDVI falls in the early stage (especially useful for wheat).

A Phenology-Based Decision Tree Approach

The selected phenology metrics were put in a decision tree to identify the five crop types in the tallgrass prairie (Figure 3). Winter wheat was first identified because of its early peak season (early spring) and large summer dry-down (dramatic decrease of NDVI) leading to harvest in early summer. Because spring wheat production was limited in the tallgrass prairie, it was grouped into winter wheat and was not further considered in this study. Spring wheat was more commonly grown in northern states, however, and could potentially be misclassified due to temporal shifts in peak greenness. Similar to winter wheat, some CSG as well as a limited number of WSG fields also reached early peak season and were characterized with reduced NDVI after haying or grazing of pastures in late spring to early summer. These grass fields were delineated from winter wheat with a shallower summer dry-down and larger cumulative NDVI ($\Sigma NDVI > 9.0$). Corn and soybean in the prairie often had an early end of season (before DOY 305), shorter season length (< 180 days) and less cumulative NDVI (< 9.0). Corn and soybean were not further separated because the primary concern was native perennial grasses in this study. After annual crops were extracted, WSG was delineated with a shorter growing length (< 257 days) than CSG. The thresholds in the decision tree were selected based on statistical analysis of training data collected in the Cherokee Plain as marked in Figure 1.

Temporal shifts of crop calendars in northern and southern states also need to be considered. It was reported in Zhang et al. (2003) that the onset of greenness for both natural vegetation and agricultural lands displayed a shift of around two days per latitude degree along the 40°–45°N latitude transect. To test temporal shifts in the tallgrass prairie, we randomly selected forty fields for each category (corn, soybean, and grass) from the CDL map. Both WSG and CSG fields were covered in the grass category because they were not specified in the CDL data. In a range of 32°–48°N in the prairie, the peak season had about two days per degree shift for corn (Figure 4A) and soybean (Figure 4B). The shift was slightly larger for grasslands (Figure 4C). There was no apparent shift for end of season of annual crops, whereas a slight shift of 1.7 days was observed in grasslands. The season length did not show apparent shift for all crop types. To be consistent with Zhang et al. (2003), a lag factor was added to peak season and end of season in the decision tree in Figure 3. Based on training data in the Cherokee Plain (around 38°N), the *lag* at each pixel was roughly calculated:

$$\text{Lag} = 2.0 \times (\text{latitude} - 38.0) \quad (1)$$

Results and Discussion

Crop Map and Validation

The MODIS-derived crop map (Figure 5) agreed with the CDL product that corn and soybean were major annual crops, especially in the Corn Belt covering the northern states of the prairie. Clustered wheat fields were mostly found in the southwest (e.g., winter wheat in Kansas and Oklahoma) and the northwest (e.g., spring wheat in Nebraska and South Dakota). For perennial crops, both CSG and WSG grew in grasslands designated in the CDL map. Warm-season native prairie grasses dominated in the south central part of the prairie whereas CSG were in the western and southeastern parts of the prairie.

The Flint Hills grassland, the largest remnant of unplowed tallgrass prairie, is marked in Figure 5. Native prairie grasses covered more than 80 percent of the Flint Hills, which was readily identified by the phenology-assisted approach in this study. Limited corn and soybean acreages were found along local river channels in the north of the Flint Hills grassland. This pattern agreed with the CDL map.

Total cropping acreages in the tallgrass prairie were compared between the MODIS-derived and CDL products. The MODIS estimation of corn and soybean was about 3.43 million ha lower than the CDL product (5.61 percent of the tallgrass prairie). The CDL product estimated 25.79 million ha (34.21 percent) of grasslands. In the MODIS crop map, CSG covered 11.82 million ha (15.65 percent), whereas warm-season native grass reached 20.73 million ha (27.45 percent). Combining CSG and WSG, the MODIS map overestimated grasslands by an area of 7.19 million ha. Part of the overestimation might come from acreages of other crops (3.21

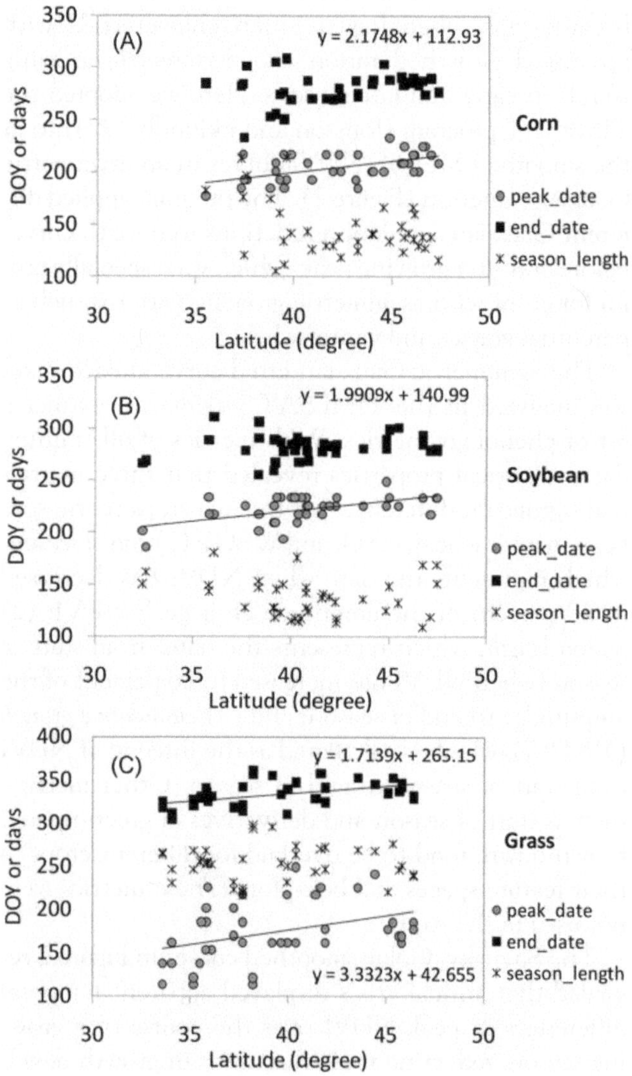

Figure 4. Temporal shifts of phenology metrics of (A) corn, (B) soybean, and (C) grass from randomly selected fields in the tallgrass prairie.

million ha) identified in the CDL product. These crops, such as sorghum, rice, cotton, and vegetables, only covered a small portion of the region and were not specified in this study. In the MODIS crop map, they were often grouped into perennial crops, especially warm-season native grasses because of their short season length. The discrepancy might also come from mixed pixels. Except for the vast landscape of cultivated lands in the Corn Belt, annual crop fields were often small and scattered with grass fields. Although they could be identified in fine-resolution CDL products, these fields were smaller than the 500-m pixel size in the MODIS-derived map. Mixed pixels tend to be classified as grasses because of grass-like phenology metrics such as extended end of season and season length. Therefore, annual crops were underestimated and perennial crops were overestimated in the MODIS-derived map.

Due to the large pixel size of MODIS imagery (500 × 500 m^2), it was difficult to perform ground-based accuracy assessment. This study applied an error matrix approach (Congalton 1988) to compare the accuracies of the MODIS-derived classes against the published CDL products. To be spatially comparable, the CDL map was down-sampled to 500-m resolution with a majority filtering process. Corn and soybean in the CDL map were compared with the class of corn/soybean in the MODIS map, whereas WSG and CSG in the MODIS map (Figure 5) were combined to match the class of grassland in the CDL map. Winter wheat was not considered in the assessment as it covered limited area of the tallgrass prairie. With a stratified random sampling design, a number of 1,448 and 1,077 sample points were selected for corn/soybean and grassland, respectively. The numbers represented 0.1 percent of total pixels for each category in the degraded CDL map. In a two-category error matrix approach, it was found that 428 out of the 1,448 corn/soybean points were misclassified as grass in the MODIS-derived map, whereas 134 out of the 1,077 grass points were misclassified as corn/soybean. The overall accuracy reached 77.74 percent when both categories were considered. For perennial crops, the producer's accuracy was 87.56 percent and the user's accuracy was only 68.78 percent, resulting from misclassifying annual crops into perennial crops in the MODIS-derived map. It should be noted, however, that the "accuracies" discussed here actually represented the agreement between the MODIS-derived map and the published CDL product, which had about 10 percent of uncertainties for classification of major crop types from fine-resolution satellite imagery (NASS 2010a). Further uncertainties were inevitably introduced when the 56-m CDL map was degraded to 500 m for purposes of comparison (Pontius and Cheuk 2006). More ground data will be collected in the future to assess real accuracies of MODIS-derived energy crop map in this study.

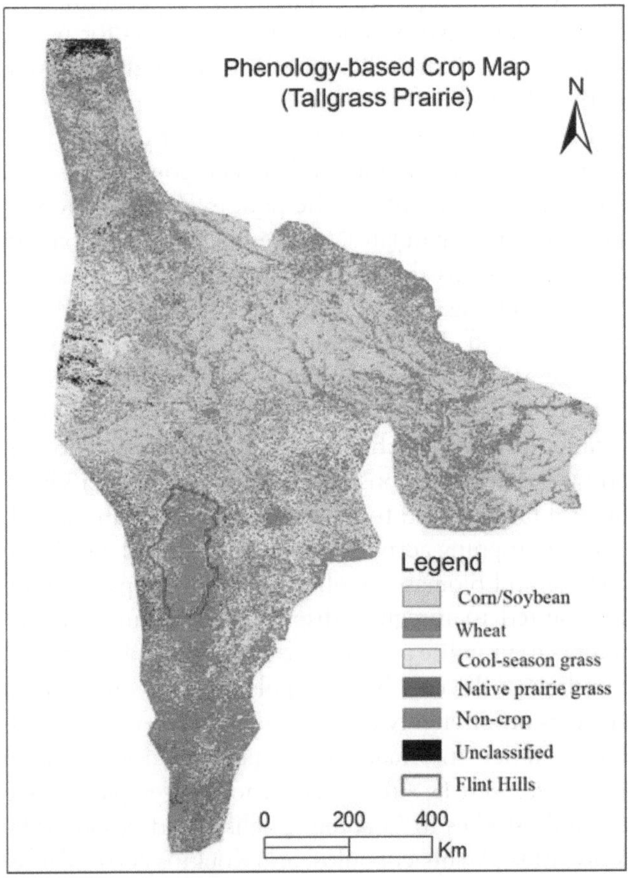

Figure 5. The 2007 crop map of the tallgrass prairie extracted from time series of Moderate Resolution Imaging Spectroradiometer (MODIS) imagery. Outlined in the map is the Flint Hills grassland, the largest validation site of native prairie grass.

Specifically for native prairie grasses, Wang, Jamison, and Spicci (2010) recorded relative abundance of WSG (%) and CSG (%) of the nineteen prairie remnants managed by MDC. In the MODIS-derived map, the percentages were calculated by counting the numbers of WSG and CSG pixels of each prairie. Due to large size of MODIS pixels, the eight prairies less than 100 ha (four MODIS pixels) were not considered. Two large ground truth sites, the Flint Hills grassland (WSG% = 80.00) and the TPNP (WSG% = 90.05) in the grassland, were also examined. Ground-observed and MODIS-calculated relative abundances of these prairies were compared in Figure 6, in which each prairie had two points (WSG% and CSG%) in the scatterplot. It agreed

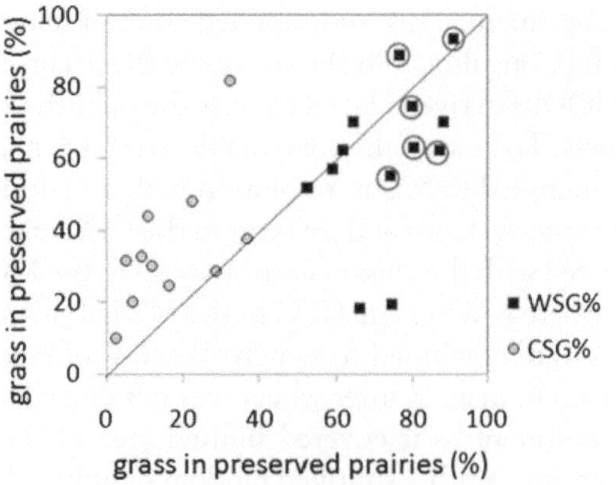

Figure 6. Comparison of ground-observed and Moderate Resolution Imaging Spectroradiometer (MODIS)-derived relative abundance of cool-season grasses (CSG) and warm-season grasses (WSG) in the preserved prairie remnants. The circled points represent the six largest prairies (> 500 ha) in the region.

with Wang, Jamison, and Spicci (2010) that native prairie grasses dominated these prairies. The six largest prairies were circled in the scatterplot: Prairie State Park, Osage Prairie, Wah-Kon-Tah Prairie, Taberville Prairie, TPNP, and Flint Hills grassland. Among the thirteen prairies, only two prairies had apparently different results from ground observations (the two WSG% points in the lower right of the scatterplot). One was the Stony Point Prairie (388.64 ha) in Dade County, Missouri. Upon ground observations it was primarily composed of WSG and CSG, whereas in the CDL map the prairie was largely classified as wetland vegetation, which was masked out in this study. The other one was the Monegow Prairie (108.12 ha) in Cedar County, Missouri. It had much lower WSG% (the lower right in Figure 6) and correspondingly higher CSG% (the upper left in Figure 6) in the MODIS-derived results. One possible reason was the relatively small area of this prairie and its mixed growth of WSG (66.23 percent) and CSG (32.43 percent) species, which resulted in large misclassification in the MODIS-derived map. With all points in Figure 6, the root mean square error (RSME) was 23.47 percent, which explained the overall discrepancy between the MODIS-derived results in this study and the published ground records. When the two outliers were removed, the RMSE dropped to 13.76 percent.

The phenology-based crop mapping approach was inevitably affected by the shift of crop calendar in a large region. Although the shift was compensated for by a lag factor in the decision tree, a large discrepancy was observed in the northern and southern states of the tallgrass prairie. For example, there were considerable amounts of spring wheat acreages in North Dakota (in the very north of the prairie). These fields tend to be misclassified as corn/soybean because they could not meet the decision rule of early peak season in Figure 3. Similarly, some grasslands in Texas (in the very south of the prairie) were misclassified as wheat because of their early peak season. Crops in these states were not further examined because they only cover limited area in the tallgrass prairie.

Biomass Potentials and Environmental and Economic Impacts

Corn has been commercially adopted as a major source of biofuel (e.g., corn ethanol), and native prairie grasses could have high potential as lignocellulosic bioenergy feedstock. The Oak Ridge National Laboratory found that approximately 10 percent of the total suitable land in the Midwest could potentially be used for energy crops without affecting food production (Wright 2007). This study recorded the geographic context of energy crops (an estimation of 26.77 million ha for corn and soybean and 20.73 million ha for native prairie grass) in the tallgrass prairie, a dominant region in the Midwest.

Based on current land use patterns of energy crops, economic and environmental impacts could be assessed for bioenergy viability in the prairie. Economic forces, such as biofuel mandates, tax credits, and tariffs explain the mechanisms through which policy changes cause land use change of energy crops (Thompson, Meyer, and Westhoff 2009). Meanwhile, potential land use change associated with energy crops is determined by crop profitability (net returns) and site-specific environmental factors such as soil quality, water availability, and weather conditions. These economic and environmental factors need to be interactively examined with current cropping activities to project potential land use change and biomass supplies of energy crops. GIS are well suited to integrate disparate data sources in attempts to visualize and analyze geographic phenomena. Similarly, GIS can be manipulated to evaluate land conversion of energy crops in a region to provide spatially explicit information to support decision-making processes such as site selection for power plants (see example studies in Graham, English, and Noon 2000; Voivontas, Assimacopoulos, and Koukios 2001; Vainio et al. 2009).

This study initiated an effort of regional bioenergy land use by taking advantage of coarse-resolution, frequent satellite observations. Only one-year data sets

(2007) were selected because of smooth climate conditions in this year. To continue this investigation, multiyear time-series observations will be applied to explore climate change in agricultural regions as extreme climate conditions and phenology abnormalities can severely affect cropping activities and biomass production. At coarse resolution, MODIS pixels are often a mixture of different fields in agricultural lands. With various linear and nonlinear unmixing techniques, mixed pixels could be decomposed to different crop types at certain percentages, which improves accuracies in estimating energy crop distributions.

Remote sensing of energy crops enhances the documentation and understanding of regional biomass feedstock production. In a long run, other large-scale efforts of phenology studies such as the U.S. National Phenology Network (http://www.uwm.edu/Dept/Geography/npn) and the European Phenology Network (http://www.dow.wau. nl/msa/epn/index.asp) could also be integrated to improve understanding in phenology of land surfaces and to provide a better basis for comparing in situ measurements against remotely sensed observations. Such spatially explicit biomass information is critical for a broad spectrum of planning activities, ranging from policy development, marketing, and siting necessary infrastructure to a variety of security and logistical issues. Therefore, the geospatial approaches initiated in this study could potentially take an important and cross-cutting role in bioenergy policy decision making for federal and state agencies.

Conclusion

This study explored a geospatial assessment of bioenergy land use in the tallgrass prairie. Time series of satellite imagery were analyzed to extract distinct phenology metrics of annual (corn, soybean, wheat) and perennial crops (CSG, WSG) over the course of the growing season in 2007. Five metrics were found to be important in delineating these crops: end of season, season length, peak season, summer dry-down, and cumulative growth. A phenology-based decision tree was developed to identify the five major crops from satellite images. The study revealed the geographic distributions of native prairie grasses, the primary perennial energy crop in the United States. The total acreage of this perennial energy crop was estimated at 20 million ha in the prairie, covering about one quarter of the region. It was demonstrated in this study that time-series MODIS products provided an efficient data source for regional bioenergy studies. If more ground data are available to better understand the uncertainties of MODIS-derived energy crop mapping, the geospatial approach developed and the distributions of energy crops extracted in this study could provide essential information for bioenergy decision making in the prairie region.

References

Allen, R., G. Hanuschak, and M. Craig. 2002. *History of remote sensing for crop acreage*. Washington, DC: USDA-NASS. http://www.nass.usda.gov/Surveys/Remotely_Sensed_Data_Crop_Acreage/index.asp (last accessed 12 August 2010).

Boatwright, G. O., and V. S. Whitehead. 1986. Early warning and crop condition assessment research. *IEEE Transactions on Geoscience and Remote Sensing* 24:54–64.

Boryan, C., R. Mueller, M. Craig, and Z. Yang. Forthcoming. Monitoring US cropland: USDA National Agricultural Statistics Service Cropland Data Layer program. *Geocarto International*.

Chang, J., M. C. Hansen, K. Pittman, M. Carroll, and C. DiMiceli. 2007. Corn and soybean mapping in the United States using MODIS time-series data sets. *Agronomy Journal* 99:1654–64.

Congalton, R. G. 1988. Using spatial autocorrelation analysis to explore the errors in maps generated from remotely sensed data. *Photogrammetric Engineering and Remote Sensing* 54:587–92.

Cropland Explorer. 2010. Cropland Explorer: USDA NASS CDL program. http://nassgeodata.gmu.edu/CDLExplorer/ (last accessed 6 September 2010).

de Beurs, K. M., and G. M. Henebry. 2004. Land surface phenology, climatic variation, and institutional change: Analyzing agricultural land cover change in Kazakhstan. *Remote Sensing of Environment* 89:497–509.

———. 2010. Spatio-temporal statistical methods for modeling land surface phenology. In *Phenological research*, ed. I. L. Hudson and M. R. Keatley, 177–208. New York: Springer Science and Business Media.

Economic Research Service, U.S. Department of Agriculture. 2010. *Feed grains data: Yearbook tables*. Washington, DC: USDA. http://www.ers.usda.gov/data/feedgrains/Table.asp?t=31 (last accessed 11 February 2010).

Farrell, A. E., R. J. Plevin, B. T. Turner, A. D. Jones, M. O'Hare, and D. M. Kammen. 2006. Ethanol can contribute to energy and environmental goals. *Science* 311:506–08.

Foreign Agricultural Service (FAS), U.S. Department of Agriculture. 2009. *USDA Crop Explorer: Global crop condition and commodity production analysis from the USDA/Production Estimates and Crops Assessment Division (PECAD)*. Washington, DC: USDA. http://gcmd.nasa.gov/records/GCMD_USDA0557.html (last accessed 12 August 2010).

Gehrung, J., and Y. Scholz. 2009. The application of simulated NPP data in improving the assessment of the spatial distribution of biomass in Europe. *Biomass and Bioenergy* 33:712–20.

Graham, R. L., B. C. English, and C. E. Noon. 2000. A geographic information system-based modeling system for evaluating the cost of delivered energy crop feedstock. *Biomass and Bioenergy* 18:309–29.

Gu, Y., S. Bélair, J. Mahfouf, and G. Deblonde. 2006. Optional interpolation analysis of leaf area index using MODIS data. *Remote Sensing of Environment* 104:283–96.

Gu, Y., J. F. Brown, J. P. Verdin, and B. Wardlow. 2007. A five-year analysis of MODIS NDVI and NDWI for grassland drought assessment over the central Great Plains of the United States. *Geophysical Research Letters* 34:L06407. DOI:10.1029/2006GL029127.

Hitchcock, A. S. 1935. *Manual of the grasses of the United States*. Washington, DC: U.S. Department of Agriculture.

Jonsson, P., and L. Eklundh. 2004. TIMESAT—A program for analyzing time-series of satellite sensor data. *Computers and Geosciences* 30:833–45.

Kort, J., M. Collins, and D. Ditsch. 1998. A review of soil erosion potential associated with biomass crops. *Biomass and Bioenergy* 14:351–59.

Loveland, T. R., J. W. Merchant, J. F. Brown, D. O. Ohlen, B. C. Reed, P. Olson, and J. Hutchinson. 1995. Map supplement: Seasonal land-cover regions of the United States. *Annals of the Association of American Geographers* 85:339–55.

Lovell, J. L., and R. D. Graetz. 2001. Filtering pathfinder AVHRR land NDVI data for Australia. *International Journal of Remote Sensing* 22:2649–54.

McLaughlin, S. B., and L. A. Kszos. 2005. Development of switchgrass (*Panicum virgatum*) as a bioenergy feedstock in the United States. *Biomass and Bioenergy* 28: 515–35.

National Agricultural Statistics Service (NASS), U.S. Department of Agriculture. 2010a. CDL metadata. http://www.nass.usda.gov/research/Cropland/metadata/meta.htm (last accessed 8 September 2010).

———. 2010b. Cropland data layer. http://www.nass.usda.gov/research/Cropland/SARS1a.htm (last accessed 1 March 2010).

Paine, L. K., T. L. Peterson, D. J. Undersander, K. C. Rineer, G. A. Bartelt, S. A. Temple, D. W. Sample, and R. M. Klemme. 1996. Some ecological and socioeconomic considerations for biomass energy crop production. *Biomass and Bioenergy* 10:231–42.

Perlack, R. D., L. L. Wright, A. F. Turhollow, R. L. Graham, B. J. Stokes, and D. C. Erbach. 2005. *Biomass as feedstock for a bioenergy and bioproducts industry: The technical feasibility of a billion-ton annual supply*. Oak Ridge, TN: U.S. Department of Energy and U.S. Department of Agriculture, Oak Ridge National Laboratory.

Pontius, R. G., Jr., and M. L. Cheuk. 2006. A generalized cross-tabulation matrix to compare soft-classified maps at multiple resolutions. *International Journal of Geographical Information Science* 20:1–30.

Reed, B. C., J. F. Brown, D. VanderZee, T. R. Loveland, J. W. Merchant, and D. O. Ohlen. 1994. Measuring phonological variability from satellite imagery. *Journal of Vegetation Science* 5:703–14.

Risser, P. G. 1988. Diversity in and among grasslands. In *Biodiversity*, ed. E. O. Wilson, 176–80. Washington, DC: National Academy of Sciences.

Savitzky, A., and M. J. E. Golay. 1964. Smoothing and differentiation of data by simplified least squares procedures. *Analytical Chemistry* 36:1627–39.

Schroeder, W. A. 1983. *Presettlement prairie*. Natural History Series, Vol. 2. Jefferson City: Missouri Department of Conservation.

Schwartz, M. D. 1999. Advancing to full bloom: Planning phenological research for the 21st century. *International Journal of Biometeorology* 42:113–18.

Thompson, W., S. Meyer, and P. Westhoff. 2009. How does petroleum price and corn yield volatility affect ethanol markets with and without an ethanol use mandate? *Energy Policy* 37:745–49.

Vainio, P., T. Tokola, T. Palander, and A. Kangas. 2009. A GIS-based stand management system for estimating local energy wood supplies. *Biomass and Bioenergy* 33:1278–88.

Voivontas, D., D. Assimacopoulos, and E. G. Koukios. 2001. Assessment of biomass potential for power production: A GIS based method. *Biomass and Bioenergy* 20:101–12.

Wang, C., B. Jamison, and A. Spicci. 2010. Trajectory-based warm season grass mapping in Missouri prairies with multi-temporal ASTER imagery. *Remote Sensing of Environment* 114:531–39.

Wardlow, B. D., and S. L. Egbert. 2008. Large-area crop mapping using time-series MODIS 250 m NDVI data: An assessment for the U.S. Central Great Plains. *Remote Sensing of Environment* 112:1096–116.

Wardlow, B. D., S. L. Egbert, and J. H. Kastens. 2007. Analysis of time-series MODIS 250 m vegetation index data for crop classification in the U.S. Central Great Plains. *Remote Sensing of Environment* 108:290–310.

Wright, L. L. 1994. Production technology status of woody and herbaceous crops. *Biomass and Bioenergy* 6:191–209.

———. 2007. Historical perspective on how and why switchgrass was selected as a "model" high-potential energy crop. OENL/TM-2007/109. Oak Ridge, TN: Office of Scientific and Technical Information. http://www.osti.gov/bridge (last accessed 11 February 2010).

Zhang, X., M. A. Friedl, and C. B. Schaaf. 2006. Global vegetation phenology from Moderate Resolution Imaging Spectroradiometer (MODIS): Evaluation of global patterns and comparison with situ measurements. *Journal of Geophysical Research* 111 (G4): G04017-30. DOI:10.1029/2006JG000217.

Zhang, X., M. A. Friedl, C. B. Schaaf, A. H. Strahler, J. C. F. Hodges, F. Gao, B. C. Reed, and A. Huete. 2003. Monitoring vegetation phenology using MODIS. *Remote Sensing of Environment* 84:471–75.

A Geographic Approach to Sectoral Carbon Inventory: Examining the Balance Between Consumption-Based Emissions and Land-Use Carbon Sequestration in Florida

Tingting Zhao,* Mark W. Horner,* and John Sulik†

*Department of Geography and Institute for Energy Systems, Economics, and Sustainability, Florida State University
†Department of Geography, Florida State University

Carbon accounting is an important analytical task that provides baseline information to assist in establishing emissions targets, developing market-based carbon trading programs, and facilitating sustainable carbon management at the regional to international scales. Although a substantial amount of research has focused on carbon emissions inventory, limited studies have been conducted to estimate consumption-based emissions and their spatial distribution in relation to vegetation carbon sinks. In this article, we develop a new approach to model the spatially detailed consumption-based carbon emissions from the household energy and transportation sectors. Emissions were in turn integrated with vegetation carbon sequestration rates that were modeled through biophysical remote sensing techniques. This enables carbon balance analysis at detailed geographic locations. To illustrate our approach for carbon balance analysis, we present a case study in Florida. Results indicate that, in 2001, Florida was able to self-assimilate residential carbon emissions from energy and transportation fuel consumption. Estimates indicate that net carbon sources (i.e., household emissions exceeding vegetation carbon assimilation) are associated with urban and suburban densities and net sinks with exurban and rural densities. In sum, the research approach can be extended to household energy consumption and carbon assessment for other geographies at alternative scales. *Key Words: carbon emissions inventory, GIS, household energy consumption, household transportation fuel consumption, vegetation carbon sinks.*

碳核算是一个重要的分析任务，它提供基线资料以协助建立减排目标，发展以市场为基础的碳交易计划，促进从区域到国际级的可持续发展的碳管理。虽然已有大量的研究集中于碳排放清单，但是以消费为基础的排放及其空间分布对有关的植被碳汇的研究还是有限的。在这篇文章中，我们发展一种新的方法来模拟空间详细的，来自家庭能源和运输部门的消费型的碳排放量。排放量反过来与通过生物物理学的遥感技术模拟的植被碳隔离率集成。这使详细地理位置的碳平衡分析成为可能。我们以佛罗里达州为例，说明我们的碳平衡分析方法。结果表明，在2001年，佛罗里达州能够自我吸收来自能源和运输燃料消耗的住宅碳排放。估计显示，净碳源（即家庭的排放量超过植被碳同化）与农村城市和郊区密度和净水槽联系在一起的远郊和郊区密度有关。总之，研究方法可以推广到家庭能源消费和在其他地区替代规模的碳评估。*关键词：碳排放清单，地理信息系统，家庭能源消费，家庭运输燃料消耗，植被的碳汇。*

La contabilidad del carbono es una tarea analítica importante que da información de base para ayudar a establecer metas de emisiones, mediante el desarrollo de programas comerciales de carbono basados en el mercado, y facilitar el manejo sostenible del carbono a escalas regionales e internacionales. Si bien una sustancial cantidad de investigación se ha enfocado al inventario de las emisiones de carbono, también se han llevado a efecto limitados estudios para calcular las emisiones basadas en consumo y su distribución espacial en relación con los sumideros de carbono de la vegetación. En este artículo desarrollamos un nuevo enfoque para modelar las emisiones de carbono del consumo de los sectores energéticos de hogares y transporte, detalladas espacialmente. A su turno las emisiones fueron integradas con tasas de secuestro del carbono por la vegetación, modeladas a través de técnicas biofísicas de percepción remota. Esto permite el análisis del balance de carbono en localizaciones geográficas detalladas. Para ilustrar nuestro enfoque de análisis del balance del carbono, presentamos un estudio de caso en la Florida. Los resultados indican que, en 2001, la Florida pudo auto-asimilar emisiones residenciales de carbono provenientes del consumo de combustibles para energía y transporte. Los estimativos indican que las fuentes netas de carbono (esto es, las emisiones de hogares que exceden la asimilación de carbono por la vegetación) están asociadas con la densidad urbana y suburbana y con sumideros netos con densidades transurbanas y rurales. En suma, el enfoque de la investigación puede extenderse al consumo familiar de energía y a la evaluación del carbono para otras geografías a escalas alternativas. *Palabras clave: inventario de emisiones de carbono, SIG, consumo de energía familiar, consumo de combustibles para transporte familiar, sumideros de carbono de la vegetación.*

Anthropogenic climate change challenges societies throughout the world. To slow the increase in global temperature, curbing emissions of carbon dioxide (CO_2), which account for nearly 75 percent of global annual greenhouse gas emissions, will be required worldwide (Intergovernmental Panel on Climate Change [IPCC] 2007). The global reduction of CO_2 emissions relies heavily on localized mitigation efforts (Knight et al. 2003) for which future targets are determined mainly by a country or region's emissions history. For example, the Kyoto Protocol established different rates of emissions reduction among industrialized countries based on their development activities since the industrial revolution. The Protocol also required that each participating country provides data to estimate its respective 1990 emission levels to designate their future emission targets (United Nations 1998).

A carbon emissions inventory, or the estimation of a country or region's historical emissions, quantifies sources and sinks of anthropogenic CO_2 over a certain geographic area and time period (U.S. Environmental Protection Agency [EPA] 2009a). It requires approaches that capture the spatial distribution of CO_2 sources associated with human activities as well as vegetation carbon sinks, which can be either anthropogenic (e.g., planted trees) or natural (e.g., undisturbed rainforests). An assessment of the literature suggests limitations with regard to several critical issues involved in the integrated assessment of historical emissions.

First, the calculation of carbon emissions is often based on production instead of consumption (Blasing, Broniak, and Marland 2005; Pétron et al. 2008). In this study we refer to *consumption* as consumer-based energy and material use. Traditional carbon emissions inventories have been performed based on combustion (e.g., electricity generation) or end use (e.g., iron and steel production) of fossil fuels, which are both on the side of industrial production. The setting of future emission targets according to production-based inventories has its pitfalls, because the consumption and production of fossil fuel–related energy and manufacturing products are often spatially disparate (Rose et al. 2005; Weber and Matthews 2008). This spatial mismatch between emissions producers and end users might have impaired international agreements on carbon emissions allowances (Helm 2008).

Second, emissions estimates are usually performed at coarse geographic scales such as by city, state, or country boundaries, with no disclosure of the spatial variation within those units (Gregg et al. 2009; Sovacool and Brown 2010). The development of continuous data sets that describe intraunit geographic and temporal patterns of carbon emissions is still in an exploratory stage. This has been performed for the globe at $1° \times 1°$ spatial resolution by allocating a country's total CO_2 emissions based on distribution of that country's population (Andres et al. 1996). More recently, emissions from fossil fuel–based combustion, known as the Vulcan inventory, were generated for the United States at a 10-km spatial resolution by incorporating emissions data available at electrical generating units and from vehicle miles traveled (Gurney et al. 2009). No spatially explicit accounts of consumption-based emissions have yet been reported.

Third, the integrated assessment of balance between anthropogenic emissions and natural carbon sinks is still limited to a few urban areas (Wentz et al. 2002; Pataki et al. 2009) as opposed to an inventory that captures carbon balance differentiation along a gradient of settlement densities. For the accounting of vegetation carbon sinks, a recent trend is to produce spatially detailed estimates from syntheses of ground surveys and ecological models with the addition of remote sensing observations (Cramer et al. 1999). These inventories, however, exist predominantly for spatially continuous vegetation communities (Imhoff et al. 2004). The accounting of carbon balance in human-dominated environments (i.e., cities, suburbs, and exurbia) requires interdisciplinary approaches that combine traditional socioeconomic and ecological studies (Grimm et al. 2008; Churkina, Brown, and Keoleian 2009; Wise et al. 2009).

In this article, we present an integrated research approach to model the CO_2 balance between consumption-based emissions and vegetation carbon sinks. Our study targets current research needs in emissions inventories with respect to the three points previously discussed. Consumption in this analysis includes the household usage of energy and transportation fuels, which account for approximately 40 percent of the carbon emissions in the United States (Brown, Southworth, and Sarzynski 2008). This inventory excludes energy and fuel consumption and associated carbon emissions from the commercial sector. The research approach presented in this article captures the spatial heterogeneity of energy and fuel consumption, consumption-based CO_2 emissions, and vegetation carbon sinks at the U.S. Census tract scale. The spatially explicit accounting of household consumption and carbon emissions or sinks were further examined among different settlement types characterized by their residential densities. Our approach integrates carbon emissions estimates based on socioeconomic data with ecological modeling of vegetation carbon sequestration potentials based on biophysical remote sensing methods. All data used in this study are publicly available in the United States and the research methods can be extended to other geographies, including the possibility of nationwide analysis at the Census tract and comparable scale.

Integrated Spatially Explicit Carbon Balance Analysis

This article develops a carbon accounting approach that (1) investigates sources of carbon emissions at the consumer end, (2) captures the spatial heterogeneity of energy and fuel consumption and related carbon emissions, and (3) provides a profile of carbon balance between emissions from consumption activities and sequestration by natural or managed vegetation. To illustrate our methods, we demonstrate an application involving the state of Florida, which is one of the largest states (by size) in the United States. In 2007, an executive order was issued on carbon emissions reduction targets for this state (State of Florida 2007).

In the United States, consumption-based emissions inventories are challenging to conduct at the substate scale. Although the U.S. Energy Information Administration (EIA) has collected point-location data about residential energy consumption since 1978, these data are distributed in aggregated form at the scale of Census regions, Census divisions, and states (EIA 2009). It is necessary to consider consumption patterns associated with households and their characteristics to estimate energy-related residential consumption at finer scales. Our methods make use of average consumption levels that vary by household size, as households form the basic measuring unit of home energy use. When this is combined with Census counts of households by household size, we are able to produce spatial estimates of residential energy consumption at the Census tract scale.

The different orientations of the social and natural sciences might explain the limited number of spatially oriented carbon balance analyses that account for both anthropogenic emissions and vegetation sequestration (Pickett et al. 2001). In fact, there are relatively few such endeavors. Among the limited efforts focusing on modeling carbon balance across heterogeneous landscapes, some scholars simulate the spatial distribution of the net carbon flow based on actual measures of CO_2 concentration in the atmosphere without partitioning the respective contribution of carbon sources and sinks. For example, in one study the spatial and temporal patterns of the atmospheric CO_2 were derived based on statistical regression of measured CO_2 concentration and variables such as traffic flow, population density, employment density, and vegetation density (Wentz et al. 2002). Other efforts tend to model carbon emissions and sinks separately and then derive the balance between those two. A recent study conducted in Salt Lake City, Utah, demonstrated an integrated approach that inventoried CO_2 emissions based on consumption records from local utility companies and estimated CO_2 sequestration rates based on a model of urban forest growth (Pataki et al. 2009). In this article, we aim to inventory emissions and sequestration separately to identify spatial locations associated with carbon sources and sinks. The vegetation carbon sinks are estimated based on photosynthesis activities of all green vegetation (including forests, shrub, grass, agriculture, and wetlands) using a modeling technique of biophysical remote sensing. This approach provides a continuous surface of vegetation carbon sink estimates at the spatial resolution of one kilometer.

Study Area

Florida is located in the southeast of the United States (Figure 1). It ranks as the fourth most populous state in this country (U.S. Census Bureau 2009). The climate of Florida ranges from subtropical in the north and central parts of the state to tropical in the southern part of the Keys (Box, Crumpacker, and Hardin 1993). Although the per capita total energy consumption of Florida ranked low (forty-fourth in the country), Florida's per capita residential electricity demand is relatively high. This could be attributed to the high cooling demand during the hot summer

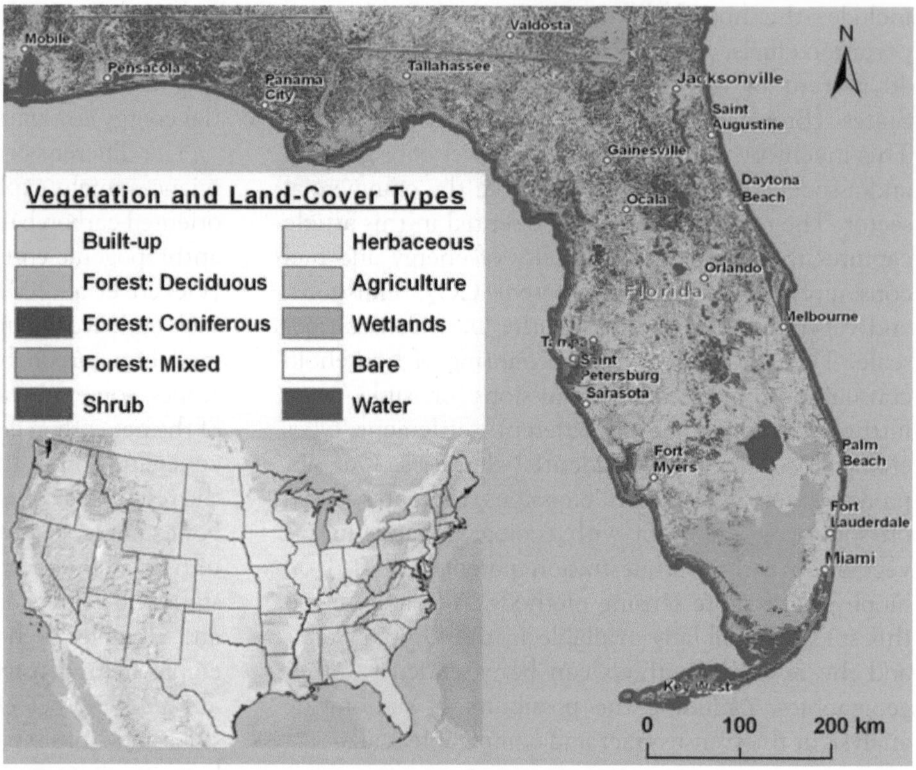

Figure 1. Florida (highlighted in yellow in the insert map) is located in the southeast of the United States. Vegetation and land-cover types were adapted from the 2001 National Land Cover Database (Homer et al. 2007). U.S. topographic map and place name reference came from ArcGIS Resource Centers.

months and the extensive use of electricity (87 percent of share in 2000) for home heating during the winter months (EIA 2010). Florida's per capita carbon emissions, calculated based on fossil fuels used for energy production and manufacturing products, were below the national average during the time period between 1960 and 2000 (Blasing, Broniak, and Marland 2005). This might be due partly to Florida's focus on tourism and agriculture instead of energy-intensive manufacturing.

Data and Methods

The research approach on carbon accounting for Florida consists of two parts. First, we investigated CO_2 emissions released from residential household energy and transportation fuel consumption at the U.S. Census tract scale. Second, we estimated vegetation carbon sinks measured by plant net photosynthesis at 1-km resolution using remote sensing and ecological approaches, the results of which were aggregated to Census tracts. The household energy and transportation fuel consumption, consumption-based carbon emissions, vegetation carbon sinks, and carbon balance were then connected to four settlement categories as measured by Census housing unit densities.

Estimating Household Energy and Fuel Consumption and Carbon Emissions

In this study, household carbon emissions were measured as CO_2 emitted through household energy use and transportation fuel. The total annual CO_2 emissions from household energy consumption in 2001 (E_e; kg) were estimated for each Census tract in Florida using the following equation:

$$E_e = \left(\sum_{i=1}^{6} HH_i \times Btu_i\right) \times 74.046 \times 0.697 \quad (1)$$

where $\sum_{i=1}^{6} HH_i \times Btu_i$ (million Btu) is the annual total household energy consumption for a given Census tract. HH_i is the Census 2000 total number of households by household size (e.g., one-person household vs. two-person household) within that Census tract, i ranges from one to six (with six representing households of six or more residents). Btu_i (million Btu) is the average per household energy consumption by household size in 2001 (EIA 2001a). In Florida, approximately 97 percent of the total household energy consumption came from electricity; consumption of 1 million Btu was equivalent to 74.046 kWh generated through electricity (EIA 2001b). The CO_2 emission coefficient was

0.697 kg/kWh for electricity generation in Florida (EIA 2002). Therefore, the total annual CO_2 emissions (kg) from household energy use were approximated based on energy consumption multiplied by the two scalars in Equation 1.

We used a standard method published by the U.S. EPA to estimate CO_2 emissions from household transportation (U.S. EPA 2005). According to this method, a typical passenger vehicle was modeled as emitting 8.877 kg of CO_2 per gallon of gas used. The total annual CO_2 emissions from household transportation in 2001 (E_t; kg) were obtained for each Census tract in Florida using Equation 2:

$$E_t = \frac{VMT}{MPG} \times 8.877 \times 365 \qquad (2)$$

where *VMT* is the daily vehicle miles of travel (VMT) per Census tract calculated based on the National Household Travel Survey (NHTS) transferability data set (Oak Ridge National Laboratory 2007). The NHTS collected daily commuting and noncommuting trips of all members in the sampled households. Those nationwide point samples were transferred to the aggregated travel behavior by Census tract through integrating settlement patterns and other socioeconomic data (Hu et al. 2007). Households were categorized into clusters based on settlement patterns, economic characteristics, and demographic characteristics of the Census tracts where they were located. The sample-based VMT values were assigned to all households within the same clusters and then summed up by Census tracts (Andrews 2008). *MPG* is the average miles per gallon (MPG) estimated based on the national average fuel economy for passenger vehicles (20.3 mpg) used in the EPA Mobile6.2 emissions modeling program (U.S. EPA 2005, 2009b).

Calculating Vegetation Carbon Assimilation Capacity

The potential vegetation carbon sinks were measured as gross primary production (GPP) of land multiplied by a scalar 0.5. GPP refers to the total amount of CO_2 entering an ecosystem through photosynthesis during a year (Chapin, Matson, and Mooney 2002). Approximately half of the total GPP is used for plant respiration (DeLucia et al. 2007). GPP minus plant respiration (i.e., net photosynthesis) indicates the maximum potential vegetation assimilation of CO_2 from the atmosphere.

Estimates of GPP were obtained using the light-use efficiency approach (Running et al. 2004; Zhao, Brown, and Bergen 2007). Our analysis includes deriving the actual light-use efficiency (ε) parameters and estimating the absorbed photosynthetically active radiation (APAR) for different types of vegetation and land cover (Zhao 2011). The 2001 National Land Cover Database (Homer et al. 2007) was used to generate ten vegetation and land-cover types throughout Florida (Figure 1). ε was derived for each vegetation and land-cover type based on documented empirical measures (Yang et al. 2007) multiplied by temperature and vapor pressure deficit (VPD) scalars (unpublished data). APAR was modeled based on the 2001 biweekly Advanced Very High Resolution Radiometer (AVHRR) normalized difference vegetation index (NDVI; U.S. Geological Survey EROS Data Center 2006) and solar radiation (Mitchell et al. 2004). The estimates of GPP, calculated at 1-km resolution throughout Florida, were aggregated to Census tracts for further analyses.

Linking Energy and Fuel Consumption and Carbon Balance to Settlement Patterns

The tract-level household energy and fuel consumption and carbon emissions and sinks were compared across four settlement categories including urban, suburban, exurban, and rural densities. Following the definition of Theobald (2001), urban densities are defined as settlement at 1 or more housing units per acre, suburban densities at 0.1 to 1 units per acre, exurban densities at 0.025 to 0.1 units per acre, and rural densities at less than 0.025 units per acre. Univariate analysis of variance (ANOVA) was constructed to detect effects of settlement category on household energy and fuel consumption, consumption-based carbon emissions, vegetation carbon sequestration, and carbon balance, respectively. The dependent variable was the tract-level per unit area or per capita measure of consumption or CO_2, and the factor variable was settlement category. Mean values for energy and fuel consumption and CO_2 estimates were derived and reported by settlement categories.

Results and Discussion

The State-Level Estimates

Florida, as a whole, appears to contain sufficient carbon sinks to offset all CO_2 emitted through household energy and fuel consumption in 2001. The estimated total annual household energy consumption was 0.58

quadrillion Btu for the entire state, which would produce approximately 30 million tons of CO_2 emissions. The total annual household vehicle travel was estimated to be approximately 74 billion miles for the entire state, which was responsible for approximately 32 million tons of CO_2 emissions. Therefore, the total annual CO_2 emitted through household energy and transportation fuel consumption in 2001 was approximately 62 million tons, which is less than the vegetation carbon assimilation capacity (88 million tons) as measured by plant net photosynthesis during the same year.

According to the last decennial Census, Florida was home to approximately 16 million residents and 6.3 million households in 2000. The per capita annual household energy and transportation fuel emissions were estimated to be 3.9 tons of CO_2 in 2001, whereas each Florida household accounted for 9.8 tons of CO_2 emissions annually on average. According to our estimates, the potential vegetation carbon sinks in Florida were 5.5 tons per person and approximately 14 tons per household.

Energy and Fuel Consumption and Carbon Balance by Census Tract

The spatial variation in household energy and transportation fuel consumption was high at the Census tract scale (Figure 2). Tracts with high energy consumption ranks might not be associated with high levels of transportation fuel usage considering that Pearson's correlation is 0.3 (significance = 0.01) between the tract-level total household energy consumption and VMT. The consumption-based CO_2 emissions normalized by Census tract area (i.e., density of emissions) followed roughly the distribution of population centers. The correlation between emissions from energy use and emissions from transportation fuel is relatively high (Pearson's correlation = 0.774, significance = 0.01) according to the tract-level density-based emissions measures.

The density-based carbon sinks estimate, or plant net photosynthesis per unit area, also varied significantly throughout Florida (Figure 3). At the Census tract scale, plant net photosynthesis correlated negatively (Pearson's correlation = –0.506, significance = 0.01) with the energy-based emissions measured by density. This indicates that tracts with higher levels of CO_2 emissions are less likely to be home to vegetation carbon sinks, which is due possibly to the reduction in greenspace associated with residential and commercial development.

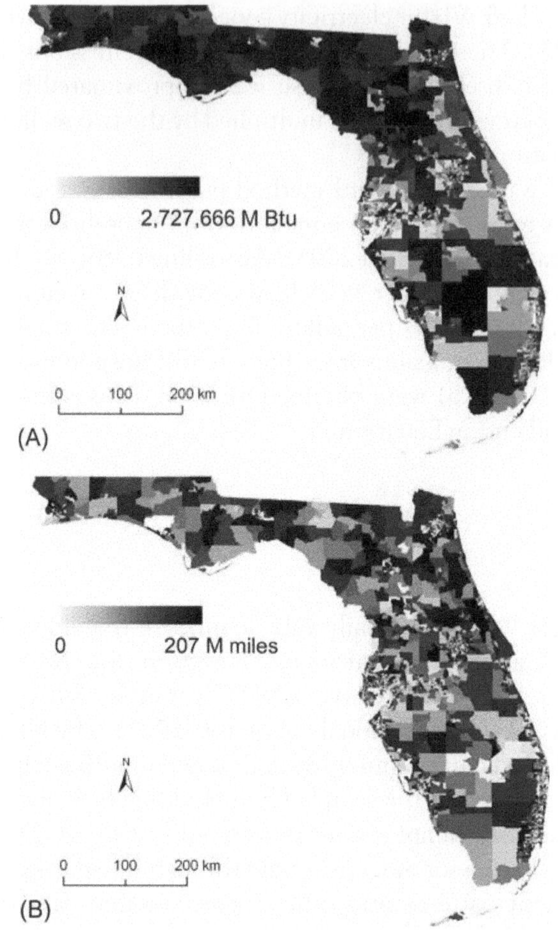

Figure 2. The Florida household energy consumption (A) and vehicle miles of travel (B) by Census tract in 2001.

Carbon balance, calculated as the total emissions from energy and transportation fuel consumption subtracting plant net photosynthesis, was generated statewide by Census tract. The spatial pattern of the estimated carbon balance varies significantly when measured by the total, per unit area, and per capita estimates (Figure 4). When measured by per unit area CO_2 balance (Figure 4B), densely populated cities such as Miami, Fort Lauderdale, Tampa, St. Petersburg, Orlando, and Jacksonville were found to be associated with net CO_2 sources (i.e., emissions exceeding vegetation carbon assimilation). Measured by per capita CO_2 balance (Figure 4C), the net carbon sources tend to extend away from the densest settlement areas.

Energy and Fuel Consumption and Carbon Balance by Settlement Densities

Four settlement categories were generated for Florida based on Census 2000 housing-unit densities (Figure 5).

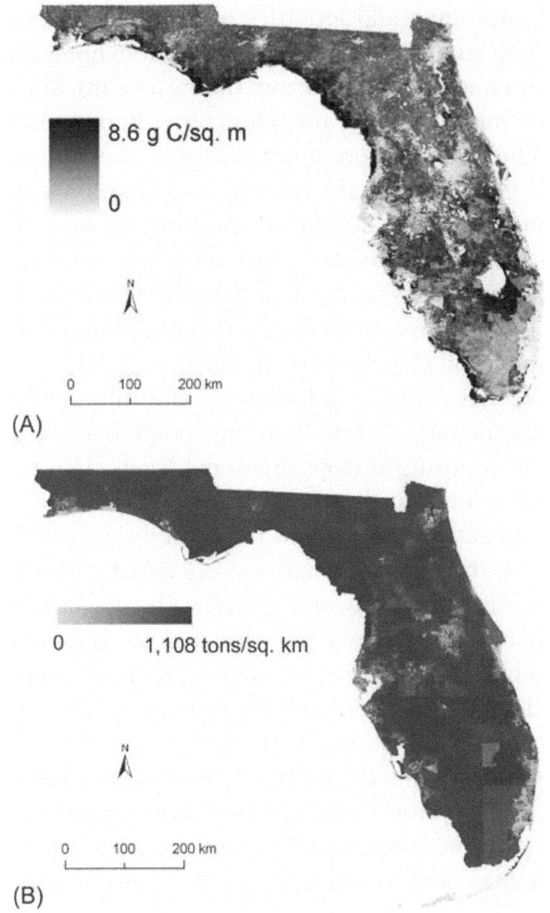

Figure 3. Vegetation carbon assimilation potential measured as gross primary production (GPP) at 1-km resolution (A) and as plant net photosynthesis aggregated by Census tract (B).

Approximately half of the state's total population and households resided in urban areas, which accounted for less than 4 percent of the total area of Florida. The urban and suburban areas together were home to nearly 90 percent of the total state population and households. Exurban densities accounted for approximately 30 percent of the state's total land and 8 percent of the state's total population and households. Rural densities occupied the largest proportion (45 percent) of the state's total land, with less than 4 percent of the total population and households distributed within those areas.

According to an ANOVA (Table 1), effect of settlement category on consumption and CO_2 levels is greater when measured by per unit area estimates than by per capita measures. The household energy consumption over per unit area was significantly higher at urban and suburban densities than the same measure at exurban and rural densities (Table 2), whereas the per capita energy consumption was less distinctive among the four settlement categories (Table 3). No variation in

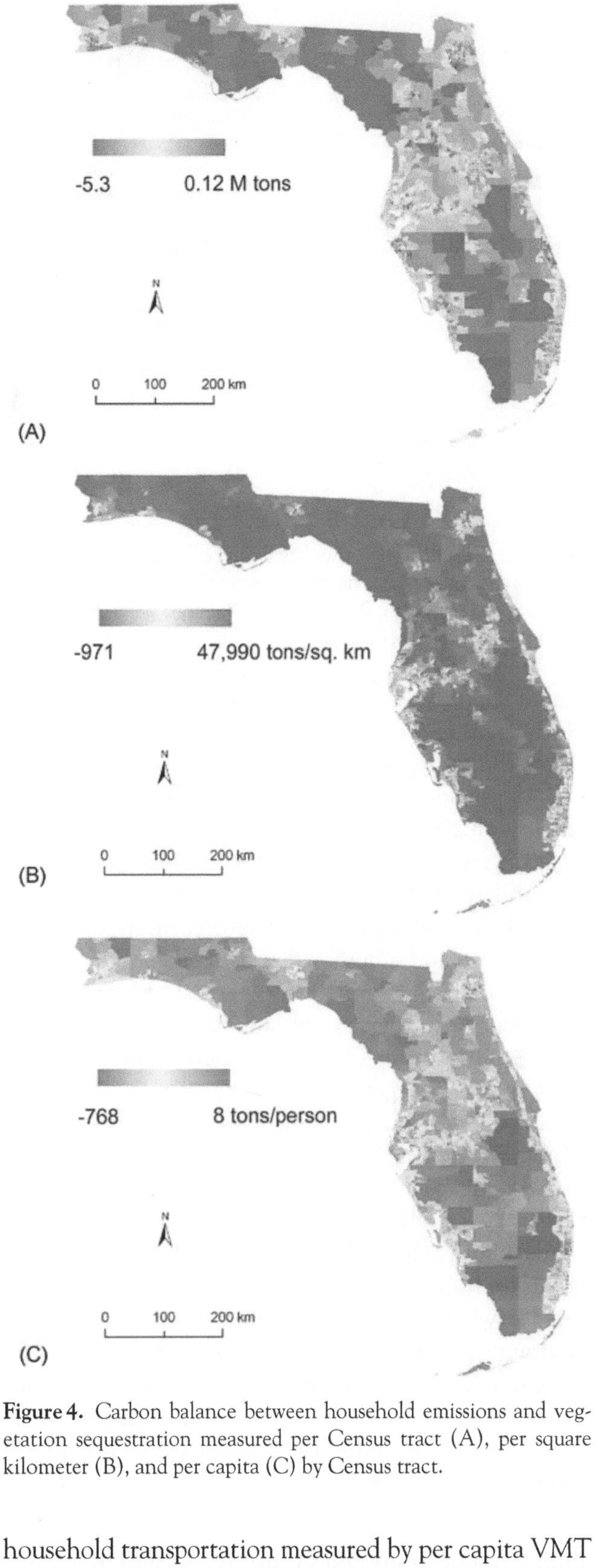

Figure 4. Carbon balance between household emissions and vegetation sequestration measured per Census tract (A), per square kilometer (B), and per capita (C) by Census tract.

household transportation measured by per capita VMT can be attributed to settlement category (Table 1). Exurban densities, however, were found to be associated with the highest per capita VMT at a rate significantly

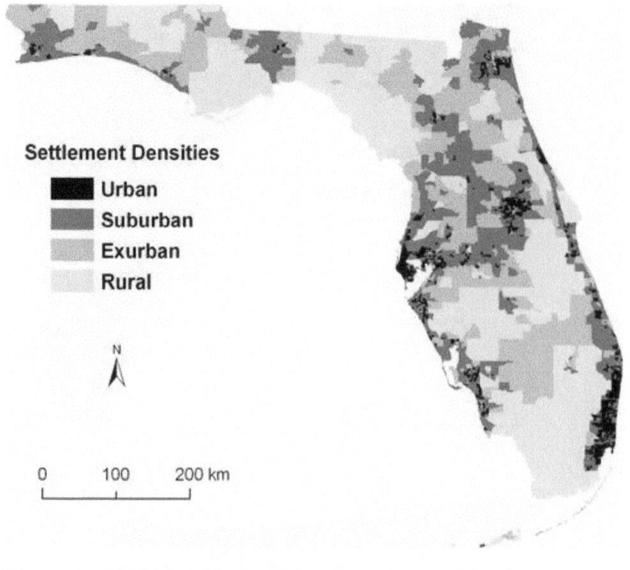

Figure 5. Settlement densities based on Census 2000 housing-unit density (Urban: ⩾ 1 housing unit per acre; Suburban: 0.1–1 housing units per acre; Exurban: 0.025–0.1 housing units per acre; Rural: < 0.025 housing units per acre).

higher than the same estimate at urban densities (Table 3). This result corroborates earlier research findings about the increasing transportation needs within the periurban zones (Ewing and Rong 2008; Brownstone and Golob 2009).

With regard to vegetation carbon sinks, both exurban and rural densities were associated with the increased amount of plant net photosynthesis, compared to the urban and suburban densities (Table 2 and Table 3). The carbon assimilation potential of exurban and rural densities appears to be similar when measured by carbon flux density (Table 2), whereas rural densities are superior for vegetation carbon assimilation than exurban densities when measured by per capita estimates (Table 3).

Settlement category accounted for the balance between household emissions and vegetation carbon assimilation (Table 1). Estimated by carbon flux density, urban and suburban densities appear to be net carbon sources, whereas exurban and rural densities are associated with slight carbon sinks (Table 2). Estimated by per capita measures, urban and suburban densities are shown to be net carbon sources at a much smaller magnitude compared to the size of net sinks at exurban and rural densities (Table 3).

Policy Implications and Future Research Directions

Consumption-based emissions analyses offer an approach to capture the carbon sources principally associated with household activities, a key driver of fossil fuel use. This study investigates emissions from home energy use and transportation sectors. Going forward, future research might consider including other household-level fossil fuel–related consumption items. These might include goods and products such as food and apparel as reported through consumer spending surveys. These analyses would provide further information to support emission trading programs, especially those involving household emission credit transfers based on fossil fuel consumption (Niemeier et al. 2008).

Another avenue could incorporate information on the commercial sectors for more comprehensive emissions accounting at (supra)national levels. This type of analysis would further contribute to helping set emissions targets and distribute responsibilities, which are top priorities for policymakers concerned with carbon emissions reductions and climate change mitigation (Hepburn and Stern 2008). Related to this, time series analysis of energy consumption and consumption-based emissions would provide insights for the efficacy of emissions reduction practices.

Per our results, the highly concentrated energy and fuel consumption levels recorded in urban and suburban areas, accompanied by lower amounts of vegetation carbon sinks, suggest the need for emissions reduction in high-density areas through cutbacks in energy use (possibly through renewable resources or new forms of mass transportation) and the utilization of carbon-efficient plant species. Exurban densities are characterized by highly efficient vegetation carbon assimilation, yet they are associated with the highest per capita household transportation demand. This suggests that lowering transportation emissions in low-density areas is a need, although the strategies for doing so are not clear. These environments tend to be unsupportive of travel reduction initiatives such as carpooling campaigns, because the complex spatio-temporal patterns of individual movement among disparate activity locations limit cooperative transport opportunities (Horner 2004). At the same time, mass transit options in these locations are also very limited.

Besides examining the spatial and temporal patterns of carbon emissions and sequestration, future research might wish to explain these patterns as a function of socioeconomic variables such as housing-unit characteristics (Newton, Tucker, and Ambrose 2000), household income levels (Druckman and Jackson 2008), or even the lifestyle characteristics of household members themselves (Lutzenhiser 1993). This contributes to a better understanding of social and economic drivers of

Table 1. Effects of settlement category (consisting of urban, suburban, exurban, and rural densities) on consumption and CO_2 estimates according to the univariate analysis of variance

Type of measures	Dependent variable	Adjusted R^2	*df*	*F*	Significance
Per unit area	Energy consumption	0.377	3	420.252	0.000
	Vehicle miles of travel	0.404	3	470.183	0.000
	CO_2 emissions	0.418	3	498.143	0.000
	Vegetation CO_2 assimilation	0.469	3	613.584	0.000
	CO_2 balance	0.430	3	523.918	0.000
Per capita	Energy consumption	0.023	3	17.436	0.000
	Vehicle miles of travel	0.001	3	1.645	0.177
	CO_2 emissions	0.002	3	2.385	0.067
	Vegetation CO_2 assimilation	0.352	3	376.186	0.000
	CO_2 balance	0.353	3	378.564	0.000

Table 2. The per unit area household energy and fuel consumption and carbon emissions and sinks by settlement densities in Florida, 2001

Settlement densities	Urban	Suburban	Exurban	Rural
Energy consumption (billion Btu/km^2)	64.26 (1.37)	9.68 (0.22)	1.12 (0.04)	0.27 (0.02)
Vehicle miles of travel (million miles/km^2)	9.84 (0.19)	1.62 (0.05)	0.20 (0.01)	0.05 (0.00)
CO_2 emissions (thousand tons/km^2)	7.62 (0.15)	1.21 (0.03)	0.15 (0.00)	0.03 (0.00)
Vegetation CO_2 assimilation (thousand tons/km^2)	0.14 (0.00)	0.31 (0.00)	0.51 (0.01)	0.51 (0.03)
CO_2 balance (thousand tons/km^2)	7.48 (0.15)	0.90 (0.03)	–0.36 (0.02)	–0.47 (0.03)

Note: Numbers in parentheses are standard error of mean.

Table 3. The per capita household energy and fuel consumption and carbon emissions and sinks by settlement densities in Florida, 2001

Settlement densities	Urban	Suburban	Exurban	Rural
Energy consumption (million Btu/person)	36.91 (0.14)	36.53 (0.17)	35.85 (0.33)	32.36 (1.05)
Vehicle miles of travel (miles/person)	5,862.85 (58.91)	5,865.64 (110.13)	6,314.17 (274.53)	5,608.84 (481.39)
CO_2 emissions (kg/person)	4,468.87 (28.05)	4,450.45 (49.66)	4,611.58 (122.45)	4,123.03 (234.11)
Vegetation CO_2 assimilation (kg/person)	1,192.00 (128.21)	2,057.51 (90.84)	19,408.97 (1,041.00)	91,689.51 (17,085.57)
CO_2 balance (kg/person)	4,340.66 (28.04)	2,392.94 (108.51)	–14,797.40 (1,037.89)	–87,566.48 (17,051.87)

Note: Numbers in parentheses are standard error of mean.

consumption and carbon emissions for effective emissions reductions.

Summary and Conclusions

In this study, we presented a new approach for spatially detailed carbon balance accounting and applied this approach to investigate the carbon balance between household emissions and vegetation assimilation by Census tract for the state of Florida. Household carbon emissions in the presented Florida study were composed of CO_2 released through residential consumption of energy and transportation fuels. We excluded carbon emissions from the commercial sector for this inventory. The tract-level residential energy and fuel consumption estimates were constructed based on Census household characteristics and the U.S. Department of Energy household consumption survey. Vegetation carbon sinks were measured as plant net photosynthesis using biophysical remote sensing techniques.

The analysis showed that Florida was able to self-assimilate its residential energy and transportation fuel-related carbon emissions (62 million tons) through the presence of vegetation (88 million tons) according to estimates in 2001. Household energy and fuel consumption, carbon emissions, and vegetation carbon sinks, however, vary significantly by Census tract across the state. The consumption-based carbon emission sources tend to be separated from vegetation carbon sinks. Consumption measures aggregated by settlement densities showed that the urban and suburban densities (settled at $\geqslant 0.1$ housing units per acre) were associated with the highest per capita energy consumption, whereas the per capita consumption of household transportation fuels tended to be the highest in exurban densities (settled at 0.025–0.1 housing units per acre). Both exurban and rural densities were associated with a significant amount of vegetation carbon sinks. The CO_2 balance indicates net carbon sources occurring in the urban and suburban densities, with net sinks in the exurban and rural densities.

This study contributes a spatial inventory of energy consumption and carbon emissions and sinks (Zimmerer 2011). The spatial inventory approach is used to provide baseline carbon fluxes estimates for establishing emission targets, distributing emission responsibilities, and establishing carbon trading programs associated with specific geographic regions or locations. Our methods are broadly applicable and can be integrated with other socioeconomic metrics for sustainability assessments. For example, future efforts might focus on the energy and carbon consequences of various urban development patterns, in particular, the location of individualized communities as well as the constituent densities of residential and commercial development (Horner 2008). This will lead to new insights with respect to sustainable low-carbon energy consumption and transportation behavior.

Acknowledgment

Part of this research was sponsored by grants received from the Institute for Energy Systems, Economics, and Sustainability (IESES) at Florida State University.

References

Andres, R. J., G. Marland, I. Fung, and E. Matthews. 1996. A 1 degrees × 1 degrees distribution of carbon dioxide emissions from fossil fuel consumption and cement manufacture, 1950–1990. *Global Biogeochemical Cycles* 10 (3): 419–29.

Andrews, C. J. 2008. Greenhouse gas emissions along the rural–urban gradient. *Journal of Environmental Planning and Management* 51 (6): 847–70.

Blasing, T. J., C. Broniak, and G. Marland. 2005. State-by-state carbon dioxide emissions from fossil fuel use in the United States 1960–2000. *Mitigation and Adaptation Strategies for Global Change* 10:659–74.

Box, E. O., D. W. Crumpacker, and E. D. Hardin. 1993. A climatic model for location of plant species in Florida, U.S.A. *Journal of Biogeography* 20 (6): 629–44.

Brown, M., F. Southworth, and A. Sarzynski. 2008. *Shrinking the carbon footprint of metropolitan America*. Washington, DC: The Brookings Institution.

Brownstone, D., and T. F. Golob. 2009. The impact of residential density on vehicle usage and energy consumption. *Journal of Urban Economics* 65:91–98.

Chapin, F. S., P. Matson, and H. A. Mooney. 2002. *Principles of terrestrial ecosystem ecology*. New York: Springer.

Churkina, G., D. G. Brown, and G. Keoleian. 2009. Carbon stored in human settlements: The conterminous United States. *Global Change Biology* 16 (1): 135–43.

Cramer, W., D. W. Kicklighter, A. Bondeau, B. Moore, G. Churkina, B. Nemry, A. Ruimy, and A. L. Schloss. 1999. Comparing global models of terrestrial net primary productivity (NPP): Overview and key results. *Global Change Biology* 5 (Suppl. 1): 1–15.

DeLucia, E. H., J. E. Drake, R. B. Thomas, and M. Gonzalez-Meler. 2007. Forest carbon use efficiency: Is respiration a constant fraction of gross primary production? *Global Change Biology* 13 (6): 1157–67.

Druckman, A., and T. Jackson. 2008. Household energy consumption in the UK: A highly geographically and socio-economically disaggregated model. *Energy Policy* 36:3177–92.

Energy Information Administration (EIA). 2001a. Total energy consumption and expenditures by household member and demographics, 2001. http://www.eia.doe.gov/emeu/recs/recs2001/detailcetbls.html (last accessed 29 July 2010).

———. 2001b. Total energy consumption in U.S. households by four most populated states, 2001. http://www.eia.doe.gov/emeu/recs/recs2001/detailcetbls.html (last accessed 29 July 2010).

———. 2002. Updated state-level greenhouse gas emission coefficients for electricity generation 1998–2000. http://www.eia.doe.gov/pub/oiaf/1605/cdrom/pdf/e-supdoc.pdf (last accessed 29 July 2010).

———. 2009. Residential energy consumption survey (RECS) 2009 update. http://www.eia.doe.gov/emeu/recs/contents.html (last accessed 29 July 2010).

———. 2010. State energy profiles: Florida. http://tonto.eia.doe.gov/state/state_energy_profiles.cfm?sid=FL (last accessed 29 July 2010).

Ewing, R., and F. Rong. 2008. The impact of urban form on U.S. residential energy use. *Housing Policy Debate* 19 (1): 1–30.

Gregg, J. S., L. M. Losey, R. J. Andres, T. J. Blasing, and G. Marland. 2009. The temporal and spatial distribution of carbon dioxide emissions from fossil fuel use in North America. *Journal of Applied Meteorology and Climatology* 48:2528–42.

Grimm, N. B., D. Foster, P. Groffman, J. M. Grove, C. S. Hopkinson, K. J. Nadelhoffer, D. E. Pataki, and D. P. C. Peters. 2008. The changing landscape: Ecosystem response to urbanization and pollution across climatic and societal gradients. *Frontiers in Ecology and the Environment* 6 (5): 264–72.

Gurney, K. R., D. L. Mendoza, Y. Zhou, M. L. Fischer, C. C. Miller, S. Greethakumar, and S. D. Du Can. 2009. High resolution fossil fuel combustion CO_2 emission fluxes for the United States. *Environmental Science & Technology* 43 (14): 5535–41.

Helm, D. 2008. Climate-change policy: Why has so little been achieved? *Oxford Review of Economic Policy* 24 (2): 211–38.

Hepburn, C., and N. Stern. 2008. A new global deal on climate change. *Oxford Review of Economic Policy* 24 (2): 259–79.

Homer, C., J. Dewitz, J. Fry, M. Coan, N. Hossain, C. Larson, N. Herold, A. McKerrow, J. N. VanDriel, and J. Wickham. 2007. Completion of the 2001 National Land Cover Database for the conterminous United States. *Photogrammetric Engineering and Remote Sensing* 73 (4): 337–41.

Horner, M. W. 2004. Spatial dimensions of urban commuting: A review of major issues and their implications for future geographic research. *The Professional Geographer* 56 (2): 160–74.

———. 2008. “Optimal” accessibility landscapes? Development of a new methodology for simulating and assessing jobs–housing relationships in urban regions. *Urban Studies* 45 (8): 1583–602.

Hu, P. S., T. Reuscher, R. L. Schmoyer, and S.-M. Chin. 2007. *Transferring 2001 National Household Travel Survey*. Oak Ridge, TN: Oak Ridge National Laboratory.

Imhoff, M. L., L. Bounoua, R. DeFries, W. T. Lawrence, D. Stutzer, C. J. Tucker, and T. Ricketts. 2004. The consequences of urban land transformation on net primary productivity in the United States. *Remote Sensing of Environment* 89 (4): 434–43.

Intergovernmental Panel on Climate Change (IPCC). 2007. *Climate 2007: Synthesis report*. Geneva, Switzerland: IPCC.

Knight, C. G., S. Cutter, J. DeHart, A. Denny, D. Howard, S.-L. Kaktins, D. E. Kromm, S. E. White, and B. Yarnal. 2003. Reducing greenhouse gas emissions: Learning from local analogs. In *Global change in local places: Estimating, understanding, and reducing greenhouse gases*, ed. Association of American Geographers Global Change in Local Places Research Group, 192–213. Cambridge, UK: Cambridge University Press.

Lutzenhiser, L. 1993. Social and behavioral aspects of energy use. *Annual Review of Energy and the Environment* 18:247–89.

Mitchell, K. E., D. Lohmann, P. R. Houser, E. F. Wood, J. C. Schaake, A. Robock, B. A. Cosgrove et al. 2004. The multi-institution North American Land Data Assimilation System (NLDAS): Utilizing multiple GCIP products and partners in a continental distributed hydrological modeling system. *Journal of Geophysical Research—Atmospheres* 109 (D7), D07S90. doi: 10.1029/2003JD003823.

Newton, P., S. Tucker, and M. Ambrose. 2000. Housing form, energy use and greenhouse gas emissions. In *Achieving sustainable urban form*, ed. K. Williams, E. Burton, and M. Jenks, 74–83. London: E & FN Spon.

Niemeier, D., G. Gould, A. Karner, M. Hixon, B. Bachmann, C. Okma, Z. Lang, and D. H. Del Valle. 2008. Rethinking downstream regulation: California’s opportunity to engage households in reducing greenhouse gases. *Energy Policy* 36:3436–47.

Oak Ridge National Laboratory. 2007. 2001 NHTS transferability national files. http://nhts-gis.ornl.gov/transferability/ (last accessed 29 July 2010).

Pataki, D. E., P. C. Emmi, C. B. Forster, J. I. Mills, E. R. Pardyjak, T. R. Peterson, J. D. Thompson, and E. Dudley-Murphy. 2009. An integrated approach to improving fossil fuel emissions scenarios with urban ecosystem studies. *Ecological Complexity* 6:1–14.

Pétron, G., P. Tans, G. Frost, D. Chao, and M. Trainer. 2008. High-resolution emissions of CO_2 from power generation in the USA. *Journal of Geophysical Research-Biogeosciences* 113 (G4): G04008. doi: 10.1029/2007JG000602.

Pickett, S. T. A., M. L. Cadenasso, J. M. Grove, C. H. Nilon, R. V. Pouyat, W. C. Zipperer, and R. Costanza. 2001. Urban ecological systems: Linking terrestrial ecological, physical, and socioeconomic components of metropolitan areas. *Annual Review of Ecology and Systematics* 32:127–57.

Rose, A., R. Neff, B. Yarnal, and H. Greenberg. 2005. A greenhouse gas emissions inventory for Pennsylvania. *Journal of the Air and Waste Management Association* 55:1122–33.

Running, S. W., R. R. Nemani, F. A. Heinsch, M. Zhao, M. Reeves, and H. Hashimoto. 2004. A continuous satellite-derived measure of global terrestrial primary production, *BioScience* 54:547–60.

Sovacool, B. K., and M. A. Brown. 2010. Twelve metropolitan carbon footprints: A preliminary comparative global assessment. *Energy Policy* 38 (9): 4856–69.

State of Florida. 2007. Executive Order 07-127. Office of the Governor.

Theobald, D. M. 2001. Land-use dynamics beyond the American urban fringe. *Geographical Review* 91:544–64.

United Nations. 1998. *Kyoto Protocol to the United Nations Framework Convention on Climate Change*. New York: United Nations.

U.S. Census Bureau. 2009. Annual estimates of the resident population for the United States, regions, states, and Puerto Rico: April 1, 2000 to July 1, 2009 (NST-EST2009–01). http://www.census.gov/popest/states/NST-ann-est.html (last accessed 29 July 2010).

U.S. Environmental Protection Agency. 2005. Emission facts: Greenhouse gas emissions from a typical passenger vehicle. EPA420-F-05-004. http://www.epa.gov/OMS/climate/420f05004.htm (last accessed 29 July 2010).

———. 2009a. Inventory of U.S. greenhouse gas emissions and sinks: 1990–2007. http://www.epa.gov/climatechange/emissions/usinventoryreport.html (last accessed 29 July 2010).

———. 2009b. MOBILE6 vehicle emission modeling software. http://www.epa.gov/OMS/m6.htm (last accessed 29 July 2010).

U.S. Geological Survey EROS Data Center. 2006. The conterminous U.S. and Alaska weekly and biweekly AVHRR composites. Sioux Falls, SD: EROS Data Center.

Weber, C. L., and H. S. Matthews. 2008. Quantifying the global and distributional aspects of American household carbon footprint. *Ecological Economics* 66:379–91.

Wentz, E. A., P. Gober, R. C. Balling, and T. A. Day. 2002. Spatial patterns and determinants of winter atmospheric carbon dioxide concentrations in an urban environment. *Annals of the Association of American Geographers* 92 (1): 15–28.

Wise, M., K. Calvin, A. Thomson, L. Clarke, B. Bond-Lamberty, R. Sands, S. J. Smith, A. Janetos, and J. Edmonds. 2009. Implications of limiting CO_2 concentrations for land use and energy. *Science* 324:1183–86.

Yang, F., K. Ichii, M. A. White, H. Hashimoto, A. R. Michaelis, P. Votava, A. Zhu, A. Huete, S. W. Running, and R. R. Nemani. 2007. Developing a continental-scale measure of gross primary production by combing MODIS and AmeriFlux data through Support Vector Machine approach. *Remote Sensing of Environment* 110:109–22.

Zhao, T. 2011. Impacts of urban growth on vegetation carbon sequestration. In *Urban remote sensing: Monitoring, synthesis and modeling in the urban environment*, ed. X. Yang, 275–86. Hoboken, NJ: Wiley.

Zhao, T., D. G. Brown, and K. M. Bergen. 2007. Increasing gross primary production (GPP) in the urbanizing landscapes of southeastern Michigan. *Photogrammetric Engineering & Remote Sensing* 73 (10): 1159–67.

Zimmerer, K. S. 2011. New geographies of energy: Introduction to the special issue. *Annals of the Association of American Geographers* 101 (4): 705–11.

Toward an Integrated GIScience and Energy Research Agenda

Mark W. Horner,* Tingting Zhao,* and Timothy S. Chapin†

*Department of Geography and the Institute for Energy Systems, Economics and Sustainability, Florida State University
†Department of Urban and Regional Planning and the Institute for Energy Systems, Economics and Sustainability, Florida State University

The growing evidence indicating that climate change is real and accelerating, coupled with a host of interrelated energy sustainability questions, has fostered increased interdisciplinary research on improving energy efficiency and reducing per capita energy consumption, as well as better understanding the sources of pollution emissions and possible policy options for limiting permanent environmental damage. Increasingly, geographic information systems, remote sensing, and other spatial technologies are being leveraged by researchers when analyzing these problems. There is, however, limited discourse regarding the possible synergies that could result from sustained engagement between those interested in geographic information science (GIScience) and researchers tackling energy issues. In this article, we outline an integrated research agenda for GIScience and energy studies that focuses on the prospects for making new contributions to the growing literature on energy sustainability. We identify three critical issues that offer substantial opportunities for new synergistic research at the nexus of GIScience and energy sustainability, including (1) the problem of carbon estimation and inventory, (2) questions of new energy infrastructure placement and transition, and (3) household energy conservation and efficiency. We lay out substantive energy considerations within each problem area and discuss possible new contributions involving GIScience. Our analysis suggests that issues of scale, representation, complexity, and several other core GIScience themes underpin these energy research needs. This article is intended to foster new dialogue between GIScience and energy studies. *Key Words: carbon, energy, GIScience, household consumption, sustainability.*

越来越多的证据表明，气候变化是真实并加快的，它与相互关联的能源可持续发展问题的主机相结合，增加了促进提高能源效率和减少人均能源消费量的跨学科研究，也使我们更好地理解污染排放源和限制永久环境破坏的可能的政策选择。越来越多的地理信息系统，遥感和其他空间技术在分析这些问题时被研究人员运用。然而，针对那些来自对地理信息科学（GIScience）感兴趣者和应付能源问题的研究者之间的持续参与的可能的协同作用的话语是有限的。在这篇文章中，我们列出了地理信息科学和能源研究的一个综合性研究议程，其焦点在于对在能源的可持续性领域增长的文献作出新的贡献的远景。我们确定了给在地理信息科学和能源可持续性关系中新的协同研究提供大量机会的三个关键问题，包括（1）碳估算和库存的问题，（2）新能源基础设施的安置和过渡问题以及（3）家庭能源的节约和效率。我们在每个问题领域计划实质性的能量考虑和讨论涉及地理信息科学的可能的新贡献。我们的分析表明，规模，代表性，复杂性，和其他几个核心地理信息科学专题问题支撑这些能源研究的需要。本文的目的是促进地理信息科学和能源研究之间新的对话。*关键词：碳，能源，地理信息科学，家庭消费可持续性。*

La creciente evidencia de que el cambio climático es real y que se está acelerando, junto a un buen número de cuestiones interrelacionadas de sustentabilidad energética, ha fomentado un mayor volumen de investigación interdisciplinaria para mejorar la eficiencia en el uso de la energía y reducir el consumo energético per cápita, al igual que para entender mejor las fuentes de emisiones de polución y posibles opciones de políticas para limitar el daño ambiental permanente. De manera creciente, los sistemas de información geográfica, la percepción remota y otras tecnologías espaciales están siendo aprestigiadas por los investigadores al analizar estos problemas. Sin embargo, poco se ha discutido en lo que concierne a las posibles sinergias que podrían resultar de un continuado compromiso entre quienes se interesan en información geográfica científica (IGciencia) y los investigadores empeñados en asuntos energéticos. En este artículo, bosquejamos una agenda de investigación integrada para IGciencia y estudios energéticos, que centra su atención en la posibilidad de hacer nuevas contribuciones a la creciente literatura sobre sustentabilidad de la energía. Identificamos tres asuntos críticos que ofrecen sustanciales oportunidades de nueva investigación sinergística en el nexo de la IGciencia y la sustentabilidad de la energía, a saber: (1) el problema del cálculo del carbono y su inventario, (2) cuestiones relacionadas con los nuevos emplazamientos de infraestructura energética y transición, y (3) conservación y eficiencia de la energía para uso familiar. Presentamos sustanciales consideraciones sobre energía en cada uno de esos problemas y discutimos posibles nuevas contribuciones que impliquen IGciencia. Nuestro análisis sugiere que asuntos relacionados con escala, representación, complejidad y varios otros temas centrales de IGciencia apoyan estas necesidades de investigación energética. *Palabras clave: carbono, energía, IGciencia, consumo doméstico, sustentabilidad.*

Because much of the world's energy supplies are derived from limited fossil fuels, there are pressing questions about the long-term environmental, social, and economic sustainability of the current global energy regime (InterAcademy Council 2007). There is a growing sense in the scientific community (Intergovernmental Panel on Climate Change [IPCC] 2007), which also has been documented in popular outlets (Friedman 2008), that humanity is on the cusp of damaging and potentially irreversible environmental changes. The adverse impacts of these changes have been linked to energy consumption and the negative externalities of the primary energy sources used (Woodcock et al. 2007). Among these, the continued release of carbon and other greenhouse gases (GHGs) is thought to be at odds with the sustainability of Earth's physical and human systems (Willson and Brown 2008).

The convergence of scientific evidence strongly points to the need to explore and pursue new ways of meeting our current and future energy needs, as well as to mitigate the impacts of our energy consumption on the environment (IPCC 2007; National Academy of Sciences 2010). In the coming years, geographic information systems (GIS) are poised to play an essential role in the design and implementation of new energy infrastructure systems (Prest, Daniell, and Stendorf 2007), as well as to help us understand the impacts of new energy policies aimed at both the macro (Stone 2008) and microscales (Cidell 2009). Many basic research issues must be confronted if we are to successfully adopt a more sustainable energy system and minimize the associated long-term environmental impacts.

A review of the recent scientific literature involving applications of GIS to energy issues, however, suggests limited participation by geographers and spatial scientists. Moreover, scholars interested in the basic science supporting GIS, GIScience, have yet to develop substantial linkages with researchers interested in energy issues. This lack of interaction represents a missed opportunity for the energy community to benefit from some of the foundational insights that could come from GIScience. From issues of spatial scale and representation (Miller and Wentz 2003; Goodchild, Yuan, and Cova 2007), to computational limits and advances with spatial data (Armstrong 2000; Xie, Batty, and Zhaoz 2007; Wang and Liu 2009), there are multiple GIScience knowledge areas underpinning potentially innovative energy research. At the same time, the diverse realm of energy studies offers GIScientists new venues in which to contemplate the boundaries of spatio-temporal analysis (Peuquet 2004; Yu and Shaw 2008), explore theories of complex systems (Evans, Sun, and Kelley 2006; Manson 2006), and advance other core research aims. More broadly, the benefits accrued from new exchanges between these areas have real potential to contribute to fostering long-term energy sustainability and aid in international efforts to mitigate climate change.

This article lays out possibilities for new interactions between GIScience and energy researchers within the context of selected ongoing research streams focused on energy sustainability issues. We first provide a brief background on GIScience and energy studies, highlighting their respective interdisciplinary foci. From there we identify three broad energy problems to which new research could contribute and provide examples of how these integrated contributions might proceed. We then close with summary statements and suggestions for further work. In short, our goal is to reflect on the opportunities for new GIScience–energy synergies and to outline a new research agenda that would support energy sustainability.

GIScience and Energy Studies

Although the term *GIScience* was coined early in the 1990s (Goodchild 2010), the field to which this term refers began decades prior, having core interests in issues such as spatial data handling, modeling, geographical representation, and so on (Mark 2003; Caron et al. 2008). It remains a vibrant interdisciplinary field drawing from computer science and engineering as well as the cognitive, physical, policy, and social sciences. Presently, there are well-attended conferences featuring GIScience research (e.g., GIScience, GeoComputation) as well as several journals devoted to GIScience content (e.g., *International Journal of Geographic Information Science*, *GeoInformatica*; Caron et al. 2008). For a fuller treatment of the history and research scope of GIScience, readers are referred to Goodchild (2010).

There have been successful efforts to integrate GIScience with other substantive domains to examine research synergies and explore specific societal issues. Some of these include emergency management (Cutter 2003), human rights (Madden and Ross 2009), and environmental justice assessment (Higgs and Langford 2009). With respect to emergency management and GIScience, synthesis has occurred with respect to topics such as human evacuation behavior (Chen, Meaker, and Zhan 2006), data dissemination during extreme events (Mills et al. 2008), and crisis management systems (MacEachren and Cai 2006). Extending this line of thought to GIScience and energy matters could yield productive new interactions.

The study of energy issues similarly draws from a diverse suite of researchers. Naturally, there are those scientists focused on understanding the physical and chemical properties of energy sources or how energy is used by biotic organisms and machines. Researchers have addressed topics ranging from where new forms of energy might be produced (Agugliaro 2007), to the viability of these alternative energy sources (Akinci et al. 2008). The implications of energy usage are increasingly important, especially in the area of GHG emissions (Andrews 2008) and policies for their management (Zegras 2007; Ewing and Rong 2008; Mack and Endemann 2010). Within this nexus, there are ample cases of GIS technology being used to address energy issues (Simao, Densham, and Haklay 2009; Velazquez-Marti and Annevelink 2009), largely because many energy problems are effectively informed by geospatial concepts. Geospatial analysis techniques and supporting theories can be brought to bear on their inherent spatial (and related temporal) dimensions.

Synthesizing Energy Issues, GIS, and GIScience

In making the case for increasing synergies with energy and GIScience, we focus our discussion on three areas within the energy sphere that seem to offer the greatest opportunities for integrated geospatial research. These areas are not only fertile ground for making new intellectual contributions to GIScience and energy studies but they also figure prominently in societal efforts to become more energy sustainable and to minimize the impacts of climate change on natural environments. These areas are (1) carbon estimation and inventory, (2) energy infrastructure placement and transition, and (3) household energy conservation and efficiency. There are two points that preface this discussion. First, we again acknowledge that GIS technology and applications are not new to any of these research areas (Saunders and da Silva 2009; Velazquez-Marti and Annevelink 2009). What is generally missing from this literature, however, is a more systematic engagement between energy and GIScience principles, including widespread participation from geographers and GIScientists. Second, we note that our focus is on research possibilities in GIScience and broader energy matters, which include but are not limited to issues related to climate change (Skole 2004). Although our motivation stems from the interactive linkages between energy and climate change (Hepburn and Stern 2008; McDonald et al. 2009), our emphasis is on examining the generation and use of energy and its impacts on climate change through GHG emissions.

Carbon Estimation and Inventory

To stabilize GHG concentrations in the atmosphere due to petroleum-based energy consumption, international collaboration on emissions reductions will be required. Some current policy ideas under consideration involve complex and difficult questions, including setting future global emissions targets, distributing emissions target responsibilities among nations or to metropolitan areas within nations, and establishing emissions trading systems (Hepburn and Stern 2008). In the United States, the federal government has examined the feasibility of cap-and-trade programs to reduce GHGs (U.S. House of Representatives 2009). There are opportunities for GIScientists and energy researchers to work together to inform these emissions management programs through innovations in spatially explicit carbon source inventory methodologies.

Carbon emissions inventories require quantitative assessments of the sources and sinks of anthropogenic GHGs during a certain time period. Carbon sources are associated with anthropogenic activities such as burning fossil fuels and deforestation, whereas carbon sinks can be either natural vegetation or manmade such as croplands (IPCC 2007). Inventories of carbon sources and sinks provide the baseline information necessary to establish future emissions targets for a region. Emissions inventories also provide support for market-based emissions reduction programs. For example, some of the economic models used to determine carbon credits require information on target GHG concentration levels that are calculated based on historical emissions (Metcalf and Weisbach 2009). This raises important questions regarding the functionality and availability of spatio-temporal databases for such applications, a long-standing concern in GIScience (Peuquet 2004; Stell and Worboys 2008). Previous work has pointed out the need for new spatio-temporal database capabilities in the context of managing climate change data (Abraham and Roddick 1999). New exchanges between energy researchers and GIScientists with respect to spatial data handling and database infrastructure issues, within the context of temporal dynamics, could lead to mutually beneficial developments.

Spatial and temporal scales are both considerations in carbon inventories. Depending on the necessary level of spatial detail, carbon estimates might be produced as finely as the household (Niemeier et al. 2008). Temporally, there is the possibility of mapping daily, monthly, or annual household carbon emissions associated with energy use. With recent advances in real-time data acquisition capacities based on environmental sensor networks, such as smart grid systems (Worboys and Duckham 2006), we are likely to see even greater temporal flexibility with respect to the availability of spatial data for carbon emissions inventories. Going forward, we will need new dialogue to address how complex spatial-temporal queries of such data are best constructed (Lopez, Snodgrass, and Bongki 2005).

Spatially disaggregated estimates can be aggregated to coarser geographic scales for the inventory of emissions over a residential neighborhood, Census geography, or municipality, which raises questions about appropriate aggregation strategies (Mu and Wang 2008). Further, highly spatially disaggregate estimation will require a large amount of data and processing effort for large study areas, which presents challenges when working at certain scales. As such, GIScience has great potential to contribute to large-scale emissions inventories through the development of new spatial interpolation and extrapolation approaches that can reduce sampling burdens and increase the spatial extensibility of results (Komarov et al. 2007; Andrews 2008). Commonly collected public spatial data sets are also likely to be of use in these large-scale efforts. For example, it is well known that the U.S. Census Bureau provides detailed data on households and housing unit characteristics at regular intervals. Also, the U.S. Department of Energy has conducted household energy and fuel consumption surveys every five years since 1993. Although these data represent a starting point for inventories, new techniques will be needed to ensure maximally accurate estimation procedures. Here, insights from the branches of GIScience related to spatial analysis and geostatistics (e.g., the use of spatial regression techniques) could be helpful, particularly given the propensity for spatial autocorrelation in the requisite socioeconomic data (Fotheringham 1997).

GIScience concepts also weigh on inventories of vegetation carbon sinks over large geographic extents with respect to scaling and remote sensing techniques. Ecosystem productivities, or measures of the carbon dioxide absorbed by plants through photosynthesis, have been mapped from stands to global levels based on ground measurements, ecological models, or both. For example, sampled crop harvest yields (kg carbon per acre) can be associated with cultivation areas to derive spatial rates of agricultural carbon assimilation (Prince et al. 2001). Advances in satellite sensors with large image swaths and rapid repeat cycles, such as the Terra Moderate Resolution Imaging Spectroradiometer (MODIS), allow the spatial heterogeneity of carbon fluxes to be captured at 1-km resolution for regional and global carbon accounting (Running et al. 2004; Zhao, Brown, and Bergen 2007).

The integrated spatial assessment of carbon balance that connects carbon sinks estimated using biophysical remote sensing to emissions accounting based on socioeconomic data is still at an early stage of development. GIScience can help support these activities, as there are opportunities for comprehensive analyses leading to detailed profiles of human-related carbon activities (Wise et al. 2009), including identifying source and sink hotspots and examining relationships between landscape dynamics and carbon sustainability. In this context, one of the most prominent trends is the rapid growth of urban populations and settlement areas (Theobald 2001). Although the impacts of these changes on carbon emissions and sinks are still inconclusive, it is widely held that the ongoing growth in urban populations will have profound impacts on global energy consumption and emissions. The vast urban modeling expertise within the GIScience literature is relevant (Batty 2005; Zhang and Guindon 2006; Horner 2010), as ongoing work seeking to measure urban morphology (Herold, Scepan, and Clarke 2003; Ji et al. 2006), for example, can inform the development of new carbon balance estimation methods. In sum, several GIScience knowledge areas underpin key aspects of the emerging area of spatially coupled carbon balance estimation and landscape evaluation.

Energy Infrastructure Placement and Transition

As the world moves toward adopting alternative means of producing energy, new infrastructure systems will likely need to be integrated into existing landscapes (Zimmerer 2011). The placement of production, transmission, and distribution facilities is subject to a variety of complex considerations ranging from minimizing possible impacts on the environment and biological habitats, particularly when converting undeveloped land (McDonald et al. 2009), to addressing population equity concerns (Marsh and Schilling 1994). There are examples in the literature where GIS and spatial modeling approaches are being applied to problems involving energy infrastructure, including the placement of wind farms for electricity production (Rodman and Meentemeyer 2006; Simao, Densham, and Haklay 2009) and siting alternative fuel refueling stations for vehicles (Nicholas, Handy, and Sperling 2004; Kuby and Lim 2007). In this context, spatial modeling implies using quantitative methodologies including spatial optimization techniques to address various location-oriented questions in the face of scarce resources or conflicting interests (ReVelle and Eiselt 2005).

Spatial modeling has an important role in research aimed at designing the energy infrastructures of tomorrow. Beyond the nominal questions of where infrastructures should be placed, subject to resource and financial constraints (DeVerteuil 2000), there are inherent social and environmental questions underlying such decisions (Talen 2001; Horner and Downs 2008). These considerations will be paramount in future infrastructure location projects as planners simultaneously seek to minimize impacts to disadvantaged population groups, sensitive habitats, water systems, and the like (McDonald et al. 2009). Moreover, residents often protest new energy infrastructure projects due to NIMBY (not in my backyard) concerns that usually involve perceived negative impacts on property values, community health, or both. These efforts can derail or slow the development of new energy facilities, particularly in the cases of wind farms, where there is a clear visual impact (Rodman and Meentemeyer 2006), or biomass plants, where residents might object to these facilities being proximal to their neighborhoods (Upreti 2004).

Given that the siting of infrastructure elements is informed by spatial analyses, GIScience is particularly well suited to contribute to research and policymaking in these areas. There will be opportunities for designing new model structures that better capture the many competing needs in infrastructure siting. In this way, researchers will not only have to think critically about representation and characterizations of "space" in model constructs (Miller and Wentz 2003; Goodchild, Yuan, and Cova 2007), but they will also need to push the envelope in terms of developing policy-applicable methodologies.

Focusing on representation in GIScience and spatial models, there has been substantial effort devoted to this issue. From relaxing the binaries inherent to discrete space (Murray 2005), to producing detailed assessments of the impacts of aggregate spatial data on model solutions (Francis et al. 2004), there are several representational issues that bear on future energy model development. As these models are typically expected to reinforce positive societal outcomes (e.g., maximizing equal access to a facility), their results are potentially compromised by poorly chosen spatial inputs. Going forward, productive synthesis could entail more discussion about how space manifests itself in spatial models of energy facility location.

Future energy work involving spatial modeling also offers GIScience a new venue in which to tackle traditional theory–practice divides in terms of getting meaningful information to practitioners and people who need it (Ligmann-Zielinska and Jankowski 2007). There is a role for GIScientists interested in facilitating public participation and collaboration through geospatial technologies (Elwood 2008; Sidlar and Rinner 2009). Many new energy facilities and infrastructures are perceived as noxious or undesirable (Berman and Huang 2008), so the challenge for modeling efforts will be to propose solutions that can be implemented given social and environmental justice concerns (Mennis and Jordan 2005). This process will involve incorporating stakeholder involvement as well as communicating key geospatial issues (Elwood 2008). The potential interface between spatial modeling efforts, geospatial collaboration, and participatory GIS in the context of energy studies can be mutually beneficial. This could spur new model developments that are more attuned to issues of representation, social and environmental context, real-world implementation, and so on, as well as offer new territory for theory and application with regard to participatory GIS. So, too, would energy facility location studies gain from the availability of more holistic and representative modeling methodologies.

Household Energy Conservation and Efficiency

We previously mentioned prospective large-scale policy efforts intended to curb and regulate GHG emissions (Ciocirlan 2008; Niemeier et al. 2008; White House 2010). These are usually aimed at private entities, including industrial sites and large corporations. Although households are typically not targeted directly, they would bear some burden as private entities pass along some of their compliance costs to them. As such, questions about how to encourage more sustainable energy use at the household level are likely gain traction in coming years. This raises two interrelated possibilities for future research involving GIScience. One emanates from attempts to model household behavior given various energy conservation and efficiency policy proposals, whereas the second examines household energy consumption behavior specifically related to transportation expenditures.

Taxes, user charges, fees, and other pricing mechanisms are among the most common policy tools for affecting individual behavior, but owing to the political challenges faced when pursuing these strategies, pricing mechanisms have been employed sparingly to reduce carbon emissions and promote energy conservation. For example, in North America, there was only one case of carbon taxes being paid by households, as a tax was established in 2008 in British Columbia on virtually all household fuels consumed, including those used in transportation (Litman 2009). In the United States, policy remedies for encouraging more sustainable household energy consumption have largely consisted of incentives or "carrots" (Coad, de Haan, and Woersdorfer 2009), such as reductions in taxes for purchasing a more energy-efficient or alternative fuel vehicle, upgrading household appliances to more energy-efficient models, or pursuing energy-sustainable housing improvements, including better insulation, multipane windows, or solar panels on rooftops. Assuming that future household-level policy instruments will take a similar incentive-based form, more research will be needed to understand behavioral responses to these policy prescriptions and the subgroups that choose to adopt (or ignore) incentives.

Long-standing geographical research foci in innovation and information diffusion (Hagerstrand 1967; Brown 1999) have bearing on these energy issues (McEachern and Hanson 2008; Cantono and Silverberg 2009) in terms of the factors that might explain why households make certain energy-saving adoptions (Zahran et al. 2008). Moreover, the pervasiveness of digital social media technologies portends a continuing reexamination of information diffusion processes themselves (Monclar et al. 2009), and household energy decision making is ripe for such analysis. More broadly, within the span of factors affecting household efficiency decisions, understanding the influence of social networks is a potential area for increased research. It is recognized that social networks play an important role in consumer choices, such as how people choose a travel mode (Dugundji and Walker 2005) or structure their daily activities (Carrasco and Miller 2009). When viewed in this fashion, households become agents within complex systems shaped by both interactions with top-down policies as well as internal interactions and lateral connections to other individuals, households, and system actors. Thus, behavioral responses to energy policies are contingent on emergent elements of the system including outcomes of information diffusion through social networks. During the last several years there have been significant advances in the modeling of complex spatial systems using agent-based models (Torrens 2006; Xie, Batty, and Zhaoz 2007; Li and Liu 2008). Using agent-based models to explore household reactions to energy policies (Nannen and van den Bergh 2010), such as incentives to purchase a fuel-efficient vehicle (Mueller and de Haan 2009), is an area for future syntheses. Opportunity for even deeper engagements could also come in the form of exchanges between energy researchers and those with interests in complex systems theories (Manson 2001; Ligmann-Zielinska and Jankowski 2007). As discussions of geographic complexity have been situated within the literature on land-use and land-cover change processes (Evans, Sun, and Kelley 2006; Manson 2006), prospects exist for new extensions such as examining the urban form dynamics associated with energy issues.

Household transportation behavior is another area for increased attention. Approximately 25 percent of energy consumed in the United States in 2006 was from motor vehicles (Akinci et al. 2008). Moreover as much as 33 percent of CO_2 emissions in the United States are from the transport sector (Ewing et al. 2008). One set of ideas proposed to reduce household travel looks at whether it is possible to save transportation energy through changes to land use (Horner 2008). Although there is inherently a question of choice in terms of where people live and where they choose to partake in activities (Levine 1998), "smart growth" advocates argue that the mixing of desirable destinations across a heterogeneous urban landscape can yield shorter trips and energy savings (Ewing et al. 2008; Stone 2008). Conversely, more segregated land-use patterns are thought to yield more energy-intensive travel because there is an inherent "built-in" distance that must be overcome to link people to destinations. Transportation specialists have sought to develop more theoretically sound models of transportation behavior to better address such questions (Miller and Shaw 2001). Moving forward, increasingly detailed simulations of human activities could allow researchers to examine how well particular land-use changes would produce improvements in transportation resource consumption (Waddell et al. 2007). Historically, there have been strong links between the GIScience and transport fields (Thill 2000). These should be expanded to explore new computational approaches that investigate the energy-saving benefits of development patterns.

Methods aimed at exploring household energy consumption are data-hungry by nature. For example, agent-based models require empirical data for their construction and validation. Information on individuals is necessary to understand decisions such as solar energy adoption rates (Sidiras and Koukios 2004) and transportation choices (Boussauw and Witlox 2009). Also necessary are data on the availability of infrastructure and the physical environment (Beccali et al. 2009). As we suggested earlier, how these data are captured spatially has important implications for how they can be used and the questions energy researchers are able to ask. Researchers have lamented the difficulties of combining the diverse spatial data sets necessary to comprehensively analyze energy questions within GIS (Ewing, Pendall, and Chen 2003), particularly for large-scale analyses (Parshall et al. 2010). As such, mutual interests in matters of data fusion (Beccali et al. 2009) as well as the aforementioned areas of spatial interpolation (Goodchild, Anselin, and Deichmann 1993) and spatial representation issues (Miller and Wentz 2003) could generate new dialogue. There has been a trend in spatial analysis toward using more disaggregate, individualized information, especially with regard to households and transportation questions (Kwan and Weber 2003). How new individual-level disaggregate data sources can be incorporated into household energy studies to enhance their representativeness, while living within the limits of computational tools (Wang and Liu 2009), will be a critical issue for future work.

Toward Research Opportunities in Energy and GIScience

We have argued that although there are examples of researchers using GIS and other geospatial technologies to address energy issues, as well as many possible areas of shared research interest, to date there has been limited engagement between those interested in GIScience and those interested in energy research, broadly defined. Energy issues offer significant research challenges for GIScience, particularly within the context of spatio-temporal inventories and models of energy behavior and emissions production, decisions regarding the placement of facilities that are environmentally viable and socially equitable, and the assessment of energy savings and land-use policies' efficacy and political feasibility. There are many potential opportunities for future synergies to develop, synergies that can both inform the knowledge base in energy systems and push the boundaries for GIScience research. Perhaps more important, work at the nexus of energy and GIScience can contribute to longer run efforts to become more energy sustainable.

We have identified three broad problems to which geospatial technologies are already contributing solutions, but where new interactions between energy researchers and those interested in GIScience seem particularly well-suited to yield deeper insights. Naturally, future GIScience contributions will not be limited to these topics. Within these subject areas we noted that good solutions are contingent on the basic science underlying GIS. From scale and representation in computational models to data fusion, interpolation, and extrapolation techniques, GIScience should be a more prominent part of new research efforts seeking to foster energy sustainability. No doubt as others contemplate future possible integration between energy and GIScience, they will see other opportunities we did not detail. Besides other energy problem areas on which we might have focused (e.g., the environmental impacts of certain types of energy uses), there are additional cross-cutting GIScience themes that could bear relevance in the future (e.g., spatial uncertainty). Our hope is that this work helps to propel further discussion and action regarding the opportunities that exist at the intersection of energy studies and GIScience.

Acknowledgments

Portions of this research were supported by Florida State University's Institute for Energy Systems, Economics and Sustainability (IESES). The views represented here are those of the authors and not those of IESES.

References

Abraham, T., and J. F. Roddick. 1999. Survey of spatio-temporal databases. *GeoInformatica* 3 (1): 61.

Agugliaro, F. M. 2007. Gasification of greenhouse residues for obtaining electrical energy in the south of Spain: Localization by GIS. *Interciencia* 32 (2): 131–36.

Akinci, B., P. G. Kassebaum, J. V. Fitch, and R. W. Thompson. 2008. The role of bio-fuels in satisfying US transportation fuel demands. *Energy Policy* 36 (9): 3485–91.

Andrews, C. J. 2008. Greenhouse gas emissions along the rural–urban gradient. *Journal of Environmental Planning and Management* 51 (6): 847–70.

Armstrong, M. P. 2000. Geography and computational science. *Annals of the Association of American Geographers* 90 (1): 146–56.

Batty, M. 2005. *Cities and complexity: Understanding cities with cellular automata, agent-based models, and fractals*. Cambridge, MA: MIT Press.

Beccali, M., P. Columba, V. D'Alberti, and V. Franzitta. 2009. Assessment of bioenergy potential in Sicily: A GIS-based support methodology. *Biomass & Bioenergy* 33 (1): 79–87.

Berman, O., and R. Huang. 2008. The minimum weighted covering location problem with distance constraints. *Computers & Operations Research* 35 (2): 356–72.

Boussauw, K., and F. Witlox. 2009. Introducing a commute-energy performance index for Flanders. *Transportation Research Part A-Policy and Practice* 43 (5): 580–91.

Brown, L. A. 1999. Change, continuity, and the pursuit of geographic understanding. *Annals of the Association of American Geographers* 89 (1): 1–25.

Cantono, S., and G. Silverberg. 2009. A percolation model of eco-innovation diffusion: The relationship between diffusion, learning economies and subsidies. *Technological Forecasting and Social Change* 76 (4): 487–96.

Caron, C., S. Roche, D. Goyer, and A. Jaton. 2008. GIScience journal ranking and evaluation: An international delphi study. *Transactions in GIS* 12 (3): 293–321.

Carrasco, J. A., and E. J. Miller. 2009. The social dimension in action: A multilevel, personal networks model of social activity frequency between individuals. *Transportation Research Part A* 43 (1): 90–104.

Chen, X. W., J. W. Meaker, and F. B. Zhan. 2006. Agent-based modeling and analysis of hurricane evacuation procedures for the Florida Keys. *Natural Hazards* 38 (3): 321–38.

Cidell, J. 2009. Building green: The emerging geography of LEED-certified buildings and professionals. *Professional Geographer* 61 (2): 200–15.

Ciocirlan, C. E. 2008. Analyzing preferences towards economic incentives in combating climate change: A comparative analysis of US states. *Climate Policy* 8 (6): 548–68.

Coad, A., P. de Haan, and J. S. Woersdorfer. 2009. Consumer support for environmental policies: An application to purchases of green cars. *Ecological Economics* 68 (7): 2078–86.

Cutter, S. L. 2003. GIScience, disasters, and emergency management. *Transactions in GIS* 7 (4): 439–45.

DeVerteuil, G. 2000. Reconsidering the legacy of urban public facility location theory in human geography. *Progress in Human Geography* 24 (1): 47–69.

Dugundji, E. R., and J. L. Walker. 2005. Discrete choice with social and spatial network interdependencies—An empirical example using mixed generalized extreme value models with field and panel effects. *Transportation Research Record* 1921:70–78.

Elwood, S. 2008. Grassroots groups as stakeholders in spatial data infrastructures: Challenges and opportunities for local data development and sharing. *International Journal of Geographical Information Science* 22 (1): 71–90.

Evans, T. P., W. J. Sun, and H. Kelley. 2006. Spatially explicit experiments for the exploration of land-use decision-making dynamics. *International Journal of Geographical Information Science* 20 (9): 1013–37.

Ewing, R., K. Bartholomew, S. Winkelman, J. Walters, and D. Chen. 2008. *Growing cooler: The evidence on urban development and climate change*. Washington, DC: Urban Land Institute.

Ewing, R., R. Pendall, and D. Chen. 2003. Measuring sprawl and its transportation impacts. *Transportation Research Record* 1831:175–83.

Ewing, R., and F. Rong. 2008. The impact of urban form on US residential energy use. *Housing Policy Debate* 19 (1): 1–30.

Fotheringham, A. S. 1997. Trends in quantitative methods: 1. Stressing the local. *Progress in Human Geography* 21 (1): 88–96.

Francis, R. L., T. J. Lowe, A. Tamir, and H. Emir-Farinas. 2004. Aggregation decomposition and aggregation guidelines for a class of minimax and covering location models. *Geographical Analysis* 36 (4): 332–49.

Friedman, T. 2008. *Hot, flat, and crowded: Why we need a green revolution—And how it can renew America*. New York: Farrar, Straus and Giroux.

Goodchild, M. F. 2010. Twenty years of progress: GIScience in 2010. *Journal of Spatial Information Science* 1:3–20.

Goodchild, M., L. Anselin, and U. Deichmann. 1993. A framework for the interpolation of areal data. *Environment and Planning A* 25:383–97.

Goodchild, M. F., M. Yuan, and T. J. Cova. 2007. Towards a general theory of geographic representation in GIS. *International Journal of Geographical Information Science* 21 (3): 239–60.

Hagerstrand, T. 1967. *Innovation diffusion as a spatial process*. Chicago: University of Chicago Press.

Hepburn, C., and N. Stern. 2008. A new global deal on climate change. *Oxford Review of Economic Policy* 24 (2): 259–79.

Herold, M., J. Scepan, and K. C. Clarke. 2003. The use of remote sensing and landscape metrics to describe structures and changes in urban land uses. *Environment and Planning A* 34 (8): 1443–58.

Higgs, G., and M. Langford. 2009. GIScience, environmental justice, & estimating populations at risk: The case of landfills in Wales. *Applied Geography* 29 (1): 63.

Horner, M. W. 2008. "Optimal" accessibility landscapes? Development of a new methodology for simulating and assessing jobs—Housing relationships in urban regions. *Urban Studies* 45 (8): 1583–1602.

———. 2010. Exploring the sensitivity of jobs-housing statistics to imperfect travel time information. *Environment and Planning B: Planning and Design* 37 (2): 367–75.

Horner, M. W., and J. A. Downs. 2008. An analysis of the effects of socio-economic status on hurricane disaster relief plans. *Transportation Research Record* 2067:1–10.

InterAcademy Council. 2007. Lighting the way: Towards a sustainable energy future. http://www.interacademycouncil.net/CMS/Reports/11840.aspx (last accessed 12 January 2011).

Intergovernmental Panel on Climate Change (IPCC). 2007. Climate change 2007: Synthesis report. http://www.ipcc.ch/publications_and_data/ar4/syr/en/contents.html (last accessed 12 January 2011).

Ji, W., J. Ma, R. W. Twibell, and K. Underhill. 2006. Characterizing urban sprawl using multi-stage remote sensing images and landscape metrics. *Computers, Environment and Urban Systems* 30 (6): 861.

Komarov, V. S., A. V. Lavrinenko, A. V. Kreminskii, N. Y. Lomakina, Y. B. Popov, and A. I. Popova. 2007. New method of spatial extrapolation of meteorological fields on the mesoscale level using a Kalman filter algorithm for a four-dimensional dynamic–stochastic model. *Journal of Atmospheric and Oceanic Technology* 24:182–93.

Kuby, M., and S. Lim. 2007. Location of alternative-fuel stations using the flow-refueling location model and dispersion of candidate sites on arcs. *Networks & Spatial Economics* 7 (2): 129–52.

Kwan, M. P., and J. Weber. 2003. Individual accessibility revisited: Implications for geographical analysis in the twenty-first century. *Geographical Analysis* 35 (4): 341–53.

Levine, J. 1998. Rethinking accessibility and jobs–housing balance. *Journal of the American Planning Association* 64 (2): 133–49.

Li, X., and X. P. Liu. 2008. Embedding sustainable development strategies in agent-based models for use as a planning tool. *International Journal of Geographical Information Science* 22 (1): 21–45.

Ligmann-Zielinska, A., and P. Jankowski. 2007. Agent-based models as laboratories for spatially explicit planning policies. *Environment and Planning B: Planning & Design* 34 (2): 316–35.

Litman, T. 2009. Evaluating carbon taxes as an energy conservation and emission reduction strategy. *Transportation Research Record* 2139:125–32.

Lopez, I. F. V., R. T. Snodgrass, and M. Bongki. 2005. Spatiotemporal aggregate computation: A survey. *IEEE Transactions on Knowledge and Data Engineering* 17 (2): 271.

MacEachren, A. M., and G. R. Cai. 2006. Supporting group work in crisis management: Visually mediated human–GIS–human dialogue. *Environment and Planning B: Planning & Design* 33 (3): 435–56.

Mack, J., and B. Endemann. 2010. Making carbon dioxide sequestration feasible: Toward federal regulation of CO_2 sequestration pipelines. *Energy Policy* 38 (2): 735–43.

Madden, M., and A. Ross. 2009. Genocide and GIScience: Integrating personal narratives and geographic information science to study human rights. *The Professional Geographer* 61 (4): 508–26.

Manson, S. 2001. Simplifying complexity: A review of complexity theory. *Geoforum* 32 (3): 405–14.

———. 2006. Land use in the southern Yucatan peninsular region of Mexico: Scenarios of population and institutional change. *Computers Environment and Urban Systems* 30 (3): 230–53.

Mark, D. M. 2003. Geographic information science: Defining the field. In *Foundations of geographic information science*, ed. M. Duckham, M. Goodchild, and M. Worboys, 3–18. New York: Wiley.

Marsh, M. T., and D. A. Schilling. 1994. Equity measurement in facility location analysis—A review and framework. *European Journal of Operational Research* 74 (1): 1–17.

McDonald, R. I., J. Fargione, J. Kiesecker, W. M. Miller, and J. Powell. 2009. Energy sprawl or energy efficiency: Climate policy impacts on natural habitat for the United States of America. *Plos One* 4 (8): e6802. doi: 10.1371/journal.pone.0006802.

McEachern, M., and S. Hanson. 2008. Socio-geographic perception in the diffusion of innovation: Solar energy technology in Sri Lanka. *Energy Policy* 36 (7): 2578–90.

Mennis, J., and L. Jordan. 2005. The distribution of environmental equity: Exploring spatial nonstationarity in multivariate models of air toxic releases. *Annals of the Association of American Geographers* 95 (2): 249–68.

Metcalf, G. E., and D. Weisbach. 2009. The design of a carbon tax. *Harvard Environmental Law Review* 33:499–556.

Miller, H. J., and S.-L. Shaw. 2001. *Geographic information systems for transportation*. New York: Oxford University Press.

Miller, H. J., and E. A. Wentz. 2003. Representation and spatial analysis in geographic information systems. *Annals of the Association of American Geographers* 93 (3): 574–94.

Mills, J. W., A. Curtis, J. C. Pine, B. Kennedy, F. Jones, R. Ramani, and D. Bausch. 2008. The clearinghouse concept: A model for geospatial data centralization and dissemination in a disaster. *Disasters* 32 (3): 467.

Monclar, R., A. Tecla, J. Oliveira, and J. de Souza. 2009. Using spatial-temporal information to improve social networks and knowledge dissemination. *Information Sciences* 179 (15): 2524–37.

Mu, L., and F. Wang. 2008. A scale-space clustering method: Mitigating the effect of scale in the analysis of zone-based data. *Annals of the Association of American Geographers* 98 (1): 85–101.

Mueller, M. G., and P. de Haan. 2009. How much do incentives affect car purchase? Agent-based microsimulation of consumer choice of new cars: Part I. Model structure, simulation of bounded rationality, and model validation. *Energy Policy* 37 (3): 1072–82.

Murray, A. T. 2005. Geography in coverage modeling: Exploiting spatial structure to address complementary partial service of areas. *Annals of the Association of American Geographers* 95 (4): 761.

Nannen, V., and J. van den Bergh. 2010. Policy instruments for evolution of bounded rationality: Application to climate-energy problems. *Technological Forecasting and Social Change* 77 (1): 76–93.

National Academy of Sciences. 2010. Real prospects for energy efficiency in the United States. http://books.nap.edu/openbook.php?record_id=12621&page=1 (last accessed 12 January 2011).

Nicholas, M. A., S. L. Handy, and D. Sperling. 2004. Using geographic information systems to evaluate siting and networks of hydrogen stations. *Transportation Research Record* 1880:126–34.

Niemeier, D., G. Gould, A. Karner, M. Hixson, B. Bachmann, C. Okma, Z. Lang, and D. Heres Del Valle. 2008. Rethinking downstream regulation: California's opportunity to engage households in reducing greenhouse gases. *Energy Policy* 36 (9): 3436.

Parshall, L., K. Gurney, S. A. Hammer, D. Mendoza, Y. Zhou, and S. Geethakumar. 2010. Modeling energy consumption and CO_2 emissions at the urban scale: Methodological challenges and insights from the United States. *Energy Policy* 38 (9): 4765–82.

Peuquet, D. J. 2004. *Representations of space and time*. New York: Guilford.

Prest, R., T. Daniell, and B. Stendorf. 2007. Using GIS to evaluate the impact of exclusion zones on the connection cost of wave energy to the electricity grid. *Energy Policy* 35 (9): 4516–28.

Prince, S. D., J. Haskett, M. Steininger, H. Strand, and R. Wright. 2001. Net primary production of U.S. Midwest croplands from agricultural harvest yield data. *Ecological Applications* 11 (4): 1194–1205.

ReVelle, C. S., and H. A. Eiselt. 2005. Location analysis: A synthesis and survey. *European Journal of Operational Research* 165 (1): 1.

Rodman, L. C., and R. K. Meentemeyer. 2006. A geographic analysis of wind turbine placement in Northern California. *Energy Policy* 34 (15): 2137–49.

Running, S. W., R. R. Nemani, F. A. Heinsch, M. Zhao, M. Reeves, and H. Hashimoto. 2004. A continuous satellite-derived measure of global terrestrial primary production. *BioScience* 54 (6): 547–60.

Saunders, M. J., and A. N. R. da Silva. 2009. Reducing urban transport energy dependence: A new urban development framework and GIS-based tool. *International Journal of Sustainable Transportation* 3 (2): 71–87.

Sidiras, D., and E. Koukios. 2004. Simulation of the solar hot water systems diffusion: The case of Greece. *Renewable Energy* 29 (6): 907–19.

Sidlar, C. L., and C. Rinner. 2009. Utility assessment of a map-based online geo-collaboration tool. *Journal of Environmental Management* 90 (6): 2020–26.

Simao, A., P. J. Densham, and M. Haklay. 2009. Web-based GIS for collaborative planning and public participation: An application to the strategic planning of wind farm sites. *Journal of Environmental Management* 90 (6): 2027–40.

Skole, D. L. 2004. Geography as a great intellectual melting pot and the preeminent interdisciplinary environmental discipline. *Annals of the Association of American Geographers* 94 (4): 739–43.

Stell, J., and M. Worboys. 2008. A theory of change for attributed spatial entities. In *Geographic information science—Lecure notes in computer science*, ed. T. Cova, H. Miller, K. Beard, A. Frank, and M. Goodchild, 308: Berlin: Springer.

Stone, B. 2008. Urban sprawl and air quality in large US cities. *Journal of Environmental Management* 86 (4): 688–98.

Talen, E. 2001. School, community, and spatial equity: An empirical investigation of access to elementary schools in West Virginia. *Annals of the Association of American Geographers* 91 (3): 465–86.

Theobald, D. M. 2001. Land-use dynamics beyond the American urban fringe. *Geographical Review* 91:544–64.

Thill, J. C. 2000. Geographic information systems for transportation in perspective. *Transportation Research Part C: Emerging Technologies* 8 (1–6): 3–12.

Torrens, P. M. 2006. Simulating sprawl. *Annals of the Association of American Geographers* 96 (2): 248–75.

Upreti, B. R. 2004. Conflict over biomass energy development in the United Kingdom: Some observations and lessons from England and Wales. *Energy Policy* 32 (6): 785.

U.S. House of Representatives. 2009. American Clean Energy and Security Act of 2009. 112th U.S. Congress, H.R. 2454.

Velazquez-Marti, B., and E. Annevelink. 2009. GIS application to define biomass collection points as sources for linear programming of delivery networks. *Transactions of the ASABE* 52 (4): 1069–78.

Waddell, P., G. F. Ulfarsson, J. P. Franklin, and J. Lobb. 2007. Incorporating land use in metropolitan transportation planning. *Transportation Research Part A* 41 (5): 382–410.

Wang, S., and Y. Liu. 2009. TeraGrid GIScience gateway: Bridging cyberinfrastructure and GIScience. *International Journal of Geographical Information Science* 23 (5): 631–56.

White House. 2010. *Federal agency strategic sustainability performance plans*. http://www.whitehouse.gov/administration/eop/ceq/sustainability/plans (last accessed 12 January 2011).

Willson, R. W., and K. D. Brown. 2008. Carbon neutrality at the local level: Achievable goal or fantasy? *Journal of the American Planning Association* 74 (4): 497.

Wise, M., K. Calvin, A. Thomson, L. Clarke, B. Bond-Lamberty, R. Sands, S. J. Smith, A. Janetos, and J. Edmonds. 2009. Implications of limiting CO_2 concentrations for land use and energy. *Science* 324:1183–86.

Woodcock, J., D. Banister, P. Edwards, A. M. Prentice, and I. Roberts. 2007. Energy and health 3: Energy and transport. *Lancet* 370 (9592): 1078–88.

Worboys, M., and M. Duckham. 2006. Monitoring qualitative spatiotemporal change for geosensor networks. *International Journal of Geographical Information Science* 20 (10): 1087–1108.

Xie, Y. C., M. Batty, and K. Zhaoz. 2007. Simulating emergent urban form using agent-based modeling: Desakota in the Suzhou-Wuxian region in China. *Annals of the Association of American Geographers* 97 (3): 477–95.

Yu, H., and S.-L. Shaw. 2008. Exploring potential human activities in physical and virtual spaces: A spatio-temporal GIS approach. *International Journal of Geographical Information Science* 22 (4): 409–30.

Zahran, S., S. D. Brody, A. Vedlitz, M. G. Lacy, and C. L. Schelly. 2008. Greening local energy: explaining the geographic distribution of household solar energy use in the United States. *Journal of the American Planning Association* 74 (4): 419–34.

Zegras, P. C. 2007. As if Kyoto mattered: The clean development mechanism and transportation. *Energy Policy* 35 (10): 5136–50.

Zhang, Y., and B. Guindon. 2006. Using satellite remote sensing to survey transport-related urban sustainability: Part 1. Methodologies for indicator quantification. *International Journal of Applied Earth Observation and Geoinformation* 8 (3): 149–64.

Zhao, T., D. G. Brown, and K. M. Bergen. 2007. Increasing gross primary production (GPP) in the urbanizing landscapes of southeastern Michigan. *Photogrammetric Engineering & Remote Sensing* 73 (10): 1159–67.

Zimmerer, K. S. 2011. New geographies of energy: Introduction to the special issue. *Annals of the Association of American Geographers* 101 (4): 705–11.

The Role of Climate Change Litigation in Establishing the Scale of Energy Regulation

Hari M. Osofsky

University of Minnesota Law School

This article argues that U.S. litigation over climate change shapes the scale at which energy regulation takes place and, in so doing, influences the sustainability of energy production and consumption. The resolution of these disputes and the resulting regulatory changes impact the scale of energy regulation because most of the cases focus on emissions from transportation or power plants, industries deeply involved in the production and consumption of energy. As these legal actions have grown from a few cases widely regarded as boundary-pushing oddities to over one hundred lawsuits under many different legal theories, their collective outcome plays an important role in the ways smaller scale regulatory action takes place. In so doing, these actions "rescale" energy regulation by mandating or preventing smaller scale regulatory action—local, state, or national—as part of overall approaches to addressing climate change. This article contributes to both the geography and legal literatures by considering how climate change litigation in the United States helps to address the multiscalar challenges facing the energy industry, and the ways in which fixed judicial structures provide a space for needed fluidity in the construction of multilevel approaches to sustainable energy production and consumption. In so doing, it connects the legal dynamic federalism scholarship with the geography literature on scale to demonstrate the constructive role that climate change litigation plays in crafting multilevel energy regulation and involving a wide range of multiscalar interests. *Key Words: climate change, energy, federalism, law and geography, litigation, scale.*

本文认为，美国对气候变化形态的的诉讼塑造了能量调节发生的规模层次，这样做也影响了能源生产和消费的可持续性。这些纠纷和由此产生的监管变化影响能源管理的规模，因为大多数情况下着重于运输或发电厂，这些深入参与了能源生产和消费的工业的排放。由于这些法律行动已经从被广泛认为是推挤边界怪行的少数情况成长为在许多不同的法律理论之下的超过一百多个的诉讼案，他们的集体成果对较小规模管制行动的发生起着重要作用在这样做时，这些行动通过授权或防止较小规模的监管行动—地方，州，或节国家—作为整体解决气候变化问题办法的一部分，来“重新调整”能源监管。本文通过考虑美国气候变化诉讼如何有助于解决能源产业所面临的多尺度的挑战，和固定的司法体制以何种方式为建设多层次的可持续能源生产和消费方法所需的流动性提供一个空间，从而对地理和法律文献作出贡献。通过这样做，把地理文献和法律动态联邦制奖学金相连接，以展示气候变化诉讼在制作多级能源章程和介入大范围多尺度利益方中所扮演的建设性角色。*关键词：气候变化，能源，联邦制，法律和地理，诉讼，规模。*

En este artículo se arguye que el litigio de los EE.UU. sobre cambio climático da forma a la escala en que tiene lugar la regulación energética y, al hacerlo, influencia la sustentabilidad de la producción y consumo de energía. La resolución de estas disputas y los cambios regulatorios que resulten impactan la escala de la regulación energética porque la mayoría de los casos se enfocan en emisiones del transporte o de las plantas generadoras de energía, industrias profundamente involucradas en la producción y consumo de energía. En la medida en que estas acciones legales han aumentado de unos pocos casos que eran siempre considerados rarezas para empujar límites, hasta cien pleitos respaldados por muy diversas teorías legales, sus resultados colectivo juegan un papel importante en la manera como tienen lugar acciones regulatorias a escalas más pequeñas. Procediendo así, estas acciones "re-escalan" la regulación energética al ordenar o prevenir acciones regulatoria a escalas más pequeña—local, estatal, o nacional—como parte de los enfoques generales con los que se está enfrentando el cambio climático. Este artículo es un aporte tanto a la literatura geográfica como a la legal, al considerar cómo los litigios por cambio climático en Estados Unidos ayuda a abocar los cambios multiescalares que enfrenta la industria de la energía, y el modo como estructuras judiciales fijas proveen un espacio para la necesaria fluidez en la construcción de enfoques de nivel múltiple para la producción y consumo sustentables de energía. Al hacerlo, se conecta la sabiduría del federalismo dinámico legal con la literatura geográfica sobre escala, para demostrar el papel constructivo que el litigio del cambio climático juega para construir regulación energética a nivel múltiple y comprometer un amplio margen de intereses multiescalares. *Palabras clave: cambio climático, energía, federalismo, ley y geografía, litigio, escala.*

As a scientific and legal matter, energy regulation is spatially and temporally multiscalar. For example, U.S. power plants, which are located in particular places and often use fossil fuels extracted from other particular places, operate under local, state, national, and, at times, international law, as well as under interstitial regional regimes. Moreover, the multiple time scales of energy resource development and of resulting externalities require regulatory efforts to interlink the past and the future in ascribing responsibility and in determining appropriate mandates (Osofsky 2005; Wiens and Bachelet 2009).

An examination of these scalar complexities in the context of climate change reinforces the daunting legal challenge facing efforts to green energy. Like the energy industry itself, climate change is both scientifically and legally multiscalar (Sayre 2005). Scientific uncertainty is greatest at the smaller spatial and temporal time scales that matter in the particular places where energy regulation is taking place, and emissions, impacts, and capacity to adapt are spread unequally (National Research Council 2007). The largest consumers of energy, who generally are the most significant greenhouse gas emitters, tend to be the richest places with the greatest capacity to adapt to climate change. Unfortunately, the physical impacts of climate change are likely to be most severe in poor places, which consume much less energy and lack the economic, political, social, and legal capacity to adapt (Intergovernmental Panel on Climate Change 2007; Osofsky 2009).

Furthermore, even though the conversation over climate change often takes place in a traditional international legal framework of treaties among nation-states, crucial decisions are being made at smaller scales. The current legal framework, however, has limited capacity for integration across scales and is often stalled at the larger scales. For example, the Copenhagen Accord does not reference the agreements among localities, states, and provinces at the Copenhagen side meetings, and the commitments taken pursuant to the Accord are mostly conditional on action by other nation-states that might or might not happen (Osofsky 2010b).

The complexities and slow pace of international and national action have led to proregulatory subnational governments, individuals, and nongovernmental organizations (NGOs) using courts to target key governmental decision makers and corporate emitters, even as antiregulatory subnational governments, national governments, and corporations use those same courts to push against action. Over the past several years, these legal actions have grown from a few cases widely regarded as boundary-pushing oddities to over one hundred lawsuits that have become part of the mainstream conversation of academics, practitioners, and policymakers. Especially after the Supreme Court found in favor of proregulatory petitioners in *Massachusetts v. EPA*, despite a standard of review very deferential to the U.S. Environmental Protection Agency (EPA), law firms and NGOs around the world have increasingly begun to incorporate climate change issues into their practice (Osofsky 2010a).

These lawsuits and petitions, which are filed in state and national courts and international tribunals, use a wide variety of legal theories and vary in whether they target governmental regulation or corporate action. What unifies them is their focus on the state–corporate relationship at the core of both energy regulation and climate change mitigation and adaptation. The regulatory suits generally consider whether or not the government should be taking more significant action to track or limit greenhouse gas emissions resulting, at least in part, from choices by major companies involved in energy production and consumption. The other, less frequent type of suits target choices by power, petrochemical, and automobile industry companies directly through claims that focus on harm to those impacted by climate change (Osofsky 2005, 2010a).

The cross-cutting nature of energy regulation raises fundamental questions for those attempting to use lawsuits over climate change to influence the energy choices of major corporations and of the governmental entities that regulate them. First, at what scales and in what spaces should this type of energy regulation take place? Second, what role does and should litigation play in establishing appropriate regulatory scale and strategies?

Both legal and geography scholarship grapple with issues at the core of these questions, and this article focuses on two streams of scholarship that are particularly relevant to them. In law, the dynamic federalism literature recognizes the need for complex governance structures and explores such structures in different variations, sometimes in combination with other theoretical approaches such as adaptive management. In so doing, it engages the multiscalar interactions described earlier (Engel 2006; Ruhl and Salzman 2010; Osofsky forthcoming). The geographical debate over scale's fixity and fluidity can add to this analysis by helping to unpack these multiple regulatory scales. The legal landscape simultaneously involves both fixed structures and ever-shifting, socially constructed arrangements, which complicate the descriptive and normative role of these lawsuits (Delaney & Leitner 1997; Swyngedouw 1997; Brenner 1998; Cox 1998; Judd 1998; Smith 1998; Martin 1999; Paasi 2004; Wood 2005; Osofsky 2010a).

Specifically, the tribunals adjudicating climate change litigation and the laws on which they rely are constituted at specific, fixed levels of governance, which, despite all of the shifts wrought by globalization, stay relatively stable most of the time. The fluidity in the scales of this litigation comes not from the tribunals themselves but from the multiscalar nature of energy regulation. Tribunals operating as "fixed" entities provide a framework in which contestation across scales takes place. As petitioners and respondents debate emissions limits and redress for those harmed, they attempt to rescale regulatory mechanisms related to the energy industry's greenhouse gas emissions and impacts by claiming that regulation at a particular level of government is appropriate or inappropriate based on the scalar nature of climate change (Osofsky 2009, 2010a).

This article considers how the rescaling function of climate change litigation should impact its role in energy regulation. The litigation itself and the legal forums hearing it not only provide spaces for scalar and policy contestation but facilitate more cross-cutting regulatory approaches that help to bridge the divide between formal law and informal sociocultural spaces. The article provides a law and geography analysis of this litigation to map how such interdisciplinary approaches can support more effective energy regulation.

The first section examines energy regulation's spaces and scales and the challenges that this structure poses for effective legal intervention. The next section considers how lawsuits over climate change interact with these challenges by comparing the rescaling function of four U.S. domestic court cases that engage different levels of government using varying legal theories. The final section brings together the geography literature on fixity and fluidity with the legal dynamic federalism literature to analyze the way in which these lawsuits' rescaling role can help to address the challenges posed by the complex geography of energy regulation.

Scalar and Spatial Challenges for Efforts at Green Energy Regulation

The simultaneously multiscalar and public–private nature of energy production and consumption creates intertwined scalar and spatial challenges for efforts at green energy regulation. The scalar challenge comes from the law's inability to cross-cut levels of governance in a fluid fashion that matches the problem. The spatial difficulty results from the boundaries of law and its subjects and objects. This section discusses both issues to frame the law and geography analysis of the rest of the article.

Energy production and consumption, by their nature, defy easy scalar compartmentalization. Energy resources are extracted and used in particular places, but a web of subnational, national, and transnational laws govern those activities. Moreover, the corporate entities involved in this industry generally have a complex institutional and spatial form, such that their parent corporations and subsidiaries span multiple nationalities (Fatouros 1994; Eisenberg 2005; Osofsky 2005).

Because law's sticky nature ties it to a particular governmental level and place, responding to the fluidity of the energy industry causes dilemmas of regulatory overlap and appropriateness. Even dynamic federalism approaches struggle with how to structure multiscalar regulatory authority in a fashion that incentivizes green choices and limits externalities like pollution, greenhouse gas emissions, and impacts on vulnerable populations. These models also have to encounter the reality of existing laws, regulations, and bureaucratic structure that constrain flexibility in any of these overlapping jurisdictions (Osofsky 2005; Engel 2006; Ruhl and Salzman 2010).

This scalar problem is compounded by a spatial one because the legal structure's limited form cannot capture the socioeconomic, political, and legal dynamics of energy production and consumption. On the production side, governmental entities, corporate actors, and impacted populations form an uneasy triad, especially because natural resource extraction often takes place within poor countries that grapple with political instability and dictatorial regimes. The limited applicability of international law to corporations and the principle of state sovereignty over natural resources mean that their activities in these unstable states can enmesh them with unsavory governments in situations in which law serves as a poor mechanism for controlling behavior (Klare 2001; Deva 2003; Dufresne 2004; Osofsky 2005; O'Lear and Diehl 2007).

With respect to consumption, the socio-legal overlaps across scales are, if anything, greater due to the carbon-based economy manifest at every level around the world. People make individual choices about their consumption, which are influenced by sublocal community efforts on sustainability and local land use planning decisions about incentivizing green choices and siting or expanding power plants. Those smallest scale decisions are influenced in turn by state and national regulatory systems, which in their interaction with transnational corporations, NGOs, and local communities help to determine what kinds of power plants have which emissions where. The legal spaces for regulating energy production and consumption simply do not overlap in a sufficiently nuanced way with key actors and their activities to address social and environmental costs effectively (Osofsky 2009).

The Rescaling Function of Climate Change Litigation

Although climate change litigation takes place in a wide range of forums under many different legal theories, all of these cases engage the scalar-spatial dilemmas discussed in the previous section. Specifically, the petitioners and respondents debate the scale and spaces for regulation of these energy-related companies' greenhouse gas emissions by presenting varying views on whether climate change is too big a problem to regulate at national, state, and local levels. Those wanting more regulation tend to scale down, arguing that emissions and impacts take place at state and local levels and that smaller scale regulation is therefore appropriate. In contrast, those wanting to block the regulation at issue generally try to scale up, claiming that climate change is a global problem for which larger scale solutions are appropriate (Osofsky 2009).

These rescaling debates have different nuances depending on the level of government at issue and whether or not they take place in a regulatory or other context. This section uses four types of disputes taking place in the U.S. domestic law context to illustrate this variation and its implications for energy regulation. The first three examples are regulatory suits focusing on different levels of government, and the final one targets major power companies directly through public nuisance law.

Disputes Over National-Level Regulation

Massachusetts v. EPA is the most prominent of the numerous cases brought under federal environmental law to force regulatory action for good reason. It resulted in the U.S. Supreme Court ordering a federal agency to justify its choices better or act, and the resulting EPA endangerment finding is serving as a primary driver of U.S. greenhouse gas emissions regulation in the face of failed comprehensive climate change legislation. This case substantively focused on whether the EPA abused its discretion in refusing to regulate motor vehicle greenhouse gas emissions under the Clean Air Act (CAA; Osofsky 2007).

The scalar disputes in the briefing, oral argument before the Supreme Court, and majority and dissenting opinions in *Massachusetts v. EPA* had two main components. Some of the debate focused on national versus international scales, with a specific emphasis on the question of whether nation-level regulation should move forward in the absence of international-level coordination. Because CAA regulatory implementation relies on state implementation plans, however, the claimants also raised the appropriateness of treating climate change as a problem that can be managed at a state level and that has impacts at that level (Osofsky 2007).

The opinion's resolution of these disputes in favor of the states, cities, territory, and NGOs pushing for greater regulation included these scalar issues. It established, subject to better developed opposing argumentation by the EPA, the national scale as appropriate for regulatory action on climate change and the harm to states as significant enough to give them standing. The opinion did not treat the state implementation issues as a barrier to further action. The opinion thus scaled down to find climate change a problem appropriate for national regulatory action, which likely will involve states heavily in implementation (Osofsky 2007).

In the broader energy regulation context that is the focus of this article, this opinion opened the door for EPA action under the CAA that will force companies that produce and use energy to reduce their greenhouse gas emissions. The EPA's decision, under President Obama, to issue an endangerment finding, followed by regulation of emissions by motor vehicles and power plants, was the first step in such regulation. Assuming this regulatory action survives judicial and legislative challenges, *Massachusetts v. EPA* has helped to move green energy regulation forward at the national scale (Osofsky forthcoming).

Disputes Over State-Level Regulation

To illustrate the similarities and differences in the scale debates at different levels and the complex interactions among levels, the second case also involves motor vehicle greenhouse gas emissions regulation under the CAA but focuses on a different provision of this national-level law that deals with state regulatory behavior. Specifically, although in general federal motor vehicle emissions regulation preempts state efforts under the CAA, the statute grants California an exception to this rule, which allows it to exceed federal standards when granted an EPA waiver. On the grant of this waiver, other states are allowed to choose between California and federal standards, which creates a dual regulatory system in which federal standards serve as a floor (Osofsky 2009, forthcoming).

The controversy over the Bush administration EPA's denial of California's request to regulate motor vehicle greenhouse gas emissions more stringently than the federal government resulted in several lawsuits. The Obama administration, within a week of taking office, began reconsidering the waiver denial and within a few months granted that waiver, which concluded the lawsuits. This waiver grant, together with the endangerment finding, paved the way for a historic compromise that included major corporate emitters and leader states. Under this compromise, the Obama administration is regulating tailpipe emissions and fuel efficiency together under the National Program and California agreed to adopt the less stringent federal standards for model years 2012 to 2016 (Osofsky 2009, forthcoming).

The focus of this analysis, however, is not simply on the policy outcome, which was influenced by both the change in administration and the lawsuits, but rather on the way in which legal scale was debated through the waiver denial and legal challenges to it. Because the substantive issue was whether to allow regulatory action by California and other states that wished to follow its higher standards, the debates centered on state versus national regulation (Osofsky 2009, 2010).

Unlike *Massachusetts v. EPA*, these suits never reached resolution in a court because the Obama administration took over and changed course, but the arguments made prior to that point illustrate the scalar dynamics. The EPA administrator at the time argued that given the global scale of the problem, national and international action should preclude state action. California and other states, in contrast, highlighted the impact of climate change on them, and motor vehicle emissions as a major driver of climate change. By resolving this dispute in favor of California and other leader states on climate change and then bringing them into the National Program, the Obama administration first supported the scaling down and then scaled the overall regulatory approach back up (Osofsky 2009, forthcoming).

As with the first example, the process and resolution of this dispute has broader implications for the scale of efforts to green energy regulation. The explicit scalar references by those on both sides of this dispute reinforce the rescaling implications of lawsuits over the future of energy production and consumption. In addition, the conflict's resolution provides a model for including bottom-up, smaller scale governmental action in a larger scale, national-level plan.

Disputes Over Local-Level Regulation

The third case example, the lawsuit between California and San Bernardino County over the county's failure to address greenhouse gas emissions in its general plan, differs from the first two examples in four ways. First, its substantive focus is broader—the suit concerns the entirety of a county's land use planning process. Second, the suit was brought in a state court under a state law, the California Environmental Quality Act (CEQA), and so has a smaller scale character than the CAA suits brought in federal court. Third, its focus is on an even smaller scale, on the responsibility of localities under state environmental law to bring greenhouse gas emissions into their planning processes. Finally, it resulted in a settlement rather than a court decision or political resolution outside of the court process, and the county has been taking steps since to green its planning processes (Osofsky 2009).

After finalizing its general plan in March 2007, San Bernardino County faced two similarly framed lawsuits—the first by NGOs that have local, state, national, and international ties and the second by the California Attorney General—under CEQA. Both petitions requested that the county disclose and analyze the plan's climate change impacts adequately and prepare a mitigation plan. The second suit settled in August 2007, with the county agreeing to amend its general plan and prepare a greenhouse gas emissions reduction plan and with the state agreeing to allow the current plan to stay in place temporarily and to provide financial support (Osofsky 2009).

In the settlement recitals, the county and state continued to express divergent views on the appropriate scale of climate change regulation. The county asserted that further discussion of climate change in its general plan would be too speculative, whereas the state reiterated that the county's plan would contribute to climate change and that the county would be affected by climate change. The resolution of the settlement has resulted in a scaling down, however, as the county engages in greenhouse gas emissions regulation under the auspices of its Green County San Bernardino initiative. Moreover, the initiative itself involves multiple substate scales as the county collaborates with other local governmental entities and supports efforts by cities, businesses, and individuals within it (Osofsky 2009).

This dispute is important to this article's broader exploration of spatial-scalar relationships in energy regulation because it reinforces the local nature of many critical decisions about energy. This lawsuit is one of many around the world involving energy, climate change, and local planning processes, which range from broader processes like this one to specific decisions to allow coal plants to expand. Furthermore, as indicated by the Copenhagen Climate Communiqué, which was signed by mayors and governors from around the world at a December 2009 meeting, urban areas are responsible for up to three quarters of the world's emissions (Osofsky 2009, 2010b). The scaling down of the climate change conversation thus serves as an important piece of greening energy locally and globally.

Disputes Over Failing to Control Emissions Adequately as Public Nuisance

The prior three cases focus on governmental behavior. The final example, *Connecticut v. AEP*, involves a lawsuit directly targeting major power companies. In this suit—one of several pending actions of its kind against major corporate greenhouse gas emitters—eight states, a city, and three land trusts sued six major electric power companies that operate plants in twenty states. The petitioners claim that these companies represent approximately one quarter of total emissions from the U.S. electric power sector and 10 percent of total U.S. emissions from human activities and that climate change is causing and will cause significant adverse impacts in their states. As a result, petitioners allege, these corporations should be held jointly and severally liable for their creation, contributions to, and maintenance of a public nuisance and should be required to cap their emissions and then reduce them. The Second Circuit ruled in September 2009 against a number of challenges to the petitioners' ability to bring the suit—most significantly finding that they had standing, had stated a claim under common law public nuisance, and did not raise a nonjusticiable political question—and could move forward to the merits (*Connecticut v. AEP* 2009). The U.S. Supreme Court is currently reviewing an appeal of this decision.

The scale issues in this case vary from the preceding three regulatory ones. The main similarity comes with respect to standing, where the focus is on whether sufficient evidence exists of current and future small-scale harms to allow the suit to continue. By finding standing, the Second Circuit was willing to scale down, at least in the limited preliminary procedural posture of the case (*Connecticut v. AEP* 2009).

The political question doctrine, however, which has been at the heart of the public nuisance climate change judicial decisions to date, has a complex relationship to scale. Each time a district court has considered a public nuisance claim to date, it has found that the political question doctrine barred the claim from proceeding. Specifically, they have all found that the cases cannot be decided without making an initial policy determination that should be left to the political branches. The two appellate panels from the Second and Fifth Circuits, that have considered appeals from these decisions thus far have both reversed the district courts and found that the petitioners did not pose a nonjusticiable political question. The Fifth Circuit, sitting *en banc*, however, vacated that decision but then realized it lacked a quorum to proceed. Those petitioners' efforts to obtain U.S. Supreme Court intervention have failed, and so the Second Circuit opinion on appeal to the Supreme Court remains the lead climate change public nuisance case (*Connecticut v. AEP* 2009; Osofsky forthcoming).

On its face, the political question doctrine is more about space than about scale; it focuses on which branch of government should resolve an issue. In this case, however, the district court found a political question problem based on the global nature of climate change and the resulting domestic policy issues that its global scale poses. The appellate court reversed the district court and scaled back down to the federal level as relevant. Focusing on the gap-filling role of common law at that level, the appellate court found that the petitioners did not have to wait for a comprehensive global solution to act, especially because the other two branches have not indicated their support for increasing greenhouse gases (*Connecticut v. AEP* 2009).

This scaling down has a very direct impact on energy regulation. The court's allowing this case to proceed to the merits, if upheld by the U.S. Supreme Court, could result in major energy corporations being forced to change the way in which they do business significantly. At the very least, it has put these corporations on notice that they are vulnerable to lawsuits if they do not reduce their greenhouse gas emissions (Osofsky forthcoming).

Overall, these case examples reflect climate change litigation's role in rescaling energy regulation. These cases impact who can regulate when and what will be required of companies that produce and consume fossil fuels. As a result, they have the potential to assist in addressing energy regulation's scalar and spatial challenges, which is the focus of the next section.

Implications of Rescaling Through Litigation for the Challenges of Energy Regulation

This section draws from the legal dynamic federalism literature and the geography literature on scale to explore how the rescaling function of climate change litigation can assist with energy regulation's challenges. These two streams of scholarship have significant synergies. The dynamic federalism literature argues that models that create more fluid allocation of regulatory authority grapple with complex problems like energy and climate change more effectively than rigid demarcations do (Osofsky 2005; Engel 2006; Ruhl and Salzman 2010). The geography literature focusing on issues of fixity and fluidity in scale engages the complex relationships that cross-cut particular levels (Delaney & Leitner 1997; Swyngedouw 1997; Brenner 1998; Cox 1998; Judd 1998; Smith 1998; Martin 1999; Paasi 2004; Wood 2005; Osofsky 2009, 2010a). Cox's exploration of a network theory of scale, in particular, helps to capture the dynamic nature of relationships among the ties to a particular level and movement among levels (Cox 1998; Osofsky 2009).

These dual insights regarding the need for fluid regulatory approaches and the cross-scalar quality of behavior taking place at particular levels illuminate the role that this litigation does and can play in advancing more effective energy regulation. First, litigation provides a mechanism for bringing greater fluidity into legal interactions, which helps create needed movement across scales and spaces (Osofsky 2009, 2010a). Although the four case examples varied significantly, they all provided opportunities for scalar contestation among a wide range of actors and a moment at which legal scale disputes were resolved. Although this resolution might not always favor the proregulatory stance, as it did in these four cases, litigation's opportunity for contestation helps to overcome the scalar and spatial fixity of law that makes effective energy regulation so difficult.

Second, the lawsuits allow for simultaneous interactions within and across levels of governance. In so doing, they create regulatory diagonals that assist in reordering a landscape dominated by horizontal interactions at a particular level and vertical interactions among key actors and institutions at different levels. As the four case examples demonstrate, these diagonal interactions vary in different lawsuits and over time along the dimensions of size (large vs. small), axis (vertical vs. horizontal), hierarchy (top-down vs. bottom-up), and cooperativeness (cooperation vs. conflict). For example, the first two conflicts evolved from small-scale actors uniting horizontally and pushing in a conflictual fashion vertically from the bottom up for legal change into larger scale, top-down, cooperative policy scheme (Osofsky forthcoming). This evolving diagonal regulatory function makes these lawsuits a helpful tool in crafting more effective cross-cutting regulation.

Moreover, the fluidity and diagonal interactions that these lawsuits bring to energy regulation have broader implications for future executive and legislative action. The Obama administration has continuing opportunities as it attempts to green energy policy to create lawmaking processes that maximize fluidity and possibilities for diagonal interactions. It has already made strides on this score, such as through the process it used to craft the National Program or through the Clean Energy Leadership Group the EPA has established, but further opportunities abound (Osofsky forthcoming).

Law and geography analyses, like the one in this article, help to frame how such approaches might be crafted most effectively within the constraints of law's fixity. Geography's deep engagement of fixity and fluidity, together with dynamic federalism's rigorous exploration of institutional possibilities, can provide the basis for creative, cross-cutting policy approaches that engage the complexities of scale. Such creativity is needed in the face of the significant challenges facing green energy regulation.

Acknowledgments

I would like to thank Alexander Murphy, the reviewers, and Karl Zimmerer for their helpful editorial input that improved the piece greatly. I also would like to thank Josh, Oz, and Scarlet Gitelson for their love, support, and patience.

References

Brenner, N. 1998. Between fixity and motion: Accumulation, territorial organization and the historical geography of spatial scales. *Environment and Planning D: Society and Space* 16 (4): 459–81.

Connecticut v. AEP, 582 F.3d 309–93 (2d Cir. 2009).

Cox, K. 1998. Spaces of dependence, spaces of engagement and the politics of scale, or: Looking for local politics. *Political Geography* 17 (1): 1–23.

Delaney, D., and H. Leitner. 1997. The political construction of scale. *Political Geography* 16 (2): 93–97.

Deva, S. 2003. Human rights violations by multinational corporations and international law: Where from here? *Connecticut Journal of International Law* 19 (1): 1–57.

Dufresne, R. 2004. The opacity of oil: Oil corporations, internal violence, and international law. *New York University International Law and Politics* 36:331–94.

Eisenberg, M. A. 2005. The architecture of American corporate law: Facilitation and regulation. *Berkeley Business Law Journal* 2 (1): 167–84.

Engel, K. H. 2006. Harnessing the benefits of dynamic federalism in environmental law. *Emory Law Journal* 56 (1): 159–88.

Fatouros, A. A., ed. 1994. *Transnational corporations: The international legal framework*. London: Routledge.

Intergovernmental Panel on Climate Change 2007. *Climate change 2007: Impacts, adaptation and vulnerability*. New York: Cambridge University Press.

Judd, D. R. 1998. The case of the missing scales: A commentary on Cox. *Political Geography* 17 (1): 29–34.

Klare, M. T. 2001. *Resource wars: The new landscape of global conflict*. New York: Metropolitan Books.

Martin, D. G. 1999. Transcending the fixity of jurisdictional scale. *Political Geography* 18 (1): 33–38.

National Research Council. 2007. *Evaluating progress of the U.S. climate change science program: Methods and preliminary results*. Washington, DC: National Academies Press.

O'Lear, S., and P. F. Diehl. 2007. Not drawn to scale: Research on resource and environmental conflict. *Geopolitics* 12 (1): 166–82.

Osofsky, H. 2005. The geography of climate change litigation: Implications for transnational regulatory governance. *Washington University Law Quarterly* 83: 1789–1855.

———. 2007. The intersection of scale, science, and law in Massachusetts v. EPA. *Oregon Review of International Law* 9: 233–60.

———. 2009. Is climate change "international"?: Litigation's diagonal regulatory role. *Virginia Journal of International Law* 49 (3): 585–650.

———. 2010a. The continuing importance of climate change litigation. *Climate Law* 1 (1): 3–29.

———. 2010b. Multiscalar governance and climate change: Reflections on the role of states and cities at Copenhagen. *Maryland Journal of International Law* 25: 64–85.

———. Forthcoming. Diagonal federalism and climate change: Implications for the Obama administration. *Alabama Law Review*.

Paasi, A. 2004. Place and region: Looking through the prism of scale. *Progress in Human Geography* 28 (4): 536–46.

Ruhl, J. B., and J. Salzman. 2010. Climate change, dead zones, and massive problems in the administrative state: A guide for whittling away. *California Law Review* 98 (1): 59–120.

Sayre, N. F. 2005. Ecological and geographical scale: Parallels and potential for integration. *Progress in Human Geography* 29 (3): 276–90.

Smith, M. P. 1998. Looking for the global spaces in local politics. *Political Geography* 17 (1): 35–40.

Swyngedouw, E. 1997. Neither global nor local: "Glocalization" and the politics of scale. In *Spaces of globalization: Reasserting the power of the local*, ed. K. R. Cox, 137–66. New York: Guilford Press.

Wiens, J. A., and D. Bachelet. 2009. Matching the multiple scales of conversation with the multiple scales of climate change. *Conservation Biology* 24 (1): 51–62.

Wood, A. 2005. Comparative urban politics and the question of scale. *Space and Polity* 9 (3): 201–15.

Energy and Identity: Imagining Russia as a Hydrocarbon Superpower

Stefan Bouzarovski* and Mark Bassin†

*GEES, *University of Birmingham, and Department of Social Geography and Regional Development, Charles University, Prague*
†*Center for Baltic and East European Studies, Södertörn University*

The relationship between energy systems, on the one hand, and narratives and practices of identity building at different scales, on the other, has received little attention in the mainstream human geography and social science literature. There is still a paucity of integrated theoretical insights into the manner in which energy formations are implicated in the rise of particular cultural self-determinations, even though various strands of work on energy and identity are frequently present throughout the wide—and rather disparate—corpus of social science energy research. Therefore, this article explores the manner in which the exploitation and management of energy resources is woven into discourses and debates about national identity, international relations, a nation's path of future development, and its significance on the global arena using the case of Russia. We investigate some of the policies, narratives, and discourses that accompany the attempt to represent this country as a global "energy superpower" in relation to the resurrection of its domestic economy and material prosperity, on the one hand, and the restoration of its global status as a *derzhava* (or "Great Power"), on the other. Using ideas initially developed within the field of critical discourse analysis, we pay special attention to the national identity-building role played by geographical imaginations about the country's past and present energy exports to neighboring states. We argue that they have created a hydrocarbon landscape in which the discursive and material have become mutually entangled to create an infrastructurally grounded vision of national identity. *Key Words: energy, hydrocarbons, identity, nation, Russia.*

一方面，能源系统之间的关系，另一方面，建于不同尺度的身份叙述和实践，在主流人文地理和社会科学文献中受到很少的重视。其中能源的形成与特定的文化自决上升相牵连，把理论见解与做法联合的研究是很少的，尽管有关能源和身份的各项工作在广泛的—而不是异类的—社会科学能源研究中频繁地展现。因此，本文用俄罗斯为例，探讨了其与国家身份，国际关系，国家的未来发展路径，和其对全球舞台的意义的演说和辩论编织在一起的能源资源开发和管理的方式。一方面，我们调查伴随把这个国家描述为一个全球的“能源超级大国”的试图，关于它的国内经济和物质繁荣复活的政策，叙述，和演说。另一方面，恢复其作为杰尔扎瓦（或“大国”）的全球地位。使用最初在批评话语分析领域发展的思想，我们着重于因国家的过去和现在的能源对邻国出口的地域想象而生的国家认同建设的作用。我们认为，他们已经创造了一个话语和材料相互纠缠的，以发展国家认同的基础的牢固的视觉碳氢化合物景观。*关键词：能源，碳氢化合物，身份，民族，俄罗斯。*

La relación que existe, por una parte, entre sistemas energéticos y las narrativas y prácticas de construcción de identidad a diferentes escalas, por la otra, ha recibido poca atención en las corrientes principales de la literatura de geografía humana y las ciencias sociales. Se nota todavía la escasez de entradas teóricas importantes en la manera como las formaciones energéticas tienen algo que ver con la elevación de autodeterminaciones culturales particulares, aunque algunas sartas de trabajo sobre energía e identidad frecuentemente están presentes a través del amplio—y muy desigual—*corpus* de investigación energética en las ciencias sociales. En consecuencia, usando el caso de Rusia, este artículo explora la manera como la explotación y manejo de los recursos energéticos se entreteje en discursos y debates acerca de identidad nacional, relaciones internacionales, la senda de una nación hacia el desarrollo futuro y su significancia en el escenario global. Investigamos algunas de las políticas, narrativas y discursos que acompañan el intento por representar a este país como una "superpotencia energética" global en relación con la resurrección de su economía doméstica y prosperidad material, por un lado, y la restauración de su estatus global como una *derzhava* ("Gran Potencia"), por el otro. Utilizando ideas desarrolladas inicialmente dentro del campo del análisis del discurso crítico, ponemos especial atención al papel de constructoras de identidad que juegan las imaginaciones geográficas acerca de las exportaciones pasadas y presentes del país hacia los estados vecinos. Sostenemos que ellos han creado un paisaje de hidrocarburos en el que lo discursivo y lo material han llegado a estar mutuamente enredados para crear una visión de la identidad nacional infraestructuralmente encallada. *Palabras clave: energía, hidrocarburos, identidad, nación, Rusia.*

Human geographers—and social scientists more generally—are becoming increasingly interested in the political dimensions of energy flows at different scales, partly thanks to the expanding body of research into the "ongoing global, yet highly differentiated, struggle for sustainability against the hegemony of fossil fuels" (Jiusto 2009, 535). There is a growing theoretical recognition that the politics of energy conversions and circulations are central to the operation of contemporary societies, involving the embedding of a wide array of power relations, institutional regulations, and collective decisions (Jones 1979; Nye 1999; Högselius 2006; Klare 2008; Poputoaia and Bouzarovski 2010). The competing global interests of different nations, expressed through the geopolitical relations underlying international energy flows and exchanges—especially hydrocarbons—have received particular attention in this regard, not the least due to their ramifications on a much broader array of state policies, behaviors, and decisions (see, for example, Carmody and Owusu 2007; Bradshaw 2009). Even though a great deal remains to be done in terms of integrating the scholarship in this field into a coherent set of epistemologies, there is little doubt that the spatialities of energy politics are gradually moving out of the margins in social science.

One of the domains that has received comparatively less attention in this context, however, is the manner in which energy systems shape, and are shaped by, narratives and practices of identity building at different scales. This links into a broader deficiency of scholarship regarding the multiple social and cultural interdependencies between energy and identity per se. Although it is widely recognized that notions of belonging and territorial affiliation are mediated, inter alia, through the realities—material as well as imagined—of energy infrastructure networks and projects (Frankel 1981; Hughes 1993; Banerjee 2003), there is a lack of clear conceptual frameworks for understanding the manner in which energy formations are implicated in processes of territorially bound cultural self-determination. Even though work on energy and identity is implicitly present throughout the wide corpus of social science energy research, it has not been theorized through an explicit conceptual lens. This is particularly true when one considers the growing literature on the geopolitics of energy, where issues of identity construction are ever-present but have rarely been linked to contemporary geographical debates regarding the production of discourse, imagination, and scale. As a result, the mechanisms through which state-level actors create particular visions of national identity with the aid of, and in relation to, energy infrastructure remain largely unexamined.

To address these gaps, this article explores some of the ways in which the exploitation, management, and transport of energy resources is woven into discourses and debates about a nation's identity, its image of its significance on the global arena, and its visions of its own future development. As a case study we draw on evidence gathered in Russia, which occupies a very special geopolitical position with respect to the flows of hydrocarbon-derived energy from the resource-rich areas of the former Soviet Union to the resource-poor regions of western and central Europe and eastern Asia. The article investigates the discourses that accompany the attempt to represent Russia as a global energy superpower and the relationship of the energy superpower image to Putinist concerns with the resurrection of Russia's domestic economy and the restoration of its global status as a *derzhava* (or "Great Power"). By examining some of the public debates around these issues, it aims to shed further light on the interaction between the discursive geopolitical spaces of energy superpower and Russia's real, existing energy landscapes. In particular, we are interested in the role played by energy in geopolitical visions about the country's past and present links with neighboring states across post-Soviet space, in what is commonly termed the *blizhnee zarubezhe* or "Near Abroad."

Restrictions on space do not allow for a separate discussion of the concept of national identity as such, especially insofar as we are concerned with geographical imaginations and discourses about identity within the specific parameters we have indicated. Although mindful of the complex controversies involving identity formation per se (see Wenger 1998; Butler 1999), most of the article is focused on exploring the attempt to build "discourse coalitions" (Bulkeley 2000; Szarka 2004; Mander 2008) that reflect the aspirations of Russia's political elite to achieve global and regional hegemony via the strategic use of energy resources and flows. The tenets of "critical discourse analysis" (Fairclough 1995, 2001, 2005) have proven particularly useful in this regard, especially thanks to its systematic overview of the different ways in which power struggles are produced and reproduced through texts, practices, and sociopolitical events. We have also relied on Star's (1999) claim that infrastructure is "learned as part of membership" via the operation of particular organizational arrangements. Before exploring the specifics of the energy superpower discourse, however, we would like to briefly outline some of the principal ways in which theorizations of energy technologies and flows have engaged with the politics of identity formation at different geographical scales.

Connecting Energy, Discourse, and Identity

Although geographers and other social scientists have written relatively little about the representational underpinnings of energy infrastructures as they relate to national identity formation, these themes have received a good deal of attention in several strands of work within the fields of the history and sociology of science. Much of this scholarship highlights the different ways in which the past development of networked energy infrastructures has been contingent on the articulation of particular political narratives and socio-technical projects (see, for example, Hughes 1993; Star 1999; Graham and Marvin 2001). In particular, investigations of the histories and politics of electrification in the developed world emphasize the importance of "communities of practice" in the electricity industry's diffusion of meaning and identity across various geographical contexts (Wenger 1998; Högselius 2009). Conceptualizations of the relationship between energy and social change have thus been enriched with an improved understanding of the multiplicity of political practices, worldviews, and cultural understandings that are implicated in the expansion and development of energy technologies (Frankel 1981; Dooley 2006; Montgomery 2010).

The historical expansion of electricity infrastructures in the developed world during the first part of the twentieth century provides vivid illustrations of the extensive symbolic and cultural ramifications of energy technology diffusion for the development of urban and regional landscapes. One example would be the extensive body of work (e.g., Nasaw 1992) that highlights the "decisive impact" of electric lighting on the psychogeography of urban space (McQuire 2005). It is argued that electric illumination helped forge a particular sense of identity and self-recognition in American and European cities, thanks to its central role in the newly created phantasmagoria of urban architectural sublimes at the onset of early modernity. But the ability of electrification to generate meaning and mold the national consciousness expands beyond its visible, material significance in everyday cityscapes. In Russia, for instance, the symbolic messages embodied in government-led mass electrification projects—embodied in Lenin's historic credo that "Communism is Soviet power plus the electrification of the entire country," prominently displayed on the central power station in Moscow—reflected long-standing national anxieties about Russia's backwardness and underdevelopment, at the same time that they served as the state's primary technological and political "instrument for modernizing the country" (Banerjee 2003, 49).

Nye's (1990) seminal exploration of the social and cultural implications of early twentieth-century electrification in America emphasizes that the spatial implications of this process were "culturally determined, as Americans used the flexibility of electrical power to atomize society rather than integrate it" (Nye 1990, 384). In making this argument, he contrasts the American experience to that of Denmark. Whereas in the former, the use of electricity was implicated in the rejection of "centralized communal services in favor of personal control over less efficient but autonomous appliances" (384), the construction of cogeneration plants in the latter had the "secondary effect of binding the community more tightly together" (384). Glaser (2009) took some of these debates outside the urban context, pointing out that "rural communities in America maintained their sense of identity and place by accessing electricity in ways that allowed them to integrate these changes on their own terms" (12).

The relationships between identity formation and energy networks are also clearly evident in the organizational and procedural capacities associated with national decision making regarding the construction and expansion of energy supply technologies (Jones 1979; Sahr 1985; Gamson and Modigliani 1989). Nuclear power provides a classic illustration of some of the ways in which "mega energy ideas," requiring centralized, national-scale, and corporate-led control networks have been advanced in relation to energy technologies (Hecht 1998; Byrne and Toly 2006). Jasanoff and Kim (2009) argued that the employment of different "sociotechnical imaginaries" in the context of nuclear power development and opposition in Korea and the United States has had "the power to influence technological design, channel public expenditures, and justify the inclusion or exclusion of citizens with respect to the benefits of technological progress" (120). A similar claim has been advanced in relation to large-scale hydropower dams, which, to cite Byrne and Toly (2006, 1), represent an "attempt at a techno-fix of the democratic-authoritarian variety."

It is worth noting that the incorporation of identity narratives and meanings in the articulation of energy technologies is not limited to state-led projects aimed at harnessing natural resources. A wide sociological literature—otherwise poorly connected with the field of energy studies—focuses on the symbolic meanings created, appropriated, and communicated through the consumption of energy. One of the largest bodies of research in this domain pertains to the social and cultural implications of automobile ownership, which has gradually come to embody combinations of meanings that were previously not associated with this technology (see Gjøen and Hård 2002; Paterson and Dalby 2006; Heffner, Kurani, and Turrentine 2007; Luedicke, Thompson, and Giesler 2010).

Debates about the cultural significance of energy consumption patterns often have a normative dimension to them, because, as argued by Perelman (1980), the "coming transition" to sustainable energy will entail a "radical transformation" in the "theory, philosophy, values and goals that define the direction of social behavior" (392). Many authors working in this vein emphasize the need for appreciating the existence of multiple identities in the process of environmental governance, to move decision making from "technical reason to political reason" while exploring deliberation and participation in politics as a way of "extending the public sphere" and providing a "normative basis of democracy" (Murphy 2007, 7). The role of environmental worldviews in the formation of the values and assumptions that underpin public preferences is a frequent theme in this literature (Kuhn 2008), as is the importance of scale and participation in alternative policy commitments to energy (Morrison and Lodwick 1981; Devine-Wright 2007).

The changing biopolitics of personal identity associated with the introduction of energy efficiency and decarbonization measures at the scale of the home and household (see, for instance, Lovell 2008) prompted Potter (2009) to claim that the emphasis on the individual as an agent of self-monitoring in this context allows the facilitation of government agendas at a distance, by introducing various self-policing measures such as energy meters, carbon accounting, and the retrofitting of energy-efficient technologies. The ontological fixity brought about by the notion of the "carbon footprint" that often underpins such interventions, she argued, brings into light the multiple intersections and entanglements of life politics created by decarbonization discourses.

Goldblatt (2005) also charted the identity underpinnings of the evolving discourse about sustainable consumption, but rather than looking at the scale of the individual, he focused on the differences among nations in the instigation of an institutional framework to create an international environmental agenda with respect to issues such as global warming. As such, his work has helped highlight the manner in which concerns about the social and environmental implications of energy use are bound up with national politics and cultural self-determination. This interdependence between energy and identity becomes even more apparent in the case of the often conflictual relationship between green politics and nationalism in Europe and America (Hamilton 2002; King 2007; Galbreath and Auers 2009; Kopeček 2009) as well as the cultural underpinnings of global discourses about environmental policy per se (E. R. A. N. Smith 2002; Dryzek 2005).

Walker and Cass (2007) pointed to the discursive basis of the modes of renewable energy implementation, thus opening the path for considering the ideologies and narratives of energy transition. Significant work has been done in this field, particularly with respect to different representations of climate change (see Boykoff 2007). Cohen et al. (1998) have emphasized that climate change discourses in the developed world are characterized by a reductionist logic based on a technical and instrumental rationality, in addition to moral-liberal politics. The discursive production of current energy transitions has been explored by authors such as A. Smith and Kern (2009), who stressed the contradictory political and institutional contexts in which socio-technical change takes place.

Narratives and representations of energy relations are no less central to the geopolitical articulation of international energy flows. This is demonstrated, for example, by the manner in which different geographical imaginations and displays of technological superiority have been used to legitimize territorial claims on the Arctic seabed (Jessup 2008). In the European context, discourses relating to the construction of a new Baltic undersea pipeline linking Russia and Germany (also known as "Nord Stream") have been used as a conduit for the articulation of particular national identity narratives and energy network development visions (Bouzarovski and Konieczny 2010). Despite being an imaginary object—the pipeline still only exists on the drawing boards held by its managing company, based in Switzerland—Nord Stream has managed to project itself into the material landscapes of the Baltic region through a variety of associated infrastructural undertakings (liquified natural gas terminals, nuclear power stations, high-voltage power lines) that have drawn their legitimacy from the discourses of fear associated with it.

Although, as was noted previously, there is a paucity of direct geographical explorations of the relationship between identity establishment and the energy sector, it is also true that a number of geographers have dealt with questions of identity and energy in their work, despite not theorizing them explicitly. Two principal themes can be discerned throughout much of this scholarship—as well as the literature surveyed earlier, some of which has resonated within the disciplinary setting of geography as a discipline. In the first instance, there is a common, if somewhat implicit, claim that the spatial inequalities created by the uneven distribution of hydrocarbon resources and their associated circulations are helping provide the symbolic and material basis for the differential formation of national and political identities (see also Watts 2004; Huber 2009). The second theme centers on the close spatial linkages between identity mobilization on the one hand and energy infrastructures nested within particular scales on the other. The question as to how and whether these spatially "heterogeneous entanglements" (Bouzarovski 2010) of energy and identity can be extended to the social, technical, and spatial imaginaries associated with identity building and energy relations among and within nation-states remains open. In particular, there is a need to investigate the manner in which different conceptions of nationhood and international politics have been implicated in this process. To explore these issues, we turn to the Russian case.

Russia as an "Energy Superpower"

The key underlying feature that defines the energy relations between Russia and most of its neighboring states is the massive hydrocarbon endowment of the former as opposed to the clear dependency of the latter on energy imports. Russia contains approximately 35.4 percent and 4.5 percent of the world's proven gas and oil reserves, respectively. Having the world's largest proven gas reserves has also made Russia the world's largest gas producer and exporter: about 53 percent of the 522 million tonnes of oil equivalent (mtoe) of gas that it produced in 2007 were destined for exports (International Energy Agency 2010). Russia is also a significant oil producer and exporter, sometimes exceeding the equivalent volumes reached by Saudi Arabia: in 2007, its oil production averaged 9.4 million barrels per day, over 70 percent of which were exported (Energy Information Administration 2010). According to the U.S. Energy Information Administration (2010),

> roughly 1.3 million barrels per day (bbl/d) were exported via the Druzhba pipeline to Belarus, Ukraine, Germany, Poland, and other destinations in Central and Eastern Europe (including Hungary, Slovakia, and the Czech Republic), around 1.3 million bbl/d via the new flagship Primorsk port near St. Petersburg, and around 900,000 bbl/d via the Black Sea.

Although the country's oil consumption remained more or less constant until 1996, its production grew rapidly after this year, averaging 700,000 barrels per day in the early 2000s. It should also be pointed out that Russia has the world's second largest recoverable coal reserves, at 173 short tons (Energy Information Administration 2010).

Despite being rather sparsely inhabited, at approximately 8.5 persons per square kilometer (Central Intelligence Agency 2008), Russia lies adjacent to some of the most populous regions of the world—western Europe and east Asia—which themselves lack indigenous hydrocarbon resources despite rising levels of domestic energy use. The dramatic spatial differential between these geographical realms provides the main supporting factor for Russia's dominant role as an energy exporter with respect to its neighboring countries. This is especially true in the case of the European Union (EU), which relies on Russia for approximately a quarter of its gas and oil alike (Euractiv 2010). The EU's energy linkages with Russia are partly the result of the infrastructural legacies of Soviet domination in eastern Europe, which today continue to cement the import dependence of the former satellite states (Bouzarovski 2009, 2010).

Although Russia's energy assertiveness in eastern Europe and the Caspian region thus has a long historical record (see, for instance, Laurila 2003; O'Lear 2004), it has only been in the last decade—since Vladimir Putin's ascension to power—that this has acquired an explicitly formulated political connotation. Over the eight years of his presidency, Putin's political vision crystallized into a cluster of doctrines that have come to be called—awkwardly but quite aptly—Putinism (Migranyan 2004, Beichman 2007; Rahn 2007; Whitmore 2007; Aron

2008). The overriding priority was to reverse the decline the country had undergone over the preceding decade and reestablish Russia's national greatness. This project of revival involved two dimensions: the resurrection of Russia's domestic economy and material prosperity on the one hand and the restoration of its global status as a *derzhava* or Great Power on the other. From the beginning, moreover, Putin was convinced that these ambitious goals could only be achieved on the material basis of the country's energy sector. Putin came to power with a specialist knowledge of the energy industry—the subject of his PhD thesis in the 1990s—and from his first days in office he made his views on its critical importance to Russia's future development eminently clear (Putin 2000, 2005).

It was not until the middle of the decade, however, that an ideological formula was devised that captured Putinism's belief in the existential dependency of Russia's national revival on the development of the energy sector. This was the novel concept of Russia as an *energeticheskaia sverkhderzhava*, or energy superpower (a formulation that had already been used for some years by Western commentators; e.g., Hill 2002, 2004). In its Putinist version, the energy superpower concept was proposed by the energy analyst and Putin minion, Dmitrii Orlov (2006a, 2006b), and it quickly caught on to become one of the most recognizable catchwords of Putinism (Irgunov 2006; Kokoshin 2006; Leont'ev 2006). It was enthusiastically taken up by leading ideologues of the regime, not least of all the influential director of Russia's National Energy Security Foundation, Konstantin Simonov, who provided the most comprehensive elaboration of the concept in a book-length manifesto simply entitled *Energy Superpower* (Simonov 2006).

The conviction that energy represents a *conditio sine qua non* for Russia's healthy national development in the twenty-first century ensures that in all of its manifestations—as a natural resource, an industrial activity, a commercial transaction, and a subject of international relations—the energy superpower discourse is implicated in an assortment of fundamental questions relating to contemporary Russia's domestic affairs, its global profile, and its national identity and destiny (Saivetz 2007; Rutland 2008). As it turns out, these patterns of implication are in and of themselves quite revealing of the complex and in certain respects contradictory aspirations for Russia's future development associated with the Putin project (Legvold 2008). The reattainment of *derzhava* involves, above all, the reestablishment of Russia's former status as a global political power. Although this is understood most immediately in terms of the global profile of the Soviet superpower after 1945—in 2005 Putin provocatively declared that the collapse of the USSR was the "greatest geopolitical catastrophe of the twentieth century" ("Putin deplores collapse of the USSR" 2005)—the aspiration for *derzhava* had been a prime motivating factor in Russian attitudes and policy since the eighteenth century.

This historical legacy plays an important role in Putin's vision for Russia's future. In the present day, however, it is Russia's potential as an energy producer and supplier that can transform the aspiration into a reality, to the extent that the Russian global superpower of the future must be precisely an "energy superpower." As Simonov (2006) explained in the introductory pages of his text,

> Every person wants to live in a great country. Citizens of the Russian Federation are no exception to this. Many people call on us to abandon our ambitious plans, to make peace with the fate of a small state, to whose opinion the rest of the world has no need to pay attention. . . . But in our case this is hardly possible, if only for the reason that the largest country in the world in territorial terms cannot remain at the periphery of world politics. We want our country's voice to be weighty and significant. (5)

The country needs projects, he continued, that will allow it "to pull Russia by its hair out of the swamp, and return it to its former status." To achieve this, Russia must develop its competitive advantages precisely as an energy superpower. This is not merely a plan for economic development and national enrichment, he emphasized, but also a "vital geopolitical task for the country, the solution of which could [help us] regain our lost status and with it our role in global political processes" (Simonov 2006, 6). As an energy superpower, Russia would necessarily be at the very heart of a new global regime of energy security, in which capacity it would act as the leading guarantor of international development and stability (Simonov 2006).

Energy and the Vision of Regional Domination

The Putinist discourse of energy superpower maintains that Russia's energy resources provide an opportunity for the country to reshape balances of power on a global scale. Although Russia currently directs much of its fossil fuel export toward the West, there is no inherent need for it to indulge in this preferential

"Eurocentrism," for there are rich and beckoning markets in Asia as well (Simonov 2006). The improvement of Russia's political and diplomatic relations with these latter regions was an important and highly visible aspect of Putin's international strategy from the outset, and Russia's energy potential plays a vital role in this endeavor. The Russian energy sector, reckoned one loyal analyst, could provide the basis for a "strategic triangle," in other words an anti-Western block between Russia and the up-and-coming—and energy-hungry—powers of India and China (Buszynski 2006; Pant 2006). "A union of the bear, the elephant, and the dragon could become a real nightmare for the West" observed Simonov (2006, 131, 138). The rapid economic growth of China and India, he argued, suggests that in future there will be insufficient energy resources for these leading global powers. This, in turn, ensures that the struggle between the United States and China (or, more broadly, between the United States and the EU on the one hand and India and China on the other hand) for world domination will be highly oriented toward energy. In these terms, he concluded that energy has become the supreme strategic weapon, for once a competitor is denied the energy resources it needs for its development, "the geopolitical struggle is won." The position of China is particularly vulnerable in this regard, insofar as without a guaranteed supply of fossil fuels its very survival is doomed. In view of these circumstances, "*Russia's significance is sharply enhanced. The development of the global political process depends on its decision as to which direction to send its oil and gas*" (Simonov 2006, 10; emphasis in original).

The great-power aspirations of Putinism are refracted regionally as well, through the relationship of Russia (as the Russian Federation) to the "Near Abroad." The breakup of the Soviet Union into sovereign independent states shattered the geopolitical unity of the traditional space of the Russian imperial state, over which Russians had exercised authority for many centuries. It is thus unsurprising that the project of resurrecting Russia's great-power status is linked in principle to reestablishing Russia's dominating position in this particular geographical arena. Once again, energy in its various manifestations is directly implicated in this striving for regional hegemony. Ironically, the deployment of energy in this manner today is only possible using a material infrastructure and set of relationships that were developed in the Soviet period for the very different purposes of integrated and "balanced" economic development of Soviet economic space. Thus, although newly independent states such as Kazakhstan, Turkmenistan, and Azerbaijan have become major energy producers in their own right, they remain dependent on the old Soviet pipeline infrastructure—based largely on Russian territory and under Russian control—to export their products. This dependency of the post-Soviet periphery on the Russian center can also be seen in the reliance of Ukraine, Belarus, Georgia, Armenia, and other states on Russia for their energy supplies—a pattern of dependency that carries over from Soviet times. Significantly, however, Russian dominance is now challenged by the fact that Russia's energy exports to Europe make heavy use of pipelines that, because they run across the territory of certain newly independent states, are subject to the jurisdiction and control of the latter.

In view of all this, the energy superpower discourse is emphatic in its insistence on the imperative for Russia to reassert its "historical rights" and prerogatives over all, or at least most, of the spaces of the former Soviet Union. On high diplomatic levels, this aspiration can be modulated and legitimated in the benign terms of a putative primordial fraternity and "commonality" ('*obshchnost*') between the Russians and the other post-Soviet nations—a rhetoric that not insignificantly finds an echo in the Eurasiaism that has served as official state doctrine in resource-rich Kazakhstan since the 1990s (Nazarbaev 1996; Kokoshin 2006). For most commentators, however, these sorts of rationalizations are dismissed in favor of more hard-headed assertions of exclusive Russian national interest. The establishment of Russian control over the energy infrastructures of the former Soviet Union is a vital "step toward achieving energy superpower status," affirmed Dmitrii Orlov. He described this as a matter of *Realpolitik* that cannot be achieved through idle "conversations about 'Slavic brotherhood' [between Russia, Ukraine and Belarus] and 'historical commonality'" (Orlov 2006a; see also Markedonov 2006). Simonov comes down yet more emphatically on this point. "Russia should clearly indicate its political interests in Central Asia . . . and in the European parts of post-Soviet space, because these territories are critical" if Russia is to achieve the sort of leading position on world energy resource markets that it seeks. Specifically, this involves Russian control over the pipelines that deliver its supplies to European markets—"even if they run across the territory of neighboring states"—together with Russia's "[complete] political domination" over Central Asia. "If we are to call things by their real names, this is one of the necessary conditions if Russia is to become an energy superpower" (Simonov 2006, 9, 26, 71).

It is well appreciated that Russia's great-power ambitions cannot be met without a thorough overhaul of the decrepit industrial infrastructure inherited from the Soviet Union. Thus—and again echoing a theme that has been part of Russian thinking about national development for over two centuries—the project of economic and industrial modernization forms a central element of the Putinist vision. As part of this, energy is now invoked as a vital and necessary foundation. The entire question as to the role energy provision should play in the modernization project is a major point of political debate in Russia, with highly influential economists and others arguing that the country's national development risks being undermined by the notorious "Dutch disease"; that is, an overdependence on the easy profits of a single resource that ultimately works to distort and undermine economic and industrial modernization rather than foster it (Gaidar 2006; Iavlinskii 2006; Shevtsova 2006). These critics emphasize the immediate threat of this syndrome—renamed by the former presidential adviser Andrei Illarionov using a more relevant contemporary geographical reference as the "Venezuelanization" of the Russian economy—and insist that in terms of it Russia's energy resources can be seen to represent a profound vulnerability rather than a national strength. The prioritization of energy production, they maintain, will convert Russia not into a "petroleum heaven" (*neftianyi rai*) but rather a resource "fringe" that is weak and dependent on the leading world economies ("Plius venesuelizatsiia vsei strany" 2005; Milov 2006).

The proponents of energy superpower, however, remain undaunted by the grim implications of the Dutch disease, and depict a very different scenario. This begins with Vladimir Putin himself, who declared that a dominant energy sector will work to "reanimate" Russian industrial growth and "become the major catalyst for the modernization and qualitative development of the entire economy of the Russian Federation" (Putin 2005). Specifically, the argument is that over a certain period of time (generally counted in decades), abundant revenues from energy sales can be used to stimulate the creation of "growth poles" that will provide the foundation for a more general transition. Leonid Grigor'ev, the articulate director of a leading think-tank devoted to energy and economic development, dismissed apprehensions about overreliance on the energy sector, maintaining that, to the contrary, energy exports represented the only available "source of assets" that can support Russia's necessary "switch to European-style capitalism," which the country had failed to achieve in the 1990s (Simonov 2006, 7; "How Russia's energy superpower status can bring supersecurity and superstability" 2006). Indeed, for some adherents, precisely this anticipated "reanimation" of Russian industry represents "the most powerful argument for the intensification of the role of the energy sector" and the creation of an energy superpower (Kokoshin 2006; Veletminskii 2006).

There is, however, a darker side to the Putin project of national rejuvenation, a subtheme of peril and danger. In the final analysis, Putinism sees international relations as driven not by international cooperation but rather by antagonism and rivalry between contending states. This means that the Russian *derzhava* of the future will operate in a hostile global environment, where powerful competitors seek constantly to undermine it and secure their own advantage at its cost. Here once again, the energy superpower discourse is implicated in a fundamental manner, both to illustrate the nature of the challenge as well as to identify a means of overcoming it. It depicts the international energy economy as a zero-sum game, indeed, a "state of war" driven by competition between the producers of energy on the one hand and between producers and consumers on the other (Simonov 2006, 9, 123, 130). The overriding significance of energy, moreover, means that this competition will not necessarily be played out in accordance with mutually accepted rules of international commerce and arbitration.

Thus, along with the advantages of Russia's abundant energy resources come risks that are "extremely high, namely that all other powers will use all means at their disposal to try to snatch away our 'oil barrel'" (Simonov 2006, 124–25). Russia has most to fear from being "torn apart" by its traditional opponents. The most acute challenge comes from the east, as the positive picture we saw earlier of an anti-Western block between Russia and China is turned upside down with a characterization of the latter as an uncouth and resource-starved superpower, intent not merely on curtailing Russia's independence but actually conquering and annexing its territory. The means for this conquest need not be military, for it can quite easily be accomplished by "creeping demographic expansion," using masses of Chinese agriculturalists to occupy the sparsely settled expanses of eastern Siberia and the Russian Far East and establish effective control. These remote territories, which contain immense stores of energy resources, are today "only nominally part of Russian economic space," and in the future they "risk being taken away (*vyvedennyi iz sostava*) from the Russian Federation" altogether (Siminov 2006, 9–10).

This prospect of an imminent "surrender of sovereignty" is not mere alarmism, moreover, as the experience of the breakup of the USSR—also deemed delusional to many people until it actually happened—clearly indicates (Simonov 2006, 56–57, 132). The only way to avoid this threat is for Russia to treat its energy resources not merely as a commercial commodity but also explicitly as a *geopoliticheskoe oruzhie* or "geopolitical weapon." In this spirit, one highly placed official in the energy industry claimed in conversation with a Western specialist that the country's energy resources play the same role for Russia today that nuclear weapons played for the Soviet Union (Legvold 2008). This weapon needs to be deployed in two ways: the urgent expansion of energy production, primarily through the development of reserves in the Russian East and the continued determination that all aspects of Russia's distribution strategy will be dictated single-mindedly by Russia's national interest. Nothing less than Russia's "national survival" is at stake (Simonov 2006, 11, 122).

Conclusion

In this article, we have highlighted the manner in which discourses associated with particular state-building initiatives and projects are grounded in the infrastructural realities associated with energy production patterns and hydrocarbon export dependencies across national boundaries. An exploration of the Putinist vision of Russia as an "energy superpower" has helped us pinpoint some of the practices that allow notions of national identity to be socially produced—and reproduced—through geographical imaginations and infrastructural materialities in the given context. By constructing a discourse coalition predicated on particular notions of national identity, Putinism has bound the "energy superpower" narrative, with the physical disposition of socio-technical networks for the conversion and circulation of energy. We would argue that this entanglement has created a hydrocarbon landscape with discursive and material aspects that are simultaneously parts of an infrastructurally grounded vision of national identity.

The existence of this spatial and political link, we would argue, points to the need for a continued critical engagement with the multiple material sites where energy and identity interact and are coproduced. Geographers have taken only a marginal interest in these questions to date, and a common framework to study them is still lacking. The broader purpose of our contribution, therefore, has been to underscore the need for the creation of such a theoretical matrix, while pointing to the importance of the relationship between the material and the imagined in the context of energy flows and infrastructure.

Acknowledgments

Stefan Bouzarovski wishes to gratefully acknowledge the support provided by the Ministry of Education, Youth and Sports of the Czech Republic, Project No. MSM0021620831, "Geographic Systems and Risk Processes in the Context of Global Change and European Integration." He is also an External Professor at the Department of Economic Geography, Institute of Geography, Faculty of Oceanography and Geography, Bażyńskiego 4, 80–952 Gdańsk, Poland.

References

Aron, L. 2008. Putinism. Washington, DC: American Enterprise Institute for Public Policy Research. http://www.aei.org/outlook/27958 (last accessed 9 January 2010).

Banerjee, A. 2003. Electricity: Science fiction and modernity in early-twentieth-century Russia. *Science Fiction Studies* 30:49–71.

Beichman, A. 2007. The perils of Putinism. *Hoover Digest* 2. http://www.hoover.org/publications/digest/7468137.html (last accessed 2 March 2010).

Bouzarovski, S. 2009. East-central Europe's changing energy landscapes: A place for geography. *Area* 41:452–63.

———. 2010. Entangled boundaries, scales and trajectories of change: Post-communist energy reforms in critical perspective. *European Urban and Regional Studies* 17:167–82.

Bouzarovski, S., and M. Konieczny. 2010. Landscapes of paradox: Public discourses and state policies in Poland's relationship with the Nord Stream pipeline. *Geopolitics* 15:1–21.

Boykoff, M. 2007. Flogging a dead norm? Newspaper coverage of anthropogenic climate change in the United States and United Kingdom from 2003 to 2006. *Area* 39:470–81.

Bradshaw, M. 2009. The geopolitics of energy security. *Geography Compass* 3:1920–37.

Bulkeley, H. 2000. Discourse coalitions and the Australian climate change policy network. *Environment and Planning C: Government and Policy* 18:727–48.

Buszynski, L. 2006. Oil and territory in Putin's relations with China and Japan. *The Pacific Review* 19:287–303.

Butler, J. 1999. *Gender trouble: Feminism and the subversion of identity*. London and New York: Routledge.

Byrne, J., and N. Toly. 2006. Energy as a social project: Recovering a discourse. In *Transforming power: Energy, environment, and society in conflict*, ed. J. Byrne, N. Toly and L. Glover, 1–34. New Brunswick, NJ: Transaction.

Carmody, P. R., and F. Y. Owusu. 2007. Competing hegemons? Chinese versus American geo-economic strategies in Africa. *Political Geography* 26:504–24.

Central Intelligence Agency. 2008. *CIA world factbook 2009*. Washington, DC: CIA.

Cohen, S., D. Demeritt, J. Robinson, and D. Rothman. 1998. Climate change and sustainable development: Towards dialogue. *Global Environmental Change* 8:341–71.

Devine-Wright, P. 2007. Energy citizenship: Psychological aspects of evolution in sustainable energy technologies. In *Governing technology for sustainability*, ed. J. Murphy, 63–88. London: Earthscan.

Dooley, B. 2006. Introduction. In *Energy and culture: Perspectives on the power to work*, ed. B. Dooley, xv–xxiv. Aldershot, UK: Ashgate.

Dryzek, J. 2005. *The politics of the earth: Environmental discourses*. Oxford, UK: Oxford University Press.

Energy Information Administration. 2010. Russia energy data, statistics and analysis—Oil, gas, electricity, coal. http://www.eia.doe.gov/cabs/russia.html (last accessed 20 August 2010).

Euractiv. 2010. EU-Russia energy dialogue. http://www.euractiv.com/en/energy/eu-russia-energy-dialogue/article-150061 (last accessed 7 March 2010).

Fairclough, N. 1995. Critical discourse analysis. In *How to analyse talk in institutional settings: A casebook of methods*, ed. A. W. McHoul and M. Rapley, 25–38. London: Continuum International.

———. 2001. Critical discourse analysis as a method in social scientific research. In *Methods in critical discourse analysis*, ed. R. Wodak and M. Meyer, 121–38. London: Sage.

———. 2005. Discourse in processes of social change: "Transition" in central and eastern Europe. *British and American Studies* 9:9–34.

Frankel, E. 1981. Energy and social change: An historian's perspective. *Policy Sciences* 14:59–73.

Gaidar, E. 2006. Neftianoe prokliatie [The petroleum curse]. *Ezhednevyi Zhurnal* 26 February. http://ej.ru/experts/entry/3114/ (last accessed 4 February 2010).

Galbreath, D. J., and D. Auers. 2009. Green, black and brown: Uncovering Latvia's environmental politics *Journal of Baltic Studies* 40:333–48.

Gamson, W. A., and A. Modigliani. 1989. Media discourse and public opinion on nuclear power: A constructionist approach. *American Journal of Sociology* 95:1–38.

Gjøen, H., and M. Hård. 2002. Cultural politics in action: Developing user scripts in relation to the electric vehicle. *Science, Technology, and Human Values* 27:262–81.

Glaser, L. S. 2009. *Electrifying the rural American West: Stories of power, people, and place*. Lincoln: University of Nebraska Press.

Goldblatt, D. L. 2005. *Sustainable energy consumption and society*. Dordrecht, The Netherlands: Springer.

Graham, S., and S. Marvin. 2001. *Splintering urbanism: Networked infrastructures, technological mobilities and the urban condition*. London and New York: Routledge.

Hamilton, P. 2002. The greening of nationalism: Nationalising nature in Europe. *Environmental Politics* 11:27–48.

Hecht, G. 1998. *The radiance of France: Nuclear power and national identity after World War II*. Cambridge, MA: MIT Press.

Heffner, R. R., K. S. Kurani, and T. S. Turrentine. 2007. Symbolism in California's early market for hybrid electric vehicles. *Transportation Research Part D: Transport and Environment* 12:396–413.

Hill, F. 2002. Russia: The 21st century's energy superpower? Washington, DC: Brookings Institution. http://www.brookings.edu/articles/2002/spring_russia_hill.aspx (last accessed 3 February 2010).

———. 2004. *Energy empire: Oil, gas, and Russia's revival*. London: The Foreign Policy Centre.

Högselius, P. 2006. Connecting East and West: Electricity systems in the Baltic Region. In *Networking Europe: Transnational infrastructures and the shaping of Europe, 1850–2000*, ed. E. van der Vleuten and A. Kaijser, 245–75. Cambridge, MA: Science History Publications.

———. 2009. The internationalization of the European electricity industry: The case of Vattenfall. *Utilities Policy* 17:258–66.

How Russia's energy superpower status can bring supersecurity and superstability. 2006. *Civil G8* 31 May. http://en.civilg8.ru/priority/energy/2054.php (last accessed 8 April 2009).

Huber, M. 2009. Energizing historical materialism: Fossil fuels, space and the capitalist mode of production. *Geoforum* 40:105–15.

Hughes, T. P. 1993. *Networks of power: Electrification in Western society, 1880–1930*. Baltimore: Johns Hopkins University Press.

Iavlinskii, G. 2006. Rossiia otvechaet luchshim dvoinym mirovym standartom [Russia meets the best double world standard]. *Novaia Gazeta* 17 July. http://www.yabloko.ru/Publ/2006/2006_07/060717_novg_yavl.html (last accessed 2 March 2009).

International Energy Agency. 2010. IEA energy statistics—Energy balances for Russian Federation. http://www.iea.org/stats/index.asp (last accessed 20 August 2010).

Irgunov, V. 2006. Energeticheskaia sverkhderzhava ili ekonomicheskii lider? [An energy superpower or an economic leader?] *Vremia Novostei* 24 January. http://www.vremya.ru/2006/10/4/143607.html (last accessed 3 April 2009).

Jasanoff, S., and S.-H. Kim. 2009. Containing the atom: Sociotechnical imaginaries and nuclear power in the United States and South Korea. *Minerva* 47:119–46.

Jessup, D. E. 2008. J. E. Bernier and the assertion of Canadian sovereignty in the Arctic. *American Review of Canadian Studies* 38:409–27.

Jiusto, S. 2009. Energy transformations and geographic research. In *A companion to environmental geography*, ed. N. Castree, D. Demeritt, D. Liverman, and B. Rhoads, 533–51. Oxford, UK: Blackwell.

Jones, C. O. 1979. American politics and the organization of energy decision making. *Annual Review of Energy* 4:99–121.

King, L. 2007. Charting a discursive field: Environmentalists for U.S. population stabilization. *Sociological Inquiry* 77:301–25.

Klare, M. T. 2008. *Rising powers, shrinking planet*. New York: Metropolitan Books.

Kokoshin, A. 2006. Rossiia: Energeticheskaia sverkhderzhava [Russia: An energy superpower]. http://www.edinros.ru/text.shtml?4/9394 (last accessed 24 April 2008).
Kopeček, L. 2009. The Slovak Greens: A complex story of a small party. *Communist and Post-Communist Studies* 42:115–40.
Kuhn, R. G. 2008. Canadian energy futures: Policy scenarios and public preferences. *Canadian Geographer/Le Géographe canadien* 36:350–65.
Laurila, J. 2003. Transit transport between the European Union and Russia in light of Russian geopolitics and economics. *Emerging Markets Finance and Trade* 39:27–57.
Legvold, R. 2008. Russia's strategic vision and the role of energy. *NBR Analysis* 19:9–22.
Leont'ev, M. 2006. Kontsept "Rossiia kak energeticheskaia sverkhderzhava" [The "Russia as an energy superpower" concept]. *Russkii Zhurnal* 27 October. http://www.russ.ru/pole/Koncept-Rossiya-kak-energeticheskaya-sverhderzhava (last accessed 3 April 2009).
Lovell, H. 2008. Discourse and innovation journeys: The case of low energy housing in the UK. *Technology Analysis and Strategic Management* 20:613–32.
Luedicke, M. K., C. J. Thompson, and M. Giesler. 2010. Consumer identity work as moral protagonism: How myth and ideology animate a brand-mediated moral conflict. *Journal of Consumer Research* 36:1016–32.
Mander, S. 2008. The role of discourse coalitions in planning for renewable energy: A case study of wind-energy deployment. *Environment and Planning C: Government and Policy* 26:583–600.
Markedonov, S. 2006. Energeticheskaia sverkhderzhava: Nastuplenie v nikuda [An energy superpower: An offensive leading nowhere]. *APN Kazakhstan*. http://www.apn.kz/news/print7338.htm (last accessed 30 December 2009).
McQuire, S. 2005. Immaterial architectures. *Space and Culture* 8:126–40.
Migranyan, A. 2004. What is "Putinism"? *Russia in Global Affairs* 2:28–45.
Milov, V. 2006. Mozhet li Rossiia stat' neftianym raem? [Can Russia become a "petroleum heaven"?] *Pro et Contra* March–June:6–15.
Montgomery, S. 2010. *The powers that be: Global energy for the twenty-first century and beyond*. Chicago: University of Chicago Press.
Morrison, D. E., and D. G. Lodwick. 1981. The social impacts of soft and hard energy paths: The Lovins' claims as a social science challenge. *Annual Review of Energy* 6:357–78.
Murphy, J. 2007. Sustainability: Understanding people, technology and governance. In *Governing technology for sustainability*, ed. J. Murphy, 3–24. London: Earthscan.
Nasaw, D. 1992. Cities of light, landscapes of pleasure. In *The landscape of modernity: Essays on New York City, 1900–1940*, ed. D. Ward and O. Zunz, 273–86. New York: Russell Sage Foundation.
Nazarbaev, N. 1996. Evraziiskii soiuz: strategiia integratsii [The Eurasian Union: The strategy of integration]. *Evraziia* 1:3–8.
Nye, D. E. 1990. *Electrifying America: Social meanings of a new technology, 1880–1940*. Cambridge, MA: MIT Press.
———. 1999. *Consuming power: A cultural history of American energies*. Cambridge, MA: MIT Press.
O'Lear, S. 2004. Resources and conflict in the Caspian Sea. *Geopolitics* 9:161–86.
Orlov, D. 2006a. Byt' li Rossii "energeticheskoi sverkhderzhavoi"? [Is Russia destined to become an "energy superpower"?] *Izvestiia* 17 January. http://www.izvestia.ru/comment/3054583_print (last accessed 1 May 2008).
———. 2006b. Neft' v obmen na demokratiiu [Petroleum in exchange for democracy]. *Finansovye Izvestiia* 14 July. http://www.finiz.ru/cfin/tmpl-art/id_art-1032607 (last accessed 15 December 2010).
Pant, H. V. 2006. The feasibility of the Russia–China–India "strategic triangle": An assessment of theoretical and empirical issues. *International Studies* 43:51–72.
Paterson, M., and S. Dalby. 2006. Empire's ecological tyreprints. *Environmental Politics* 15:1–22.
Perelman, L. J. 1980. Speculations on the transition to sustainable energy. *Ethics* 90:392–416.
Plius venesuelizatsiia vsei strany [Plus the Venezuelaization of the entire country]. 2005. *Vremiia Novostei* 58 (6 April). http://www.vremya.ru/2005/58/4/122066.html (last accessed 2 January 2010).
Poputoaia, D., and S. Bouzarovski. 2010. Regulating district heating in Romania: Legislative challenges and energy efficiency barriers. *Energy Policy* 38:3820–29.
Potter, E. 2009. Calculating interests: Climate change and the politics of life. *M/C Journal* 12. http://journal.media-culture.org.au/index.php/mcjournal/article/view/182 (last accessed 4 April 2011).
Putin, V. V. 2000. Mineral'no-syrevye resursy v strategii razvitiia Rossiiskoi ekonomiki [Mineral and raw material resources in the strategy for the development of Russia's economy]. In *Rossiia v okruzhaiushchem mire*, ed. N. N. Marfenin, 72–79. Moscow: Azist.
———. 2005. Vstupitel'no slovo na zasedanii Soveta Besopastnosti po voprosu o roli Rossii v obespechenii mezhdunarodnoi energeticheskoi besopastnosti [An introductory word at the security council's meeting on the question of Russia's role in providing international energy security]. 22 December. http://archive.kremlin.ru/text/appears/2005/12/99294.shtml (last accessed 2 January 2010).
Putin deplores collapse of the USSR. 2005. *BBC News* 25 April. http://news.bbc.co.uk/1/hi/4480745.stm (last accessed 3 February 2010).
Rahn, R. 2007. From communism to Putinism. *The Brussels Journal* 21 September. http://www.brusselsjournal.com/node/2501 (last accessed 3 February 2010).
Rutland, P. 2008. Russia as an energy superpower. *New Political Economy* 13:203–10.
Sahr, R. 1985. *The politics of energy policy change in Sweden*. Ann Arbor: University of Michigan Press.
Saivetz, C. R. 2007. Russia: An energy superpower? Cambridge, MA: MIT Center for International Studies. http://web.mit.edu/cis/editorspick_saivetz07_audit.html (last accessed 3 February 2010).

Shevtsova, L. 2006. Bessilie Putina [Putin's weakness]. *demokratiia.ru* 12 July. http://www.democracy.ru/library/newsarchive/article.php?id=1131 (last accessed 2 January 2010).

Simonov, K. V. 2006. *Energeticheskaia sverkhderzhava* [An energy superpower]. Moscow: Algoritm.

Smith, A., and F. Kern. 2009. The transitions storyline in Dutch environmental policy. *Environmental Politics* 18:78–98.

Smith, E. R. A. N. 2002. *Energy, the environment, and public opinion*. Lanham, MD: Rowman and Littlefield.

Star, S. L. 1999. The ethnography of infrastructure. *American Behavioral Scientist* 43:377–91.

Szarka, J. 2004. Wind power, discourse coalitions and climate change: Breaking the stalemate? *European Environment* 14:317–30.

Veletminskii, I. 2006. S TEKom napereves [Forward with TEK]. *Rossiiskaia gazeta* 4124 (21 July). http://www.rg.ru/2006/07/21/eksperty-energoderzhava.html (last accessed 1 May 2008).

Walker, G., and N. Cass. 2007. Carbon reduction, "the public" and renewable energy: Engaging with sociotechnical configurations. *Area* 39:458–69.

Watts, M. 2004. Resource curse? Governmentality, oil and power in the Niger Delta, Nigeria. *Geopolitics* 9:50–80.

Wenger, E. 1998. *Communities of practice: Learning, meaning and identity*. Cambridge, UK: Cambridge University Press.

Whitmore, B. 2007. Putin may go, but can Putinism survive? *Radio Free Europe/Radio Liberty* 29 August. http://www.rferl.org/content/article/1078413.html (last accessed 3 February 2010).

The Changing Structure of Energy Supply, Demand, and CO_2 Emissions in China

Michael Kuby,* Canfei He,† Barbara Trapido-Lurie,* and Nicholas Moore‡

*School of Geographical Sciences and Urban Planning, Arizona State University
†College of Urban and Environmental Sciences, Peking University
‡Department of Geography, Hunter College–CUNY

Because of its enormous population, rapid economic growth, and heavy reliance on coal, China passed the United States as the world's largest source of CO_2 emissions in 2006. China is also becoming a major factor in the global oil market. This article analyzes China's energy production and consumption, with a focus on the energy and CO_2 emissions per capita and per unit of gross domestic product (GDP) and the mix of energy sources and end uses. Energy flow diagrams for 1987 and 2007 make it possible to visualize the allocation of energy from sources through energy transformation to final uses in units of metric tons of coal equivalent. Declining coal use by residences, agriculture, and transportation has been more than offset by a massive increase in electricity and industry usage. The article places these changes in political–economic context and helps illustrate and explain the difficulties China faces in trying to reduce its absolute CO_2 emissions and why it instead proposes to reduce its CO_2 per unit of GDP. *Key Words: China, coal, electricity, energy flow diagram, energy intensity.*

由于其人口众多，快速的经济增长，和对煤炭的严重依赖，中国在 2006 年超过美国成为世界上最大的二氧化碳排放源。中国也正成为一个全球石油市场的主要因素。本文分析了中国的能源生产和消费，着重于人均和单位国内生产总值（GDP）的能源和二氧化碳排放量，以及能源的来源和最终用途的混合。1987 和 2007 年的能流图使通过能量转换到最终用途的，以吨标煤为单位的能源配置可视化成为可能。住宅，农业，和运输煤炭下降的使用已被在电力和工业界巨型的增量使用过度抵消了。文章把这些变化放置于政治和经济的背景，并有助于说明和解释中国在试图减少其绝对二氧化碳排放量面临的困难，和为什么它建议减少其单位 GDP 的，而不是绝对的二氧化碳排放量。*关键词：中国，煤炭，电力，能流图，能源强度。*

Debido a su enorme población, rápido crecimiento económico y alta dependencia en carbón mineral, China sobrepasó a Estados Unidos en 2006 como la fuente más grande de emisiones de CO_2. China también se está convirtiendo en un factor mayor del mercado global del petróleo. Este artículo analiza la producción y consumo de energía en China, enfocando el tema de la energía y las emisiones de CO_2 per cápita y por unidad del producto nacional bruto (PNB) y la mezcla de fuentes de energía y usos finales. Los diagramas de flujos de energía para 1987 y 2007 hacen posible visualizar la asignación de energía desde las fuentes, a la transformación de la energía, hasta los usos finales en unidades equivalentes a toneladas métricas de carbón. La disminución del uso de carbón en residencias, agricultura y transporte ha sido poco menos que opacado por un incremento masivo en su uso para electricidad e industria. El artículo coloca estos cambios en contexto político–económico y ayuda a ilustrar y explicar las dificultades que enfrenta China al tratar de reducir sus emisiones de CO_2 en términos absolutos y por qué en su defecto propone reducir su CO_2 por unidad de PNB. *Palabras clave: China, carbón, electricidad, diagrama del flujo de energía, intensidad energética.*

Economic reforms have liberalized the Chinese economy, resulting in remarkable economic growth, structural transformation, and energy consumption since the late 1970s. China's energy consumption doubled twice to 2.65 billion tons of coal equivalent (tce) in the last twenty-five years (National Statistical Bureau [NSB] 2008). China has long been the world's largest producer and consumer of coal and now uses 39 percent of the world's total. In 2006, China passed the United States as the world's top CO_2 emitter, with 6.1 billion tons of annual emissions, and by 2008 had already outdistanced the United States by 1.5 billion tons (Figure 1). Then, in July 2010, the International Energy Agency (IEA) announced that China had, in 2009, passed the United States to become the world's largest energy consumer. China's National Energy Administration disputed the claim, citing a lack of knowledge "about China's latest developments of

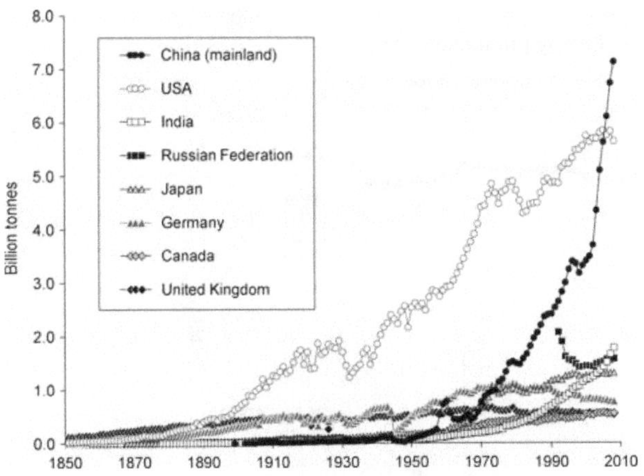

Figure 1. Top eight countries in annual CO_2 emissions, in order of 2008 emissions. Rounding out the top ten are Italy and Mexico. *Source:* Carbon Dioxide Information Analysis Center (CDIAC): http://www.cdiac.ornl.gov/ftp/ndp030/nation1751_2006.ems.

energy conservation and renewable energy" (Smith and Schmollinger 2010). Indeed, China has managed to use energy more efficiently, and its economy has become less energy intensive (He and Wang 2007; C. Ma and Stern 2008), but its phenomenal economic growth has made it one of only two countries in the world (along with Qatar) whose energy use doubled in the 2000s (British Petroleum 2010). Furthermore, China, a net oil importer since 1993, has become a force in the global energy market.

With the global focus on climate change, the energy consumption–CO_2 nexus has politicized China's energy issue. This nexus, however, depends critically on the structural change of energy consumption. The purpose of this study is to explore the structural change since the dawn of the People's Republic of China and especially over the last twenty years. We highlight changes in energy sources, consumption, scale, imports and exports, intensity, and emissions. These changes are presented using energy flow diagrams that allow us to visualize the entire energy system at a glance. These diagrams are available for some countries in some years (e.g., Lawrence Livermore National Laboratory 2010; U.S. Energy Information Administration [EIA] 2010; U.K. Department of Business Innovation and Skills 2010), but we know of none published for China for the 1990s or 2000s. This article supplements flow diagrams for 1987 and 2007 with other graphs and tables describing China's changing energy system.

Background on China's Energy

Many studies have been written on China's energy system. The main themes in the literature are energy efficiency and intensity, structural change within the energy sector, environmental impacts, policy reforms, and data problems. These studies have attributed China's energy-efficiency improvements to structural, technological, and institutional changes in the economy and, in particular, the energy sector (Kambara 1992; Huang 1993; World Bank 1994; Lin and Polenske 1995; Garbaccio, Mun, and Jorgenson 1999; Sinton and Fridley 2000; Thomson 2003; Z. X. Zhang 2003; Wu 2005; Fan, Liao, and Wei 2007; He and Wang 2007; H. Ma et al. 2008). H. Ma and Oxley (2009) reviewed the literature and concluded that changes in economic activities across sectors of the economy (structural change) was the larger cause before the 1990s, whereas declining energy intensity within a sector (technological change) has had greater impact more recently. Additional debate has centered on whether the fall in energy intensity stagnated or reversed in the 2000s (Fisher-Vanden et al. 2004; Liao, Fan, and Wei 2007; C. Ma and Stern 2008; Chai et al. 2009).

Many studies have highlighted interfuel shifts in the energy system, including the shift away from—and back toward—coal, largely driven by expansion of electricity use (Thomson 2003; H. Ma et al. 2008; H. Ma, Oxley, and Gibson 2009; Q. Wang, Qiu, and Kuang 2009). The growth in oil due to expanded road transport has attracted much attention (Kuby and Cook 1997; Downs 2000; IEA 2000; Cornelius and Story 2007; Konan and Zhang 2008). H. Ma, Oxley, and Gibson (2009) and Shi (2009) have looked at renewable energy's potential to increase supply without detrimental environmental impacts. Numerous studies have also highlighted the domestic and global environmental impacts of China's coal usage (World Bank 1994; Smil 1997; C. Wang, Chen, and Zou 2005).

Another theme has been reforms in the energy sector since Deng Xiaoping began to dismantle the Mao-era command economy and open China to the outside world (Thomson 2003; Wu 2003; Fan, Liao, and Wei 2007; He and Wang 2007). Levine (2005) divided China's energy policies into a pre-1980 Soviet-style period, an early reform period from 1980 to 1992 emphasizing price reforms and energy-use monitoring, and a transition period from 1993 to 2001 with rapid price increases and marketization. Glomsrød and Wei (2005) highlighted 1994, when the dual-price system was abandoned. Osnos (2009) called China's launch of the 863 Program its "Sputnik moment." The program, named after the year and month of its launch, boosted research and development in key sectors of the Chinese economy, including energy. Studies have highlighted

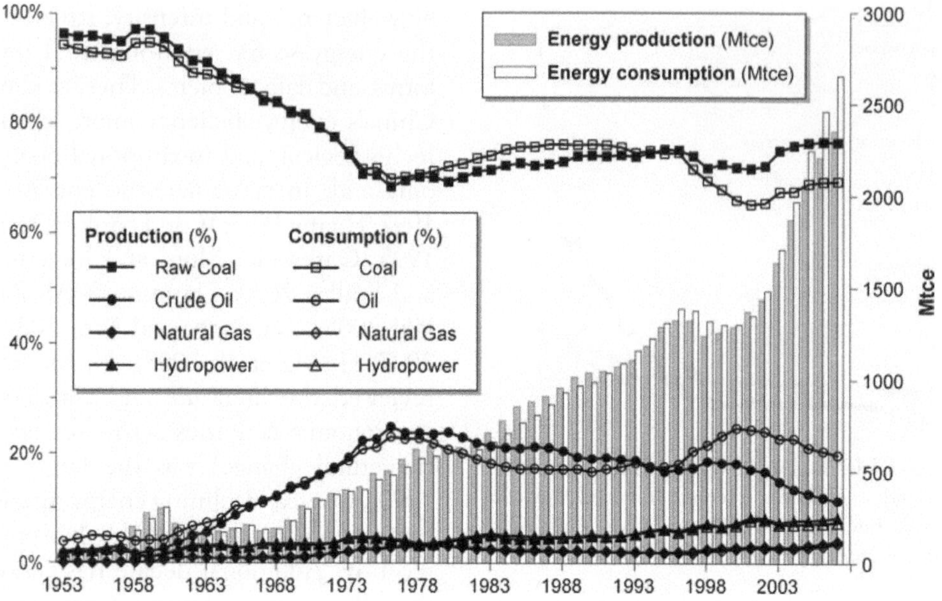

Figure 2. Energy production and consumption trends. Mtce = millions of tons of (standard) coal equivalent. *Source: China Energy Statistical Yearbook* (NSB 2008).

reforms related to regulation (Andrews-Speed, Dow, and Gao 2000), prices (Hang and Tu 2007), and renewable energy (Cherni and Kentish 2007; P. Zhang et al. 2009). H. Ma and Oxley (2009, 7) described energy reforms as "slow, incremental, and gradual" compared to other sectors. Other studies focused on causes of and solutions to energy shortages and transport bottlenecks (Kuby et al. 1995; Xie and Kuby 1997) from the late 1980s to the mid-1990s, and again more recently (Levine 2005).

Many data problems face researchers studying energy in China, but such problems are not unique to China (Farla and Blok 2001). Shi (2009) argued that Chinese energy statistics exclude biomass consumption and thus underestimate renewable energy. Chinese statistics might have overestimated economic growth and therefore underestimated energy intensity (Rawski 2001). Sinton (2001) found that data accuracy declined in the late 1990s, when economic growth skyrocketed without growth of energy consumption, and underreporting of coal production by small, nonstate mines was rampant. Sinton and Fridley (2002) explained that data for oil, gas, and hydro production are more reliable than for coal production and most energy consumption because the former are collected from full reports from large government-owned enterprises, whereas the latter also rely on surveys and estimates. In addition, up to half of transportation consumption data might be buried in the data for other sectors that use their own vehicles. Data issues could be one reason researchers have not developed energy flow diagrams for China until now.

Structural Changes of China's Energy System

Figure 2 portrays the growing totals and shifting shares of energy sources in production and consumption. Commercial production grew forty-five-fold and consumption forty-nine-fold from 1953 to 2007. In the reform era since 1978, consumption grew 5.4 percent annually, whereas production grew at 4.7 percent. Accession to the World Trade Organization (WTO) in 2001 led to an explosive expansion of industrial exports and greatly increased energy demand. Exponential growth of motor vehicles was another driving force: From 2003 to 2007, privately owned cars increased from 12.2 million to 28.8 million, an annual increase of 24 percent (NSB 2009).

Accompanying overall growth is structural change within China's energy system. Although coal dominates both production and consumption, its share has fluctuated, with 1976—the year Mao died—marking a watershed. In the 1950s, coal consistently accounted for over 95 percent of total energy production and over 92 percent of consumption, gradually dropping below 70 percent of both by 1976. Due to rapid economic growth since 1976, however, China has not been able to sustain this decline in coal's share, because coal is the

Table 1. Main sources of crude oil imports: 2007

Region/country	Million tons of oil	%	Region/country	Million tons of oil	%
Asia	84.0	51.50	Africa	53.0	32.51
Saudi Arabia	26.3	16.14	Angola	25.0	15.32
Iran	20.5	12.59	Sudan	10.3	6.32
Oman	13.7	8.38	Congo	4.8	2.94
Kazakhstan	6.0	3.68	Eq. Guinea	3.3	2.01
United Arab Emirates	3.7	2.24	Libya	2.9	1.78
Kuwait	3.6	2.22	South Africa	2.3	1.43
Yemen	3.2	1.99	Latin America	10.3	6.32
Indonesia	2.3	1.40	Venezuela	4.1	2.53
Europe	14.9	9.10	Brazil	2.3	1.42
Russia	14.5	8.90			

Source: Ministry of Commerce (2008).

one source that China can easily increase in the short term. Coal is available in many regions (although not in equal quality and quantity) and can be mined using labor-intensive methods. The 1971–1978 and 2001–2007 increases in coal's share correspond to periods of the most rapid total energy growth. In addition, the post-WTO economic growth generated huge demands for electricity, most of which was supplied by coal.

Since 1953, China's efforts to diversify its energy supply have met with mixed results. Hydropower grew from 2 percent of energy production in 1953 to 4.1 percent in 1976 to 8.2 percent in 2007.[1] Natural gas use in China began in 1957, and its shares of production and consumption, although still below 4 percent, have doubled since the mid-1990s.

As China modernizes and urbanizes, it increasingly depends on crude oil. In absolute terms, consumption has doubled almost seven times over since 1953, and oil has absorbed most of the share lost by coal. As Figure 2 shows, oil and coal have been mirror images of each other in terms of self-sufficiency. From the mid-1970s to the mid-1990s, China was a net coal importer and a net oil exporter, but since the mid-1990s it has become a net coal exporter and net oil importer. Oil production has been limited by a lack of economically feasible oil reserves (IEA 2007), and oil imports have grown at 23 percent annually since 1980. By 2007, China imported 211 million tons of oil and has diversified its foreign suppliers. Specifically, Saudi Arabia, Angola, and Iran top the list, followed by a second tier consisting of Russia, Oman, and Sudan (Table 1). An increasing dependence on foreign oil has made China central in the global energy market, leading to its attempt to purchase Marathon Oil, construction of an entire port city in Angola, and concerns about weapons sales to oil-producing countries (Downs 2000; Hanson 2008).

In 2007, fossil fuels as a group (coal, crude oil, natural gas) accounted for 93 percent of China's energy consumption, followed by hydropower and nuclear. Increasing importance of renewable energy from sunlight, biomass, wind, geothermal heat, and tidal and wave sources is anticipated in the near future as the Chinese government promotes renewable energy development (H. Ma, Oxley, and Gibson 2009; Shi 2009).

Although consumption has expanded rapidly, the Chinese economy has become less energy intensive (Figure 3). Energy intensity, defined as the ratio of total energy consumption to gross domestic product (GDP; computed here at constant 1952 prices), dropped

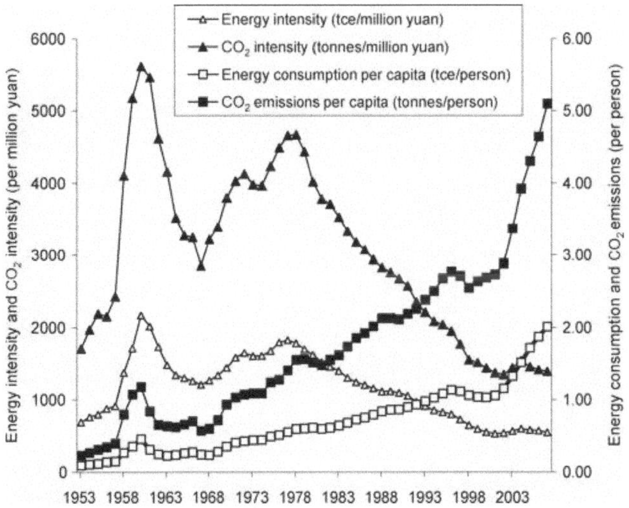

Figure 3. Energy consumption and CO_2 emissions per capita and energy and CO_2 intensity, 1953–2007. Tce = tons of coal equivalent. *Source:* Original data from *China Energy Statistical Yearbook* (NSB 2008) and Carbon Dioxide Information Analysis Center.

from 1,785 to 549 tce/million yuan from 1978 to 2007, for three main reasons. First, China's industrial structure has shifted from heavy industries toward export-oriented light industries (Kambara 1992; World Bank 1994). Second, technological change has lowered the energy required to produce a particular product of a particular sector (Huang 1993; Sinton and Levine 1994; Lin and Polenske 1995; Garbaccio, Mun, and Jorgenson 1999; Z. X. Zhang 2003). Third, market-oriented reforms (Sinton and Fridley 2000; Fisher-Vanden et al. 2006; Fan, Liao, and Wei 2007; He and Wang 2007) in the energy sector (He and Wang 2007) have incentivized energy efficiency. The slight increase in energy intensity after 2002 is associated with the heavy industrialization strategy in coastal provinces and a nationwide construction boom, which boosted production of energy-intensive steel, aluminum, and cement (Liao, Fan, and Wei 2007).

As energy intensity has declined since 1978, per capita consumption has increased. The opposite trends of these two indicators in Figure 3 are not contradictory because while energy per unit of GDP dropped threefold, GDP per capita increased almost fivefold.

Trends in CO_2 intensity and CO_2 emissions per capita mirror those of energy intensity and energy per capita (Figure 3). In percentage terms, the fall in CO_2 intensity and rise in CO_2 per capita are very similar to those for the equivalent energy curves. This confirms M. Zhang et al.'s (2009) CO_2 decomposition analysis, which showed that CO_2 emissions are driven largely by GDP and energy intensity. Figures 1 and 3 together show that a dramatic reduction in CO_2 emissions intensity can occur simultaneously with continued large increases in total CO_2 emissions, as long as the economy continues to grow faster than energy use.

Energy Flow Diagrams

Figures 4 and 5 show the energy flows from sources through several energy transformation processes to final end uses for 1987 and 2007. The latter year is the most recent for which data were available, whereas the former is twenty years earlier and representative of the early stages of reform. To put 1987 in historical context, China established the first Special Economic Zones in 1980, opened fourteen coastal cities to foreign investment in 1984, and expanded reforms to broader coastal zones by 1988. The first *China Energy Statistical Yearbook* was published in 1986, and the Ministry of Energy was established in 1988.

Figures 4 and 5 are drawn to the same scale, illustrating the growth in the size of China's energy system. Amounts of energy are shown in millions of tons of (standard) coal equivalent (Mtce). Although coal varies widely in heat content, standard coal is considered to contain 7,000 kcal/kg, or 29 GJ/t. In comparison, U.S. energy statistics report energy in quadrillion BTU,

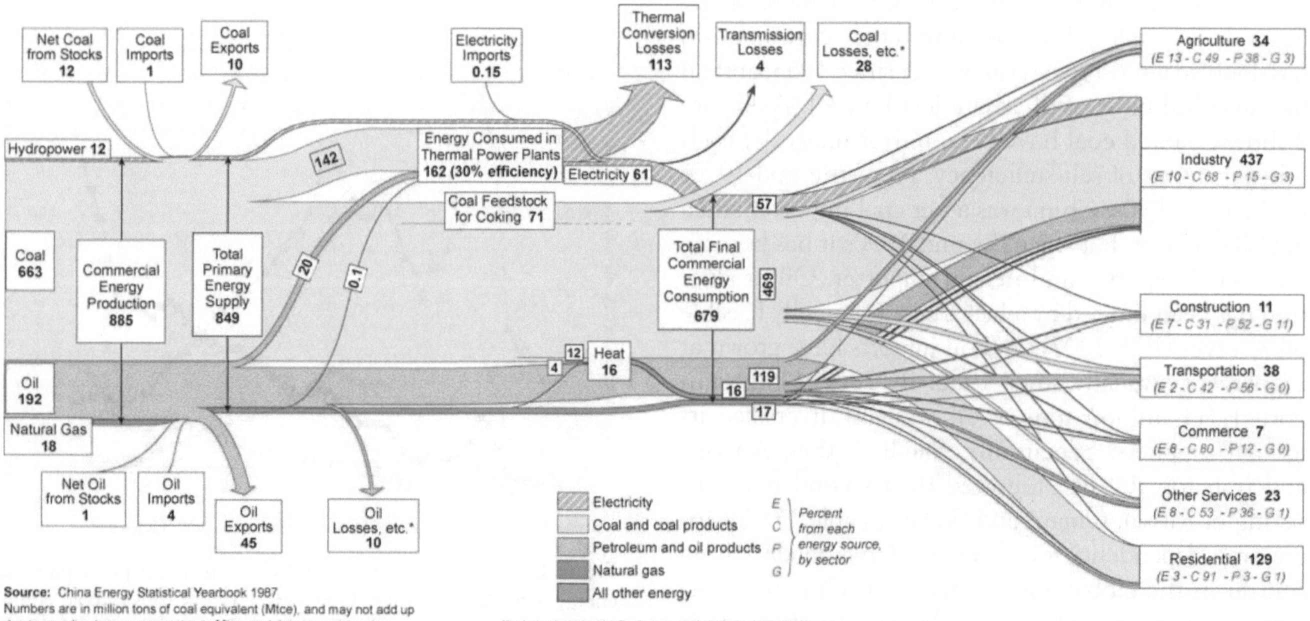

Source: China Energy Statistical Yearbook 1987
Numbers are in million tons of coal equivalent (Mtce), and may not add up due to rounding losses, conversion to Mtce, and data inconsistencies.

*Includes losses, nonfinal uses, and statistical inbalance

Figure 4. Energy flow diagram for 1987.

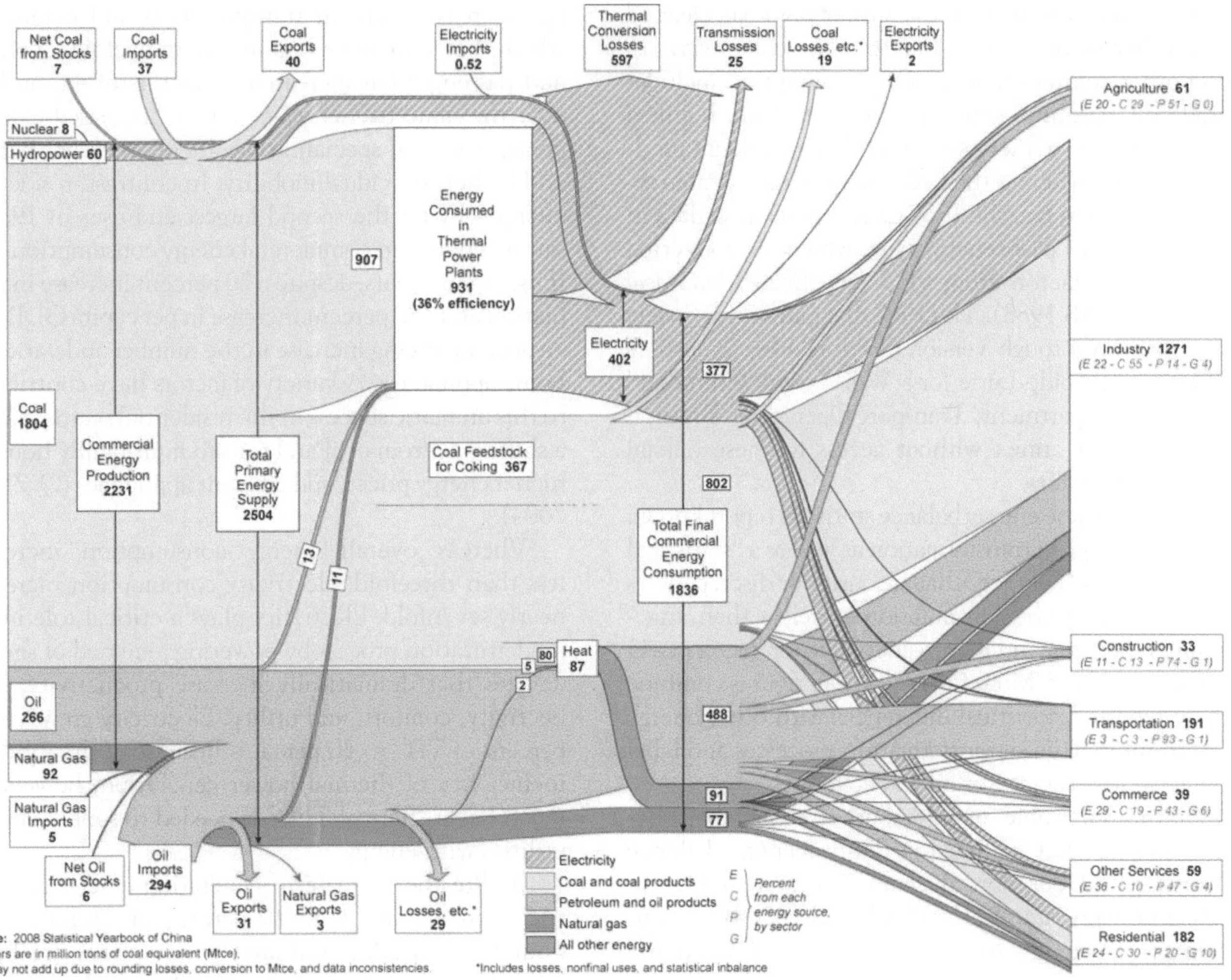

Figure 5. Energy flow diagram for 2007 (same scale as Figure 4).

where 1 quad = 36 Mtce. Note that most coal in China is not standard quality. Raw coal is considered to average 0.71 tce (or 5,000 kcal/kg), washed coal 0.90 tce, and coke 0.97 tce (NSB 2009).

At the far left of each figure is domestic commercial energy production, by resource type. Noncommercial energy, such as firewood or agricultural waste, is not included. Next comes imports, exports, and net change in stocks (long-term storage), resulting in a total amount of primary (raw) energy available for domestic use, or total primary energy supply (TPES). Next, raw energy is transformed into usable forms, such as electricity, refined petroleum products, coal products, and usable heat, with substantial energy losses. The resultant usable energy delivered to the end users is known as total final consumption (TFC).

The energy flow diagrams are based on data collected by township and county authorities, aggregated at the provincial level, and reported by the NSB (1988, 2008). Although known to suffer from some deficiencies, these are the same data on which the IEA and World Bank rely (Sinton and Fridley 2002).

Many assumptions had to be made to create these figures. First, on the production side, we show hydropower and nuclear power in terms of their electrical output. In some Chinese statistics, much higher numbers are given representing the energy value of the coal needed to generate an equivalent amount of electricity. Second, Chinese energy statistics include many transformation processes: thermal-power production, heating supply, coal washing, coking, petroleum refining, gas works, and briquetting, with complex flows among them such as raw coal to washed coal to coke to coke-oven gas to heat. To avoid cluttering the diagram, we show only thermal power, heat, and coking. Third, we included all products made from coal and oil as forms of coal and oil

TFC. Coal is consumed in the form of raw coal, cleaned coal, other washed coal, briquettes, coke, coke-oven gas, and other coking products. Oil consumption includes crude oil, gasoline, kerosene, diesel, fuel oil, liquified petroleum gas, refinery gas, and other petroleum products. Fourth, whereas the 2007 energy balance table reports all data in tce, the 1987 table reports most data in units native to each energy type, which we converted to tce using factors from the *China Energy Statistical Yearbook* (NSB 1988). The 1987 diagram in this article improves on a rough version developed by co-authors Kuby and Trapido-Lurie for a World Bank China and Mongolia Department, Transport Operations Division (1995) report annex without access to these official conversion factors.

Fifth, Chinese energy balance statistics report known losses of energy in transformation as well as a "statistical difference" or "balance" that accounts for discrepancies between supply and consumption as well as theft, inaccurate reporting, and incorrect conversion factors (–45 Mtce for 2007; 5 Mtce for 1987). Our figures combine these reported statistical differences with other energy lost in intermediate transformation processes and label them as "losses, etc." for oil and coal.

Sixth, renewable energy sources other than hydropower are not included in China's energy balance statistics and our flow diagrams. Separate tables in *China's Energy Statistical Yearbook* (NSB 2008) for noncommercial energy attribute 93 Mtce of firewood and 160 Mtce of plant stalks to rural residential consumption. Other sources list China's installed windpower capacity at 5.9 GW in 2007 (Xia and Song 2009). IEA (2007) estimated China's nonhydro, nonbiomass renewables at 0.17 percent of total consumption in 2005.

Discussion

Comparison of Figures 4 and 5 shows that total energy production grew 250 percent from 1987 to 2007, and TFC grew 270 percent. Within these totals, however, coal production nearly tripled, hydropower and natural gas production quintupled, and nuclear power came on line. Constrained by limited domestic reserves, oil production increased by only 38 percent, whereas oil consumption quadrupled. Increased oil imports made up the difference: From 1987 to 2007, China went from net exports of over 40 Mtce to net imports of 263 Mtce.

On the consumption side, industrial energy use nearly tripled. Industry consumes more than two thirds of TFC. From 1987 to 2007, the largest percentage increases were in transportation and commerce, which both increased more than fivefold. Ton-km and passenger-km increased by factors of 4.6 and 4.0 over the same twenty years, due to increased exports, greater regional specialization, increased urbanization, and higher individual mobility. In contrast, residential energy—by far the second largest end user in 1987—saw no change in commercial energy consumption over these twenty years, despite a 20 percent increase in population and 480 percent increase in per capita GDP, and an accompanying increase in the number and variety of home appliances. A variety of factors have contributed to this dramatic success in the residential sector, such as a shift away from coal and toward multifamily housing, higher energy prices, and efficient appliances (Q. Zhang 2004).

Whereas overall energy consumption increased less than threefold, electricity consumption increased nearly sevenfold. Electricity plays a critical role in the modernization process by powering a myriad of smaller devices that dramatically increase productivity, connectivity, comfort, and utility. Electricity grew from 8 percent of TFC to 20 percent. Because of the inherent inefficiency of thermal power generation, however, a much larger share of TPES is needed to supply electric utilities with energy.

Despite the increase in generating efficiency, electricity grew from 20 to 40 percent of China's TPES from 1987 to 2007. Oil-fired power plants are being phased out, as power from natural gas, which is better suited to serving peak-period loads and urban areas, has increased more than 100-fold. Despite the large percentage increases in electricity from natural gas, nuclear, and hydropower, it is clear from Figure 5 that electricity drives China's coal consumption and CO_2 emissions. From 1987 to 2007, the percentage of primary coal supply burned in thermal power plants grew from 21 to 50 percent. In comparison, the United States used 86 percent of coal for electricity in 1987 and 91 percent in 2007, with coke-making for steel production using most remaining coal (U.S. EIA 2009).

Most modern economies today have all but eliminated direct burning of coal in ovens, kilns, boilers, and residential stoves and heaters because of the health impacts and inconvenience of delivery and handling. Coal use in these smaller scale energy applications often has no particulate filters and precipitators typical of large point source users such as power and coking plants. In 1987, China was still getting 91 percent of residential commercial energy from household stoves and heaters and apartment building boilers. Commercial establishments (restaurants, hotels, shops) derived 79 percent of their energy from coal. Even transportation in 1987 burned coal for 42 percent of its energy, mainly in steam locomotives. Note that these percentages of TFC do not include their consumption of electricity derived from coal.

Comparing Figures 4 and 5 shows the progress China made from 1987 to 2007 in reducing coal burning by end users. The percentage of TFC from coal fell from 49 to 29 percent in agriculture, 31 to 13 percent in construction, 79 to 19 percent in commerce, and 36 to 6 percent in other services. The most dramatic decline was in the transport sector, which slashed coal use from 49 to 3 percent by replacing steam locomotives with cleaner and more efficient diesel and electric ones and rapid growth of road and air transport, which rely mainly on liquid hydrocarbons. The residential sector cut direct coal burning from 91 to 30 percent by substituting a diverse mix of electricity, petroleum products, natural gas, and steam for heating, cooking, lighting, and appliances. A deeper look into the residential sector (not shown in the figures) reveals an urban–rural split: Coal accounts for only 19 percent of urban use but 61 percent of rural needs. Meanwhile, natural gas and district heating provide 33 percent of urban home-energy use but less than 0.1 percent of rural use.

Coke production, which grew fivefold over these twenty years, is another driving factor behind the dramatic growth of coal in China. Metallurgical coke, produced by pyrolysis of low-ash, low-sulfur bituminous coals, is a nearly smokeless fuel for smelting pig iron. Steel for construction, infrastructure, rail, and vehicles is driving the need for coke. China's consumption of metallurgic coal for coking is higher than the total coal consumption of every single country in the world except the United States and India (U.S. EIA 2010).

Not shown in Figures 4 and 5 is coal preparation, a relatively inexpensive procedure that reduces the ash content from over 30 percent in raw coal to around 10 percent by grinding, sifting, floating, and other mechanical means. In 2007, China treated 415 Mtce mostly for coking (318 Mtce), industry (46 Mtce), and electricity (17 Mtce). Coal preparation is necessary for coking coal and optional for steam coal. If the ash content and the distance shipped are both high enough, preparation of steam coal can pay for itself by lowering transport costs while benefitting the environment by reducing particulates and acid rain (but not CO_2; Kuby et al. 1995; Glomsrød and Wei 2005).

Summary and Future Prospects

This article presented energy flow diagrams to visualize China's complex energy system. It highlighted the extent to which China has modernized an inefficient coal-based system and how much it has yet to accomplish. Great strides have been made in reducing energy intensity and eliminating direct coal burning in transport, commerce, services, construction, residential, and agriculture. The explosion of coal use for electricity, coking, and industry, however, has more than offset those gains. Whenever the Chinese economy has heated up and energy consumption has accelerated, China has turned to its most plentiful source of energy—coal—to meet its needs. Now its middle class and export industries are demanding the convenience and mobility of road transport, and oil imports are skyrocketing.

As China looks to move 1.3 billion people toward Western living standards, its energy system will likely become the world's largest, continue to depend on coal, put enormous pressure on world oil markets, and accelerate global climate change. China's leadership is acting boldly on all four of these, and although it has been reluctant to commit to CO_2 reduction terms that might limit economic growth, it has agreed to pursue a carbon intensity reduction target. The IEA's (2007) "reference case" predicts that China's net oil imports will grow from 3.5 million barrels per day (273 Mtce) in 2006 to 13.1 million barrels per day (1,022 Mtce) by 2030, surpassing current U.S. imports of 12.9 million barrels per day. The same report predicted that China will account for over 60 percent of the world's increase in coal use by 2030, largely for electricity generation, and that by 2030 China's CO_2 emissions will surpass those of the United States and European Union combined.

China's leadership is aware of these trends and has the political will to invest heavily, subsidize liberally, and direct national research and development to greater efficiency and toward renewable technologies—for domestic purposes and to dominate green-technology export markets globally. China's days of thinking that they can solve their energy problems solely by supply-side measures are long gone (Hang and Tu 2007; Xie and Kuby 1997). China is making the United States and Japan nervous with its efforts to take the world lead in battery electric vehicles and has aggressively expanded mass transit, high-speed rail, and electric bicycles (Osnos 2009). Yet, as the energy flow diagrams have shown, transportation actually accounts for a small part of total energy use, and greater reliance on electric vehicles will translate back to more coal-fired electricity. Thus, China has been pouring resources into becoming a world leader in clean-coal technology. By 2006, 7 percent of China's coal power capacity employed highly efficient supercritical combustion technology, and they are now building integrated gasification combined cycle (IGCC) power plants (IEA 2007). Carbon capture and storage—in which CO_2 is separated from emissions, compressed, and sent by pipeline to be injected and stored long term underground in geologic formations or in depleted oil fields for enhanced oil recovery—is likely to be central to China's efforts to reduce greenhouse gases (Zheng et al., 2010). A silver lining in the continued expansion of coal-fired generating capacity is that IGCC makes it easier to separate the CO_2 emissions prior to combustion (IEA 2007). With their vast workforce of lower-cost young engineers, China appears able to build such plants at a fraction of the U.S. cost (Osnos 2009). In addition, China leads the world in solar-panel production, produces more than half of the world's solar water heaters, and doubled its installed wind-power capacity every year from 2006 to 2008 (Osnos 2009).

As the IEA (2007, 216) put it, "environmental concerns have come more to the fore in China and India in recent years, but they remain subordinate to the demands of economic development and poverty alleviation." China will continue heavy industrialization and rapid urbanization, with associated growth of energy and CO_2 emissions. On climate change, China supports the Kyoto Protocol principle of "common and differentiated responsibilities" (Xinhua News Agency 2009) and argues that responsibility should be based on historical cumulative CO_2 emissions (the areas under the curves in Figure 1) and per capita emissions (Figure 3).

Although China has surpassed the United States in annual emissions (Figure 1), its share of global cumulative emissions is only 8 percent versus 30 percent for the United States, and its per capita emissions are less than one quarter those of the United States (IEA 2007). Differing viewpoints on the basis for CO_2 reductions were a primary reason why the Copenhagen Climate Summit of 2009 stalemated with no internationally binding agreement for reductions. China has promised to reduce CO_2 emission intensity (per yuan of GDP) by 40 to 45 percent by 2020 relative to 2005 (Xinhua News Agency 2009), in part by increasing renewable energy to 15 percent of total consumption (Shi 2009). As Figures 1 and 3 have shown, however, dramatic reductions in CO_2 intensity are not at all inconsistent with continued escalation of total CO_2 emissions. In fact, in the two years from 2005 to 2007, China already achieved 4.5 percent of their pledged 40 to 45 percent drop in CO_2 intensity, even as its total CO_2 emissions rose by 19.7 percent. The graphs and energy flow diagrams presented here give the reader a holistic perspective on the role that coal continues to play in China's energy economy and the opportunities and limits for further restructuring and progress toward CO_2 goals.

Acknowledgments

Canfei He would like to acknowledge funding from the National Natural Science Foundation of China (#41071075) and thanks Yan Yan for her excellent research assistance.

Note

1. As the world's largest power plant of any kind, the Three Gorges Dam rightly attracts a great deal of attention, but its total contribution to China's energy needs must be put into perspective. In 2007, hydropower accounted for 14 percent of electricity generation in China, and the Three Gorges complex, then running at 61 percent of its planned 100TwH production (and 22.5 GW capacity), accounted for 14 percent of hydropower generation. Thus, the Three Gorges accounted for about 2 percent of China's electricity production in 2007. By the time it reaches full capacity in 2011, China's total generating capacity will have grown similarly, from 624 GW to over 1,000 GW, and the Three Gorges contribution will remain about 2 percent (U.S. EIA 2009).

References

Andrews-Speed, P., S. Dow, and Z. Gao. 2000. The ongoing reforms to China's government and state sector: The case of the energy industry. *Journal of Contemporary China* 9:5–20.

British Petroleum (BP). 2010. *Statistical review of world energy 2010*. London: British Petroleum.

Chai, J., J. Guo, S. Wang, and K. Lai. 2009. Why does energy intensity fluctuate in China? *Energy Policy* 37:5717–31.

Cherni, J. A., and J. Kentish. 2007. Renewable energy policy and electricity market reforms in China. *Energy Policy* 35:3616–29.

Cornelius, P., and J. Story. 2007. China and global energy markets. *Orbis* 51:5–20.

Downs, E. S. 2000. *China's quest for energy security*. Monograph MR-1244-AF. Santa Monica, CA: Rand Corporation.

Fan, Y., H. Liao, and Y. Wei. 2007. Can market-oriented economic reforms contribute to energy efficiency improvement? Evidence from China. *Energy Policy* 35: 2287–95.

Farla, J. C. M., and K. Blok. 2001. The quality of energy intensity indicators for international comparisons in the iron and steel industry. *Energy Policy* 29:523–43.

Fisher-Vanden, K., G. H. Jefferson, H. Liu, and Q. Tao. 2004. What is driving China's decline in energy intensity? *Resource and Energy Economics* 26:77–97.
Fisher-Vanden, K., G. H. Jefferson, J. Ma, and J. Xu. 2006. Technology development and energy productivity in China. *Energy Economics* 28:690–705.
Garbaccio, R. F., S. Mun, and D. W. Jorgenson. 1999. Why has the energy-output ratio fallen in China. *The Energy Journal* 20:63–91.
Glomsrød, S., and T. Wei. 2005. Coal cleaning: A viable strategy for reduced carbon emissions and improved environment in China? *Energy Policy* 33:525–42.
Hang, L., and M. Tu. 2007. The impact of energy prices: Evidence from China. *Energy Policy* 35:2978–88.
Hanson, S. 2008. *China, Africa, and oil.* New York: Council on Foreign Relations. http://www.cfr.org/publication/9557/ (last accessed 16 December 2009).
He, C., and J. Wang. 2007. Energy intensity in light of China's economic transition. *Eurasian Geography and Economics* 48:439–68.
Huang, J. 1993. Industry energy use and structural change: A case study of the People's Republic of China. *Energy Economics* 15:131–36.
International Energy Agency (IEA). 2000. China's worldwide quest for energy security. http://www.iea.org/textbase/nppdf/free/2000/china2000.pdf (last accessed 16 December 2009).
———. 2007. World energy outlook 2007: China and India insights. http://www.iea.org/textbase/nppdf/free/2007/weo_2007.pdf (last accessed 18 February 2010).
Kambara, T. 1992. The energy situation in China. *China Quarterly* 131:608–36.
Konan, D., and J. Zhang. 2008. China's quest for energy resources on global markets. *Pacific Focus* 23:382–99.
Kuby, M., and P. Cook. 1997. The implications of structural change, reform and congestion for freight traffic in China. *Asian Profile* 25:1–21.
Kuby, M., Q. Shi, W. Thawat, X. Sun, W. Cao, Z. Xie, D. Zhou, et al. 1995. Planning China's coal and electricity delivery system. *Interfaces* 25:41–68.
Lawrence Livermore National Laboratory. 2010. Energy flow charts. https://flowcharts.llnl.gov/# (last accessed 28 December 2010).
Levine, M. 2005. *Energy efficiency in China: Glorious history, uncertain future.* Berkeley: University of California Energy Resources Group Colloquium.
Liao, H., Y. Fan, and Y. Wei. 2007. What induced China's energy intensity to fluctuate: 1997–2006? *Energy Policy* 35:4640–49.
Lin, X., and K. R. Polenske. 1995. Input–output anatomy of China's energy use changes in the 1980s. *Economic System Research* 7 (1): 67–84.
Ma, C., and D. I. Stern. 2008. China's changing energy intensity trend: A decomposition analysis. *Energy Economics* 30:1037–53.
Ma, H., and L. Oxley. 2009. China's energy economy: A survey of the literature. Working Paper No. 02/2009, Department of Economics, University of Canterbury, New Zealand.
Ma, H., L. Oxley, and J. Gibson. 2009. China's energy situation in the new millennium. *Renewable and Sustainable Energy Reviews* 13:1781–99.
Ma, H., L. Oxley, J. Gibson, and B. Kim. 2008. China's energy economy: Technical change, factor demand and interfactor /interfuel substitution. *Energy Economics* 30:2167–83.
Ministry of Commerce. 2008. *China commerce yearbook.* Beijing: Ministry of Commerce.
National Statistical Bureau (NSB). 1988. *China energy statistical yearbook 1987.* Beijing: China Statistical Publishing House.
———. 2008. *China energy statistical yearbook 2007.* Beijing: National Statistical Bureau.
———. 2009. *Statistical yearbook of China 2008.* Beijing: National Statistical Bureau.
Osnos, E. 2009. Green giant: Beijing's crash program for clean energy. *The New Yorker* December 21–28:54–63.
Rawski, T. G. 2001. What is happening to China's GDP statistics? *China Economic Review* 12:347–54.
Shi, D. 2009. Analysis of China's renewable energy development under the current economic and technical circumstances. *China & World Economy* 17:94–109.
Sinton, J. E. 2001. Accuracy and reliability of China's energy statistics. *China Economic Review* 12:373–83.
Sinton, J. E., and D. G. Fridley. 2000. What goes up: Recent trends in China's energy consumption. *Energy Policy* 28:671–87.
———. 2002. A guide to China's energy statistics. *Journal of Energy Literature* 8:22–35.
Sinton, J. E., and M. D. Levine. 1994. Changing energy intensity in Chinese industry: The relative importance of structural shift and intensity change. *Energy Policy* 22:239–55.
Smil, V. 1997. China shoulders the cost of environmental change. *Environment* 39:33–37.
Smith, G., and C. Schmollinger. 2010. China passes U.S. as world's biggest energy consumer, IEA says. Bloomberg News July 20. http://www.bloomberg.com/news/2010-07-19/china-passes-u-s-as-biggest-energy-consumer-as-oil-imports-jump-iea-says.html (last accessed 27 August 2010).
Thomson, E. 2003. *The Chinese coal industry: An economic history.* London and New York: RoutledgeCurzon.
U.K. Department of Business Innovation and Skills. 2010. Energy flow chart. http://webarchive.nationalarchives.gov.uk/+/http://www.berr.gov.uk/energy/statistics/publications/flowchart/page37716.html (last accessed 28 December 2010).
U.S. Energy Information Administration (EIA). 2009. Country analysis briefs: China, electricity. http://www.eia.doe.gov/cabs/China/Electricity.html (last accessed 23 July 2010).
———. 2010. Annual energy review. http://www.eia.doe.gov/aer (last accessed 28 December 2010).
Wang, C., J. Chen, and J. Zou. 2005. Decomposition of energy-related CO_2 emissions in China, 1957–2000. *Energy* 30:73–83.
Wang, Q., H. Qiu, and Y. Kuang. 2009. Market-driven energy pricing necessary to ensure China's power supply. *Energy Policy* 37:2498–2504.
World Bank. 1994. *China: Issues and options in greenhouse gas emissions control.* Washington, DC: The World Bank.
World Bank, China and Mongolia Department, Transport Operations Division. 1995. Investment strategies for

China's coal and electricity delivery system. Report No. 12687-CHA, World Bank, Washington, DC.
Wu, Y. 2003. Deregulation and growth in China's energy sector: A review of recent development. *Energy Policy* 31:1417–25.
Xia, C., and Z. Song. 2009. Wind energy in China: Current scenario and future perspectives. *Renewable and Sustainable Energy Reviews* 13:1966–74.
Xie, Z., and M. Kuby. 1997. Supply side-demand side optimization and cost-environment tradeoffs for China's coal and electricity delivery system. *Energy Policy* 25:313–26.
Xinhua News Agency. 2009. China to cut 40 to 45 percent GDP unit carbon by 2020. *China Daily* 26 November:1.
Zhang, M., H. Mu, Y. Ning, and Y. Song. 2009. Decomposition of energy-related CO_2 emissions over 1991–2006 in China. *Ecological Economics* 68:2122–28.
Zhang, P., Y. Yang, J. Shi, Y. Zheng, L. Wang, and X. Li. 2009. Opportunities and challenges for renewable energy policy in China. *Renewable and Sustainable Energy Reviews* 13:439–49.
Zhang, Q. 2004. Residential energy consumption in China and its comparison with Japan, Canada, and USA. *Energy and Buildings* 36:1217–25.
Zhang, Z. X. 2003. Why did the energy intensity fall in China's industrial sector in the 1990s? The relative importance of structural change and intensity change. *Energy Economics* 25:625–38.
Zheng, Z., E. D. Larson, Z. Li, G. Liu, and R. H. Williams. 2010. Near-term mega-scale CO_2 capture and storage demonstration opportunities in China. *Energy and Environmental Science* 3:1153–69.

Mountaintop Removal and Job Creation: Exploring the Relationship Using Spatial Regression

Brad R. Woods* and Jason S. Gordon†

*Office for Research Protections, The Pennsylvania State University
†Department of Forestry, Mississippi State University

This project focused on a new and increasingly contested method of coal extraction, mountaintop removal (MTR), and its effects on central Appalachian residents' quality of life vis-à-vis increased employment. Attention is given to central Appalachia because its fossil fuel landscapes have undergone major changes as a result of two interrelated forces: (1) a national push for energy independence that led to the region's all-time high production of coal (supplying over half of the nation's coal); and (2) changes in mining technology that allowed for increased production. Such transitions have led to widespread use of MTR mining, a method that entails removal of extensive land area to expose coal seams. Although policymakers are aware of the negative environmental effects of MTR, its continued use is primarily rationalized using the argument that it contributes to local economies, especially job retention and development. MTR proponents argue that, without MTR, other regions and countries more competitively extract coal. Opponents counter that MTR fails to substantially contribute to employment due to efficiencies in mechanization. This study used socio-spatial analysis to understand MTR's impact on employment in southern West Virginia populated places. We integrated coal mining permit boundaries with employment indicators obtained from the U.S. Census. Contrary to pro-MTR arguments, we found no supporting evidence suggesting MTR contributed positively to nearby communities' employment. Implications for economic development are discussed. *Key Words: Appalachia, coal, economic development, mountaintop removal, spatial regression.*

该项目着重于一个新的越来越有争议的煤炭开采方法，即山顶搬迁（MTR），和其对中央阿巴拉契亚居民生活的质量相对于增加就业的影响。注重中央阿巴拉契亚是因为它的化石燃料的景观因为两个相互关联的力量已发生了主要变化：（1）国家对能源独立的推动，导致该地区空前的高煤炭生产（供应过半的全国煤炭），而且（2）在开采技术上的改变使产量增加。这种过渡导致 MTR，一个需要除去大量的土地表面以暴露煤层的开采方法的广泛使用。尽管政策制定者意识到 MTR 的负面影响，它的继续使用因其有利于当地经济，特别是保留工作和发展的论点而主要地被合理化。MTR 支持者认为，没有 MTR，其他地区和国家将更竞相地提取煤。反对者则反驳说，由于机械化效率，MTR 没有实质上促进就业。本研究使用于社会空间分析，了解 MTR 在西弗吉尼亚州南部人口稠密的地方对就业的影响。我们综合了煤炭采矿许可界限和由美国人口统计局获得的就业指标。与 MTR 支持论据相反，我们没有发现任何支持证据表明 MTR 对附近社区的就业作出了积极贡献。并讨论了此结果对经济发展所含的意义。*关键词：阿巴拉契亚，煤炭，经济发展，山顶搬迁，空间回归。*

Este proyecto concentró su interés en un nuevo y cada vez más debatido método de extracción de carbón, la remoción de la cubierta montañosa (MTR, sigla en inglés), y sus efectos en la calidad de vida de los habitantes de los Apalaches centrales frente a un incremento del empleo. Se le dio atención a la parte central de los Apalaches porque sus paisajes de combustibles fósiles han experimentado cambios importantes como resultado de dos fuerzas interrelacionadas: (1) una presión nacional en pro de la independencia energética que llevó a la más alta producción de carbón de todos los tiempos en la región (suministrando más de la mitad del carbón de la nación); y (2) cambios en la tecnología minera que permitieron el aumento de la producción. Estas transiciones han conducido al amplio uso de la minería MTR, método que significa la remoción de vastas extensiones de tierra para exponer las vetas de carbón. Aunque los hacedores de políticas saben de los efectos ambientales negativos del MTR, continuar con su aplicación se racionaliza primariamente con el argumento de que así se contribuye a fortalecer las economías locales, especialmente en conservación del empleo y desarrollo. Los proponentes del MTR arguyen que sin el MTR, otras regiones y países extraen carbón más competitivamente. Los oponentes al método replican que el MTR falla en contribuir sustancialmente al empleo debido a las eficiencias en mecanización. Este estudio utilizó el análisis socio-espacial para evaluar el impacto del MTR sobre el empleo en los lugares poblados del sur de Virginia Occidental. Integramos los límites permitidos para minería del carbón con los indicadores de empleo obtenidos del Censo de los EE.UU. Contrario a lo argumentado en favor del MTR, no encontramos evidencia de apoyo que sugiera que el MTR contribuyó positivamente al empleo de las comunidades cercanas. Se discuten las implicaciones que esto tiene para el desarrollo económico. *Palabras clave: Apalaches, carbón, desarrollo económico, remoción de la cubierta montañosa, regresión espacial.*

A 1999 U.S. District Court decision affirmed waste generated from mountaintop removal coal mining (MTR) violated the Clean Water Act. This ruling was declared an economic setback by coalfield politicians, operators, and many coalfield residents. MTR supporters argued the policy would stifle economic development and emphasized the potential loss of jobs resulting from a heavy-handed regulatory environment (Ward 1999). After fierce lobbying to have this decision reversed, the Environmental Protection Agency (EPA) and U.S. Army Corps of Engineers (ACOE) redefined the meaning of "fill material" (i.e., waste) in Section 404 of the Clean Water Act, which, in turn, resulted in a dramatic increase in the number of MTR projects (Stewart-Burns 2005; Davis and Duffy 2009).

The EPA revisited the issue in 2009, threatening to use its veto authority under the Clean Water Act to reverse permits issued by the ACOE (U.S. EPA 2005). Such actions have occurred only twelve times since 1972.[1] Again, MTR advocates argued that vetoes would compromise the region's economic stability, including job retention. A recent report issued by the U.S. Senate Environment and Public Works Committee (2010) indicated, "The [EPA] has taken several actions to obstruct, delay, and shutter surface coal mining operations in Appalachia. . . . Moreover, the EPA either ignored or dismissed the fact that the projects *bring jobs* and economic growth to the Appalachian region" (emphasis added).

The "coal means jobs" mantra is clearly of vital importance for justifying the initiation and maintenance of extraction activities in coal-dependent communities. In his seminal piece, Gaventa (1980) described the evolution of a discourse focusing on job creation and the benefits of industrial society but with the latent purpose of shaping a quiescent and readily available labor supply. Gaventa further noted that prior to the ideological inception of the discourse, the capitalization of labor was minimal and work was less glorified. Local culture was shaped by its relationship to the landscape, whereas "coal means jobs" insisted on dominance of nature. In short, social and environmental policies have long been interlinked with the prevalent discourse of mining, including MTR. As well, residents at the local level consistently support pro-MTR arguments, at least outwardly. For instance, Woods (2010, 83) found, "although some [community residents] acknowledged problems [caused by MTR] still persisted, they emphasized opportunities for employment and larger economic realities, made MTR problems bearable."

This study explored the argument that coal mine employment is associated with the scale of MTR projects. A spatial model was constructed of West Virginia's fossil fuel landscape, including mine sites and adjacent human populations. Using geospatial analysis, we created buffer zones to capture MTR sites and nearby communities. An ordinary least squares (OLS) regression model was developed to examine the relationship between type of mining (MTR vs. underground), scale of mining operation, and percentage of the population employed in MTR coal mining. West Virginia was chosen due to (1) local perceptions of employment opportunities (Scott 2007; Bell and York 2010); (2) its distinction as the nation's second largest coal-producing state (U.S. Energy Information Administration 2009); and (3) the most acreage of permitted MTR projects in the country (Stewart-Burns 2005).

Our primary objective was to engage in an empirical examination, which is principally driven by policy considerations, but the research problem leads to a broader question addressing the entrenched persistence of the discourse. We utilize Gaventa's (1980) description of the processes of legitimation to link our research to broader issues of hegemony at the local level as well as at larger scales of policy formation. Gaventa's work explains the emergence and persistence of the "coal means jobs" mantra—our research questions the authenticity of the discourse. As such, our research quantitatively contributes to Gaventa's work in understanding the hydrocarbon landscape in central Appalachia.

Further, findings have implications for research that has tended to emphasize ecological problems associated with MTR operations (e.g., Negley and Eshleman 2006; Ferrari et al. 2009; Palmer et al. 2010) largely to the exclusion of socioeconomic analysis. Our study indicates the need to reexamine arguments linked to MTR using socioeconomic indicators and spatial methodologies. Finally, we present implications for policies that affect coalfield communities in West Virginia and other areas of central Appalachia where MTR is practiced. Current guidelines and regulations tend to focus on negative consequences associated with postextraction phases of fossil fuel production (e.g., carbon by-products from burning). As the need for energy increases, policy considerations must extend beyond carbon mitigation and consider the effects of extraction (Zimmerer 2011).

Background

MTR is a relatively recent form of coal extraction technology. It is distinguishable from traditional mining techniques, which include all forms of underground mining, such as continuous, room and pillar, and longwall methods. MTR begins by clear-cutting native forests. This is followed by the use of heavy explosives to loosen soil and dislodge underlying rock. Coal is then extracted, loaded onto freight haulers, and transported to energy production facilities, usually located far from the local area (Fox 1999). MTR represents a transformation in the industrial sector of mining communities. Natural resource depletion, reduced labor costs, and globalization have driven technological innovations such as MTR. In turn, innovations have led to hope for many communities in the form of much needed employment.

Despite benefits for the coal mining and energy industries, controversy surrounds MTR (Peng 2000; Lindsey 2007). For those living closest to mining operations, threat of injury looms from "flyrock."[2] Other problems from explosive blasting have included high levels of respirable dust, disruptive noise, and cracks in the walls of residents' homes (Fox 1999; Woods and Meyer 2006; Woods 2010; Zeller 2010). Further hazards include flooding, drinking water contamination, and stream degradation.[3] The U.S. EPA (2005) estimated that over 1,200 miles of streams have been directly impacted by the creation of valley fills.

Industry representatives and many state officials have acknowledged the ecological disruptions caused by MTR, but they have argued such "temporary" effects are offset by benefits from MTR. For example, proponents claim that flat land created after mining benefits the region by creating space for development, resulting in further job growth (Mountaintop Mining 2008).

An even more common argument is that direct employment by MTR projects is a practical solution to the area's historic impoverishment (cf. McGinley 2004; Mountaintop Mining 2008; Zeller 2010). Increasing demand for low-cost energy and global competition have forced mining firms to adopt larger scale, more cost-efficient extraction practices than those used in underground mining (Dunn 1999). Decreased costs lead to increased numbers of mining projects. In turn, low-skilled job opportunities increase, which would otherwise be eliminated due to competition of low-cost coal production in other locations. In a recent study, funded by the West Virginia Coal Association, researchers at West Virginia and Marshall Universities found employment in all coal mining sectors has recently increased (Bureau of Business and Economic Research and Center for Business and Economic Research 2010). However, their study further noted, "West Virginia coal mining employment of 20,454 in 2008 can be divided into two major categories: underground and surface mining. Underground mining employment was the highest in the state in each quarter accounting for 66 to 67 percent of total employment" (21).

It is notable that, despite widespread use of the "jobs" argument in media and public relations outlets, we found little empirical evidence supporting the localized job creation claim. To the contrary, Figure 1 illustrates a consistent and alarming decline in West Virginia coal mining employment. From 1900, employment in the mining industry rose steadily to a peak of 130,457 employees in 1940. In 1952, direct employment in the mining industry declined rapidly with the diffusion and wide acceptance of continuous mining technology (Lewis 1992; West Virginia Coal Association 2009).

In more recent years, the technology used in MTR surpassed continuous mining in its ability to minimize reliance on human labor while increasing rates of coal extraction. Between the early 1980s and early 2000s, total production increased as direct employment in coal mining declined (Woods and Meyer 2006; West Virginia Coal Association 2009). During the same period, income inequality increased dramatically across the state. Qualifying as the lowest in the nation, the bottom 20 percent of families earned an average of $13,208 annually (Center on Budget and Policy Priorities and the Economic Policy Institute 2006).[4] In the meantime, West Virginia's wealthiest families were able to increase their average incomes to seven times that of the state's poorest families, representing the sixth highest increase in income inequality in the nation.

Although our study is largely driven by a policy-based question, Gaventa's (1980) work serves as a theoretical foundation from which to revisit the ubiquity of the "coal means jobs" mantra. Gaventa described the evolution of the discourse as a basic feature of central Appalachian life since the early days of coal production in the nineteenth century. The "coal means jobs" mantra emerged out of the boomtown enthusiasm surrounding the transformation of central Appalachia from small farm communities to an industrial society (see also Salstrom 1994). Residents and mine operators, regardless of social class, believed that they stood an equal chance to lose or gain in the endeavor to industrialize. The mutual needs of miners and operators meant dedication to the mine was vital to societal advancement. Eventually, consumptive values and almost total dependency on the wealth and power of the mining companies solidified the belief in "economic oneness" (Gaventa 1980, 67). The nexus between the job and the community further shaped residents' values and wants, uncertainties, and collective action.

The discourse is a fundamental component of the way power (as wielded by those who control the resource) has influenced perceptions of needs, capacities, and collective agency in situations of latent conflict (Gaventa 1980). The discourse shaped competition between labor groups (i.e., union and nonunion coal miners), which further reinforced the notion that failure to identify with the coal industry was inconsistent with supporting local livelihoods. Although Gaventa focused on the quiescence of local residents, his thesis applies to the processes legitimating dominant meanings and patterns beyond the local scale of analysis. Fundamentally, these processes advanced the notion

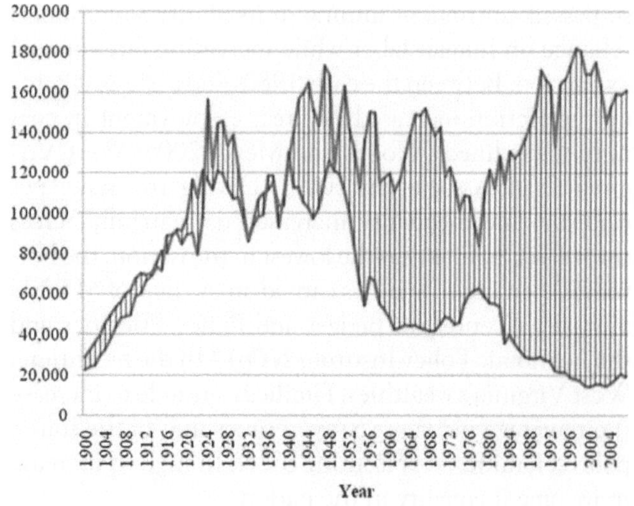

Figure 1. West Virginia mining employment and production (1900–2005).

that coal mining was the only employment option leading to the rewards of industrialized society. As a result, policy discussions focus on employment because it has been shaped as the dominant concern among various levels of society across the fossil fuel landscape.

To complement Gaventa's qualitative work, we quantitatively explored the assumption that "coal means jobs" is but a legitimation of discourse intended to justify the retention of a low-cost labor force. This research problem is driven by recent policy formulations but has roots in the processes of hegemony that have long characterized the region. Specifically, we explored the argument that MTR improves community well-being vis-à-vis job creation within communities adjacent to mine sites. The lack of such a relationship would suggest the need to consider the major arguments surrounding MTR from a new perspective.

Study Area

Active MTR mining occurs in ten southern West Virginia counties (see Figure 2). Communities in these counties have historically relied on coal mining as a source of employment as well as cultural identification (Billings 1974; Lee 2002).

Table 1 presents selected socioeconomic and biophysical characteristics of the ten counties in this study. Logan County had the most acres permitted for MTR (18,301 acres), and Raleigh had the fewest (1,314 acres; West Virginia Department of Environmental Protection 2008). However, when Census figures were reviewed for the number of individuals employed in coal mining, Raleigh County had the most coal miners (1,693), whereas Logan County only had 1,400.

Methods

Figure 3 illustrates populated places in the study area's ten counties. These were identified using U.S. Geological Survey (USGS) Populated Places data for West Virginia, which identified residential areas not included in the U.S. Census Populated Places data.[5] Based on the need to concatenate coal mining employment attributes to place data, U.S. Census ZIP code files were added to the USGS data. Any place without a ZIP code designation was excluded from the analysis. The resultant data set contained 291 populated places with amended employment data.

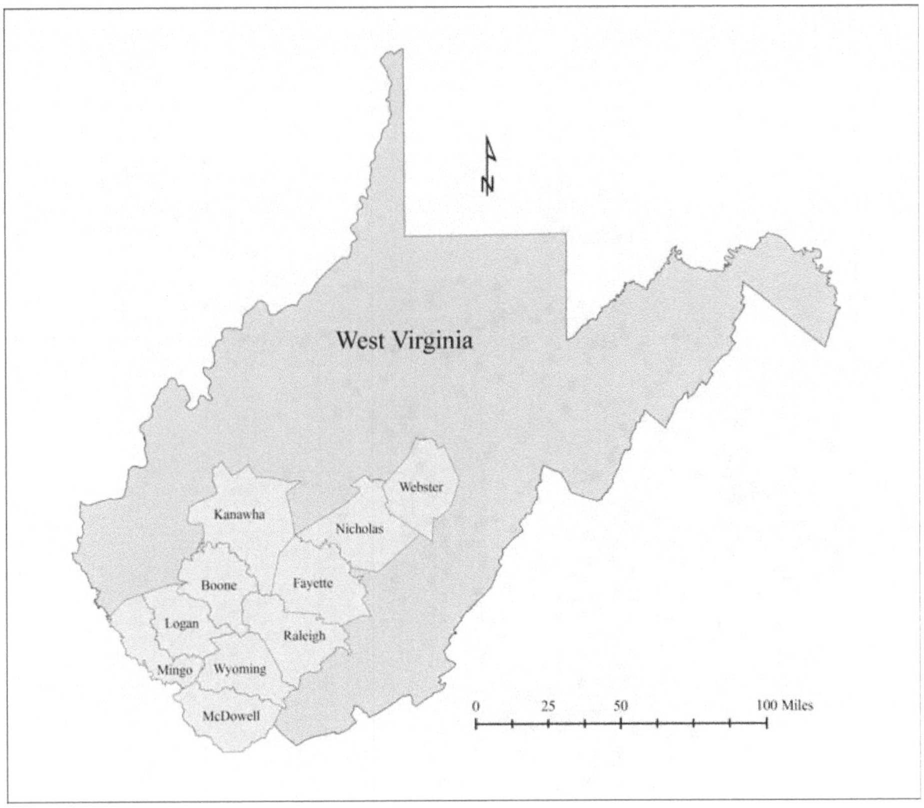

Figure 2. Mountaintop removal (MTR) counties (1:2,844,473).

Data on MTR and traditional mining were obtained from the West Virginia Department of Environmental Protection. The data set contained information on all current permitted mines as of February 2008 (both underground mining and surface mining). For each identified place, the number of individuals employed in coal mining was obtained from the U.S. Census Bureau (2000).[6] The dependent variable was percentage of people employed in coal mining. The independent variable was MTR size (extraction site area in acres). Because we wanted to focus on a possible association between MTR and employment, underground mine size (in acres) was used as a control variable, keeping in mind that MTR occupies a larger area.

MTR and underground mining boundary files, obtained from the West Virginia Department of

Table 1. Descriptive statistics for MTR counties in southern West Virginia

County	2000 population	Median household income (1999 dollars)	Number of individuals employed in coal mining	Total MTR permitted acres (active permits)[a]	2008 total coal production in West Virginia (by rank)[b]
Boone	25,535	25,669	1,353	10,899	1
Fayette	47,579	24,788	752	6,767	—
Kanawha	200,073	33,766	1,184	13,095	3
Logan	37,710	24,603	1,400	18,301	2
McDowell	27,329	16,931	802	2,063	10
Mingo	28,253	21,347	1,426	13,468	4
Nicholas	26,562	26,974	763	1,776	—
Raleigh	79,220	28,181	1,693	1,314	7
Webster	9,719	21,055	226	1,565	9
Wyoming	25,708	23,932	1,373	2,661	—

Note: MTR = mountaintop removal. *Source:* U.S. Census Bureau (2000).
[a]West Virginia Department of Environmental Protection (2008).
[b]West Virginia Coal Association (2009).

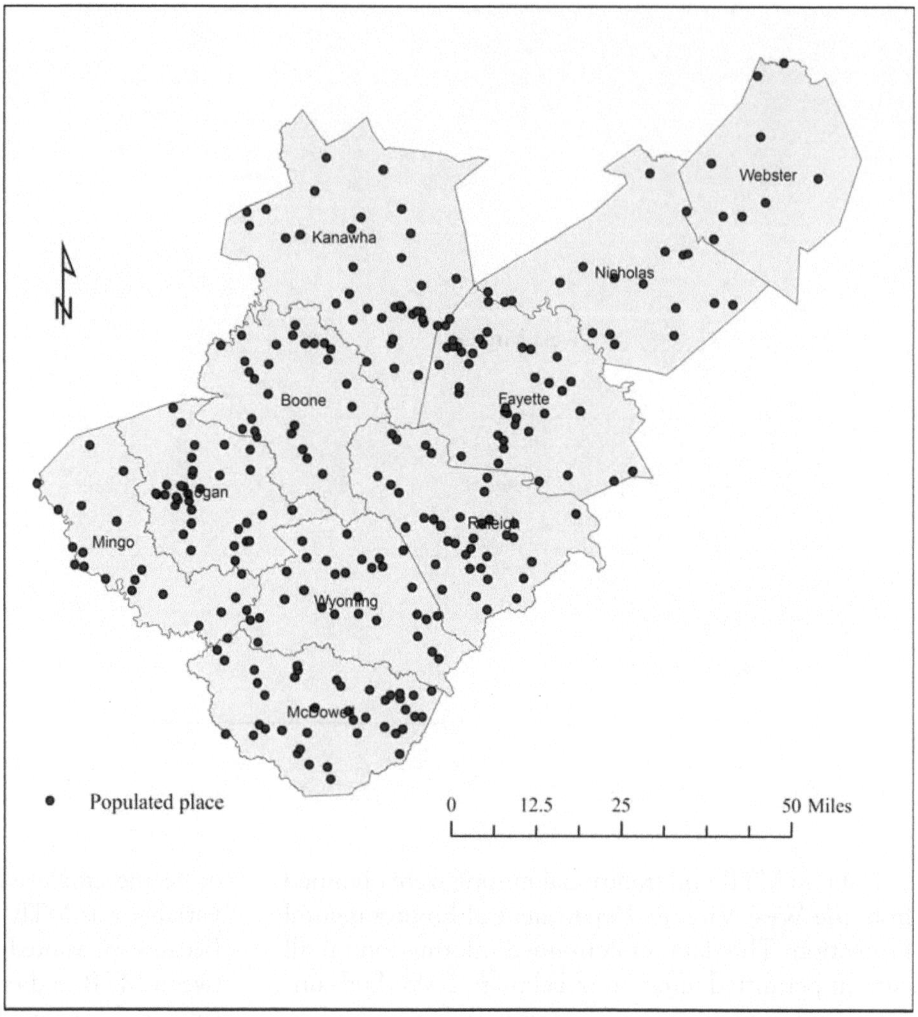

Figure 3. U.S. Geological Survey Population Places (1:1,306,618).

Environmental Protection's Division of Mining and Reclamation permit maps, were amended to the county boundaries. Mining permit status (open or closed) was reviewed, and inactive mines were excluded from the data set. Using the coal mine permit number as a common identifier, the permit status data were joined to the mining boundary data to exclude mining not relevant to the analysis. The resulting data were sorted based on mining type (i.e., MTR vs. underground) to differentiate between types of mining. After disaggregating the mining permit boundaries, place-point locations were added to the project data set.

Following the incorporation of the mining and place-point data, buffer zones were created with six-mile diameters and three-mile spacing between each buffer zone boundary. Such distances capture the majority of direct environmental impacts (e.g., detectable stream degradation) associated with MTR (see U.S. EPA 2005) and fall within parameters used in prior studies that have examined the spatial relationship between disadvantaged populations and their proximity to environmental hazards (Mennis 2002; Bullard et al. 2007; Maranville, Ting, and Zhang 2009; Mohai et al. 2009). According to the developer of the method, locating the place-point data and corresponding buffer zones along gridlines was a technique useful for quasi-geostatistical sampling of databases for exploratory geoinformatic analysis (W. L. Myers, personal communication, e-mail, 3 January 2011). Populated places, MTR boundaries, and underground mining boundaries were then amended to the buffer zones. Figure 4 illustrates the populated places and mining boundaries inside each buffer zone. Figure 5 uses a closer scale to better illustrate the scale of mining operations (both underground and MTR) along with adjacent populated places.

In preparation for analysis, data were linked to each uniquely numbered buffer zone. After each link was completed, the zone identifier was included for the community, underground mining, and MTR mining shape files. These amended shape files were then

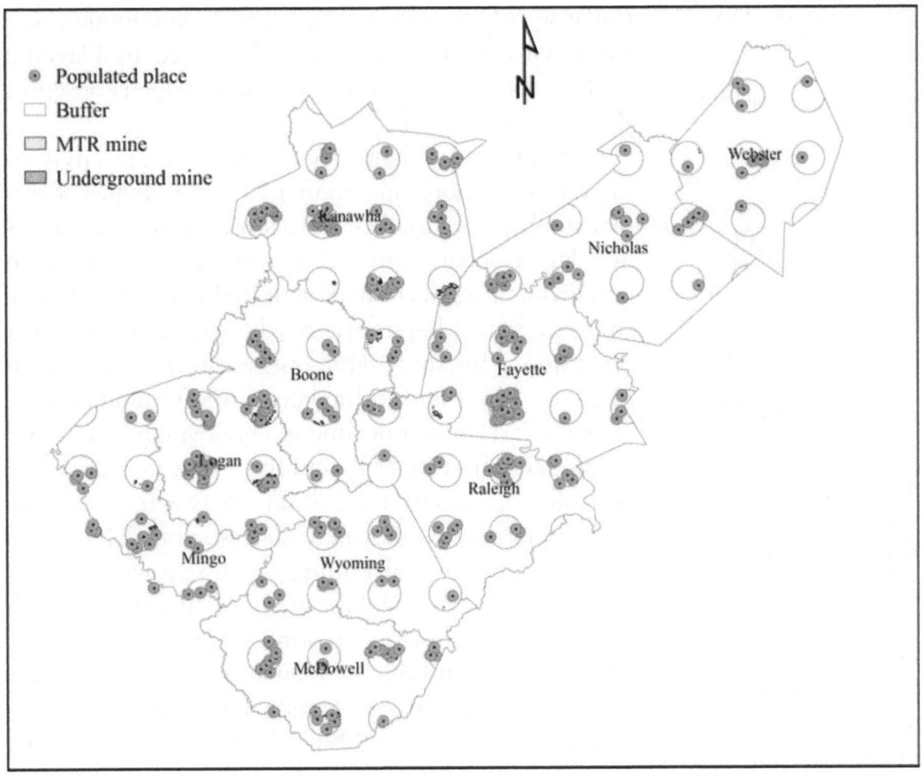

Figure 4. Mining boundaries and populated places (1:1,450,909). MTR = mountaintop removal.

joined to the buffer zone shape files by their common identifier.

Before estimating the OLS regression, several steps were taken to investigate distributional characteristics and propensity toward spatial autocorrelation of the variables. These included reviews of Moran's scatterplots, histograms, and box plots. These processes used GeoDa software to uncover spatial autocorrelation as well as run the multivariate OLS regression (Anselin 2004). Box plots and histograms were developed as indicators of distributional characteristics (Anselin 1996).

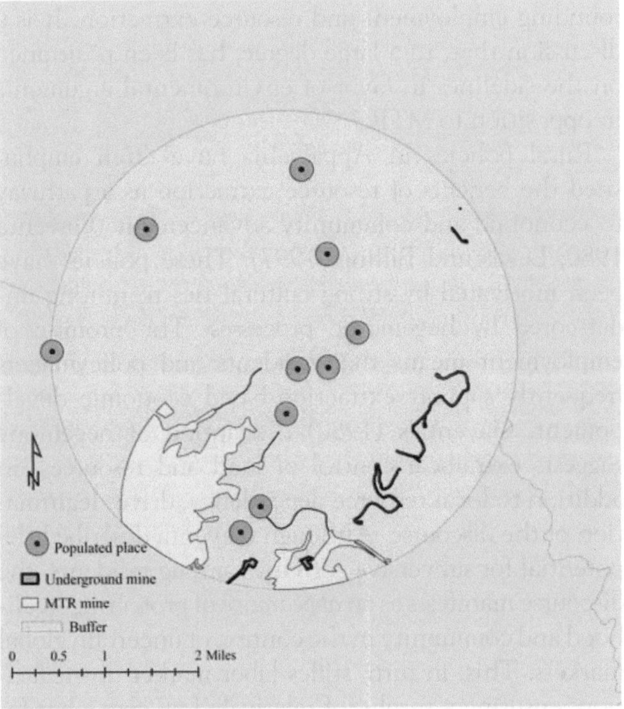

Figure 5. Mining permit boundaries and populated places, increased scale (1:64,789). MTR = mountaintop removal.

Percentage employed in mining was calculated for each community using geographic information system (GIS) spatial analysis tools. This number was then divided by the total number of persons at or above sixteen years of age (male and female) to obtain percentage of persons employed in coal mining in each place. Finally, data were compiled and added to the buffer zone shape file by creating variables of mine size (by extraction method) and percentage employed in mining. The percentage employed in coal mining box plot indicated that the majority of points fell above the 75th percentile. There were no indications of outliers.

Results

Our research question was straightforward: Is there a relationship between the size of MTR mining and employment, which justifies the "coal means jobs" mantra? Using OLS regression, we examined the effects of mine size on percentage employed in coal mining. The resultant model, including tests for spatial error and spatial

Table 2. Multivariate model of percentage employed in coal mining

Variables	Coefficient
MTR mine size (acres)	68.37
Traditional mine size (acres)	1.25
$F = 0.12$; adjusted $R^2 = .03$	
Lagrange multiplier (lag)	.01
Lagrange multiplier (error)	2.35

Note: $n = 64$. MTR = mountaintop removal.

lag, was not significant (Table 2).[7] The results of the overall model suggested insufficient evidence to support a positive relationship between mine size (either MTR mining or underground mining) and percentage of the working population employed in coal mining. This finding casts doubt on the pervasive and dominant argument of MTR advocates.

Some caveats to this finding warrant mention. First, MTR is practiced in much of central Appalachia and the finding could vary according to geographic location. Additional analysis is needed that compares findings across different geographies. Second, this study explored the possibility of a relationship between direct employment and MTR. We did not intend to measure an association between MTR and indirect mining employment, such as local occupations that service the coal mining industry. It is plausible that indirect employment might increase with the increasing size of mining operations. At any rate, pro-MTR arguments refer to increased direct employment in the mines. Finally, future research should examine other factors that might affect coal mining employment, such as the influence of shifting coal markets, which make coal from central Appalachia less attractive. As regulations associated with the extraction and burning of coal tighten, and West Virginia's most accessible seams are exhausted, larger and easier to extract coal seams in Wyoming and abroad will likely displace Appalachian coal to meet energy demands. In turn, this will likely result in shifting employment patterns in the West Virginia mining sector.

Discussion and Conclusion

Changes in coal mining extraction methods over the past two decades have resulted in conflicts among federal, state, and local decision makers and among many southern West Virginia community residents. These conflicts often emerged from differences in claims about the ways MTR affects communities, economies, and the environment. Yet, as also illustrated by Gaventa (1980), the discourse has successfully shaped consensus that "coal means jobs."

This discourse claims that, given flexible environmental regulations, more efficient technologies of surface mining potentially lead to more mining operations and larger acreage sites. Accordingly, an increase in employment results from growth in the number of mine sites and scale, despite the need for fewer workers per operation. This study represents an initial exploration into a possible relationship between MTR and jobs.

Neither a rise nor decline in employment was found for underground or MTR mining. The lack of a statistically significant relationship between coal mine size and mining employment suggests that reliance on new methods of coal mining for job growth is tenuous at best. This supports previous literature, which has rejected arguments favoring increased numbers of MTR projects and limited regulatory oversight of such projects (Mountain Association for Community Economic Development 2009). The lack of any statistical relationship between coal mining and job creation raises questions concerning coal mining's role in developing local economies.

This article does not seek to analyze policy per se, although it is motivated by a contradiction in policy. Rather, it utilizes the question of policy as a springboard to revisit Gaventa's discussion about the discourse surrounding employment and resource extraction. It is a discussion that, to a large degree, has been positioned on the sidelines in favor of environmental arguments in opposition to MTR.

Rural policies in Appalachia have often emphasized the benefits of resource extraction as a pathway to economic and community advancement (Gaventa 1980; Lewis and Billings 1997). These policies have been motivated by strong cultural ties to mining underscored by hegemonic processes. The promise of employment means that residents and policymakers frequently support extraction-based economic development. Gaventa's (1980) description of hegemony suggests extralocal control of land and resources, in addition to local resource dependency, drives legitimation of the discourse. Although Gaventa described the potential for subversive activities among residents, the discourse manifests as an *appearance* of protecting livelihood and community in the context of uncertain global markets. This, in turn, stifles labor market diversification, ensuring a surplus of relatively low-cost labor for the coal industry.

Because no significant relationship emerged from our analysis, we suggest that unsupported claims of MTR advocates are implicated in processes of hegemony as described by Gaventa. Residents' hesitancy to condemn MTR, as suggested by Woods (2010), is consistent with Gaventa's argument and reifies the discourse that "coal means jobs." Researchers and activists have attempted to break through the discourse by emphasizing the negative environmental implications of MTR. Spatial regression using sociodemographic data provides a means to better understand how the pro-MTR argument unfolds across the fossil fuel landscape. Our failure to verify a statistical relationship between job creation and MTR suggests the critical need to apply quantitative measures examining socioeconomic implications of mining technology.

Future research employing a variety of methods and study sites will further illuminate the social and ecological implications of MTR. Repetition of the study in various locations and using multiple variables will assist policymakers in critically analyzing the use of MTR coal as an energy source. Advocating a questionable extraction technology, which brings with it potentially widespread and severe environmental outcomes, fails to acknowledge long-term needs of communities, states, and nations.

Acknowledgment

The authors thank Dr. Wayne Myers for his contribution to this analysis and the useful comments of anonymous reviewers and editors on drafts of this article.

Notes

1. On 13 January 2011, the EPA announced their veto of the Spruce Number 1 Surface Mine, the largest permitted mine in West Virginia history (see http://water.epa.gov/lawsregs/guidance/cwa/dredgdis/upload/Spruce_No-_1_Mine_Final_Determination_011311_signed.pdf).
2. Flyrock "[is] the undesired propulsion of rock fragments through the air or along the ground beyond the blast zone, by the force of the explosion" (Little and Blair 2010, 641).
3. Impacts include introduction of toxins such as nickel, chromium, zinc, and other toxic compounds (Paybins et al. 2000; Lemly 2007).
4. Income figures reflect pooled data, "in order to have enough cases to generate reliable estimates of income by quintile by state we pooled income data for three consecutive years" (i.e., 1980–1982, 1990–1992, and 2001–2003 (see Center on Budget and Policy Priorities and the Economic Policy Institute 2006).
5. USGS Populated Places are places or areas with clustered or scattered buildings and a permanent human population (city, settlement, town, village).
6. These data did not differentiate between commuters and noncommuters.
7. See Anselin (1996) for a discussion on spatial dependence.

References

Anselin, L. 1996. The Moran scatterplot as an ESDA tool to assess local instability in spatial association. In *Spatial analytic perspectives on GIS*, ed. M. M. Fischer, H. J. Scholten, and D. J. Unwin, 111–26. London and New York: Taylor & Francis.

———. 2004. *GeoDa 0.9.5-i release notes*. Urbana, IL: Spatial Analysis Laboratory.

Bell, S. E., and R. York. 2010. Community economic identity: The coal industry ideology construction in West Virginia. *Rural Sociology* 75 (1): 111–43.

Billings, D. 1974. Culture and poverty in Appalachia: A theoretical discussion and empirical analysis. *Social Forces* 53:315–23.

Bullard, R. D., P. Mohai, R. Saha, and B. Wright. 2007. *Toxic wastes and race at twenty 1987–2007: Grassroots struggles to dismantle environmental racism in the United States*. Cleveland, OH: United Church of Christ Justice and Witness Ministry.

Bureau of Business and Economic Research and Center for Business and Economic Research. 2010. *The West Virginia coal economy 2008*. Morgantown: West Virginia University.

Center on Budget and Policy Priorities and the Economic Policy Institute. 2006. *Pulling apart: A state-by-state analysis of income trends*. Washington, DC: Economic Policy Institute.

Davis, C. E., and R. J. Duffy. 2009. King coal vs. reclamation: Federal regulation of mountaintop removal mining in Appalachia. *Administration and Society* 41: 674–92.

Dunn, S. 1999. King coal's weakening grip on power. *World Watch Magazine* 12 (5): 10–19.

Ferrari, J. R., T. R. Lookingbill, B. McCormick, P. A. Townsend, and K. N. Eshleman. (2009). Surface mining and reclamation effects on flood response of watersheds in the central Appalachian plateau region. *Water Resources Research* 45:W04407.

Fox, J. 1999. Mountaintop removal in West Virginia: An environmental sacrifice zone. *Organization and Environment* 12:163–83.

Gaventa, J. 1980. *Power and powerlessness: Quiescence and rebellion in an Appalachian valley*. Urbana: University of Illinois Press.

Lee, H. B. 2002. *Bloodletting in Appalachia: The story of West Virginia's four major mine wars and other thrilling incidents of its coal fields*. Parsons, WV: McClain Printing Company.

Lemly, A. D. 2007. A procedure for NEPA assessment of selenium hazards associated with mining. *Environmental Monitoring and Assessment* 125:361–75.

Lewis, R. L. 1992. Appalachian restructuring in historical perspective: Coal, culture, and social change in West

Virginia. Paper presented at the Conference on Regional Restructuring and Regeneration, Glasgow, Scotland.

Lewis, R. L., and D. B. Billings. 1997. Appalachian culture and economic development. *Journal of Appalachian Studies* 3:3–42.

Lindsey, R. 2007. Coal controversy in Appalachia. http://earthobservatory.nasa.gov/Features/MountaintopRemoval (last accessed 8 January 2011).

Little, T. N., and D. P. Blair. 2010. Mechanistic Monte Carlo models for analysis of flyrock risk. In *Rock fragmentation by blasting*, ed. A. Sanchidrián, 641–47. London and New York: Taylor & Francis Group.

Maranville, A. R., T.-F. Ting, and Y. Zhang. 2009. An environmental justice analysis: Superfund sites and surrounding communities in Illinois. *Environmental Justice* 2 (2): 49–58.

McGinley, P. C. 2004. From pick and shovel to mountaintop removal: Environmental injustice in the Appalachian coalfields. *Environmental Law* 34:21–106.

Mennis, J. 2002. Using geographic information systems to create and analyze statistical surfaces of population and risk for environmental justice analysis. *Social Science Quarterly* 83 (1): 281–97.

Mohai, P., P. M. Lantz, J. Morenoff, J. S. House, and R. P. Mero. 2009. Racial and socioeconomic disparities in residential proximity to polluting industrial facilities: Evidence from the Americans' Changing Lives study. *American Journal of Public Health* 99 (S3): S649–S656.

Mountain Association for Community Economic Development. 2009. *The economics of coal in Kentucky: Current impacts and future prospects*. Berea, KY: Mountain Association for Community Economic Development.

Mountaintop Mining. 2008. Some simple facts. http://www.mountaintopmining.com/default.htm (last accessed 28 July 2008).

Negley, T. L., and K. N. Eshleman. 2006. Comparison of stormflow response of surface-mined and forested watersheds in the Appalachian Mountains, USA. *Hydrological Processes* 20 (16): 3467–83.

Palmer, M. A., E. S. Bernhardt, W. H. Schlesinger, K. N. Eshleman, E. Foufoula-Georgiou, M. S. Hendryx, A. D. Lemly, et al. 2010. Science and regulation: Mountaintop mining consequences. *Science* 327:148–49.

Paybins, K. S., T. Messinger, J. H. Eychaner, D. B. Chambers, and M. D. Kozar. 2000. Water quality in the Kanawha–New River Basin West Virginia, Virginia, and North Carolina, 1996–98. U.S. Geological Survey Circular 1204. Denver, CO: USGS.

Peng, S. 2000. Mountaintop removal controversy slows West Virginia coal mining. *Mining Engineering* 52 (9): 53–58.

Salstrom, P. 1994. *Appalachia's path to dependency: Rethinking a region's economic history 1730–1940*. Lexington: The University Press of Kentucky.

Scott, R. R. 2007. Making sense of mountaintop removal: The logic of extraction in the southern West Virginia coalfields. PhD dissertation, Department of Sociology, University of California, Santa Cruz.

Stewart-Burns, S. L. 2005. The impact of mountaintop removal surface coal mining on southern West Virginia communities, 1970–2004. PhD dissertation, Department of History, West Virginia University.

U.S. Census Bureau. 2000. Census 2000 summary file 3 (SF3) custom table: Employed civilian population 16 years and over; Mining, by place (West Virginia). http://factfinder.census.gov (last accessed 25 April 2009).

U.S. Energy Information Administration. 2009. Annual coal report 2009: Table 7. http://www.eia.doe.gov/cneaf/coal/page/acr/acr_sum.html (last accessed 10 October 2010).

U.S. Environmental Protection Agency. 2005. *Final programmatic environmental impact statement: Mountaintop mining/valley fills in Appalachia*. Philadelphia: U.S. EPA.

U.S. Senate Environment and Public Works Committee. 2010. The Mingo Logan Spruce No. 1 Mine: EPA ignores transparency, concerns of West Virginia officials. http://www.epw.senate.gov/public/index.cfm?FuseAction=Files.View&FileStore_id=ec6c3919-4949-452d-b0d3-6ca95e441494 (last accessed 23 January 2010).

Ward, K. 1999. Logan County leaders argue against Haden mine ruling. *Charleston Gazette* March 11. http://www.wvgazette.com/static/series/mining/haden0311.html (last accessed 11 January 2011).

West Virginia Coal Association. 2009. *Coal facts 2009*. Charleston: West Virginia Coal Association.

West Virginia Department of Environmental Protection. 2008. Mining permit database. CD-ROM. Charelston: West Virginia Department of Environmental Protection.

Woods, B. R. 2010. Social well-being in the Appalachian coalfields. PhD dissertation, Department of Agricultural Economics and Rural Sociology, The Pennsylvania State University.

Woods, B. R., and A. L. Meyer. 2006. Mountaintop removal mining in Appalachia: The context of poverty, the dearth of employment, and the outrage of residents in the West Virginia coalfields. Paper presented at the annual meeting of the Rural Sociological Society, Louisville, KY.

Zeller, T. 2010. Beyond fossil fuels: A battle in mining country pits coal against wind. *New York Times* 15 August:BU1, NY edition.

Zimmerer, K. S. 2011. New geographies of energy: Introduction to the special issue. *Annals of the Association of American Geographers* 101 (4): 705–11.

Enforcing Scarcity: Oil, Violence, and the Making of the Market

Matthew T. Huber

Department of Geography, The Maxwell School of Syracuse University

Oil has always been at the center of discussions of resource scarcity. Over the last decade of volatile and often rising oil prices, a vast "peak oil" literature has emerged citing the geological finitude of petroleum as a harbinger of an era of catastrophic energy scarcity. Many analysts focused on the geopolitics of oil also presume that natural oil scarcity is the primary driver of global conflict and "resource wars." In contrast, I follow geographical discussions of the social production of scarcity, to problematize oil scarcity as not a geological fact but as a social relationship mediated by capitalist commodity relations. Specifically, I focus on the role of violence in socially producing the scarcity necessary for the oil market to function. I first discuss the broader historical and legal problems of "overproduction" in the United States. I then examine the 1931 declaration of "martial law" in the oil fields of east Texas and Oklahoma in a moment of lax depression-era demand, glut, and collapsing oil prices. I argue that violently imposing oil scarcity was not merely sectoral but a broader project of stabilizing the chaotic oil market in accordance with the reorganization of capitalism during the 1930s. Such stabilization was critical for the emergence of an oil-powered Fordism in the postwar United States responsible for the intractable patterns of oil demand so vexing to energy policymakers today. I conclude by suggesting that contemporary debates on petro-imperialism might consider questioning the role of violence not as a product but as a generator of scarcity.
Key Words: energy history, oil, scarcity.

石油一直在资源稀缺讨论的中心。在过去动荡的和石油价格屡次上涨的十年中，出现了一个巨大的“峰值石油”文献，这些文献引用石油的地质限定作为一个时代灾难性的能源紧缺预言者。许多着重于石油地缘政治分析的人士也推测天然石油短缺是全球性冲突和“资源战争”的主要决定因子。相反，我遵循社会生产稀缺的地域讨论，把石油稀缺作为一种资本主义商品关系介导的社会关系而不是地质事实来命题。具体来说，我将重点放在石油市场运行所需的社会性生产稀缺的暴力作用上。我首先讨论在美国“生产过剩”更广泛的历史和法律问题。然后我调查了在 1931 年这个宽松的大萧条时代的需求，供过于求，和油价暴跌的时刻，德克萨斯州东部和俄克拉荷马州油田的“戒严法”声明。我认为，暴力地强加石油稀缺不仅是行业的，而是一个在 20 世纪 30 年代期间与资本主义的整顿相符合的更广泛的稳定混乱石油市场的工程。这种稳定对战后美国一个以石油为动力的福特主义的出现是关键的，它也对使决策者伤脑筋的石油能源需求难以处理的形势负责。我总结建议当代的石油帝国主义论辩可以考虑质疑暴力作为一个石油稀缺的发电机而不是产品的作用。*关键词：能源的历史，油，稀缺。*

El petróleo ha estado siempre en el centro de las discusiones sobre escasez de recursos. Durante la pasada década, caracterizada por precios volátiles y a menudo en aumento, apareció una literatura del "pico del petróleo" que cita lo finito de la vigencia geológica de este hidrocarburo, como un presagio de una era de catastrófica escasez de energía. También, muchos analistas enfocados en la geopolítica del petróleo presumen que la escasez natural de este combustible es el principal motor de conflicto global y de "las guerras de los recursos". En contraste, yo sigo las discusiones geográficas sobre la producción social de escasez, para problematizar la escasez de petróleo no como un hecho geológico sino como relación social mediada por relaciones mercantiles capitalistas. Específicamente, me concentro en el papel de la violencia para producir socialmente la escasez necesaria para que el mercado del petróleo funcione. Primero discuto los más amplios problemas históricos y legales de "superproducción" en Estados Unidos. Luego examino la declaración de "ley marcial" de 1931 en los campos petroleros del oriente de Texas y Oklahoma en un momento de floja demanda asociada a la era de la depresión, la saturación y colapso de precios del crudo. Argumento que el imponer con violencia la escasez de petróleo no fue meramente sectorial sino un proyecto de mayor alcance para estabilizar el caótico mercado del petróleo, de acuerdo con la reorganización del capitalismo durante los años 1930. Tal estabilización era crítica para la emergencia de un Fordismo propulsado por petróleo en los Estados Unidos de la posguerra, responsables de los patrones inmanejables de la demanda de petróleo, tan irritantes para quienes orientan las políticas petroleras de nuestros días. Concluyo sugiriendo que los debates contemporáneos sobre petro-imperialismo podrían considerar poner en picota la discusión del papel de la violencia no como un producto sino el generador de la escasez. *Palabras clave: historia de la energía, petróleo, escasez*

For the first time since the Industrial Revolution, the geological supply of an essential resource will not meet demand.

—Kenneth Deffeyes (2005, xi)

America in the year 2001 faces the most serious energy shortage since the oil embargoes of the 1970s.

—Dick Cheney's "Energy Task Force" Report (National Energy Policy Development Group 2001, viii)

A world of rising powers and shrinking resources is destined to produce intense competition among an expanding group of energy-consuming nations.

—Michael Klare (2008, 7)

These statements come from three different perspectives—geological science, presidential power, and left-wing critique—yet converge on the same conclusion: Resource scarcity is a fact likely to produce concerning effects. After a protracted period of "resource triumphalism" (Bridge 2001), the past decade has revived predictions of scarcities for key resources such as petroleum, water, and food. In 2008, a year characterized by rising food prices and $147 per barrel oil, the *Wall Street Journal* published an article entitled, "New Limits to Growth Revive Malthusian Fears" with special attention paid to the billions of Indians and Chinese, "stepping up to the middle class" (Lahart, Barta, and Batson 2008, A1).

Because of its energetic importance in industrial life, petroleum is perhaps the most contested in discussions of resource scarcity. Over the last decade of volatile and often rising prices, peak oil arguments cite the geological finitude of petroleum as a harbinger of an era of energy scarcity (Heinberg 2003; Deffeyes 2005; Kunstler 2005). Many analysts focused on the geopolitics of oil also start from the premise that the natural scarcity of oil is the primary driver of "resource wars" (e.g., Klare 2008). What is striking is how the category of "scarcity" is the *premise* of the discussion—as a geological fact that societies confront as an external problem. It seems that little has changed in the thirty-seven years since Harvey (1974, 272) quipped, "It is often erroneously accepted that scarcity is something inherent in nature, when its definition is inextricably social and cultural in origin." Since Harvey's intervention, several geographers have deepened our understanding of the multiple social, political, and institutional relations that combine to produce scarcity (e.g., Yapa 1996; Peluso and Watts 2001; Birkenholtz 2009). From this perspective, scarcity is not the premise but the object of explanation itself. The aim of this article is to link this geographical perspective on scarcity with wider critiques of oil scarcity (e.g., Nitzan and Bichler 2002; Caffentzis 2005; Retort 2005; Le Billon and Cervantes 2009; Labban 2010). Although many processes contribute to the production of petroleum scarcity (e.g., state policies, inequality, cartels), I focus on the role of violence in producing the scarcity necessary for the operation of commodity relations. In so doing, I argue that violence is *constitutive* of the crisis-prone and unstable capitalist market.

This article contains four sections. First, I review the literature on environment, scarcity, and violence to argue that more attention should be paid to the crisis-prone nature of resource-based political economies. Second, despite repeated fears of the looming scarcity of petroleum and the violent repercussions, I review how the political economy of oil is more profoundly shaped by fears of overproduction and glut. Third, I review the history of early U.S. oil production and its endemic tendencies toward overproduction. Fourth, I examine the U.S. oil "overproduction" crisis during the 1930s and the military response.

Violence and the Production of Scarcity

Scarcity is treated as a natural fact of economic life. The discipline of economics is often defined as the study of how the forces of supply and demand combine to allocate "scarce resources." It is rarely pointed out that scarcity must itself be produced for commodity markets and the price mechanism to function (Harvey 1974). This often stands in contradiction to the gigantic capacity of capitalist industrial production to produce "an immense collection of commodities" (Marx [1867] 1976, 125). Thus, specific institutional arrangements are often constructed by the state and capital to limit the production of commodities to stabilize prices and secure profits for particular fractions of capital. In other words, scarcity has to be socially produced. Although others have focused on institutional aspects of produced scarcity (e.g., Yapa 1996; Berkenholtz 2009), less attention has examined the role of violence.

In opposition to the tendency to presume that violence is the natural product of scarcity (Kaplan 1994; Homer-Dixon 1999), several have attempted to understand violence and scarcity as coproduced through the uneven development of capitalism (Watts 1983; Keen 1998; Peluso and Watts 2001; Dalby 2002; Cramer 2007). The mutually constitutive nature of violence and capitalist market relations has been approached from a number of ways. Most notably, as Marx ([1867] 1976, 874) famously argued and Harvey (2003) rekindled interest in, capital accumulation is predicated on "conquest, enslavement, robbery, murder, in short, force." In particular, capital is made possible through the violent dispossession of the majority of producers from the means of producing their own livelihood. In discussing the displacement associated with resource extraction in Indonesia, Tsing (2004, 68) commented, "Violence became key to ownership." Moreover, as capital is reproduced through unequal access to the means of subsistence, the market itself generates a form of abstract violence where commodity relations structure everyday forms of hunger and displacement (Heynen 2009). Through the colonization of life by the market, Taussig (1980, 18) famously argued that those "neophyte proletarians" recently subsumed within commodity relations, "understand the world of market relations as intimately associated with a spirit of evil . . . barrenness and death."

Yet, these forms of market violence are operable, in fact necessary, to the stable circulation of capital and its foundation on proletarianization and uneven development. As the current conjuncture makes clear, capital accumulation is also prone to cycles of market collapse and crisis. Lurking in the instability of capitalist markets is the continual possibility of economic chaos and dysfunction. In moments of market breakdown and perceived "emergency," Benjamin (1999, 287) argued that "law preserving" forms of state violence become imagined as necessary. In this case, the deployment of state force is meant explicitly to preserve the legal-institutional frameworks of the market itself—private property, profit, and exploitation.

Although state violence in response to market collapse is often seen as emerging out of scarcity—particularly the scarcity of labor through "general strike"—crises are more often not a product of scarcity but overaccumulation, overproduction, and excess capacity (Harvey 1982, 190–203; Brenner 2002; McNally 2009). The problems of overproduction are illustrated through the political economy of boom and bust that afflicts natural resource producing sectors. Although on a regional level depletion and scarcity often come along with the "bust" cycle (Clapp 1998), at the level of the world market abundance causes commodity price collapses and crisis in resource-producing regions (Freudenberg 1992; Auty 2004; Wright and Czelusta 2007). Such crisis conditions can surely lead to different forms of social displacement, poverty, and violence (Le Billon 2004). Thus, although Fairhead (2001, 221) suggested that resource "wealth" (not scarcity) should be positioned as a key driver of violent conflict, overproduction and the consequent incapacity of resources to generate "wealth" seems just as likely to produce conditions of desperation and possible violence. Although the past decade has renewed the boom cycles of high prices and scarcity narratives (Klare 2008), the period through which Homer-Dixon (1999) and Kaplan (1994) couched their "scarcity causes violence" arguments was one characterized by glut and falling prices for several natural resources (Bridge 2001). As a finite, yet often overproduced, commodity, oil is at the center of debates over scarcity and violence.

Oil Scarcity and Violence

Since at least the 1920s petroleum geologists have warned of the looming exhaustion of U.S. petroleum supplies (Yergin 1991; Olien and Olien 2000). The most recent manifestation of this narrative is those proclaiming the imminence of "peak oil" when global production reaches its peak and goes into decline. Given the fixed nature of petroleum deposits, it is clear that petroleum production will indeed peak at some point, but *predicting* the peak is much more difficult than peak oil proponents claim (Smil 2008). As some critics point out,[1] peak oilers tend to downplay capital's dynamic capacities for innovation. From a Marxist perspective (e.g., Labban 2010), peak oil theorists fail to theorize the wider social relations through which both the "supply" and "demand" of oil are structured. Supply is certainly not the result of petroleum producers producing at the maximum capacity that geology allows,[2] and demand is not always increasing and actually declines in moments of crisis.[3] Moreover, they approach oil as simply a geological "thing" and not a *commodity* produced for exchange and profit (Caffentzis 2005, 170–76). Quite apart from geological supply, if producers do not expect the money (M) invested in producing oil (C) will produce a profit (M′) then they will not supply the market.[4] Indeed, private and national oil capitals are still investing substantial money in oil exploration and development with the expectation that the oil age will continue for years to come (Bridge and Wood 2005).[5]

Although petroleum is indeed a finite resource, and its limitation certainly should not be ignored, Labban (2008, 2) commented that history reveals "the problem of oil is not its scarcity but its abundance." Retort (2005, 59) argued that the "constant menace" threatening the oil industry is not scarcity but "falling prices . . . surplus and glut." If oil scarcity is not actively managed, prices will not be high enough for profitable accumulation. The oil market is beset by a struggle to find a balance between abundance and scarcity that both allows high enough prices for profitable accumulation and low enough prices to maintain levels of demand.

What is striking is how difficult it is to achieve the scarcity necessary for market stability and a profitable oil market. From the Texas Railroad Commission to the Organization of Petroleum Exporting Countries (OPEC), elaborate institutional mechanisms have been constructed with the purpose of *limiting* oil production. After high prices during the 1970s, non-OPEC production combined with OPEC's own failure to control each others' production levels to cause price collapses throughout the 1980s and 1990s. In 1986, Vice President George H. W. Bush was forced to visit Saudi Arabia to plead with them to limit production on behalf of U.S. independent oil producers (Yergin 1991). A glutted market during the financial crisis of 2008 forced petroleum sellers to store 80 million barrels of oil in thirty-five idling supertankers (Krauss 2009).

When scarcity is presumed as a fact of nature and not a remarkable social achievement, it is posited as the natural cause of violence and war. Over the last decade, many have argued that oil scarcity is ushering in a new era of "resource wars" and geopolitical conflict (Heinberg 2003; Kunstler 2005; Klare 2008). Yet, given that the world was "drowning in oil" as recently as a decade ago ("Drowning in oil" 1999), it is questionable whether or not oil scarcity should be so quickly assumed as the cause of recent war and conflict. Nitzan and Bichler (2002) provocatively argued that the abundance of oil (low prices) is the best predictor of violence in the Middle East. Le Billon and Cervantes (2009, 842) examine how violence triggers a "scarcity . . . narrative . . . constructed for and through prices." The appearance of scarcity and violence yields high prices and, thus, high returns for particular interests. Thus, although common sense expects violence to emerge out of oil scarcity, few have explored the possibility that violence could be a response to oil abundance and falling prices. In what follows, I examine perhaps the most spectacular case of the use of violence to produce scarcity—the declaration of martial law in the oil fields of east Texas and Oklahoma.[6] First, I provide some background on the legal geography of U.S. petroleum production and its endemic problems of overproduction.

Overproduction and the Legal Geography of U.S. Oil

For much of its history, the U.S. oil market was constantly threatened by overproduction: "Economic disaster stalked the industry with the discovery of each new field" (Mills 1960, 3). Unlike much of the rest of the world, ownership of subsurface minerals in the United States is delegated to private owners on nonpublic lands (Mommer 2002). This created a diverse geography of multiple producers dispersed through the landscape of private property parcels. Concretely, overproduction resulted from a specific legal decision in 1889 by the Pennsylvania Supreme Court known as "the rule of capture" (Zimmerman 1957, 91–100). The court likened petroleum to a "fugitive" substance that, like "wild game," moves below the surface of the earth and declared that, "If an adjoining, or even a distant, owner drills his own land, and taps your gas, so that it comes into his well and under his control, it is no longer yours, but his" (Thorton 1918, 43). Thus, often several private property owners found themselves all having reasonable "property rights" to a single oil pool beneath the surface.[7] This law corresponds to a "Lockean" notion of property that imbues property rights to any individual who "mixes their labor" with the land (Brand 1983, 106). The courts reasoned that the rule of capture would allow for more production and thus cheaper natural resources for consumers. Yet, the only rational response to this situation was to quickly pump oil for fear that your neighbor would suck the oil from underneath your property.[8] An oil discovery automatically created a chaotic rush to produce as much oil as possible from each individual "lease." This created the prototypical landscape of U.S. oil production as a "forest of derricks" with each well producing against the other in a competitive race to maximize resource wealth. The process is vividly described by a rig operator: "Boy, when you get in there with a good rig, with probably thirty or forty rigs right around you, and everyone trying to beat the other one, it's fascinating" (quoted in Boatright and Owens 1970, 157).

This chaotic race mentality often had disastrous consequences from the standpoint of conservation. Haphazard production depleted the gas or water pressure needed to push the oil above the surface and left untold barrels of oil trapped underground. Moreover, the rush to extract petroleum often took precedence over the ability to store or transport it, leaving oil evaporating in nearby dug holes and flowing through local waterways (Zimmerman 1957). Beyond conservation concerns, overproduction produced substantial market volatility and the constant threat of collapsing prices.

The obvious solution to this contradiction between property and geology was to rescale property relations in harmony with the shape of the geological pool. This was central to the "unitization" movement to rationally manage oil deposits as a single unit (Yergin 1991, 220–23). But private property was embedded in the longstanding geographies of private landownership buoyed by dreams of "magical" oil wealth (cf. Watts 1994; Coronil 1997). That wealth was often imagined to be tapped by the "independent wildcatter,"[9] who discovers and develops his own oil with entrepreneurial tenacity. As famous Texas oil regulator E. O. Thompson stated, "I am of the opinion that unitization means the extermination of the little man" (Texas State Archives 1932). Within each geological oil "unit" existed countless "little" landowners and "little" producers. The idea of centralized, collective management of oil pools as a natural unit was a "drastic and large scale a departure from the basic principles of maximum freedom for individuals" (Zimmerman 1957, 345). More important than geological integrity, the "little man" stood in for American market competition and petty commodity production (de Janvry 1981, 187). The question was how scarcity could be produced without violating the ideological conditions of American capitalism.

The Wake of the Flood: Martial Law and the "Black Giant"

As endemic as overproduction was, its market effects remained regional until the late 1920s and early 1930s and the emergence of the Great Depression (Olien and Hinton 2007). In October 1930, wildcatter C. M. "Dad" Joiner discovered the single largest lower-forty-eight oil field in the history of the United States—the Black Giant, or the "East Texas Oil Field" (Clark and Halbouty 1972). It was forty-two miles long and four to eight miles wide, containing thousands of private property tracts (Mills 1960). In a case of terrible timing, as depression-era demand for petroleum collapsed, production from the East Texas field soared. The major companies and their geologists "assumed . . . [that the discovery well] was a west edge well, and the pool lay east" (Fischer 1955, 2). As independents gathered up oil-rich leases to the west, the majors drilled dry wells to the east. The shallow depth and ease of production along with nearby cheap supplies of lumber for oil derricks made it ideal for independent producers (Olien and Hinton 2007). Thus, the field had an abnormally high level of independent operators that had leased approximately 50 to 60 percent of the acreage (Nordhauser 1979). These highly indebted independents sought to tap resource wealth quickly, but this only exacerbated the overproduction problem (Mills 1960).

By June 1931, 700 wells were drilled and the field was producing 350,000 barrels per day (bpd), or nearly 15 percent of total domestic consumption (Olien and Hinton 2007, 58). At the same time, the price of oil plummeted—15 cents per barrel nationally. In some parts of Texas, oil only cost 2.6 cents per barrel in an industry that considered $1 a bare minimum for profitable production (Nordhauser 1979, 29; Yergin 1991, 259). The discovery of the East Texas field threatened to bankrupt the entire industry. In a context where oil was increasingly seen as the "lifeblood" of society, the bankruptcy of the oil industry represented an unacceptable proposition.

Militarizing Scarcity—Martial Law in the Oil Fields

> [T]he "law" of the police really marks the point at which the state, whether from impotence or because of the immanent contradictions within any legal system, can no longer guarantee through the legal system the empirical ends that it desires at any price to attain. (Benjamin 1999, 287)

With the oil market in disarray, it was clear that some mechanisms had to be put into place to control production in East Texas. For a variety of reasons, the regulatory authority for petroleum conservation fell to the Texas Railroad Commission (TRC; Childs 2005), but the TRC had not demonstrated the ability to enforce the "proration" policies meant to set "allowables" for production in East Texas. Such policies held the possibility of stabilizing the market by coordinating production levels with projected demand. Yet, in June 1931, a federal court ruled that the TRC was attempting to regulate "economic waste"—oil produced over market demand—rather than "physical waste"—oil production that threatened the field's geological integrity. The state conservation law of 1929 only allowed the TRC to regulate physical waste. Any attempt to control production over market demand was considered an uncompetitive form of price fixing.

As the Benjamin (1999) quote suggests, the state could resort to "police" forms of violence as a result of the impotence and general contradictions of an existing legal structure. Initially, Texas Governor Ross Sterling, a former oil executive and a proponent of free markets, was confident in the legal authority of the TRC and the market to solve the problem (Mills 1960). Yet, by August 1931, East Texas production was at 1 million barrels per day and prices were still floating around 10 cents per barrel (Mills 1960). In Oklahoma, Governor Bill Murray declared martial law and closed 3,016 wells in the state, vowing not to open them until the price reached $1. Word started to circulate from East Texas that a coalition of landowners and 1,500 oil producers were organizing to violently dynamite drilling rigs, pipelines, and other property associated with illicit producers and "hot oil" transporters (Eason 2005; see also Mills 1960). The major company, Humble Oil and Refining Co. (later Exxon), hired armed security to protect its property (Mills 1960).

Sterling was forced to reverse his earlier confidence. On 17 August 1931, he declared that, "There exists an organized and entrenched group of crude petroleum and natural gas producers . . . who are in a state of insurrection against the conservation laws of the State" (Texas State Archives 1931a). Martial law was declared and 4,000 troops from the Texas National Guard were sent to enforce the "allowable" production levels (Nash 1968). Initially the "troops encountered no resistance . . . [and] . . . compliance was prompt and universal" (Mills 1960, 28). By the end of August with nearly 1 million barrels per day from Texas and Oklahoma off the market, prices started to rebound toward 70 cents per barrel (Mills 1960).

Although martial law was orderly, there was widespread belief that violence always "lies near the surface of East Texas" (Mills 1960, 25). Many oil producers and landowners "interpreted the shutdown . . . as a personal affront" (Eason 2005, 50). As the field came under martial law, General Jacob Wolters sent telegrams to Governor Sterling, with stark warnings of what would happen if martial law was lifted: "All day men have openly said that if the Governor could not hold the field down, they were going to take charge of it themselves, hang a few S.B.s and if necessary dynamite pipelines and save this field from destruction and save the oil industry" (Texas State Archives 1931b).

By sending armed troops to the oil fields, the declaration of martial law was positioned more as a *threat* of the use of violence in the face of the possible breakdown of the foundations of capitalist society based on private property, competition, and profit. Although the oil boom created an influx of law-breaking roughnecks, bootleggers, and prostitutes swelling the social infrastructures of small East Texas towns like Kilgore (Eason 2005), the real threats were the direct violent destruction of private property by those disaffected with the capacity of "law" to solve the crisis and the indirect destruction of private property and the capacity of oil to yield profit if overproduction continued. The "insurrection" was not an action of armed groups against the state but a multiplicity of individuals armed with nothing more than a drilling rig. The insurgent was oil itself. As Nordhauser (1979, 43) remarked, "the insurrection was suppressed by turning a set of valves." Precisely because oil was not scarce, such overproduction made profitable accumulation impossible. Scarcity, or the presupposition of capitalist commodity relations, had to be enforced militarily if the capitalist market was to regain functionality.

By the late fall of 1931, prices rebounded to over $1 per barrel. As Figure 1 illustrates, Sterling and Murray were credited with saving the industry by using executive power to lift prices. Yet, in early 1932, martial law was declared unconstitutional by the federal courts when independent refiners, who welcome low oil prices, sued the state of Texas, calling martial law "an arbitrary and tyrannical taking of property" (Mills 1960, 37). Sterling claimed that martial law was necessary in terms of the "law of self-defense as applied to nation or state" (Mills, 1960, 38). This defense was difficult to substantiate. After his failure to win reelection in 1932, Sterling was more candid about the real purpose of martial law: "I have no apology for martial law in East Texas. . . . It saved the state $6 million in taxes and it saved the people of Texas $40 million in the value of their products" (quoted in Mills 1960, 41). In other words, violence was necessary in the production of value and the reproduction of Texan state power. Violence, scarcity, and the market were intertwined.

Aftermath: Institutionalizing Scarcity

Once martial law was lifted, intransigence returned to East Texas and prices plummeted again. After the

Figure 1. "Said the Governor to the Governor," *Daily Oklahoman* 22 August 1931. Records of Ross S. Sterling, Texas Office of the Governor, Texas State Archives, Austin, TX.

inauguration of Franklin D. Roosevelt in 1933, the dependable production of oil was accorded national importance; it was emerging as both the lifeblood of American society and a strategic resource for national defense.[10] Thus, solving the oil crisis was not merely a sectoral issue for the oil industry but also critical for the wider New Deal project of rescaling American capitalism.

FDR directed the Secretary of Interior, Harold Ickes, to solve the crisis. Ickes quickly raised alarms, claiming that the industry was in a state of "utter demoralization" and its failure threatened "the general economic situation in the country" (Franklin D. Roosevelt Library 1933). Many believed that this problem could not be solved by the states and needed strong federal action. Figure 2 portrays Texas as a site of waste, unregulated competition, and blithe disregard for the profoundly national significance of oil resources. There was substantial momentum, even among industry leaders, to declare Ickes "oil dictator" and nationalize the oil industry as a public utility. One 1933 telegram to FDR from an oil producer in East Texas put it simply, "We want an oil dictator" (Franklin D. Roosevelt Library 1933).

Texas, by its opposition to "federal encroachment" in the oil fields, recklessly squanders the nation's resources

Figure 2. Political cartoon illustrated by Daniel R. Fitzpatrick, in White, O. P. 1935. Piping hot. *Collier's Weekly* 12 January 95 (2): 11.

Despite Ickes's best efforts, he could not overcome ideological stubbornness toward free enterprise and state's rights. By 1935, a complicated system was devised to limit production throughout the midcontinental states.[11] Ultimate authority remained with state conservation agencies (such as the TRC), and an Interstate Oil Compact Commission was formed where states could share information and coordinate production levels in line with demand. The Federal Bureau of Mines published monthly estimates of national demand that became the statistical groundwork for the state allowables. Production was controlled so as not to exceed what was referred to as "reasonable market demand" (Lovejoy and Homan 1967, 130). Demand's very reasonability was, of course, internally related to the emerging Fordist social formation dependent on automobility and mass oil consumption. The price level that induced this "reasonable" demand was not a natural product of market equilibrium but a concerted effort to protect oil profits from the ruinous effects of competition. Thus, this "conservation" policy actually locked in mammoth patterns of demand that have led to current concerns with the U.S. "oil addiction." That demand itself might represent a conservation problem simply never entered the minds of the relevant policymakers (cf. Zimmerman 1957, 46). Thus, Fordism was powered not by cheap oil but cheap *enough* oil to protect profits and spur mass consumption. After the exceptional period of World War II where production was ramped up to near full capacity to fuel the war effort (Yergin 1991), this scalar coordination of state and federal regulation worked well to keep nominal prices stable, if not increasing, throughout the postwar period up until the crisis of the 1970s (Libecap 1989). In fact, the state conservation schemes were the model studied in the formation of OPEC in their efforts to limit their own production levels (Yergin 1991).[12]

Conclusion

This article should not be interpreted as a denial of scarcity as a natural limit, but in capitalism in general, and the oil market in particular, commodity relations presuppose the social production of scarcity. Capitalism is characterized by its capacity for abundance amidst enforced scarcity. Moreover, too many analysts take natural scarcity as the premise of their critique of U.S.

petro-imperialism. Most critics of the Iraq war assume that oil scarcity drove those in power, like Dick Cheney, to seek control of Iraqi oil. Yet, a look at the evidence does not support this view. Several years after the U.S. invasion, Iraq remains mired below prewar production levels (Walt 2009). Iraq still has the second or third[13] highest proven oil reserves in the world, but constant violence over the last thirty years—much as a direct result of U.S. policy—has prevented production, let alone exploration. Indeed, it was Saddam Hussein's determination to increase Iraqi production in awarding contracts to Russian and French oil firms that presaged the invasion (Banerjee 2003). Although there is still much uncertainty about what exactly drove the U.S. invasion, the question should at least be raised (e.g., Labban 2008, 163) as to whether or not the motives mirrored those of Governor Sterling in Texas: Was violence used to *prevent* the production of oil that threatened to return oil capital into the less profitable eras of the 1980s and 1990s?

Acknowledgments

The author is grateful to the Franklin D. Roosevelt Presidential Library for their support of this research. He would also like to thank Karl Zimmerer for his editorial guidance and Jody Emel, Deborah Martin, Dick Peet, James McCarthy, Jesse Goldstein, George Caffentzis, and three anonymous reviewers for helpful comments on various iterations of this article (the usual disclaimers apply).

Notes

1. The most prominent being Yergin and the Cambridge Energy Research Associates, who believe in the power of high oil prices to spur technical innovation and increased production capacity (e.g., Jackson 2009).
2. Texas and later Saudi Arabia were famous as "swing producers" who could increase production at any moment. Most estimate that Iraq is currently producing at only about 20 percent of present capacity (Walt 2009).
3. Many suggest that the United States has reached peak oil demand (e.g., Jackson 2009).
4. This is complicated with the emergence of so many state-run national oil companies, but even most of these "public" enterprises operate with a profit orientation (Mommer 2002: Labban 2008).
5. For example, Exxon/Mobil's recent $4.1 billion for a field off the coast of Ghana (Mouawad and Kouwe 2009).
6. This case study is just that. Space precludes a more general examination of the wider relations among oil, violence, and U.S. policy over the last century.
7. Similar problems beset groundwater property regimes (Emel, Roberts, and Sauri 1992).
8. A process described in the 2007 film, *There Will Be Blood*, as drinking your milkshake. This phrase, "I drink your milkshake," has become infused into pop culture, inspiring T-shirts and Web sites.
9. The term *independent* usually refers to oil producers, refiners, or marketing outlets that were nonintegrated and did not, like "major companies," control oil from the production well to the gasoline station (Olien and Hinton 2007, 1–11).
10. The importance of oil in fueling war was made abundantly clear during World War I but was emphasized even more during World War II.
11. It should be noted that California—a major oil producer—was a somewhat separate concern, as Californian oil was largely consumed within the western part of the United States (Sabin 2005).
12. OPEC has never replicated the kind of market stability enjoyed by the United States during the postwar period.
13. Depending on whether or not the vast Tar Sands of Alberta are included in Canada's reserve figures.

References

Auty, R. M., ed. 2004. *Resource abundance and economic development*. New York: Oxford University Press.

Banerjee, N. 2003. In Iraqi fields, technical and political challenges. *New York Times*, 13 April:11.

Benjamin, W. 1999. Critique of violence. In *Selected writings: Vol. 1. 1913–1926*, ed. M. Bullock and M. W. Jennings, 277–300. Cambridge, MA: Harvard Belknap Press.

Birkenholtz, T. 2009. Irrigated landscapes, produced scarcity, and adaptive social institutions in Rajasthan, India. *Annals of the Association of American Geographers* 99 (1): 118–37.

Boatright, M. C., and W. A. Owens. 1970. *Tales from the derrick floor: A people's history of the oil industry*. New York: Doubleday.

Brand, D. R. 1983. Corporatism, the NRA, and the oil industry. *Political Science Quarterly* 98:99–118.

Brenner, R. 2002. *The boom and the bubble: The U.S. and the world economy*. London: Verso.

Bridge, G. 2001. Resource triumphalism: Postindustrial narratives of primary commodity production. *Environment and Planning* A 33 (12): 2149–73.

Bridge, G., and A. Wood. 2005. Geographies of knowledge, practices of globalisation: Learning from the oil exploration and production industry. *Area* 37 (2): 199–208.

Caffentzis, C. G. 2005. *No blood for oil! Energy, class struggle and war, 1998–2004*. RadicalPolytics.org. http://www.radicalpolytics.org/caffentzis/no_blood_for_oil-entire_book.pdf (last accessed 31 March 2011).

Childs, W. R. 2005. *The Texas Railroad Commission: Understanding regulation in America to the mid-twentieth century*. College Station: Texas A&M University Press.

Clapp, R. A. 1998 The resource cycle in forestry and fishing. *The Canadian Geographer* 42 (2): 129–44.

Clark, J. A., and M. T. Halbouty. 1972. *The last boom*. New York: Random House.

Coronil, F. 1997. *The magical state: Nature, money, and modernity in Venezuela*. Chicago: University of Chicago Press.

Cramer, C. 2007. *Violence in developing countries: War, memory, progress*. Bloomington: Indiana University Press.

Dalby, S. 2002. *Environmental security*. Minneapolis: University of Minnesota Press.

Deffeyes, K. S. 2005. *Beyond oil: The view from Hubbert's Peak*. New York: Hill and Wang.

de Janvry, A. 1981. *The agrarian question and reformism in Latin America*. Baltimore: Johns Hopkins University Press.

Drowning in oil. 1999. *The Economist* 5 March:19.

Eason, A. 2005. *Boomtown: Kilgore, TX*. Kilgore, TX: East Texas Oil Museum.

Emel, J., R. Roberts, and D. Sauri. 1992. Ideology, property, and groundwater resources: An exploration of relations. *Political Geography* 11 (1): 37–54.

Fairhead, J. 2001. International dimensions of conflict over natural and environmental resources. In *Violent environments*, ed. N. Peluso and M. Watts, 83–116. Ithaca, NY: Cornell University Press.

Fischer, F. W. 1955. Oral history interview of Judge F.W. Fischer, Tape 172, 5/21/55, "Pioneers in Texas Oil," Box 3K23, First Folder, Center for American History, University of Texas-Austin, TX.

Franklin D. Roosevelt Library. 1933. FDR Official File 56, Box 1, Franklin D. Roosevelt Library and Museum, Hyde Park, NY.

Freudenberg, W. 1992. Addictive economies: Extractive economies and vulnerable localities in a changing world economy. *Rural Sociology* 57 (3): 305–32.

Harvey, D. 1974. Population, resources and the ideology of science. *Economic Geography* 50 (3): 256–77.

———. 1982. *The limits to capital*. Oxford, UK: Blackwell.

———. 2003. *The new imperialism*. New York: Oxford University Press.

Heinberg, R. 2003. *The party's over: Oil, war and the fate of industrial societies*. Gabriola Island, BC, Canada: New Society Publishers.

Heynen, N. 2009. Bending the bars of empire from every ghetto to feed the kids: The Black Panther party's radical anti-hunger politics of social reproduction and scale. *Annals of the Association of American Geographers* 99 (2): 406–22.

Homer-Dixon, T. F. 1999. *Environment, scarcity and violence*. Princeton, NJ: Princeton University Press.

Jackson, P. M. 2009. The future of global oil supply: Understanding the building blocks. Cambridge, MA: CERA.

Kaplan, R. D. 1994. The coming anarchy. *The Atlantic Monthly* February:44–76.

Keen, D. 1998. *The economic function of violence in civil wars*. New York: Oxford University Press.

Klare, M. T. 2008. *Rising powers, shrinking planet: The new geopolitics of energy*. New York: Metropolitan Books.

Krauss, C. 2009. Where is oil going next? *New York Times* 15 January:B1.

Kunstler, J. H. 2005. *The long emergency: Surviving the converging catastrophes of the twenty-first century*. New York: Atlantic Monthly Press.

Labban, M. 2008. *Space, oil, and capital*. London and New York: Routledge.

———. 2010. Oil in parallax: Scarcity, markets and the financialization of accumulation. *Geoforum* 41 (4): 541–52.

Lahart, J., P. Barta, and A. Batson. 2008. New limits to growth revive Malthusian fears. *Wall Street Journal* 24 March:A1.

Le Billon, P. 2004. The geopolitical economy of resource wars. *Geopolitics* 9 (1): 1–28.

Le Billon, P., and A. Cervantes. 2009. Oil prices, scarcity and the geography of war. *Annals of the Association of American Geographers* 99 (5): 836–44.

Libecap, G. D. 1989. The political economy of crude oil cartelization in the United States, 1933–1972. *Journal of Economic History* 49 (4): 833–55.

Lovejoy, W. F., and P. T. Homan. 1967. *Economic aspects of oil conservation regulation*. Baltimore: Johns Hopkins University Press.

Marx, K. [1867] 1976. *Capital*. Vol. I. Trans. B. Fowkes. London: Penguin.

McNally, D. 2009. From financial crisis to world slump: Accumulation, financialisation, and the global slow-down. *Historical Materialism* 17 (2): 35–83.

Mills, W. E. 1960. *Martial law in east Texas*. Birmingham: University of Alabama Press.

Mommer, B. 2002. *Global oil and the nation state*. Oxford, UK: Oxford University Press.

Mouawad, J., and Z. Kouwe. 2009. Exxon said to pay $4 billion for oil field. *New York Times* 7 October:B5.

Nash, G. D. 1968. *United States oil policy, 1890–1964: Business and government in twentieth century America*. Pittsburgh, PA: University of Pittsburgh Press.

National Energy Policy Development Group. 2001. *National energy policy*. Washington, DC: U.S. Government Printing Office.

Nitzan, J., and S. Bichler. 2002. *The global political economy of Israel*. London: Pluto.

Nordhauser, N. E. 1979. *The quest for stability: Domestic oil regulation 1917–1935*. New York: Garland.

Olien, R. M., and D. D. Hinton. 2007. *Wildcatters: Texas independent oilmen*. College Station: Texas A&M University Press.

Olien, R. M., and D. D. Olien. 2000. *Oil and ideology: The cultural creation of the petroleum industry*. Chapel Hill: North Carolina University Press.

Peluso, N. L., and M. Watts. 2001. Violent environments. In *Violent environments*, ed. N. Peluso and M. Watts, 3–38. Ithaca, NY: Cornell University Press.

Retort. 2005. *Afflicted powers: Capital and spectacle in a new age of war*. London: Verso.

Sabin, P. 2005. *Crude politics: The California oil market, 1900–1940*. Berkeley: University of California Press.

Smil, V. 2008. *Oil: A beginner's guide*. Oxford, UK: Oneworld.

Taussig, M. 1980. *The devil and commodity fetishism in South America*. Chapel Hill: University of North Carolina Press.

Texas State Archives. 1931a. Ross Sterling, Governor Records, Box 301–457, Folder 1, Oil—Martial Law in East Texas, August 15–21, 1931. Texas State Archives, Austin, TX.

———. 1931b. Wolters telegram, Gov. Sterling Records, Box 301–458, Folder 17, Production of Wells in East Texas, 10/12/31—12/31/31. Texas State Archives, Austin, TX.

———. 1932. Texas Railroad Commission, Commissioner's Records, E. O. Thompson, Box 4/3/330. Texas State Archives, Austin, TX.

Thorton, W. W. 1918. *The law relating to oil and gas*. Vol. 1. Cincinnati, OH: W. H. Anderson.

Tsing, A. 2004. *Friction: An ethnography of global connection*. Princeton, NJ: Princeton University Press.

Walt, V. 2009. U.S. companies shut out as Iraq auctions its oil fields. *Time.com* 19 December. http://www.time.com/time/world/article/0,8599,1948787,00.html (last accessed 31 March 2011).

Watts, M. 1983. *Silent violence: Food, famine, and peasantry in northern Nigeria*. Berkeley: University of California Press.

———. 1994. Oil as money: the Devil's excrement and the spectacle of black gold. In *Money, power and space*, ed. S. Corbridge, N. Thrift, and R. Martin, 406–45. Oxford, UK: Blackwell.

Wright, G., and J. Czelusta. 2007. Resource-based growth: Past and present. In *Natural resources: Neither curse nor destiny*, ed. D. Lederman and W. F. Maloney, 183–211. Palo Alto, CA: Stanford University Press.

Yapa, L. 1996. Improved seeds and constructed scarcity. In *Liberation ecologies: Environment, development, social movements*, ed. R. Peet and M. Watts, 69–85. London and New York: Routledge.

Yergin, D. 1991. *The prize: The epic quest for oil, money, and power*. New York: Free Press.

Zimmerman, E. W. 1957. *Conservation in the production of petroleum: A study in industrial control*. New Haven, CT: Yale University Press.

Constructing Sustainable Biofuels: Governance of the Emerging Biofuel Economy

Robert Bailis and Jennifer Baka

School of Forestry and Environmental Studies, Yale University

In recent decades, new modes of governance have emerged in which an array of non-nation-state actors (NNSAs) drive norms and behaviors related to the production and consumption of goods and services with potentially large environmental and social impacts. These modes of governance are evident in the governance of biofuels, where intergovernmental organizations, national and subnational governments, corporations, and civil society organizations have recently developed an array of standards, metastandards, and codes of conduct attempting to define the conditions in which crops can be grown, processed, and used as fuel. Although the field is populated by dozens of efforts, a few binding state-sponsored regulations appear to dominate the major markets for biofuels in the United States and European Union. Nevertheless, existing regulations were heavily influenced by NNSAs. Further, several state-derived modes of governing take the form of metastandards that permit standards developed by NNSAs to be used in their place, resulting in a hybrid system of state and nonstate governance regimes. The regimes taking shape attempt to minimize the negative impacts of biofuel production in numerous ways; for example, by introducing penalties for production associated with direct and indirect land use change, crediting for the coproduction of livestock feed, and encouraging biofuel production on marginal lands. However, these issues are plagued by contradictions, raising questions about how sustainability is defined and assessed in the context of biofuel governance. *Key Words: biofuels, environmental governance, land use change.*

近几十年来，新的管理模式已经出现，对于那些事关生产和消费和对环境和社会具有潜在的巨大影响的商品和服务，许多的非民族国家行为者（NNSAs）驱动了其中的规范和行为。这些管理模式在治理生物燃料方面是显而易见的，政府间组织，国家和地方政府，企业和民间社会组织最近制定了一系列标准和元标准，行为法则，试图以此界定作物被种植，加工，并作为燃料使用的条件。虽然这个领域是多方共同努力的结果，在美国和欧盟，一些具有约束力的、由国家资助的规定似乎主宰了生物燃料的主要市场。然而，NNSAs 严重影响了现行条例。此外，一些国家派生的治理模式采取了元标准的形式，即允许 NNSAs 在他们自己的地方开发制定规则，造成了一种国有和非国有混合型的治理制度。这些治理制度正在形成规模，试图尽量减少生物燃料生产在许多方面的负面影响，例如，通过引入惩罚制度，对导致直接和间接土地利用变化的生产课以罚款，奖励共同生产生物燃料和牲畜饲料，并鼓励在贫瘠的土地上生产生物燃料。然而，这些问题受到冲突矛盾的困扰，对生物燃料治理方面如何定义和评估可持续发展提出了质疑。*关键词：生物燃料，环境治理，土地利用的变化。*

En décadas recientes, han surgido nuevas formas de gobierno en las que una serie de actores no nacionales o estatales (NNSAs, acrónimo en inglés) manipulan las normas y conductas relacionadas con la producción y consumo de bienes y servicios capaces de generar impactos ambientales y sociales potencialmente grandes. Estos tipos de control son evidentes en lo que concierne a biocombustibles, caso en el que recientemente organizaciones intergubernamentales, gobiernos nacionales y sub-nacionales, corporaciones y organizaciones de la sociedad civil han desarrollado una serie de estándares, metaestándares y códigos de conducta con los cuales intentan definir las condiciones como se debe hacer el cultivo de las cosechas, procesarlas y usarlas como combustible. Aunque este campo está repleto de docenas de esfuerzos, unas cuantas normas obligatorias patrocinadas por el Estado parecen dominar los mercados más importantes de biocombustibles en Estados Unidos y la Unión Europea. No obstante, las reglas existentes fueron fuertemente influenciadas por los NNSAs. Más todavía, varios tipos de control de origen estatal toman la forma de metaestándares que permiten el uso de estándares desarrollados por los NNSAs, lo cual desemboca en un sistema híbrido de regímenes de gobierno estatal y no-estatal. Los regímenes que ya están consolidando su forma intentan minimizar los impactos negativos de la producción de biocombustibles de muy diversas maneras; por ejemplo, estableciendo castigos por la producción asociada directa o indirectamente con cambios en el uso del suelo, o dando crédito apropiado por la coproducción de concentrados para los animales y estimulando la producción de biocombustibles en tierras marginales. Sin embargo, todas estas cosas están plagadas contradicciones que inducen a cuestionamientos sobre la manera como se define y evalúa la sustentabilidad de en el contexto del manejo de biocombustibles. *Palabras clave: biocombustibles, administración ambiental, cambio de uso del suelo.*

Biofuels have been simultaneously upheld as a means of reducing the impacts of fossil use and as a profound risk to livelihoods and the environment.[1] In the last decade, numerous national and subnational policies have incentivized biofuel production in both developed and developing economies. Global production has increased by a factor of four and international trade of feedstock and fuel is rapidly expanding (Heinimö and Junginger 2009; Energy Information Administration [EIA] 2010). These policies seek to meet a range of environmental and sociopolitical objectives, including climate change mitigation and domestic energy security, as well as support for the agricultural sector in both the North and South.[2]

Large-scale biofuel production can alter landscapes and strain social-ecological relationships (Gordon 2008; Piñeiro et al. 2009; Zimmerer 2011). In response, numerous sets of actors—intergovernmental organizations, national and subnational governments, private corporations, and civil society organizations—have produced a complex mix of standards, metastandards, and codes of conduct to minimize the negative impacts associated with biofuels.

These efforts represent a large and rapidly emerging attempt to govern for sustainability (Hysing 2010). However, sustainability itself is a complex and contested concept,[3] so it should not be surprising that this complexity carries through to biofuels. The modes of governance that have taken shape range from standards addressing a single technical issue like greenhouse gas (GHG) emissions to comprehensive metastandards that incorporate multiple social issues like labor conditions, land tenure, poverty alleviation, and food security.[4]

Attempts to "govern for sustainability" raise a number of contradictions. For example, minimizing risks to food security creates an incentive to cultivate biofuel on currently uncultivated lands, which could threaten natural forests, reduce biodiversity, and increase GHG emissions. To avoid threatening natural forests, biofuel producers shift toward marginal lands with minimal vegetative cover, but this could create risks for populations dependent on marginal lands, raising questions about modes of governance that promote such production practices. In this article, we explore these and other contradictions raised by emerging modes of biofuel governance. In the following sections, we contextualize the emergence of current governance regimes and examine some of the theoretical issues that might help to clarify how biofuel governance is evolving. We then explore three issues that pose particular challenges to the governance of biofuel sustainability: land use change (LUC), the use of marginal lands for biofuel production, and the relationship between biofuel production and livestock. We conclude by summarizing these challenges and suggesting that they be made more explicit within the governance regimes that are emerging.

Biofuel Governance in Context

Recent research has raised questions about whether large-scale biofuel production can achieve the social and environmental objectives that proponents claim (Naylor et al. 2007; Fargione et al. 2008; Searchinger et al. 2008). Between 2005 and 2008, as major biofuel programs were initiated in the United States and European Union (EU), followed by similar policies in two dozen other countries (see Table 1), world food markets experienced the largest price shock in thirty years: Grain and edible oil prices increased 70 to 120 percent (Food and Agricultural Policy Research Institute 2009), setting off riots in dozens of cities worldwide (Vidal 2007; Martin 2008; "If Words Were Food" 2009). Since that time, other impacts have come to light through studies linking biofuels to deforestation (Waterfield 2007), land grabs (Actionaid 2008; Zoomers 2010), and forced labor (Welch 2006; Dos Santos 2007).

Both in anticipation of these impacts and in response, several dozen distinct efforts have emerged from national and subnational governments, intergovernmental organizations, nongovernmental organization (NGOs), and private corporations (van Dam, Junginger, and Faaij 2010). Several additional roundtables, consisting entirely of non-nation-state actors (NNSAs), have been organized for major biofuel crops like soy, oil palm, and sugarcane (Roundtable on Sustainable Palm Oil [RSPO] 2009; Round Table on Responsible Soy [RTRS] 2009; Better Sugar Cane Initiative 2010).

The range of actors attempting to define acceptable modes of biofuel production exemplifies broader shifts in environmental governance, which reveal the power of NNSAs to shape market behaviors (Bernstein and Cashore 2007). Acting with multiple agendas, NNSAs have been key participants in what Swyngedouw (2005, 1991) terms "governance beyond the state."

Table 1. Current biofuel mandates

Country	Mandate[a]
Australia	E2 in New South Wales, increasing to E10 by 2011; E5 in Queensland by 2010
Argentina	E5 and B5 by 2010
Bolivia	B2.5 by 2007 and B20 by 2015
Brazil	E22 to E25 existing (slight variation over time); B3 by 2008 and B5 by 2013
Canada	E5 by 2010 and B2 by 2012; E7.5 in Saskatchewan and Manitoba; E5 by 2007 in Ontario
Chile	E5 and B5 by 2008 (voluntary)
China	E10 in nine provinces
Colombia	E10 and B10 existing
Dominican Republic	E15 and B2 by 2015
Germany	E5.25 and B5.25 in 2009; E6.25 and B6.25 from 2010 through 2014
India	E5 by 2008; E20 and B20 by 2017 (Government of India 2009)
Indonesia	E1 and B1 in Jakarta and Surabaya. Actual mixture for biodiesel varies. It reached B5 in late 2008, but returned to B1 when feedstock prices increased in early 2009 (U.S. Department of Agriculture Foreign Agricultural Service 2009a)
Italy	E1 and B1
Jamaica	E10 by 2009
Korea	B3 by 2012
Malaysia	B5 by 2008 (currently on hold according to U.S. Department of Agriculture Foreign Agricultural Service 2009b)
Paraguay	B1 by 2007, B3 by 2008, and B5 by 2009; E18 (or higher) existing
Peru	B2 in 2009; B5 by 2011; E7.8 by 2010
Philippines	B1 and E5 by 2008; B2 and E10 by 2011
South Africa	E8–E10 and B2–B5 (proposed)
Thailand	E10 by 2007 and B10 by 2012; 3 percent biodiesel share by 2011
United Kingdom	E3.5/B3.5 by 2010/11 (Renewable Fuels Agency 2010)
United States	Nationally, 130 billion liters/year by 2022 split between first- and second-generation fuels along with numerous subnational blending targets: – E10 in Iowa, Hawaii, Missouri, and Montana – E20 in Minnesota – B5 in New Mexico – E2 and B2 in Louisiana and Washington – 3.4 billion liters/year by 2017 in Pennsylvania – California has introduced a low-carbon fuel standard that mandates a 10 percent reduction in 2005 greenhouse gas emissions from the transportation sector by 2020. Analysts estimate that this will require 10–17 billion liters of ethanol in 2020 (Crane and Prusnek 2007)
Uruguay	E5 by 2014; B2 from 2008–2011 and B5 by 2012

Note: Data from REN21 (2009), except where otherwise noted.

[a]Blends are described by the volumetric percentage of biofuel relative to fossil fuel and abbreviated with an E for ethanol and a B for biodiesel; e.g., 5 percent ethanol in gasoline is E5 and 10 percent biodiesel in fossil diesel is B10.

Importantly, in such modes of governance, the state is not necessarily absent; rather, it might function "at arm's length" (Hysing 2009, 321).[5]

As with governing for sustainability of other commodities, several of the emerging modes of biofuel governance link environmental impacts with notions of ethical consumption (see Table 2). Biofuels, however, possess several distinct characteristics. First, as feedstocks for electric power production and transportation fuels, biofuels are not typical consumer goods. They are usually blended with the "unsustainable" product they are meant to substitute for, leaving retail consumers with little choice between commodities. This is not to say that consumers have no role in building the governance regime, only that their role is not directly exercised at the point of sale, as it is with other products. Instead, it is played out through the participation of civil society organizations like environmentally and socially motivated NGOs, which play a direct role (as NNSAs) in the development of biofuel sustainability standards. These actors do not speak for all consumers but rather for the section of the public that adheres to similar ideologies as the groups themselves.[6]

Moreover, fuels are fundamental inputs into all types of productive activities. Disruptions in supply bring economies to a standstill. As a result, fuels have a

Table 2. Greenhouse gas emission reduction criteria in a selection of standards

Governance scheme	Mandated emission reduction	State, NNSA, or hybrid[a]	Includes non-GHG criteria?	Source
U.S. Renewable Fuel Standard (RFS)	Conventional biofuels: 20% reduction compared to baseline.[b] Advanced biofuels: 50 reduction compared to the baseline.[c] Cellulosic biofuels: 60% reduction compared to the baseline	State	No	U.S. EPA (2010)
EU Renewable Energy Directive (RED)	Initially a 35% reduction but increasing to 50% in 2017 and 60% for new facilities thereafter	Hybrid[d]	Biodiversity Food security	European Council (2009)
California Low Carbon Fuel Standard (LCFS)	Requires refineries and distributors to market a fuel mix that achieves a 10% reduction by 2020	State	No	CalEPA Air Resources Board (2010)
Roundtable for Sustainable Biofuels (RSB)	50% reductions relative to the fossil fuel baseline	NNSA	Biodiversity Food security Land tenure Poverty alleviation Human rights Labor laws Environmental quality	RSB (2010)
U.K. Renewable Transportation Fuel Obligation (RTFO)	Obligation to demonstrate that renewable fuel is supplied equivalent to a specified percentage of total fuel sales and meets a gradually increasing emission reduction	Hybrid	Biodiversity Food security Land tenure Environmental quality	Renewable Fuels Agency (2010)

Note: NNSA = non-nation-state actor; GHG = greenhouse gas.
[a]Hybrid describes state-led metastandards that will accept qualifying standards that originate from NNSAs.
[b]This rule applies to newly constructed facilities producing ethanol from corn. Existing facilities were "grandfathered" into the standard (U.S. EPA 2010a).
[c]Advanced biofuels are effectively defined by this criteria; any fuels that achieve 50 percent reduction are considered advanced. Biomass-based diesel is a subset of this category (U.S. EPA 2010a).
[d]Note that some analysts considerthe European Union and similar intragovernmental agencies to be nonstate actors in the sense that they differ from traditional nation-states.

strategic nature matched by few other commodities, which tends to keep them under close state oversight. In addition, this strategic importance creates a powerful incentive to prioritize security of supply over environmental or social concerns.

Second, both state and nonstate actors promulgating modes of governance over biofuels are concerned with quantifiable environmental performance in ways that actors promoting governance of sustainability for other commodities are not. Regulators of the largest markets for biofuels (the United States, the United Kingdom, and the EU) have mandated that life-cycle GHG emissions must demonstrate specific reductions relative to the fossil fuels that they are meant to replace (see Table 2). Demonstrating that emission reduction objectives are satisfied requires complex life-cycle assessment (LCA) methodologies that are absent from other governance regimes.[7]

Third, demand for biofuels has been legislated into existence in extremely large volumes, which can both affect global commodity markets and drive LUC. In 2008, global biodiesel production utilized approximately 16 percent of the world's soy crop, 36 percent of the world's rapeseed crop, and 4 percent of the global supply of palm oil; global ethanol production utilized roughly 18 percent of the world's sugarcane and 11 percent of the world's corn harvest.[8] Further, demand for biofuels is projected to double in the coming decade (Food and Agriculture Organization 2008), which creates pressure to open new land for biofuel cultivation as well to compensate for the diversion of grains, sugars, and edible oils into biofuel production.

A fourth issue relates both to LUC and the need to quantify environmental performance. Emerging governance regimes place substantial weight on the GHG balance of biofuel systems as an indicator of sustainability. LUC can be a critical determinant of GHG emissions (Fargione et al. 2008; Searchinger et al. 2008), and biofuel regimes address LUC impacts through several means. Some, like the U.S. federal and California state-level standards, account for changes in terrestrial carbon that result as both direct land use change (dLUC) and indirect land use change (iLUC; CalEPA Air Resources Board 2009; U.S. Environmental Protection Agency [EPA] 2010b). Several modes of biofuel governance attempt to minimize LUC impacts by encouraging production on marginal land (Renewable Fuels Agency [RFA] 2008; European Council 2009; Roundtable for Sustainable Biofuels [RSB] 2010). We use the term *marginal land* to mirror the language used in sustainability standards. The term describes land that has minimal productive potential. Other terms, such as *wastelands* and *degraded lands*, are also used. We acknowledge that the label is value laden and discuss the implications of its use further later.

Fifth, biofuels are closely linked to existing global commodity markets. Today's most prevalent biofuel feedstocks—maize, sugarcane, soybeans, rapeseed, and oil palm—are among the world's most widely produced and traded commodities (Statistics Division of the Food and Agriculture Organization of the United Nations [FAOSTAT] 2009). Whether produced for fuel or other applications, each crop yields multiple coproducts. For example, soy, rapeseed, corn, and, to a lesser degree, oil palm, yield a protein-rich seed meal that is readily incorporated into markets for animal feed. This partitioning of resources raises challenging contradictions and complicates attempts to achieve sustainability, a theme that we also return to in more detail later.

Thus, as multiple agro-industries attempt to meet projected biofuel demand, a wide range of actors is constructing modes of governance that seek to define sustainable practices, but the processes that these actors set into motion raise several challenges for this endeavor.

Sustainable Biofuels in Theory

Through the 1990s, as biofuels and particularly fuel-ethanol production steadily increased, numerous debates emerged about the fuel's environmental performance (Giampietro, Ulgiati, and Pimentel 1997). Simultaneously, fundamental changes were underway in society's regulation of social and environmental impacts of production and consumption. These changes saw management of many environmental problems shift away from "hierarchical state activity" (Biermann and Pattberg 2008, 278) toward a multifaceted set of institutional arrangements characterized by "new types of agency," "new mechanisms and institutions," and "increasing segmentation and fragmentation " (Biermann and Pattberg 2008, 280). This shift coincided with a broader neo-liberal turn in society (Swyngedouw 2005; Lemos and Agrawal 2006; Biermann and Pattberg 2008) that saw environmental regulations rolled back along with social safety nets, trade rules, and labor protections (Dryzek 1996; McCarthy and Prudham 2004).

Placing environmental governance in the hands of nonstate actors brings about numerous tensions (Mansfield 2004; Hughes 2006). For example, if governance regimes are perceived as skewed too much toward the interests of particular actors, they will lack political legitimacy (Bernstein and Cashore 2007). A second tension arises with the way in which scientific expertise is deployed in governing such markets. Scientific approaches to environmental management have enabled regulations that constrain the excesses of the free market. By the same token, though, scientific approaches to environmental management can be disempowering to laypersons, favoring quantifiable metrics (McCarthy and Prudham 2004). With biofuels, this is exemplified by the regulations limiting GHG emissions associated with the biofuel life cycle. Sustainability standards require biofuels to achieve a specified reduction in GHG emissions relative to the fossil fuel that they are meant to displace. LCA is used to estimate the reduction including, in most cases, an estimate of LUC impacts (see later).

A third tension emerges as a result of regulation in newly emerging systems of governance. Critical scholars of neo-liberalism have long identified the contradictory nature of regulation in the "free market," citing the ways in which deregulation of neo-liberalized markets actually requires active political intervention to be established and maintained (Mansfield 2004). When environmental and social risks arise as a result of market activity, however, the complete dismantling of environmental regulation might become untenable. Nevertheless, the underlying tenets of neo-liberal ideology are maintained; the state can no longer be the sole arbiter of environmental regulation and new forms of governance are brought to bear on the problem (Bulkeley 2005). Further, the cross-scalar nature of complex social-environmental challenges creates space for multiple (potentially overlapping) scales of governance. As Swyngedouw (2005) noted, the state remains a central figure, but a greater role exists for "private economic agents as well as more vocal civil-society-based groups," yielding a "complex hybrid form of government/governance" (2002). This interplay is made explicit with the metastandard, in which compliance can be demonstrated through the satisfaction of existing "qualifying" standards (Kaphengst, Ma, and Schlegel 2009; Endres 2010). Legislated sustainability standards for biofuels in the United Kingdom, Germany, and The Netherlands have all adopted this approach and the EU-wide Renewable Energy Directive is also considering allowing its provisions to be met through "qualifying standards" originally developed by NNSAs like the various roundtables for biofuels, palm oil, sugar, and soy.

The degree to which hybrid modes of governance achieve the social and environmental goals set by state and nonstate actors is an open question. One encouraging outcome is that NNSAs like the roundtables previously mentioned have defined stronger social principles than the state-promulgated metastandards for which they might qualify.[9] Numerous contradictions arise, however, that could prevent biofuel producers from achieving a broad mix of social and environmental goals. The following sections explore three such contradictions: LUC, marginal lands, and links between biofuels and livestock production.

Three Challenges to Biofuel Sustainability

Direct and Indirect Land Use Change

As world food prices reached their peak in 2008, several scientific assessments were published that raised (or confirmed) doubts about the ability of biofuels to meet some or all of their stated objectives by demonstrating how, under existing and proposed production systems, biofuels could contribute to large-scale LUC (Fargione et al. 2008; Searchinger et al. 2008). Such effects could occur through dLUC, which entails the replacement of natural vegetation with biofuel crops, or through iLUC, in which biofuel crops affect market conditions either by displacing crops or livestock or by diverting existing crops from one market (e.g., food or feed) into biofuel production.

Conceptualizing dLUC is straightforward. The process of converting land from some prior use to biofuel crops inevitably changes the land's numerous ecological functions. If that land had been under natural vegetation, many functions will degrade. For example, land that is rich in biodiversity or carbon stocks will likely lose that richness (Philpott et al. 2008).

Impacts associated with dLUC are likely to be most profound when biofuels are planted on land that was previously covered in natural vegetation. In contrast, iLUC impacts are largest when biofuels are grown on land that was previously used for crop or livestock production, or food or feed crops are diverted into biofuel supply chains. Under those circumstances, supplies of food or feed are reduced, which causes prices to increase. Over time, other commodity producers will respond to higher prices by increasing output. Those producers might increase output through intensification or cultivating new land (Naylor et al. 2007; Ros et al. 2010). If new land is indeed opened, then the ramifications are similar to those associated with dLUC discussed in the previous paragraph, with risks to biodiversity, loss of ecosystem function, and decreased carbon stocks. The opening of uncultivated land is displaced spatially and temporally from the site of biofuel production, however; thus, unlike dLUC, which is attributable to the biofuel producer, the outcomes associated with iLUC cannot be linked to a specific set of actors. Further, the impacts associated with iLUC might not occur until many growing seasons after the biofuel crop is planted. Hence, iLUC introduces explicit spatial and temporal components to biofuel governance by linking the actions of specific biofuel producers to the workings of a global commodity market and the producers who respond to market signals.

It is difficult, if not impossible, to attribute future changes in land use to the present-day actions of a specific set of crop producers. It should be no surprise that the role of iLUC in governance regimes has created a great deal of opposition, primarily from industry actors, who hold that they are being penalized for actions that are beyond their control. Objections have also been raised by academics and government researchers based on doubts about the validity of the concept and the inherent uncertainty in modeling iLUC.

Despite opposition, several iLUCs have been explicitly included in several biofuel governance regimes. This is done through complex modeling techniques,[10] with critical inputs from industry and other NNSAs. When it was first introduced, the U.S. EPA's Renewable Fuel Standard disallowed corn ethanol because projected iLUC impacts caused GHG emissions to exceed the threshold set by the EPA (Table 2). Brazilian cane ethanol was also penalized for iLUC, although it qualified as a "conventional biofuel." Under industry pressure, however, the EPA repeated its iLUC analyses utilizing "significant new scientific data available to the agency, rigorous independent peer review; and extensive public comments" (U.S. EPA 2010a, 2). This reanalysis estimated that iLUC emissions from corn-ethanol were 50 percent less than the original analysis, allowing it to qualify as a "conventional biofuel" by a narrow margin. UNICA (União da Indústria de Cana-de-Açúcar), the Brazilian Sugarcane Industry Association, was also vocally opposed to the EPA's initial iLUC findings (UNICA 2009a). A similar reanalysis led to a downward adjustment of iLUC associated with Brazilian sugarcane-based ethanol by 93 percent, qualifying it as an "advanced biofuel" (also see Table 2). Thus, despite arguments that the methods are inherently uncertain, industry players have been appeased because the state actors setting sustainability standards have adjusted their analyses.[11]

Biofuels and Marginal Lands

As a result of concern about iLUC, modes of biofuel governance actively encourage biofuel production in marginal lands. Although no single definition exists, marginal land typically describes land that is perceived by outsiders as unused, often governed by common property rights, and of little productive value, but this notion of marginality overlooks the deeply political meaning that marginality carries with respect to land designations in (post)colonial societies. Further, the designation is applied in a homogenizing way, obscuring the wide range of land types, tenure relations, and social-ecological interactions that characterize lands falling under this broad category.

For example, in India, such land is officially designated "wasteland," a categorization that dates to colonial land settlement schemes (Gilmartin 2003; Gidwani 2008). The designation was transferred into postcolonial Indian land policy and continues to this day. India still grapples with food security (Indo-Asian News Service 2010) and the Indian concept of wastelands is closely tied to food production. Thus, it is not surprising that the National Policy on Biofuels draws an explicit connection among food production, marginal lands, and biofuels. Passed by Parliament in December 2009, the policy specifies that the country's approach to biofuels "is based solely on non-food feedstocks to be raised on degraded or wastelands that are not suited to agriculture, thus avoiding a possible conflict of fuel vs. food security" (Government of India 2009, 3).

China and Brazil also promote biofuel production on marginal or degraded lands, but neither follows that doctrine. Brazil is the world's second largest producer of ethanol (EIA 2010). The country produces ethanol entirely from sugarcane, which is grown primarily on productive land in the south of the country (UNICA 2009b). In defense of the use of arable land for fuel production, proponents of Brazilian biofuels cite both the country's huge endowment of arable land (355 million ha) and the small fraction that is currently occupied by sugarcane for ethanol production (currently a little over 1 percent). However, industry proponents also cite the large quantities of degraded pasture available in Brazil and cite the industry's use of that land for current and future expansion (Goldemberg, Coelho, and Guardabassi 2008; UNICA 2008). Similarly, fieldwork conducted by the authors on Brazil's nascent Jatropha industry revealed that large-scale growers (those with holdings larger than 1,000 ha) are also targeting former pasturelands for the development of their plantations.[12] It is not yet clear if former pasture is actually degraded and how this status is determined.

In Tanzania, the state invited biofuel investors with favorable policies. By mid-2009, foreign investors had applied for access to nearly 4 million hectares of land.[13] Investors might be targeting lands that both they and the state perceive as marginal, but local perceptions can differ considerably (Sulle and Nelson 2009). Land available for biofuel production in Tanzania falls under two categories: general land and village land (a third category, reserved land, includes conservation areas and is off-limits to commercial development). *Village land* is land within demarcated boundaries that define each of Tanzania's roughly 10,000 villages. Property rights in that land are vested in village councils (Sulle and Nelson 2009). *General land* is any land that is not classified as reserved or village land and falls under state authority, but general land might incorporate village land that is unoccupied or unused (Sulle and Nelson 2009). In practice, biofuel producers have targeted areas that would be perceived as marginal by the broad definitions already outlined. When investors simply target "marginal land," however, it obscures critical differences between tenure categories and leads to potential conflict between different levels of decision makers.

As these examples indicate, the designation of marginal land encompasses numerous land types, under a broad range of tenure arrangements. In India alone, the wasteland designation includes thirteen distinct categories, including degraded pasture and underutilized or degraded forest land (Rao, Gautam, and Sahal 1991), both of which are being targeted for biofuel production. Underutilized or degraded forest land in particular warrants further discussion because it highlights tensions between dLUC and iLUC. Marginal lands are not, by definition, devoid of vegetation; nor are they unable to support livelihoods. For example, in India, underutilized or degraded forestland is the second largest wasteland category, covering more than 120,000 square kilometers (National Remote Sensing Agency 2005). Similarly, in Tanzania, and targeted for biofuel development, is *miombo* woodland (Sulle and Nelson 2009). In India, Tanzania, and elsewhere, forested landscapes store large quantities of carbon in biomass and soil, which is largely lost when the natural vegetation is replaced with biofuel crops.[14] These losses represent a considerable "carbon debt" that can negate most or all of the benefits of replacing fossil fuels with biofuels (Fargione et al. 2008). Moreover, so-called underutilized forest areas might host a range of socioeconomic activities, including fuel wood collection, charcoal production, and collection of nontimber forest products (Robbins 2001; Sulle and Nelson 2009), which would all be disrupted if existing vegetation were replaced by biofuel crops.

Biofuel–Livestock Linkages

As was discussed earlier, debates over iLUC and the impulse to promote biofuel production on marginal land are both grounded in concern over food security. This concern is typically voiced in terms of biofuels either diverting or replacing food crops, but the reality is somewhat more subtle. When oilseed crops are used for biofuel production, the oil itself is processed into fuel and removed from food markets, but the "seedcake" remains. Soymeal, rapeseed meal, and palm kernel meal are high in protein and are typically used for animal feed. Indeed, much of the global expansion of soy production through the last half-century was based on expanding demand for soymeal as animal feed rather than soy oil, which was treated almost as a by-product of meal production (Shurtleff and Aoyagi 2007). Similarly, when grain crops are processed into biofuels, the starch is converted into alcohol, leaving behind distillers' grains, which are also utilized as feed. Oilseed meals and distillers' grains have a nutrient value equivalent to conventional sources of feed (Owen and Larson 1991; Nuez, Waldo, and Yu 2009).

Thus, the food–fuel debate is more subtle than is usually implied. Rather than replacing food crops, the most prevalent biofuel feedstocks (soy, maize, rapeseed, and, to a lesser degree, palm oil and sugarcane) have synergistic ties with an increasingly global feed market. The international trade in oilseed meals exceeded 70 million tons in 2008 (roughly 30 percent of total production) and is growing at 7 percent annually (FAOSTAT 2009). Biofuel proponents point to this synergy as evidence of biofuel sustainability. Indeed, when oilseed meal or distillers' grain is incorporated into LCAs, estimates of energy and GHG emissions improve because biofuels are credited with avoiding the use of livestock feed produced by other means (Farrell et al. 2006a; Wang, Huo, and Arora forthcoming).

Biofuel opponents, on the other hand, typically point to the potential conflict between food and fuel but fail to acknowledge the link between most biofuel crops and feed production. This link extends from biofuel producers to scores of livestock-producing countries around the globe. For example, in 2008 Argentina and Brazil collectively produced roughly 12 percent of the world's biodiesel almost entirely from soybeans (EIA 2010). In the same year, the two exported nearly 30 million tons of soymeal to more than sixty countries. The majority went to the EU, but trade partners included South American neighbors as well as countries in South and East Asia, sub-Saharan Africa, Oceania, and the Middle East (ISTA Mielke GmbH 2009).

Governing biofuels to ensure their sustainability requires a shift from the food–fuel dichotomy to an understanding that current biofuel crops are inextricably linked to existing food and feed markets but that those markets have themselves been implicated as unsustainable (Galloway et al. 2007; McMichael et al. 2007). Most current biofuel production systems yield both fuel and feed, and the latter ultimately reaches the human food chain in the form of meat or dairy products. By current accounting metrics, this combination makes an incremental improvement over the environmental performance of both conventional fossil fuels and feed production (Kim and Dale 2005; Farrell et al. 2006b), but these incremental improvements might actually obscure the degree to which biofuel systems reinforce existing practices in industrial livestock production.

Concluding Thoughts

Biofuel governance is in its infancy. Modes of biofuel governance consist of dozens of efforts ranging from narrowly defined binding regulations promulgated by the state to broad sustainability standards promulgated by several different NNSAs that attempt to cover social impacts, environmental quality, and economic feasibility. The narrowly defined binding regulations originate from both the United States and EU; however, modes of governance in the latter region appear more flexible.

There, modes of governance have evolved as metastandards allowing voluntary schemes to substitute for binding regulations. Thus, an explicit representation of Swyngedouw's (2005) government/governance hybrid is clearly evident. Although the state-led modes of governance have failed to introduce social principles, this hybridity allows NNSAs to promote socially and environmentally responsible biofuel production, with the tacit approval of the state. Empirical evidence is not yet available to assess whether this is actually occurring, however. This is an area of active research by the author and others.

In contrast to the EU, biofuel governance in the United States, currently the largest biofuel consumer, is focused narrowly on GHGs and LUC with no attention to other dimensions of environmental and social impacts. In addition, U.S.-based modes of governance are not metastandards; standards originating from NNSAs cannot qualify as equivalent substitutes as they can in the EU. The ramifications of this regional asymmetry deserve close observation.

Biofuel governance regimes ostensibly exist to minimize the impacts of biofuel production across space and time. Several do this explicitly by adding crop-specific iLUC impact factors estimated with complex global economic models onto equally complex LCA models to delineate good biofuels from bad ones. To minimize iLUC, regimes also encourage biofuel production on marginal lands, but they do so in a way that depoliticizes the concept of marginality and risks negatively impacting the lives and livelihoods of people dependent on those lands. Further, incentivizing production on marginal lands to minimize iLUC might simply increase dLUC.

Similarly, the discourse of food versus fuel obscures the relationship between biofuels and livestock production. Most biofuel supply chains coproduce feed and fuel. When accounting for GHG emissions, the LCA methodologies embedded in sustainability standards account for this contribution by positively crediting biofuels. This accounting procedure reduces the overall negative impacts that biofuels are associated with. To date, however, analysts have failed to critically question the broader implications of this positive association. An accurate assessment of biofuel sustainability might require a politically uncomfortable interrogation of the sustainability of the global meat industry, which is absent from current attempts to govern biofuels.

As biofuel production continues to increase and new blending mandates come into force, the tensions discussed here are likely to increase. These issues deserve to be addressed directly to understand whether and under what conditions biofuels can make a sustainable contribution to global energy supplies.

Acknowledgments

The author acknowledges valuable input from three anonymous reviewers as well as Blake Harrison and Rob Fetter. In addition, the first draft of this article was written when the author was visiting the Copernicus Institute for Sustainable Development and Innovation at Utrecht University, where he benefited from frequent conversations about biofuel sustainability from students and faculty. All errors and omission are, of course, the responsibility of the author.

Notes

1. The polarized sentiments about biofuels are too numerous to review, but a sample of the most negative include "Biofuels Threaten 'Billions of Lives'" (Smith and Elliott 2008) and "Will Biofuels Trigger Genocide?" (Zeigler 2008), whereas the most positive include "As the Saudi Arabia of Ethanol, Brazil Leads by Example" (Harris, Kite-Powell, and Lyle 2009) and "Ethanol Facts" (RFA 2010a).
2. Support for farms in the industrial North is substantially different from efforts in the global South.
3. There is a rich literature exploring and critiquing the notion of sustainability. See, for example, Lélé (1991), Williams and Millington (2004), and Sneddon, Howarth, and Norgaard (2006).
4. Space does not permit a description of standards or their characteristics. For a thorough overview, see van Dam, Junginger, and Faaij (2010).
5. I am grateful to an anonymous reviewer for directing me to Hysing's contributions.
6. I am grateful to a reviewer who suggested the link between consumers and civil society. Civil society participation in biofuel governance is evident from membership lists of the RSB, RSPO, and RTRS, among others (RSPO 2009; RTRS 2009; RSB 2010).
7. Carbon credits are an exception, which must also demonstrate emission reductions.
8. Estimates by the author under a separate project (German, Bailis, and van Gelder 2010).
9. Compare the RSB principles for social impacts to those of the EU or United Kingdom (European Council 2009; RFA 2009; RSB 2010).
10. The EPA estimated the iLUC component of GHG emissions resulting from the implementation of the U.S. Renewable Fuel Standard by linking FASOM, a partial equilibrium forestry and agriculture model, with an general equilibrium model from the Food and Agricultural Policy Research Institute, plus two soil nutrient cycling models (CENTURY and DAYCENT), and an LCA model for transportation (GREET; U.S. EPA 2010).

11. U.S. corn-based ethanol is still disallowed by California (CalEPA Air Resources Board 2009), which uses different models to calculate iLUC emissions, prompting the U.S. ethanol industry to sue the state (Guerrero 2010).
12. Jatropha is a perennial shrub with an oil-rich seed promoted as a promising biofuel crop because of its ability to survive in arid conditions and poor soils (Jatropha Alliance 2009).
13. At the time of writing, more than 600,000 ha had been allocated (Sulle and Nelson 2009).
14. Tropical shrublands can contain sixty to eighty tons of dry matter per hectare. Land under annual crops typically holds five tons of dry matter per hectare and tends to lose soil carbon over time as a result of tillage (Intergovernmental Panel on Climate Change 2003).

References

ActionAid. 2008. *Food, farmers and fuel: Balancing global grain and energy policies with sustainable land use 20*. Johannesburg, South Africa: ActionAid International.

Bernstein, S., and B. Cashore. 2007. Can non-state global governance be legitimate? An analytical framework. *Regulation & Governance* 1:347–71.

Better Sugar Cane Initiative. 2010. Better Sugar Cane website. London: Better Sugar Cane Initiative. http://www.bettersugarcane.org (last accessed 12 April 2011).

Biermann, F., and P. Pattberg. 2008. Global environmental governance: Taking stock, moving forward. *Annual Review of Environment and Resources* 33:277–94.

Bulkeley, H. 2005. Reconfiguring environmental governance: Towards a politics of scales and networks. *Political Geography* 24:875–902.

CalEPA Air Resources Board. 2009. Proposed regulation to implement the Low Carbon Fuel Standard: Volume I, 374. Sacramento, CA: CalEPA Air Resources Board.

———. 2010. *Low Carbon Fuel Standard program*. Sacramento, CA: CalEPA Air Resources Board.

Crane, D., and B. Prusnek. 2007. *The role of a low carbon fuel standard in reducing greenhouse gas emissions and protecting our economy*. Sacramento, CA: Office of the Governor.

Dos Santos, S. 2007. The tainted grail of Brazilian ethanol achieving oil independence but who has borne the cost and paid the price. *N.Y. City Law Review* 61:61–93.

Dryzek, J. S. 1996. *Democracy in capitalist times: Ideals, limits, and struggles*. New York: Oxford University Press.

Endres, J. M. 2010. Clearing the air: The meta-standard approach to ensuring biofuels environmental and social sustainability. *Virginia Environmental Law Journal* 28:73–120.

Energy Information Administration (EIA). 2010. *International energy statistics: Biofuels production*. Washington, DC: National Energy Information Center.

European Council. 2009. Directive 2009/28/EC of the European Parliament on the promotion of energy from renewable sources. *Journal of the European Union* 140:16–62.

Fargione, J., J. Hill, D. Tilman, S. Polasky, and P. Hawthorne. 2008. Land clearing and the biofuel carbon debt. *Science* 319:1235–38.

Farrell, A. E., R. J. Plevin, B. T. Turner, A. D. Jones, M. O'Hare, and D. M. Kammen. 2006a. EBAMM Release 1.1. Berkeley: University of California, Berkeley, Energy and Resources Group. http://rael.berkeley.edu/ebamm/EBAMM_1_1.xls (last accessed 27 July 2010).

———. 2006b. Ethanol can contribute to energy and environmental goals. *Science* 311:506–8.

Food and Agriculture Organization. 2008. *The state of food and agriculture 2008*. Rome: United Nations Food and Agriculture Organization.

Food and Agricultural Policy Research Institute. 2009. *2009 U.S. and world agricultural outlook*. Ames: Iowa State University, Food and Agricultural Policy Research Institute.

Galloway, J. N., M. Burke, G. E. Bradford, R. Naylor, W. Falcon, A. K. Chapagain, J. C. Gaskell et al. 2007. International trade in meat: The tip of the pork chop. *AMBIO: A Journal of the Human Environment* 36:622–29.

German, L., R. Bailis, and J. W. van Gelder. 2010. *Patterns in biofuel investments in forest-rich countries of Asia, Latin America and Africa*. Bogor, Indonesia: CIFOR.

Giampietro, M., S. Ulgiati, and D. Pimentel. 1997. Feasibility of large-scale biofuel production. *BioScience* 47:587–600.

Gidwani, V. 2008. *Capital, interrupted: Agrarian development and the politics of work in India*. Minneapolis: University of Minnesota Press.

Gilmartin, D. 2003. Water and waste: Productivity and colonialism in the Indus basin. *Economic and Political Weekly* 38 (48): 5057–65.

Goldemberg, J., S. T. Coelho, and P. Guardabassi. 2008. The sustainability of ethanol production from sugarcane. *Energy Policy* 36:2086–97.

Gordon, G. 2008. The global free market in biofuels. *Development* 51:481–87.

Government of India. 2009. *National policy on biofuels*. New Delhi: Ministry of New & Renewable Energy.

Guerrero, T. J. 2010. Lawsuit: LCFS violates US Constitution. *Ethanol Producer Magazine*. http://www.ethanolproducer.com/articles/6246/lawsuit-lcfs-violates-us-constitution (last accessed 27 February 2010).

Harris, M., K. Kite-Powell, and A. Lyle. 2009. As the Saudi Arabia of ethanol, Brazil leads by example. Paper presented at BIO International Convention, Atlanta, GA.

Heinimö, J., and M. Junginger. 2009. Production and trading of biomass for energy—An overview of the global status. *Biomass and Bioenergy* 33:1310–20.

Hughes, A. 2006. Geographies of exchange and circulation: Transnational trade and governance. *Progress in Human Geography* 30:635–43.

Hysing, E. 2009. Governing without government? The private governance of forest certification in Sweden. *Public Administration* 87:312–26.

———. 2010. *Governing towards sustainability: Environmental governance and policy change in Swedish forestry and transport*. Örebro, Sweden: Örebro University.

If words were food, nobody would go hungry; Feeding the world. 2009. *The Economist*. http://www.economist.com/node/14926114 (last accessed 3 September 2010).

Indo-Asian News Service. 2010. Food security, productivity major concerns in India, says PM. *The Times of India* 1

February. http://economictimes.indiatimes.com/News/Economy/Indicators/Food-security-productivity-remain-major-concerns-in-India-PM/articleshow/5523182.cms (last accessed 12 April 2011).

Intergovernmental Panel on Climate Change. 2003. *Good practice guidance for land use, land-use change and forestry*. Hayama, Japan: Institute for Global Environmental Strategies (IGES) for the IPCC.

ISTA Mielke GmbH. 2009. *Oil world*. Hamburg, Germany: ISTA Mielke GmbH.

Jatropha Alliance. 2009. *Jatropha has the potential to reduce hunger and to fight climate change*. Berlin: Jatropha Alliance.

Kaphengst, T., M. S. Ma, and S. Schlegel. 2009. At a tipping point? How the debate on biofuel standards sparks innovative ideas for the general future of standardisation and certification schemes. *Journal of Cleaner Production* 17:S99–S101.

Kim, S., and B. E. Dale. 2005. Life cycle assessment of various cropping systems utilized for producing biofuels: Bioethanol and biodiesel. *Biomass and Bioenergy* 29:426–39.

Lélé, S. M. 1991. Sustainable development: A critical review. *World Development* 19:607–21.

Lemos, M. C., and A. Agrawal. 2006. Environmental governance. *Annual Review of Environment and Resources* 31:297–325.

Mansfield, B. 2004. Rules of privatization: Contradictions in neoliberal regulation of north Pacific fisheries. *Annals of the Association of American Geographers* 94:565–84.

Martin, A. 2008. Fuel choices, food crises and finger-pointing *The New York Times* 15 April:1.

McCarthy, J., and S. Prudham. 2004. Neoliberal nature and the nature of neoliberalism. *Geoforum* 35:275–83.

McMichael, A. J., J. W. Powles, C. D. Butler, and R. Uauy. 2007. Food, livestock production, energy, climate change, and health. *The Lancet* 370:1253–63.

National Remote Sensing Agency. 2005. *Wastelands atlas of India, 2005*. Balanagar, India: Ministry of Rural Development, Department of Land Resources and the National Remote Sensing Agency.

Naylor, R. L., A. J. Liska, M. B. Burke, W. P. Falcon, J. C. Gaskell, S. D. Rozelle, and K. G. Cassman. 2007. The ripple effect: Biofuels, food security, and the environment. *Environment* 49:30–43.

Nuez, O., G. Waldo, and P. Yu. 2009. Nutrient variation and availability of wheat DDGS, corn DDGS and blend DDGS from bioethanol plants. *Journal of the Science of Food and Agriculture* 89:1754–61.

Owen, F. G., and L. L. Larson. 1991. Corn distillers dried grains versus soybean meal in lactation diets. *Journal of Dairy Science* 74:972–79.

Philpott, S. M., W. J. Arendt, I. Armbrecht, P. Bichier, T. V. Diestch, C. Gordon, R. Greenberg, et al. 2008. Biodiversity loss in Latin American coffee landscapes: Review of the evidence on ants, birds, and trees. *Conservation Biology* 22:1093–1105.

Piñeiro, G., E. G. Jobbágy, J. Baker, B. C. Murray, and R. B. Jackson. 2009. Set-asides can be better climate investment than corn ethanol. *Ecological Applications* 19:277–82.

Rao, D. P., N. C. Gautam, and B. Sahal. 1991. IRS-1A applications for waste land mapping. *Current Science (Proceedings of the Indian Academy of Science, Section A)* 61:193–97.

REN21. 2009. *Renewables global status report: 2009 update*. Paris: REN21 Secretariat and Deutsche Gesellschaft für Technische Zusammenarbeit (GTZ) GmbH.

Renewable Fuels Agency. 2008. *The Gallagher review of the indirect effects of biofuels production*. East Sussex, UK: Renewable Fuels Agency.

———. 2009. *Carbon and sustainability reporting within the renewable transport fuel obligation*. Technical guidance Part 1, version 2.0. East Sussex, UK: Renewable Fuels Agency.

———. 2010a. *Ethanol facts: Agriculture feeding the world, fueling a nation*. Washington, DC: Renewable Fuels Agency.

———. 2010b. RTFO targets. In *About the RTFO*. London: Renewable Fuels Agency. http://www.renewablefuelsagency.org/aboutthertfo/rtfotargets.cfm (last accessed 28 February 2010).

Robbins, P. 2001. Tracking invasive land covers in India, or why our landscapes have never been modern. *Annals of the Association of American Geographers* 91:637–59.

Ros, J. P. M., K. P. Overmars, E. Stehfest, A. G. Prins, J. Notenboom, and M. v. Oorschot. 2010. *Identifying the indirect effects of bio-energy production*. Bilthoven, The Netherlands: Netherlands Environmental Assessment Agency (PBL).

Roundtable for Sustainable Biofuels (RSB). 2010. *RSB principles and criteria for sustainable biofuel production*. Lausanne, Switzerland: École Polytechnique Fédérale de Lausanne.

Roundtable on Responsible Soy (RTRS). 2009. *Round table on responsible soy*. Buenas Aires, Argentina: RTRS.

Roundtable on Sustainable Palm Oil (RSPO). 2009. *Promoting the growth and use of sustainable palm oil*. Kuala Lumpur, Malaysia: RSPO.

Searchinger, T., R. Heimlich, R. A. Houghton, F. Dong, A. Elobeid, J. Fabiosa, S. Tokgoz, D. Hayes, and T.-H. Yu. 2008. Use of U.S. croplands for biofuels increases greenhouse gases through emissions from land-use change. *Science* 319:1238–40.

Shurtleff, W., and A. Aoyagi. 2007. *History of world soybean production and trade: Part 1. History of soybeans and soyfoods: 1100 B.C. to the 1980s*. Lafayette, CA: Soyinfo Center.

Smith, L., and F. Elliott. 2008. Biofuels threaten "billions of lives." *The Australian* 8 March:17.

Sneddon, C., R. B. Howarth, and R. B. Norgaard. 2006. Sustainable development in a post-Brundtland world. *Ecological Economics* 57:253–68.

Statistics Division of the Food and Agriculture Organization of the United Nations (FAOSTAT). 2009. *FAOSTAT agricultural production data*. Rome: United Nations Food and Agriculture Organization.

Sulle, E., and F. Nelson. 2009. *Biofuels, land access and rural livelihoods in Tanzania*. London: International Institute for Environment and Development and Tanzania Natural Resource Forum.

Swyngedouw, E. 2005. Governance innovation and the citizen: The Janus face of governance-beyond-the-state. *Urban Studies* 42:1991–2006.

UNICA. 2008. *Brazilian sugarcane ethanol: Get the facts right and kill the myths*. Sao Paulo, Brazil: UNICA.

———. 2009a. *Global solutions, not immature science, needed to avoid CO2 emissions from ILUC*. Sao Paulo, Brazil: UNICA.

———. 2009b. *Quotes & stats*. Sao Paulo, Brazil: UNICA.

U.S. Department of Agriculture Foreign Agriculture Service. 2009a. *Indonesia biofuels annual: 2009*. Washington, DC: USDA.

———. 2009b. *Malaysia: Biofuels annual report 2009*. Washington, DC: USDA Foreign Agricultural Service.

U.S. Environmental Protection Agency (EPA). 2010a. EPA finalizes regulations for the National Renewable Fuel Standard Program for 2010 and beyond. Washington, DC: Office of Transportation and Air Quality, EPA.

———. 2010b. Renewable Fuel Standard Program (RFS2) regulatory impact analysis. Washington, DC: Assessment and Standards Division, Office of Transportation and Air Quality, EPA.

van Dam, J., M. Junginger, and A. P. C. Faaij. 2010. From the global efforts on certification of bioenergy towards an integrated approach based on sustainable land use planning. *Renewable and Sustainable Energy Reviews* 14:2445–72.

Vidal, J. 2007. Global food crisis looms as climate change and fuel shortages bite. *The Guardian* 3 November:27.

Wang, M., H. Huo, and S. Arora. Forthcoming. Methods of dealing with co-products of biofuels in life-cycle analysis and consequent results within the U.S. context. *Energy Policy*.

Waterfield, B. 2007. EU admits green fuel targets will damage rainforests. *The Daily Telegraph* 27 April: International News.

Welch, C. 2006. Globalization and the transformation of work in rural Brazil: Agribusiness, rural labor unions, and peasant mobilization. *International Labor and Working-Class History* 70:35–60.

Williams, C. C., and A. C. Millington. 2004. The diverse and contested meanings of sustainable development. *Geographical Journal* 170:99–104.

Zeigler, A. 2008. Will biofuels trigger genocide? Conscious Cultural Evolution Website. http://www.conev.org/biofueltriggergenocide.html (last accessed 28 February 2010).

Zimmerer, K. S. 2011. New geographies of energy: Introduction to the special issue. *Annals of the Association of American Geographers* 101 (4): 705–11.

Zoomers, A. 2010. Globalisation and the foreignisation of space: Seven processes driving the current global land grab. *Journal of Peasant Studies* 37:429–47.

Social Perspectives on Wind-Power Development in West Texas

Christian Brannstrom, Wendy Jepson, and Nicole Persons

Department of Geography, Texas A&M University

Since 2000, U.S. wind-energy capacity has increased 24 percent per year, with Texas emerging as the leading state. Multidimensionality, economic decline, and ownership-participation hypotheses dominate recent geographical research on social perspectives toward wind energy. We examine these hypotheses regarding support of wind power from the perspective of a county that leads Texas in installed capacity. Using Q-method, we present empirically determined, statistically significant social perspectives regarding wind energy. Key actors surveyed included landowners with wind turbines, elected and civil service government officials, and prominent local business and community leaders. We identified five significant clusters of opinion varying in terms of degree of support for wind energy and concern for negative impacts. Stakeholders use economic decline discursively to support wind power, but views on tax policy and distribution of costs and benefits of wind power condition the overall favorable position of key actors to wind-energy development. Specific forms of ownership and participation and positions on tax abatements for attracting wind farms frame discourses of support for wind power. *Key Words: Q-methodology, social perspective, Texas, wind energy.*

自 2000 年以来，美国的风力发电能力每年增长百分之 24，得克萨斯州是其中的新兴主导州。多维性，经济衰退和所有制参与的假说主宰了近期对风能的社会学角度的地理研究。我们从一个县的角度研究了这些假设对于风力发电的支持，该县的装机容量领先于德克萨斯州。利用 Q 方法论，我们提出了一个关于风能的经验性确定的、具有统计显著性的社会学观点。本项调查的主要参与者包括风力涡轮机的地主，民选和文职政府官员，突出的本地商界及社区领袖。我们确定了五个重要的意见集群，包括对风能的支持程度和对负面影响的关切。利益相关者利用经济衰退推断式地支持风力发电，但对税收政策和风能成本和效益分布的不同观点决定了主要参与者对风能发展的整体性支持程度。用以吸引风电场的减税政策，涉及到所有权和参与程度等具体形式，这些因素构成了支持风力发电的话语立场。*关键词：Q 方法论，社会角度，得克萨斯州，风能。*

Desde el 2000, la capacidad de generación de energía por el viento en EE.UU. se ha incrementado anualmente en un 24 por ciento, con Texas asumiendo el papel de estado líder en este campo. La multidimensionalidad, la declinación económica y las hipótesis sobre la propiedad compartida, dominan la reciente investigación geográfica sobre las perspectivas sociales que tiene la energía de origen eólico. Examinamos las hipótesis que tienen que ver con el respaldo que se ofrece a la energía derivada del viento desde la perspectiva de un condado que pone a Texas a la cabeza en términos de capacidad instalada. Utilizando el método Q, presentamos las perspectivas sociales determinadas de manera empírica y que son estadísticamente significativas en lo que concierne a la energía eólica. Se incluyeron para el estudio como actores claves a los propietarios de tierras que usan turbinas de viento, agentes gubernamentales elegidos y del servicio civil, y a destacados líderes comunitarios y comerciantes locales. Identificamos cinco grupos de opinión significativos que varían en términos del grado de apoyo a la energía eólica y en la preocupación por los impactos negativos. Los interesados utilizan discursivamente la declinación económica como argumento de apoyo a la energía generada por el viento, pero, en general, el asunto de las políticas de impuestos y distribución de costos y beneficios de la energía eólica condicionan la posición favorable de actores claves para el desarrollo de la energía eólica. Las formas específicas de propiedad y participación y las posiciones sobre reducción de impuestos para atraer granjas que manejen el viento son los marcos que orientan los discursos en apoyo de la energía eólica. *Palabras clave: metodología Q, perspectiva social, Texas, energía eólica.*

Land-based wind turbines have the potential for supplying a significant portion of electricity demand in the United States and beyond (Lu, McElroy, and Kiviluoma 2009). In 2008, wind energy supplied less than 2 percent of U.S. electricity (Bolinger and Wiser 2009); however, wind energy capacity has increased 24 percent per year since 2000, moving toward a well-publicized target for wind energy to provide 20 percent of electricity by 2030 (U.S. Department of Energy 2008). The U.S. Great Plains, sometimes described as a "wind corridor," has received major investments, with nearly one quarter of the projected 42,000 MW total U.S. capacity located in Texas.

Study of social perceptions and public opinion toward wind-power development is a major aspect of human geography research in renewable energy landscapes (Pasqualetti 2000, 2001). In Europe, a paradox has framed research on wind landscapes: Overall support for wind projects is high, but individual projects often fail because of highly visible public opposition. Three broad arguments, seeking to explain or describe support and opposition regarding wind-power development, have emerged from studies of this paradox. A multidimensionality argument developed from Devine-Wright's (2005, 129) suggestion that scholars focus on the "complex, multidimensional nature of public perceptions of wind farms," rather than simplistic "not-in-my-backyard" (NIMBY) descriptions (Wolsink 2000, 2007; Bell, Gray, and Haggett 2005; Abbott 2010; Warren and McFayden 2010). Warren and Birnie (2009, 120) argued that debates regarding wind farms "are complex, multifaceted and passionate, tapping into deeply held beliefs and value systems," and Ellis, Barry, and Robinson (2007, 537) indicated that "supporters as well as objectors display complex reasons for their respective positions."

Second, the economic decline argument derives from scholars who have associated socioeconomic factors with support for wind power. For example, Horst (2007, 2705) wrote that "residents of stigmatised places are more likely to welcome facilities that are relatively 'green,'" and Toke, Breukers, and Wolsink (2008, 1137) predicted high support in places "perceived as being in economic decline or which are not highly valued as living spaces." In addition, Devine-Wright and Howes (2010) reported significant differences in opinions for offshore wind farm development between two communities with divergent economic development trajectories.

Recent support for the ownership and participation hypothesis comes from Bohn and Lant (2009, 98), who predicted that "local investiture" and "local control over siting" wind-energy projects will lead to strong local and community support for wind power. This view echoed Pasqualetti's (2001) prediction that support for wind will increase as economic returns and property values increased. In European countries, institutional factors have been recognized as key variables in social perception of emerging wind landscapes (Wolsink 2000; Warren et al. 2005; Breukers and Wolsink 2007; Loring 2007; Toke, Breukers, and Wolsink 2008; Ellis et al. 2009). Local ownership, through cooperatives or other institutions, leads to greater social acceptance of wind power (Krauss 2010; Warren and McFayden 2010; Wolsink and Breukers 2010).

These recent contributions have exposed gaps—beyond the more obvious lack of attention to support for wind power and U.S. onshore cases—that this article aims to address. First, key determinants of multidimensionality are deeply held "foundational" moral or environmental values, but there is little attention to how place-based material practices and experiences with the wind-energy industry influence social and political perspectives of new renewable landscapes. Second, the ontological status of "declining" and "marginalized" is not defined, so that it is unclear whether reference to economic decline is interpreted as an objective reality or subjective claim by the public or stakeholders (Horst and Toke 2010). Finally, we do not yet know what "ownership and participation" mean in U.S. cases, where diverse meanings and practices of participation exist because of wide variation in permitting (Bohn and Lant 2009).

To help build a more robust theoretical framework for analyzing public and stakeholder views of wind-energy development in the United States, we interrogate these three claims using a case of rapid wind-power expansion with little apparent opposition. We use Q-method, which has been applied in the European context for studying discourses of support and opposition to wind farms (Ellis, Barry, and Robinson 2007; Fisher and Brown 2009; Wolsink and Breukers 2010), in a site that is at the epicenter of the wind-power boom in the United States. Our research aims to answer the following question informed by this literature: What ideas, claims, and experiences inform support or opposition for wind-power development? We find that stakeholder experiences with tax policies, housing, and economic decline, rather than proximity or landscape aesthetics, structure social perspectives. As wind power increases its geographical footprint, multiple factors will inform how people respond to wind farms. Our findings indicate the need to revise the content and conduct of public opinion surveys, which scholars commonly use to assess public opinion on wind-power development (Firestone, Kempton, and Krueger 2009; Swofford and Slattery 2010).

Study Region and Methods

Nolan County is the location of rapid wind-power development, dating from the 150 MW Trent Mesa site, which began operations in 2001. By 2009, the

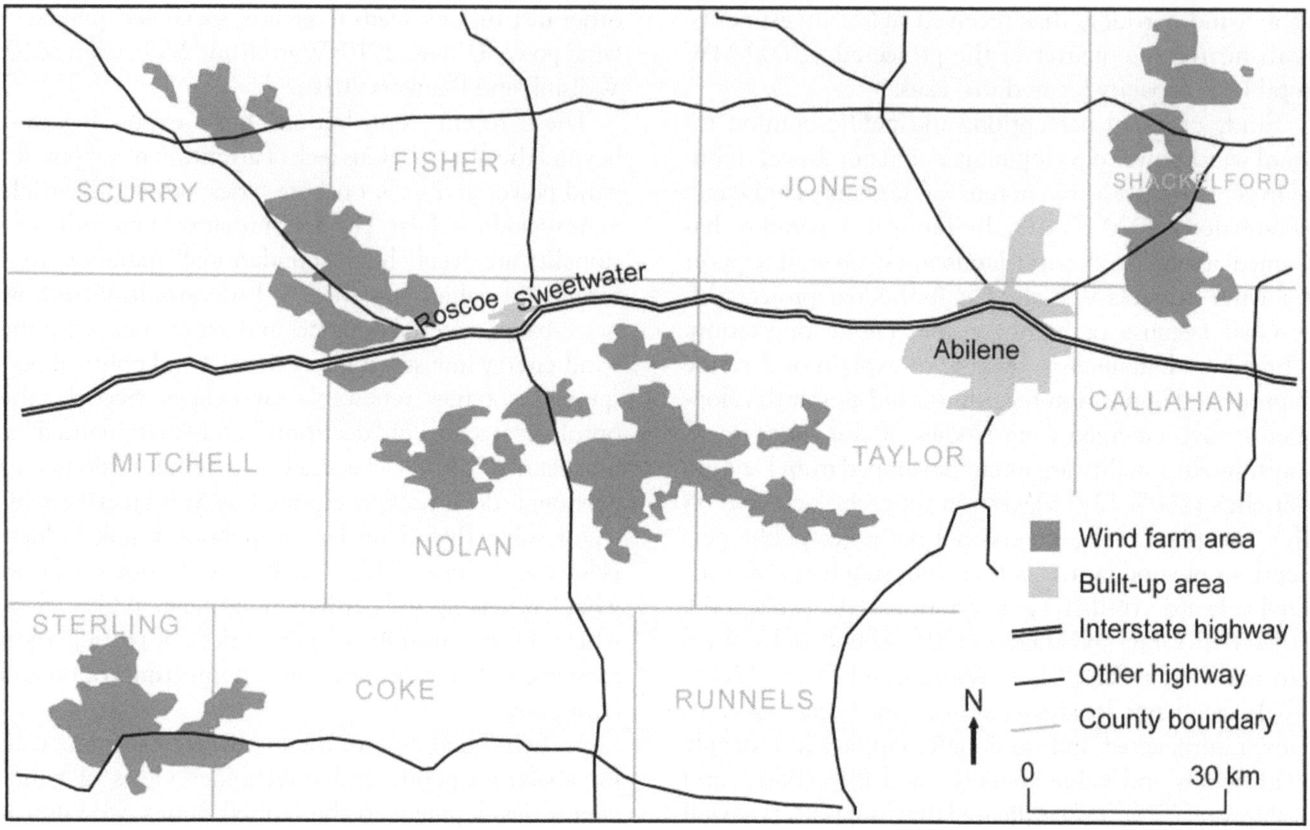

Figure 1. Study region. Nolan County and adjacent areas have approximately 3,000 MW of installed wind-power capacity. Significant clusters of wind turbines also exist northeast of Abilene, in Shackelford County, and southwest of Sweetwater, in Sterling County.

county and adjacent areas had approximately 3,000 MW, or nearly one third of the installed capacity of Texas (Figure 1). Nolan County is a microcosm of a much wider process that has made Texas the leading U.S. state in wind power. Texas has favorable wind climatology, buoyant demand for electricity, little political opposition to wind-farm construction, and the country's lowest permitting requirements. Wind farms in Texas are commonly constructed on private properties. Firms establish land-use agreements with landowners, who receive royalty payments, and seek tax abatements from county authorities. In addition, the state enacted renewable portfolio standards in 1999 and in 2008 committed to subsidizing a $4.9 billion transmission line system to deliver power from rural wind farms to urban centers (Langniss and Wiser 2003; Parker 2008a, 2008b; Bohn and Lant 2009; Wilson and Stephens 2009; Fischlein et al. 2010).

We employed Q-method, which combines qualitative and quantitative techniques, to determine subjectivities of key actors involved in wind power. Q-method allows for a systematic study of social perspectives, without reliance on group affiliation, and opens the analysis to discursive alliances, contention, and schisms across the broad spectrum of actors involved in resource management (Robbins 2006). Our reliance on Q-method is part of what Fisher and Brown (2009, 2517) called a "discursive turn" in renewable energy siting and planning studies. Scholars increasingly employ qualitative methods to elicit views of key actors or stakeholders (Breukers and Wolsink 2007; Loring 2007; Barry, Ellis, and Robinson 2008; Aitken 2009, 2010a, 2010b; Nadaï and Labussière 2009; Stevenson 2009; Zografos and Martínez-Alier 2009; Cowell 2010; Fischlein et al. 2010; Gee 2010; Jessup 2010; Krauss 2010; Wolsink 2010) and to study the statements made during public comment periods (Aitken, McDonald, and Strachan 2008; Abbott 2010).

Q-method is based on the notion that subjectivity is observable through statistical measurement (Robbins and Krueger 2000), assuming that there are limited ordered patterns of discourses or accounts that can be elicited from the data (Barry and Proops 1999). Q-method does not address the prevalence of discourses across a population. Respondents "test" statements that they and other respondents made previously about a given topic. The numerical ranking of each statement is then subjected to factor analysis, which generates factors or groups of respondents who have ranked the same statements in similar fashion.

Q-method relies on four main steps well described in the literature (Brown 1980; Robbins and Krueger 2000; Eden, Donaldson, and Walker 2005). First, we created a concourse of statements after coding semistructured interviews conducted between March and June 2009 with ten key actors representing organizations we determined from a reading of wind-farm siting literature, supplemented by a review of more than sixty articles published in two local newspapers, the *Sweetwater Reporter* and the *Abilene Reporter*, and a study on the economic impacts of wind energy in Nolan County (New Amsterdam Wind Source LLC 2008). The snowball sampling technique, which was initiated in parallel with several stakeholders, was used to identify respondents. Our final selection of twenty-seven Q-statements encompassed five domains: aesthetics, energy/environmentalism, community, landowners, and taxes (Table 1). Impacts on birds and bats were not mentioned in the semistructured interviews for the concourse, so statements to this effect were not included in the concourse.

We then moved to the second phase: Identify respondents, using purposive sampling, to sort the statements into a quasi-normal distribution. Twenty-one participants, including the original ten informants, ranked the same statements in a seminormal distribution that forced sorters to place the statements into a grid that allowed for only one statement corresponding to *most agree with my views* (+4) and only one for *most disagree with my views* (–4), but that permitted seven statements in the "neutral" category. Postsort interviews elicited the rationale for statement rankings.

The third phase of Q-method involved the correlation and factor analysis using dedicated freeware (PQMethod, version 2.11 2002) to determine the correlation matrix, extract and rotate significant factors, and calculate statement z scores for each factor. We selected a five-factor solution that accounted for 77 percent of variance (Table 2). We examined other factor solutions, but they explained little additional variance, included factors that we could not interpret, or provided excessively simplistic factors. Some factors were positively correlated with other factors, suggesting similarities in social perspectives (Table 3).

The fourth step of Q-method is to interpret factors and validate results by semistructured interviews with "loaders" on factors. In January 2010, our preliminary descriptions were used as the basis for semistructured interviews with six participants who loaded highly. After transcription and coding, we incorporated these interviews into the analysis we present here, with preliminary descriptions slightly modified. Our interpretations of these factors relied on "distinguishing" statements, which have significantly different rankings across the factors, statement rankings in the extremes (+4, +3, –3, –4) in idealized sorts for each factor, and respondent rationales for statements ranked highly or lowly. We refer to statements by number (Table 1) and respondents by number (Table 4).

Analysis

Wind Welcomers (Factor 1)

The Wind Welcomers factor describes the idea that wind-power development has had an overwhelming positive impact on the economy and community and that tax abatements are important to foment wind-energy development. Two statements were ranked differently, in a statistically significant manner, compared to other factors: the claim that farmers were major supporters of wind-power development (22; $z = 1.42$, $p < 0.01$) and the denial of negative impacts from wind energy (6; $z = 0.76$, $p < 0.05$). Respondents described how "Sweetwater was struggling. . . . We were starting to lose population, we were seeing a large decline in our enrollment in our schools" (Respondent 24; F1 loading = 0.8326), reflecting a discourse in the media that Sweetwater, the "fastest shrinking city in Texas," was beginning a revival (Myers 2009). Respondents also argued that the arrival of wind power meant that "ranchers and farmers can keep ranching and farming because there is enough confidence that the underlying amount of income will be coming" (Respondent 9; F1 loading = 0.7261).

Statement rankings in the extreme, but that were not statistically different than other factors, also help interpret this factor. For example, the claim of significant local economic impacts from wind (Statement 26) was ranked as +4 and the statement using the "turning wind into wealth" claim (Statement 19) was ranked with +3. By contrast, the lowest idealized ranking (–4) was attached to Statement 8, strongly critical of the wind industry, and a low score (–3) was given to the statement that "Nolan County, the community of Sweetwater or schools have not benefited from the wind farm tax revenue yet" (Statement 25).

Reacting to criticisms of wind-energy development in Statement 25, respondents argued that economic benefits were obvious. One respondent claimed that the wind-energy firms "are paying taxes," while also

Table 1. Scores and rank of each statement by factor

		Factor									
		1 (wind welcomers)		2 (land-based wind welcomers)		3 (disenchanted tax abatements)		4 (favorable tax abatements)		5 (community advocate)	
No.	Statement	z score	Rank	z score	Rank	z score	Rank	z score	Rank	z score	Rank
1	At first the wind turbines were trashy looking but as more turbines are built, the more they become beautiful because of what they're going to provide for the economy around here.	0.27	0	0.31	0	−0.95	−2	0.79	1	−0.51	−1
2	We have to look at these wind towers dotting our west Texas skyline yet we don't reap any of the benefits from them. The people in West Texas should be compensated for these eyesores.	−1.41	−3	−0.51	−1	−1.31	−2	−1.17	−2	0.51	1
3	In other places people complain about how the wind turbines ruin the scenery, but people here see there are just shrubs and cacti. The fact that there are so many turbines, it's almost like going to a garden and looking at something growing.	0.66	1	−0.11	0	−0.73	−2	0.79	1	−0.51	−1
4	The people that think the turbines are unattractive are the same people who do not own any turbines or will never own one.	0.87	2	−0.33	0	0.73	2	0	0	0	0
5	The people who have been neglected are the people who have been in the community before the wind industry arrived and those that are going to be here when the wind industry leaves.	−0.4	0	−0.81	−2	−0.65	−1	−0.2	−1	**1.53**	3
6	There have not been any negative impacts due to the wind industry. People have not come into the community committing crimes, for example.	**0.76**	1	−1.97	−4	0	0	**1.57**	3	−1.02	−2
7	People that used to be the main oil and gas players do not like the wind industry because it is eating into their oil and gas stronghold. The wind industry has politicized the town between wind and oil. Currently in local politics, everything is about wind (**C,** $p>$ **0.05**).	−0.11	0	−0.22	0	0.29	0	−0.2	−1	−0.51	−1
8	The wind industry has negatively impacted the community. Several businesses are failing now because the wind industry has left. In addition, the wind industry has horribly impacted the local housing market, as it impossible to rent an apartment.	−1.43	−4	**1.12**	2	−0.58	−1	−0.59	−1	−2.04	−4
9	The security of the town has diminished since the wind industry began. Before, there were not a lot of transient populations, even with Interstate 20, or when there was a prison. Some of the wind energy employees are leaving, but there are still a few people.	−1.12	−2	0.05	0	**−2.33**	−4	−0.98	−2	1.02	2
10	The town as a whole turned to pure greed during wind turbine construction. Rent tripled in price because the wind energy employees are able to afford higher rent prices, while other members of the community could not afford the new rent prices.	−1.02	−1	0.33	1	0.58	1	−0.79	−1	**2.04**	4
11	The wind energy companies do not contribute to the community and have not been involved in community events. The money is being invested in the community through the tax revenue, but it would be nice if the wind energy companies would contribute.	−1.04	−2	−1.18	−2	**1.53**	3	−1.18	−3	−1.53	−3

Table 1. Scores and rank of each statement by factor (*Continued*)

		Factor									
		1 (wind welcomers)		2 (land-based wind welcomers)		3 (disenchanted tax abatements)		4 (favorable tax abatements)		5 (community advocate)	
No.	statement	*z* score	Rank	*z* score	Rank	*z* score	Rank	*z* score	Rank	*z* score	Rank
12	When wind energy construction booms, it booms and when wind energy construction busts, it busts. The pay was great for the wind employees and construction workers but now there is nothing.	−0.55	−1	−0.41	−1	−0.07	0	−0.2	−1	1.02	2
13	There are a minimum of 300 full-time wind energy employees located in Sweetwater that were not here 8 years ago. If you do not count the employees that work for government agencies, the wind industry would be the 2nd largest employer in Sweetwater.	1.44	3	1.04	2	1.17	3	0.98	2	0	0
14	People here are not spending a lot of time thinking about how they're saving the planet. In fact, a lot of them are dubious of the whole concept of global warming.	−0.09	0	0.47	1	0.73	2	1.17	2	0.51	1
15	Wind energy does not change your view on renewable energy sources. It is not really even green energy. Even though we are not burning coal to generate the electricity, inside the wind turbines are components that are going to be thrown into a landfill **(C, $p>$ 0.01)**.	−0.97	−1	−0.68	−1	−1.38	−3	−0.19	0	0	0
16	People from here are more aware of renewable energy but are not very environmentally conscious. Wind energy is not viewed as a clean power source but just as additional revenue.	−0.42	0	0.06	0	0.8	2	0.78	1	0	0
17	Farmers may have to plow around a wind turbine, and the cattle may feed up to it, but ultimately it has not taken anything of consequence when you compare it to the pollution of a coal plant or nuclear waste **(C, $p >$ 0.05)**.	0.82	1	0.74	1	0.29	0	0.2	1	0	0
18	People from this area do not take the pledge of being green and only for renewable energy. You have to be for every type of energy source because gas pumps are not going away for decades **(C, $p>$ 0.01)**.	0.51	1	0.46	1	−0.22	−1	0.98	2	0.51	1
19	"Turning wind into wealth" is a slogan familiar to many people in this region because many landowners, businesses, school districts, and other taxing entities have seen extra wealth now that wind is being used as a resource.	1.49	3	1.36	2	0.58	1	**−0.79**	−1	0.51	1
20	Hunters from other locations see wind turbines and hunting as incompatible. They are coming here for the experience of hunting in the wilderness, and the wind turbines are taking away from that.	0.07	0	**<u>1.4</u>**	3	0	0	0.19	0	−0.51	−1
21	The difference between ranchers and farmers with wind turbines on their property is about two or three decimal places. Some ranchers have turbines producing over 100 megawatts on their property, whereas most farmers have turbines producing 3 megawatts.	0.26	0	0	0	0	0	0	0	−1.02	−2

(*Continued on next page*)

Table 1. Scores and rank of each statement by factor *(Continued)*

		Factor									
		1 (wind welcomers)		2 (land-based wind welcomers)		3 (disenchanted tax abatements)		4 (favorable tax abatements)		5 (community advocate)	
No.	Statement	*z* score	Rank	*z* score	Rank	*z* score	Rank	*z* score	Rank	*z* score	Rank
22	Farmers used to cuss the wind because it killed crops, carried moisture away, and dried out land. They now love the wind, because income from a windmill is more dependable than dry-land cotton farming, where drought and hail are constant threats.	**1.42**	2	**−1.12**	−2	0.22	0	0	0	0	0
23	The county should not lower the tax rate by 2 cents and save me 70 dollars while saving the wind farm companies millions of dollars. The county is giving away money that the county will need to fix roads some day or build a jail.	−0.47	−1	−0.52	−1	0.66	1	**−1.96**	−4	0	0
24	Tax abatements are a great way to invite wind energy companies to build in your community.	1.15	2	1.53	3	**−1.53**	−3	1.57	3	1.02	2
25	Nolan County, the community of Sweetwater, or schools have not benefited from the wind farm tax revenue yet.	−1.26	−3	−1.42	−3	**0.22**	0	**1.77**	4	−1.02	−2
26	The wind energy companies have provided jobs, use supplies, and buy gasoline from local businesses. The wind industry has been good for the merchants of Nolan County and has allowed for tax values to increase, which leads to lower tax rates.	1.79	4	1.88	4	−0.37	−1	−0.98	−2	1.53	3
27	Tax abatements and the economic development tax should be done away with all together. This land and this country were built without tax abatements and everyone paid on a level playing field. Tax abatements should be given to no one.	−1.22	−2	−1.49	−3	**2.33**	4	−1.57	−3	−1.53	−3

Note: Bold indicates significance at $p < 0.05$; bold underline identifies significance at $p < 0.01$. Nonsignificant "consensus" statements are indicated by C and associated p values.

pointing to other benefits, such as an overall decrease in the tax rate (Respondent 1; F1 loading = 0.7561). Another respondent claimed that "the revenue . . . collect[ed] for these [taxing] entities has gone up exponentially every year," although this respondent admitted that "most people do not understand the tax system and they don't see those numbers or the benefit" (Respondent 5; F1 loading = 0.8446). Another

Table 2. Statistics of factors extracted and rotated

	Factors				
Factor characteristics	1	2	3	4	5
Number of defining variables (loaders)	13	3	2	2	1
Eigenvalue	9.7545	2.5159	1.8308	1.3359	1.0279
Composite reliability	0.981	0.923	0.889	0.889	0.800
Standard error of factor scores	0.137	0.277	0.333	0.333	0.447
Percentage variance explained	37	14	10	9	7
Distinguishing statements	6, 22	8, 20, 22	9, 11, 24, 25, 27	3, 6, 19, 25	5, 10

Table 3. Correlations between factor scores with 98 percent confidence interval

	1	2	3	4	5
1	1.0000	0.4310 ± 0.1567	0.0512 ± 0.1919	0.3491 ± 0.1689	0.2284 ± 0.1823
2	0.4310 ± 0.1567	1.0000	−0.2162 ± 0.1834	0.0512 ± 0.1919	0.3452 ± 0.1695
3	0.0512 ± 0.1919	−0.2162 ± 0.1834	1.0000	−0.1406 ± 0.1886	−0.3351 ± 0.1708
4	0.3491 ± 0.1689	0.0512 ± 0.1919	−0.1406 ± 0.1886	1.0000	−0.0656 ± 0.1916
5	0.2284 ± 0.1823	0.3452 ± 0.1695	−0.3351 ± 0.1708	−0.0656 ± 0.1916	1.0000

Note: From Brown (1980, 286).

argument was that the Sweetwater Independent School District had suffered from enrollment declines for several years, but "now our enrollment has been steady for four years and it has been due to the activity from the wind farms. That has been huge for us" (Respondent 16).

In postsort interviews, loaders on the Wind Welcomers factor were strongly supportive of tax abatements, although the ranking of statements on tax abatements was not significantly different. This is expected, considering that this factor had positive correlations with the Favorable Toward Tax Abatements factor (0.3491). One government official defended abatements, arguing "We can lower our tax rate and still have more tax revenue than we ever thought possible before the wind industry came here" (Respondent 11; F1 loading = 0.6516). A prominent community member argued that tax abatements were necessary because it makes wind-farm financing "more attractive" for firms (Respondent 21; F1 loading = 0.6727).

Land-Based Wind Welcomers (Factor 2)

The Land-Based Wind Welcomers factor, positively correlated with Wind Welcomers (0.4310), represents the social perspective of landowners who were highly supportive of wind-power development and the use of tax abatements and also emphasized how wind turbines fit into the productive landscape of farming, ranching, and hunting, as Sowers (2006) reported for northern Iowa. This social perspective was skeptical of the optimistic claims for how farmers "loved" the wind,

Table 4. Factor loadings

ID	Respondent	Factor 1	Factor 2	Factor 3	Factor 4	Factor 5
25	Landowner	0.8223*	0.0068	0.1360	0.2223	−0.1130
03	Government official	0.0992	0.2805	−0.2872	−0.0578	0.8228*
07	Government official	−0.1239	0.1784	0.3119	−0.8532*	0.1242
05	Government official	0.8446*	−0.0201	−0.1814	0.2133	−0.1145
27	Business owner	−0.0134	−0.2137	0.8595*	−0.1003	0.0478
24	Business/landowner	0.8326*	0.1075	−0.2647	0.2080	−0.0655
29	Government official/landowner	0.1077	0.7502*	−0.1748	−0.0811	0.2094
30	Government official/landowner	0.1718	0.9215*	−0.0895	−0.0046	−0.0158
9	Government official/landowner	0.7621*	0.3684	0.2630	0.2911	−0.0882
26	Business owner	0.1647	0.0140	0.7765*	0.0542	−0.3339
11	Government official	0.6516*	0.4807	−0.0219	0.2923	0.3004
28	Government official/landowner	0.7136*	0.4482	−0.0891	0.1271	0.2670
19	Landowner	0.7846*	0.1729	0.2453	0.1428	0.3058
1	Government official	0.7651*	0.1706	0.2084	0.1655	0.0287
31	Landowner	0.7866*	0.4604	0.0093	−0.0777	0.0413
14	Government official	0.6529*	0.0364	0.2383	−0.2129	0.3303
16	Government official	0.8152*	0.2183	−0.0962	−0.0232	0.1414
32	Prominent community member	0.4570	0.6147*	0.1531	0.2922	0.1690
21	Prominent community member	0.6727*	0.0010	0.1918	0.4075	0.4451
17	Prominent community member	0.3122	0.3275	0.3060	0.7299*	0.1010
4	Government official	0.6515*	0.2877	0.2279	−0.0137	0.1954

Note: * = defining sort.

indicated concerns for increased crime and cost of housing, and correlated positively with the Community Advocate factor (0.3452). Statement 8 ($z = 1.12$, $p < 0.01$), critical of how the wind industry "negatively impacted the community," was significantly different from all factors, especially Wind Welcomers ($z = -1.43$). Yet, Land-Based Wind Welcomers mirrored the Wind Welcomers in giving the highest score to Statement 26, which claimed that wind firms "provided jobs, use supplies, and buy gasoline."

Respondent rationale for high rankings of statements critical of the negative impacts was more resigned than impassioned. For example, a government official and landowner (Respondent 29; F2 loading = 0.7502) argued that "people came here to work" but that they also "do what people do when they start making money—they go out to eat and go out to party," sometimes getting cited for drunk-driving violations or illegal drugs. Another government official and landowner (Respondent 30; F2 loading = 0.9215) argued that crime increases were real but somewhat justified by the scale of the overall economic impact: "Anytime you put a big industry in, it's going to have an impact. It ripples and it gets bigger and bigger, not to say it's all negative."

The experience of Land-Based Wind Welcomers in hunting explains the highest idealized score to a statement supportive of how wind turbines are compatible with the productive landscape (20; $z = 1.4$, $p < 0.01$). A government official and landowner (Respondent 30; F2 loading = 0.9215) admitted his initial skepticism that hunting and wind could coexist but went on to argue that wildlife returned "within six months of construction . . . I guide right underneath these things [turbines] . . . nature adapts and it overcomes."

Land-Based Wind Welcomers also differed significantly from other factors on Statement 22, which claimed that "Farmers used to cuss the wind" but now "love it" ($z = -1.12$, $p < 0.01$). Experience in the farming and ranching economy sustained the disagreement with this statement, as evidenced by one landowner who affirmed that "we always have cussed the wind, and we have learned to love it . . . most of us that work outside all the time have chicken-fried skin" (Respondent 19). In addition, the same respondent who noted that hunting was compatible with turbines argued for an inherent problem that the wind of west Texas offered to farmers: "the turbines have helped subsidize their income to help them to keep farming, [but the wind is] still gonna kill crops" (Respondent 30).

Disenchanted About Tax Abatements (Factor 3)

Tax abatements emerged as an issue separating two social perspectives. The Disenchanted About Tax Abatements social perspective, which was not correlated to the Wind Welcomers factor (0.0512), sustained that tax abatements should not be given to any business and that wind-energy companies have made only slight contributions to the community apart from tax revenue, strongly supporting Statement 11 ($z = 1.53$, $p < 0.01$). Stakeholders loading on this factor believe the wind-energy companies should donate directly to the community and that the wind employees should become more involved in local community events. This should not be confused with the idea that crime has increased because of wind-power development; this factor had a strongly negative reaction to Statement 9, claiming reduced security since the arrival of wind turbines ($z = -2.33$, $p < 0.01$).

The "disenchanted" view of tax abatement is sustained by two distinguishing statements, one arguing that tax abatements "should be done away with altogether" (27; $z = 2.33$, $p < 0.01$) and another supportive of tax abatements, at the opposite end of the rankings (24; $z = -1.53$, $p < 0.01$). Many arguments were made in opposition to tax revenue, but perhaps the most candid came from a government official who warned of a "day of reckoning" when funds would be necessary to repair infrastructure that wind-energy firms had used: "You need to be collecting that tax. . . . As the infrastructure degrading is going on, you got to take care of it" (Respondent 7; F3 loading = 0.3119, F4 loading = –0.8532). Another business owner (Respondent 27; F3 loading = 0.8595) expressed a similar sentiment on the basis of equitable tax treatment: "I don't understand why big companies should get tax abatements and the little guys don't." This respondent argued further that if tax abatements were eliminated, the benefits of wind-power development would be more widely distributed: "If they didn't have the tax abatements, then [wind-energy firms] wouldn't be paying those landowners so much. . . . Now if the companies had to pay taxes, the whole community would benefit, not just those people out there that are getting real wealthy."

Favorable Toward Tax Abatements (Factor 4)

The Favorable Toward Tax Abatements social perspective, which was negatively correlated with the Disenchanted About Tax Abatements factor (–0.1406) and positively correlated with the Wind Welcomers factor (0.3941), represents the view that tax abatements are legitimate instruments to attract wind-energy development into economically marginalized areas such as Nolan County. Defining statements included the idea that the county was "giving away money" (23; $z = -1.96$, $p < 0.01$), the claim that "taxing entities have seen extra wealth" from wind power (19; $z = -0.79$, $p < 0.05$), and the argument that school districts and the county "have not benefited from the wind farm tax revenue yet" (25; $z = 1.77$, $p < 0.01$). For comparison, Wind Welcomers strongly opposed Statement 25; in addition, the two factors differed strongly on Statement 26, which referred to the positive economic and tax benefits of wind firms.

Respondents on this factor opposed the idea that tax abatements meant "giving money away" and supported the claim that taxing agencies have no extra revenue because of wind-power development. In postsort interviews, respondents constructed a narrative of economic decline to justify tax abatements. Sweetwater was "on a downward spiral" before the arrival of wind firms: "We were a typical small town in Texas, with . . . a slump [or] a leveling off economic activity, until wind energy" (Respondent 7, F4 loading = 0.8532). Similarly, "cutting taxes to allow them [wind-energy firms] to operate" is "all we have to offer" (Respondent 17; F4 loading = 0.7299). This respondent went on to argue "our population stopped disappearing" because of tax abatements:

> We give tax abatements all the time. It's the only thing a town like Sweetwater has to entice people to come. We can't sell our thriving nightlife or vast cultural scene. . . . The thing is that people do not understand is that [abatements] are going to end at some point. So if we had not offered them a tax abatement, then all the wind turbines would be in a different county.

Community Advocate (Factor 5)

The Community Advocate social perspective represents the view that wind-power development generated significant negative impacts to the community; however, this perspective also supports wind-energy development and tax abatements because of what it has provided for some—but not all—in Nolan County. The Community Advocate perspective recognizes that low-income residents have been neglected prior to and after the wind energy industry. This social perspective pays attention to people who are not receiving benefits from wind-energy employment, tax breaks, or wind royalties and at the same time have to pay for increased rent and services. Postsort interviews revealed that other social perspectives might not acknowledge or know of these negative impacts to the community because they represent the direct beneficiaries of wind farms.

One significantly different statement indicated that the town "turned to pure greed" during wind turbine construction because of increasing property values and tripling of rent (10; $z = 2.04$, $p < 0.01$). The rationale illustrated the importance of personal experiences:

> I wish they [wind-energy employees] would go somewhere else. My [friend] was kicked out of a house that she was renting, and had three days to get out. Her landowner literally came and told her, "I have a wind farm person who has $1,000 in his hand and your $500 is not helping me. So you have three days to get out." . . . It's impossible to rent an apartment, a storage building, a house. . . . That doesn't help your average Joe. (Respondent 3; F5 loading = 0.8228)

In addition, a significantly different score in the Community Advocate factor was for Statement 5, which described the perception that there was a population in the community that had been neglected before the wind industry had arrived and will continue to be neglected after the wind industry leaves ($z = 1.53$, $p < 0.01$).

Other respondents, loading on different factors, shared the general belief that many residents were excluded, in particular, "low-income people, who [were] pushed out of a renting a house" (Respondent 7, F4 loading = –0.8532). Another respondent witnessed "a lot of people who have invested in a lot of real estate, and charging high prices . . . and I think it is based on greed" (27; F3 loading = 0.8595). By contrast, other respondents understood the negative impacts either as a natural process or believed that positive issues outweighed negative ones. One respondent claimed that rent increases were a natural process that came with economic growth: "It's like with any type of progress when it comes in—if you are in the position to reap the benefits of that income" (Respondent 5; F1 loading = 0.8446). The claim that few people have benefited from wind-power development represented "just a typical head in the sand" (Respondent 9; F1 loading = 0.7261). Another respondent treated any suggestion of negative issues arising from wind-energy development with scorn, arguing that if housing prices did not increase, then "the town is dying . . . you are just filling up the cemetery" (Respondent 17; F4 loading = 0.7299).

Discussion

To what degree do our findings from west Texas advance or challenge the economic decline, multidimensionality, and participation-ownership arguments? The overall support for wind farms in west Texas would advance the economic decline argument, but to accept this would be to ignore that economic decline was a narrative created by some respondents to portray wind energy as a solution to population loss and depressed property values. Decline is a relative, not an objective, description of west Texas, reflecting a particular dominant view of economic development that suits certain interests. Respondents across all factors shared a similar narrative on the recent development of Sweetwater, emphasizing a dismal future without wind energy. Yet the Community Advocate perspective indicated negative economic impacts involving housing and cost-of-living prices, which hurt Sweetwater's nonlandowning elites or boosters not directly engaged in renewable energy. In addition, the economic decline discourse obscures material processes, such as wind rents and reduced property tax rates, that underpin the Wind Welcomer perspective.

Does support for wind power confirm the multidimensionality argument? We found broad support for wind-power development because of perceived increased employment and economic activity. Job creation and economic development are important positive benefits in the eyes of key stakeholders. These place-based experiences, not foundational aesthetic or moral values, framed discourses of support. Negotiations with wind firms, experiences with the housing market, and participation in business revival informed the social perspectives. The wide disagreement on tax policy indicates concerns regarding a key institution in attracting wind-power development, and others considered tax abatements to be a highly inequitable means to pursue economic development. We found no basis to collapse different supporting arguments into a please-in-my-backyard (PIMBY) synthesis (Sowers 2006).

Do ownership and participation explain support for wind power? The west Texas case indicates the need to rethink the ownership-participation argument in more material terms. The Texas siting process removed tax abatements and landowner royalties from public discussion; in addition, large firms own the turbines, which are sited on private land. The "planning" process involves wind firms negotiating contracts, which center on royalty payments based on electricity generation, and lobbying county officials for tax abatements on the value of wind turbines. The irony is that stakeholders are intensively engaged through negotiations with wind firms, although the broader public is excluded because public meetings do not exist. Indeed, the Disenchanted About Tax Abatements social perspective indicates fissures in the argument, created by elites who receive wind-power royalties, that tax abatements benefit all landowners because overall property tax rates have decreased with the growth of wind farms.

What are the implications for future research on renewable energy landscapes? Three lines of inquiry build from our broad conclusion that place-based experiences, rather than moral or foundational values, structure social perspectives. First, our findings indicate that it is essential to focus on the material benefits arising from wind-power development, especially royalties to landowners and tax abatements extended to wind firms. Both material benefits sustain Wind Welcomer perspectives, as well as the overall view that wind power will reverse economic decline. The ownership and participation hypothesis should include specific reference to these benefit flows as the means by which ownership and participation are produced. The minimal planning process in Texas truncates political engagement evident elsewhere in the United States (Bohn and Lant 2009; Abbott 2010). We found no opposition to wind power based on aesthetics. Many respondents believed that wind turbines made the landscape more, not less, aesthetically pleasing. Second, public opinion surveys, long the standard instrument in the study of social perceptions, should incorporate opinions about tax policy and other county- or state-level economic incentives. Surveys should include groups that might be excluded from the benefits of wind power because they do not own land with turbines or businesses catering to wind firms; therefore, surveys should ensure that residents of apartments and trailer parks are included or are approached face-to-face by enumerators. Finally, our findings demonstrate the need to move beyond the headlines of conflict and resistance and to study, instead, the subtle but significant changes that wind-power development has on rural landscapes and communities. Renewable energy studies will benefit from examining the material, place-based processes of change, while emphasizing that the absence of resistance to wind-energy development does not signify complete acceptance. It is at this geographical scale that we will appreciate the challenges and new dynamics arising from rapid increases in wind energy.

Conclusion

Perceived acceptance of wind power requires analysis of the subtle discourses grounded in place-based experiences. We investigated renewable energy landscapes by focusing on social perspectives on wind-energy development in a site of rapid increase in capacity. Our study revealed that support for wind power is multidimensional, informed by place-based experiences relating to economic change, tax policy, the housing market, distribution of benefits, and costs of economic changes. Ownership and participation help inform support but in a process different than reported in the literature because the siting process limits public participation. Finally, our findings offer insights into the issues that might arise in the new renewable energy landscapes of the U.S. Great Plains. The construction of a "wind corridor" will generate debates on the distribution of costs and benefits, in addition to the policies used to attract wind-energy firms. Future research should be attentive to the unequal distribution of benefits from wind-power development and to the debates surrounding the incentives that elites offer to wind-energy firms. Social perspectives are closely linked to the political economy of wind-power growth. The study of both phenomena will lead to a more complete understanding of the production of renewable energy landscapes.

Acknowledgments

We thank the support of the Texas Christian University-Oxford-Next Era Energy Resources Wind Research Initiative for funding and anonymous reviewers for insightful comments on earlier versions of this article.

References

Abbott, J. A. 2010. The localized and scaled discourse of conservation for wind power in Kittitas County, Washington. *Society and Natural Resources* 23 (10): 696–85.

Aitken, M. 2009. Wind power planning controversies and the construction of "expert" and "lay" knowledges. *Science as Culture* 18 (1): 47–64.

———. 2010a. Why we still don't understand the social aspects of wind power: A critique of key assumptions within the literature. *Energy Policy* 38 (4): 1834–41.

———. 2010b. Wind power and community benefits: Challenges and opportunities. *Energy Policy* 38 (10): 6066–75.

Aitken, M., S. McDonald, and P. Strachan. 2008. Locating "power" in wind power planning processes: The (not so) influential role of local objectors. *Journal of Environmental Planning and Management* 51 (6): 777–99.

Barry, J., G. Ellis, and C. Robinson. 2008. Cool rationalities and hot air: A rhetorical approach to understanding debates on renewable energy. *Global Environmental Politics* 8 (2): 67–94.

Barry, J., and J. Proops. 1999. Seeking sustainability discourses with Q methodology. *Ecological Economics* 28 (3): 337–45.

Bell, D., T. Gray, and C. Haggett. 2005. The "social gap" in wind farm siting decisions: Explanations and policy responses. *Environmental Politics* 14 (4): 460–77.

Bohn, C., and C. Lant. 2009. Welcoming the wind? Determinants of wind power development among U.S. states. *The Professional Geographer* 61 (1): 87–100.

Bolinger, M., and R. Wiser. 2009. Wind power price trends in the United States: Struggling to remain competitive in the face of strong growth. *Energy Policy* 37 (3): 1061–71.

Breukers, S., and M. Wolsink. 2007. Wind power implementation in changing institutional landscapes: An international comparison. *Energy Policy* 35 (5): 2737–50.

Brown, S. R. 1980. *Political subjectivity: Applications of Q methodology in political science*. New Haven, CT: Yale University Press.

Cowell, R. 2010. Wind power, landscape and strategic, spatial planning—The construction of "acceptable locations" in Wales. *Land Use Policy* 27 (2): 222–32.

Devine-Wright, P. 2005. Beyond NIMBYism: Toward an integrated framework for understanding public perceptions of wind energy. *Wind Energy* 8 (2): 125–91.

Devine-Wright, P., and Y. Howes. 2010. Disruption to place attachment and the protection of restorative environments: A wind energy case study. *Journal of Environmental Psychology* 30 (3): 271–80.

Eden, S., A. Donaldson, and G. Walker. 2005. Structuring subjectivities? Using Q methodology in human geography. *Area* 37 (4): 413–22.

Ellis, G., J. Barry, and C. Robinson. 2007. Many ways to say "no," different ways to say "yes": Applying Q-methodology to understand public acceptance of wind farm proposals. *Journal of Environmental Planning and Management* 50 (4): 517–51.

Ellis, G., R. Cowell, C. Warren, P. Strachan, and J. Szarka. 2009. Expanding wind power: A problem of planning, or of perception? *Planning Theory & Practice* 10 (4): 523–32.

Firestone, J., W. Kempton, and A. Krueger. 2009. Public acceptance of offshore wind power projects in the USA. *Wind Energy* 12 (2): 183–202.

Fischlein, M., J. Larson, D. M. Hall, R. Chaudhry, T. Rai Peterson, J. C. Stephens, and E. J. Wilson. 2010. Policy stakeholders and deployment of wind power in the sub-national context: A comparison of four U.S. states. *Energy Policy* 38 (8): 4429–39.

Fisher, J., and K. Brown. 2009. Wind energy on the Isle of Lewis: Implications for deliberative planning. *Environment & Planning A* 41 (10): 2516–36.

Gee, K. 2010. Offshore wind power development as affected by seascape values on the German North coast. *Land Use Policy* 27 (2): 185–94.

Horst, D. 2007. NIMBY or not? Exploring the relevance of location and the politics of voiced opinions in renewable energy siting controversies. *Energy Policy* 35 (5): 2705–14.

Horst, D., and D. Toke. 2010. Exploring the landscape of wind farm developments; local area characteristics and planning process outcomes in rural England. *Land Use Policy* 27 (2): 214–21.

Jessup, B. 2010. Plural and hybrid environmental values: A discourse analysis of the wind energy conflict in Australia and the United Kingdom. *Environmental Politics* 19 (1): 21–44.

Krauss, W. 2010. The "*dingpolitik*" of wind energy in northern German landscapes: An ethnographic case study. *Landscape Research* 35 (2): 195–208.

Langniss, O., and R. Wiser. 2003. The renewables portfolio standard in Texas: An early assessment. *Energy Policy* 31(6): 527–35.

Loring, J. M. 2007. Wind energy planning in England, Wales and Denmark: Factors influencing project success. *Energy Policy* 35 (5): 2648–60.

Lu, X., M. B. McElroy, and J. Kiviluoma. 2009. Global potential for wind-generated electricity. *Proceedings of the National Academy of Sciences of the United States of America* 106 (27): 10933–38.

Myers, D. 2009. Wind energy revives the former "fastest shrinking city in Texas." *Abilene Reporter* 7 March. http://www.reporternews.com/news/2009/mar/07/sweeterwater/ (last accessed 9 May 2009).

Nadaï, A., and O. Labussière. 2009. Wind power planning in France (Aveyron), from state regulation to local planning. *Land Use Policy* 26 (3): 744–54.

New Amsterdam Wind Source LLC. 2008. Nolan County: Case study of wind energy economic impacts in Texas. http://cleanenergyfortexas.org/downloads/Nolan_County_case_study_070908.pdf (last accessed 2 March 2009).

Parker, B. D. 2008a. Capturing the wind: The challenges of a new energy source for Texas. *House Research Organization Focus Report* 80–9:8 July.

———. 2008b. Recent decisions affect wind energy. *House Research Organization, Interim News* 80–7:9 October.

Pasqualetti, M. J. 2000. Morality, space, and the power of wind-energy landscapes. *The Geographical Review* 90 (3): 381–94.

———. 2001. Wind energy landscapes: Society and technology in the California desert. *Society and Natural Resources* 14 (8): 689–99.

PQMethod, version 2.11. 2002. Munich, Germany: Peter Schmolck.

Robbins, P. 2006. The politics of barstool biology: Environmental knowledge and power in greater northern Yellowstone. *Geoforum* 37 (2): 185–99.

Robbins, P., and R. Krueger. 2000. Beyond bias? The promise and limits of Q method in human geography. *The Professional Geographer* 52 (4): 636–48.

Sowers, J. 2006. Fields of opportunity: Wind machines return to the Plains. *Great Plains Quarterly* 26 (2): 99–112.

Stevenson, R. 2009. Discourse, power, and energy conflicts: Understanding Welsh renewable energy planning policy. *Environment and Planning C: Government and Policy* 27 (3): 512–26.

Swofford, J., and M. Slattery. 2010. Public attitudes of wind energy in Texas: Local communities in close proximity to wind farms and their effect on decision-making. *Energy Policy* 38 (5): 2508–19.

Toke, D., S. Breukers, and M. Wolsink. 2008. Wind power deployment outcomes: How can we account for the differences? *Renewable and Sustainable Energy Reviews* 12 (4): 1129–47.

U.S. Department of Energy. 2008. *20% Wind energy by 2030: Increasing wind energy's contribution to U.S. electricity supply*. DOE/GO-102008-2567. Washington, DC: U.S. Department of Energy.

Warren, C. R., and R. V. Birnie. 2009. Re-powering Scotland: Wind farms and the "energy or environment" debate. *Scottish Geographical Journal* 125 (2): 97–126.

Warren, C. R., C. Lumsden, S. O'Dowd, and R. V. Birnie. 2005. "Green on green": Public perceptions of wind power in Scotland and Ireland. *Journal of Environmental Planning and Management* 48 (6): 853–75.

Warren, C. R., and M. McFayden. 2010. Does community ownership affect public attitudes to wind energy? A case study from southwest Scotland. *Land Use Policy* 27 (2): 204–13.

Wilson, E. J., and J. C. Stephens. 2009. Wind deployment in the United States: States, resources, policy, and discourse. *Environmental Science and Technology* 43 (24): 9063–70.

Wolsink, M. 2000. Wind power and the NIMBY-myth: Institutional capacity and the limited significance of public support. *Renewable Energy* 21 (1): 49–64.

———. 2007. Wind power implementation: The nature of public attitudes: Equity and fairness instead of "backyard motives." *Renewable and Sustainable Energy Reviews* 11 (6): 1188–1207.

———. 2010. Near-shore wind power—Protected seascapes, environmentalists' attitudes, and the technocratic planning perspective. *Land Use Policy* 27 (2): 195–203.

Wolsink, M., and S. Breukers. 2010. Contrasting the core beliefs regarding the effective implementation of wind power: An international study of stakeholder perspectives. *Journal of Environmental Planning and Management* 53 (5): 535–58.

Zografos, C., and J. Martínez-Alier. 2009. The politics of landscape value: A case study of wind farm conflict in rural Catalonia. *Environment and Planning A* 41 (7): 1726–44.

Farmer Attitudes Toward Production of Perennial Energy Grasses in East Central Illinois: Implications for Community-Based Decision Making

Miriam A. Cope, Sara McLafferty, and Bruce L. Rhoads

Department of Geography, University of Illinois at Urbana–Champaign

Throughout the Midwestern United States, land owners and managers, mainly farmers, are increasingly considering the possibility of transforming industrial agricultural landscapes that currently are used almost strictly for food production to landscapes that include renewable energy production. Because most land in this region is privately owned and independently farmed, transformation of the landscape will be the product of myriad decisions by individual farmers. Little is known about the geographic, environmental, and sociocultural forces that influence farmers' decisions. We use survey methods and a geographic information system (GIS)-aided focus group to elicit farmers' perspectives on growing perennial energy grasses such as switchgrass in central Illinois. Approximately one third of surveyed farmers are willing to plant energy grasses if a local market exists. Farmers' planting decisions are bound up with their understandings of land suitability for planting at the farmstead and regional scales. Through a GIS-aided focus group, participants defined lands suitable for energy grass production—marginal lands—not purely in environmental terms but in relation to existing cropping patterns, farming operations, land parcel characteristics, and the social relations of farming. We find that farmers perceive an array of economic, social, and geographic barriers to energy grass cultivation and that these perspectives deserve attention in renewable energy policy debates. *Key Words: bioenergy, farmers' decision making, GIS-aided focus group, marginal land, Midwestern United States.*

在整个美国中西部，土地的所有者和管理者，主要是农民，都越来越多地开始考虑产业景观转化的可能性，即由目前几乎严格地用于食品生产的格局向包括可再生能源生产的农业景观转换。由于该地区大多数的土地是私人拥有和独立经营，农业生产景观的转变将是众多农民个人决策的集合产物。我们对影响农民的决定的地理因素，环境和社会文化力量知之甚少。我们在研究中使用了统计调查方法和地理信息系统（GIS）辅助的焦点团体，以征求伊利诺斯州中部农民对种植多年生能源作物，例如柳枝草的观点。大约有三分之一接受调查的农民愿意种植能源草，如果当地市场确实存在的话。农民关于种植的决策与他们在农庄和区域规模上对土地适宜性的理解是相关联的。通过地理信息系统辅助的焦点团体，参与者确定了适宜能源草生产的土地—边际土地—不单纯是在环境方面，也与现有的种植模式，农业经营，地块特点，以及农业社会关系相关。我们发现，农民察觉到能源草种植的一个经济，社会和地理障碍的阵列，而且这些观点在可再生能源政策辩论中值得注意。*关键词：生物能源，农民的决策，地理信息系统辅助的焦点团体，贫瘠的土地，美国中西部。*

A través de todo el Medio-Oeste de los Estados Unidos, los propietarios de tierras y administradores, principalmente agricultores, crecientemente están considerando la posibilidad de transformar los paisajes agrícolas industriales, hasta ahora casi estrictamente utilizados para la producción de alimentos, en paisajes que que incluyan la producción de energía renovable. Debido a que la mayor parte de la tierra en esta región es de propiedad privada y se cultiva de manera independiente, la transformación del paisaje será el producto de infinidad de decisiones de granjeros individuales. Muy poco se conoce de las fuerzas geográficas, ambientales y socio-culturales que influyen en las decisiones de los agricultores. Utilizamos métodos de levantamiento de campo y un grupo focal ayudado por un sistema de información geográfica (SIG), para establecer las perspectivas que tienen los granjeros de cultivar pastos energéticos perennes, como el pasto varilla (*Panicum virgatum*), en la parte central de Illinois. Aproximadamente la tercera parte de los granjeros entrevistados tienen la disposición de cultivar tales pastos, si existe un mercado local. Las decisiones de los granjeros sobre qué cultivar están ligadas a sus conocimientos sobre las propiedades adecuadas del suelo para determinado cultivo a escalas de la finca y regional. Mediante el grupo focal ayudado por SIG, los participantes definieron la tierra apropiada para la producción de pasto energético—las tierras marginales—no meramente en términos ambientales sino en relación con los patrones de cultivo existentes, las operaciones agrícolas, las características de las parcelas de tierra y las relaciones sociales de la agricultura. Encontramos que los agricultores perciben un conjunto de barreras económicas, sociales y geográficas para el cultivo de pastos energéticos y que estas perspectivas merecen atención en los debates sobre políticas de energía renovable. *Palabras clave: bioenergía, toma de decisiones de los cultivadores, grupo focal ayudado por SIG, tierra marginal, Medio-Oeste de los Estados Unidos.*

Throughout the Midwestern United States, land owners and managers, mainly farmers, are increasingly considering the possibility of transforming industrial agricultural landscapes that currently are used almost strictly for food production to landscapes that integrate renewable energy production. Because most land in this region is privately owned and independently farmed, transformation of the landscape will be the product of myriad decisions by individual farmers who are influenced not only by markets, technologies, and policies but also by local social norms and networks and cultural factors (Atwell, Schulte, and Westphal 2010). Currently, the rationale behind farmers' decisions on whether to grow energy crops, including perennial energy grasses, and where on the landscape to grow these crops is poorly understood. Much emphasis in policy circles is placed on the utility of marginal land as an initial focal point for renewable energy production, but the notion of what constitutes marginal land in this context remains unclear. This is problematic, as conceptions of the environment can be viewed as socially constructed (Urban and Rhoads 2003; Castree 2005), and social processes strongly influence farming practices and land-use decisions (D. Wilson et al. 2003). The geography and conservation literatures are replete with examples of how social relations, cultural beliefs, and personal values influence decision making about modifications to rural landscapes (G. Wilson 1997; D. Wilson et al. 2003; Rodriguez et al. 2008). Thus, determining how farmers view bioenergy crops and how their views are bound up with conceptions of marginal land is critically important for understanding evolving landscapes of bioenergy production in the Midwestern United States.

Our study examines how farmers, who as owners and managers of private land are key stakeholders, might reconfigure extant agricultural landscapes, including land currently considered as "marginal" for corn and soybean production to support energy grass cultivation. This research is situated within the context of a community-based bioenergy initiative in Decatur, Illinois. The initiative seeks to ascertain the potential for energy grass cultivation to yield multiple local and regional benefits such as protection of soil and water resources, enhancement of biodiversity, increased and diversified farm income, and sustainable economic development. We use survey methods and a GIS-aided focus group to elicit farmers' perspectives on growing perennial energy grasses and to examine how such perspectives relate to farmers' decision making about energy grass cultivation and their conceptions of marginal land.

Our research explores fundamental questions relevant to the community-based bioenergy initiative: What are farmers' perspectives on energy grass cultivation? How do farmers conceptualize marginal land, and what associations do they make between marginal land and the potential for energy grass production? Our conceptual framework emphasizes the importance of economic and noneconomic processes, including sociocultural and biophysical processes (Zimmerer 2011), in farmers' decisions about whether and where perennial energy grasses might be planted.

Bioenergy Policy and Marginal Land

Biomass energy constitutes 3 percent of the total energy consumed in the United States (Jensen et al. 2007; U.S. Department of Energy 2008). The production of bioenergy crops will likely increase given recent federal and state renewable energy policies, such as the 2007 Energy Independence and Security Act, which established a production target of 36 billion gallons of corn-based ethanol by 2022 and 100 million gallons of cellulosic ethanol by 2010. The 2008 Farm Bill passed by Congress created a program called the Biomass Crop Assistance Program (BCAP) to subsidize the collection, harvest, storage, and transportation of eligible biomass materials and provide matching payments (up to $45 dollars per ton) for biomass sold to a certified conversion facility (U.S. Department of Agriculture 2009a).

Switchgrass and miscanthus are two perennial energy grasses included under BCAP that are considered to have significant environmental and economic benefits (Heaton, Dohleman, and Long 2008). They require fewer mechanical and chemical inputs than grain crops and their extensive root systems make them resilient and capable of growing on highly erodible land, thereby providing the potential to reduce soil erosion and improve water quality (Jensen et al. 2007). The high yield potential of miscanthus and switchgrass, along with their ability to grow in poor soil, has made them a central focus of bioenergy policy discussions in the

Midwest (Heaton, Dohleman, and Long 2008). Perennial energy grasses have also attracted attention from local and regional conservation organizations who see potential environmental benefits in these grasses.

The issue of where on the landscape perennial energy grasses should be planted is critically important as policymakers attempt to evaluate energy grass potential. A frequently cited advantage of these crops is that they can be cultivated on "marginal land," thus reducing competition with food crops (Royal Society 2008). The governments of countries such as India, Indonesia, and China are adopting policies to encourage bioenergy crop cultivation on marginal and degraded land (Plieninger and Gaertner 2011). In the United States, the suitability of bioenergy crops for marginal land is frequently cited as an advantage (Schmer et al. 2008).

Despite the emphasis on degraded and marginal lands as sites for perennial energy crop cultivation, the concept of marginal land has been neither clearly defined nor critically analyzed (Dale et al. 2010). Marginal land has been defined as "land that is of poor quality with regard to agricultural use and unsuitable for housing and other uses" (Organisation for Economic Co-operation and Development 2001). But this definition begs the question: How is quality defined and by whom?

Economists typically define marginal land as land with a low economic return, but this definition ignores the social and subsistence value of land for local populations, especially in cases where land resources are shared (Biswas 1979). In contrast, ecologists and agricultural scientists typically define marginal land in biophysical terms as land that is unproductive due to physical properties such as soil quality or slope. Such biophysical definitions have dominated assessments of bioenergy crop production in the United States. For example, in evaluating switchgrass potential, marginal land was defined as: "limited by erosiveness, excessive wetness, soil chemistry constraints, rooting constraints, or climate issues" (Wright 2007, 3). Abandoned farmland and land designated for the Conservation Reserve Program have also figured prominently in definitions of marginal land for bioenergy production in the United States (Schmer et al. 2008).

Although marginal land is relevant to geographical conceptions of landscapes, the concept has received scant attention in the geographic literature aside from a handful of studies of marginal lands' spatial distribution (e.g., Breunig-Madsen, Reenberg, and Holst 1990). We argue here that strict economic and biophysical definitions of marginal land are limited because they ignore important social and political–economic valuations and meanings of land in particular geographic contexts. Contemporary geographic perspectives emphasize the ways in which rural landscapes are socially constructed and the political and ecological contexts of landscape change (Halfacree 2001; G. Wilson 2001). In the intensively cropped landscapes of the Midwestern United States, land valuations are made by farmers whose views reflect a range of social, economic, and ecologic considerations (Figure 1). Decisions about land are based not only on economic imperatives but also pragmatic concerns related to farming practices, social relations such as tenancy, aesthetic judgments about landscape appearance, values about environmental stewardship, and attitudes toward nature, family, and community (Walter 1997; D. Wilson et al. 2003; Urban 2005; White et al. 2009).

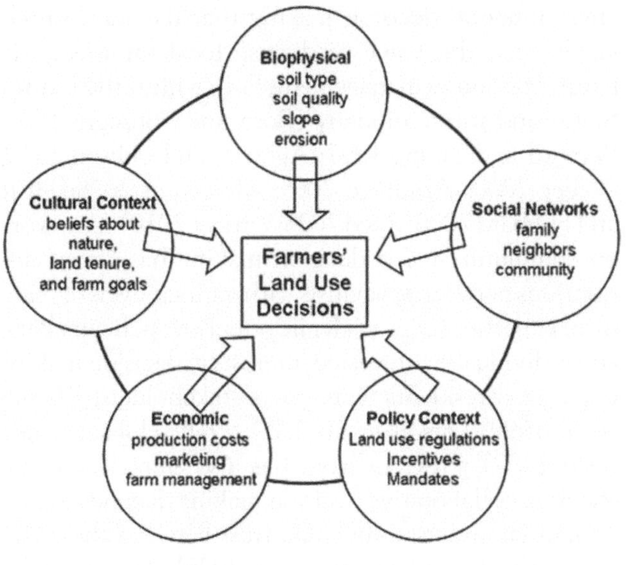

Figure 1. Factors that influence farmers' decisions concerning perennial energy crops. *Source:* Adapted from S. S. White, J. C. Brown, J. Gibson, E. Hanley, and D. Earnhardt. 2009. Planting food or fuel: An interdisciplinary approach to understanding farmers' decision to grow second-generation biofuel feedstock crops. *Comparative Technology Transfer and Society* 7 (3): 287–302.

Community-Based Natural Resources Management

Transformation of landscapes of food production into landscapes that include energy production depends strongly on the active involvement of local landowners and managers (i.e., farmers), especially when decision making occurs within a community-based resource management process and land under consideration for transformation is privately owned. Therefore, this study is situated within a community-based approach to environmental decision making that has increasingly emphasized the value of diverse local knowledge in natural resources management, both within the United States and internationally (Born and Sonzogni 1995; Bernard and Young 1997; Agrawal and Gibson 1999; Weber 2000; Armitage 2005; Margerum 2008; Lurie and Hibbard 2008; Reed 2008; Gruber 2010). Past work on community-based decision making has drawn distinctions between specialists with various levels of technical expertise (e.g., academic scientists, policymakers) and individuals with vested interests in decision making (e.g., local residents, grass-roots stakeholders; Rhoads et al. 1999; Larson et al. 2009; Prell, Hubacek, and Reed 2009). Previous work has also emphasized that effective collaborative decision making depends in part on mutual understanding and trust between these two groups (Focht and Trachtenberg 2005). This interplay between layperson knowledge and that of specialists has been critically interpreted as a complex process in which professionals ultimately benefit from the knowledge of local stakeholders (Waller 1995; Rhoads et al. 1999). Whether or not community-based decision making is the panacea originally envisioned has been questioned (Mitchell 2005; Koontz and Thomas 2006), but this new paradigm has led to an explosion of grassroots partnerships and nonprofit organizations seeking to influence management of local resources.

In many cases, knowledges that are part of a community-based decision-making process are heavily influenced by social, political, and cultural factors—a characteristic that links community-based decision making to the idea that many landscapes are socially constructed (Greider and Garkovich 1994). In this regard, community-based natural resource management directly related to farming practices hinges not only on appropriate understandings of farmers' attitudes about economic issues, educational programs, technical assistance, and assessments of risk (e.g., Napier and Tucker 2001) but also on sociocultural factors such as farmers' cultural conventions and identities, farming practices, and sense of aesthetics (Nassauer 1989; Urban 2005; Figure 1). Yet policymakers and others engaged in community-based resource management typically focus on biophysical and economic criteria that influence farmers' land use decisions at the expense of "intrinsic" sociocultural motivations, such as protecting land for future generations and assuring the visual quality of the landscape (Ryan, Erickson, and De Young 2003). These considerations might influence farmers' perspectives on energy grass cultivation in ways that are important for community-based bioenergy initiatives.

Research Setting

In the past several years, a Local Bioenergy Initiative (LBI) has emerged in Decatur, Illinois, through the combined efforts of the Agricultural Watershed Institute (AWI), a local nonprofit organization concerned with watershed protection, the Soil and Water Conservation District, city governments, and businesses. The LBI seeks to develop a profitable and environmentally beneficial energy grass market in and around Macon County. To do so, the LBI will conduct community deliberations, with significant involvement of farmers, regarding the economic, social, and environmental impacts of energy grass production in Macon County and throughout east central Illinois.

The LBI is situated in Macon County, an intensely farmed area in central Illinois covering 1,515.1 square kilometers. Macon County is home to a population of 105,044 and 708 farms (U.S. Department of Agriculture 2007; U.S. Census Bureau 2008). Almost 75 percent of the land is used for corn and soybean production (U.S. Department of Agriculture 2007). Soils in the area are mostly classified as "very suitable" for crop production with the addition of even minimal fertilizer (U.S. Department of Agriculture 2009b).

For reasons not well understood, farmer involvement in LBI deliberations has been limited. Given their key roles as landowners and managers, coupled with the goals of the LBI, the lack of farmer input signifies an important omission. The AWI, already engaged in research partnerships at the university, invited us to assist in conducting initial research on local farmer perspectives regarding renewable energy and energy grasses. We viewed this endeavor as an opportunity to contribute to geographic knowledge of landscapes of renewable energy in the Midwestern United States and to enhance geographic understandings of marginal land.

Data and Methods

The research was conducted in two phases. First, we administered a mail survey to 400 rural residents in the study region to assess farmers' knowledge of and attitudes toward perennial energy grasses. Survey recipients were chosen randomly from a database of more than 1,000 rural landowners and farmers maintained by the Macon County Soil and Water Conservation District. Although the survey explicitly targeted farmers, it was impossible to identify only farmers from the database prior to mailing, but the respondents discussed here all self-identified as being engaged in farming operations.

The questionnaire consisted of closed- and open-ended questions about farmers' knowledge of perennial energy grasses, their attitudes about trade-offs between food and biofuel production, their understanding of the environmental benefits or costs of energy grasses, and constraints on planting. We asked about farmers' willingness to plant perennial energy grasses and the criteria used in deciding where to plant these new crops. We also included a form for respondents to indicate their interest in attending a GIS-aided focus group to explore issues related to planting perennial energy grasses in central Illinois. Fifty-seven farmers responded to the survey, for a response rate of 14.25 percent. Many factors influence response rates among farmer populations, including length of survey and seasonal timing (Pennings, Irwin, and Good 2002; Morgaine et al. 2005). Our response rate is comparable to that achieved in similarly designed surveys in which farmers are not sent reminders to return a questionnaire (Pennings, Irwin, and Good 2002). The heterogeneity of our sampling frame also might have affected the response rate.

The second phase of the project comprised a GIS-aided focus group. As in a typical focus group, the session was organized as an exploratory "conversation" among a small group of farmer participants (Longhurst 2010). The session was GIS-aided in the sense that participants were able to view and manipulate maps of the study region and direct spatial queries. GIS provided a tool for encouraging participants to think about energy grass cultivation in relation to local land use and environmental conditions. This use of GIS is consistent with the sort of interactive geovisualization advanced in contemporary research on critical and qualitative GIS (Kwan 2002; Cope and Elwood 2009; Elwood 2010). The focus group centered on two questions: (1) What factors are important to you in deciding whether or not to plant perennial energy grasses? (2) If a viable market for such grasses existed, where would you be willing to plant energy grasses? The first question aimed at evaluating farmers' interest in planting energy grasses and the economic, social, and environmental constraints to planting. The second question was explicitly geographical and explored farmers' perspectives about where on the landscape energy grasses should be planted.

Twelve respondents indicated an interest in attending the energy grasses workshop, although only five actually participated. A member of the AWI who provided information for the survey also attended the session. Using a method described in Nyerges et al. (2006), one researcher operated the GIS and the others facilitated and observed the discussion. Participants directed the GIS operations, which included panning, zooming, and proposing spatial queries. The GIS contained six data layers: a 30-m crop cover image of Macon County (classified by USDA-NASS), 2008 soil data, parcel data, slope, streams and watershed boundary data from the Illinois State Water Survey, and street centerline data.

Survey Results

The farmers who responded to our survey were similar in demographic characteristics to farmers in central Illinois (U.S. Department of Agriculture 2007). Three quarters of the respondents were fifty years of age or older. Most (88 percent) considered themselves "family farmers." More than half the respondents had farmed for three decades or more, but a substantial minority (14.5 percent) had farmed for less than a decade. Tenancy was common among the respondents: The average number of acres farmed (926 acres) was more than four times the average number of acres owned (171 acres). In sum, the respondents consisted primarily of experienced farmers with long histories in central Illinois.

In general, farmers saw perennial grasses as a potentially important source of renewable energy but had limited knowledge of energy grass cultivation. Only 7 percent reported that they were "very informed" about the grasses, and almost half (40 percent) reported being "not informed," results similar to those reported in the literature (Jensen et al. 2007; Villamil, Silvis, and Bollero 2008). Thus, farmers lacked detailed knowledge about the grasses themselves and how to cultivate them. Nevertheless, respondents identified many benefits to energy grasses: Over 80 percent agreed that using the grasses for fuel and power supply helps to reduce dependence on foreign oil (Table 1). They were also generally aware of possible environmental benefits, citing improvements in water quality and wildlife habitat associated with energy grasses.

Despite recognizing the benefits of perennial energy grasses as a source of renewable energy, the respondents were hesitant about replacing current crops (primarily corn and soybeans) with energy grasses. Ten percent took an extreme view, agreeing with the statement that "energy grasses should not be grown in Illinois"; 21 percent disagreed that they "would replace some corn and soybeans with energy grasses if a local market existed." On the other hand, 35 percent indicated a willingness

Table 1. Responses to selected survey questions

Question	Agree %	Agree n	Disagree %	Disagree n	No opinion %	No opinion n
Biofuels are important for reducing dependence on foreign oil	85.7	48	8.9	5	5.4	3
Raising crops and conserving the environment are competing goals	36.4	20	54.5	30	9.1	5
Perennial energy grasses should not be grown in central Illinois	10.7	6	46.4	26	42.9	24
Planting perennial energy grasses can help improve water quality	66.7	36	0.0	0	33.3	18
Grasses benefit bird and wildlife habitat	67.3	37	0.0	0	32.7	18
Local market for energy grasses would interest me in replacing some row crops with grasses	35.7	20	21.5	12	42.9	24
Willing to plant grasses on my marginal land	47.4	27	7.0	4	45.7	26

Note: Missing values are excluded in calculating percentages.

to replace some row crops with energy grasses if a local market for the energy crop existed (Table 1). Those willing to consider planting energy grasses viewed them primarily as an "extra" crop, not a wholesale replacement for corn and soybeans. Even if market conditions were favorable, most farmers saw themselves converting less than 10 acres of corn or soy land to energy grasses. Importantly, the respondents overwhelmingly favored planting the grasses on marginal land. Responses to open-ended questions mentioned highly erodible land and land with poor soil as ideal areas for energy grass cultivation. In this sense, the survey pointed toward a biophysical definition of marginal land.

Many barriers to planting perennial energy grasses emerged from the survey responses. Economic barriers were most important, particularly the lack of a market for the grasses and the lack of profitability. Farmers also mentioned the high cost of shifting from one crop to another, including the costs of purchasing new equipment and obtaining rhizomes to establish the grasses. Changing crops also requires a substantial investment of time, as noted by several respondents, one of whom described the "many years to profitability" in the context of her or his advancing age. Lack of storage and transportation infrastructure was also cited frequently. Central Illinois includes a well-developed network of grain elevators and transportation facilities that tightly link corn and soybean production areas with markets. The absence of such a network for energy grasses was mentioned by some respondents.

GIS-Aided Focus Group

The GIS-aided focus group provided an opportunity to explore farmers' perspectives on energy grasses in detail. The first part of the session focused on knowledge and barriers, and it echoed many of the themes identified in the survey. Initially, economic concerns dominated the conversation. Participants emphasized the need for a market and expressed concerns about profitability. Getting local agribusinesses to create a market for energy grasses was seen as critically important. Referring to a large multinational agribusiness firm headquartered in the study area, a participant looked ahead and commented, "There's our market." In addition to profitability concerns, participants stressed the high initial cost of cultivating a new crop, including equipment, rhizomes, transportation, and other issues raised in the survey. Concern was also expressed about timing required to realize a cash crop. One participant described the "five-year wait" for perennial energy grasses to become fully established as a "risky and expensive" proposition.

Viewing maps of the study region sparked a more grounded discussion of energy grass cultivation. The land-use map, a map quilted with corn and soybean patches, framed participants' responses (Figure 2). While viewing this map, participants raised concerns about the fact that the grasses are perennial species and thus potentially invasive. Respondents described the need to protect their productive corn and soybean fields from invasive species. Discussions also focused on the highly productive soils in the study region and their value in producing essential "food" crops of corn and soybeans. Questions were raised about replacing food crops with fuel crops. As one farmer summarized, "I can't take corn out for something questionable." Another said, "If it was prime [farmland] I wouldn't." Thus, farmers expressed concerns about ecological impacts and conveyed an attachment to well-established cropping patterns.

We asked participants to describe suitable locations for energy grasses on thematic maps and to identify criteria for determining land suitability that were then implemented via spatial queries. Biophysical characteristics were important in participants' assessments of land suitability for energy grass cultivation; they identified

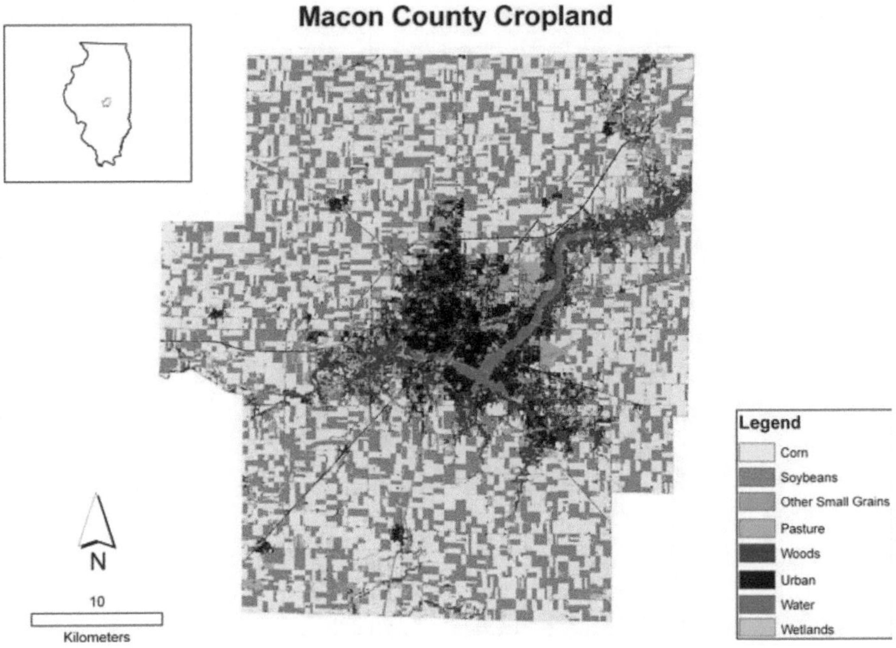

Figure 2. Land cover in Macon County.

stream and ditch corridors and areas with poorly drained and highly erodible soils as suitable areas for grass production. All of the participants viewed energy grasses as a sort of interstitial crop, not something that would supplant corn and soybeans. As in the survey responses, they emphasized "marginal land" as an appropriate place for energy grass cultivation. One participant said that he would be willing to plant the grasses on "marginal ground" such as a persistent gully at the end of a filter strip. That land "washes away bad," and perennial grasses might help reduce erosion.

In addition to well-recognized biophysical and economic criteria, characteristics of the built environment shaped farmers' conceptions of marginal land in relation to energy grasses. Areas bordering railway lines and rights of way for electric power lines attracted attention because they contain lower value farmland and because these human-built features can impede cultivation of corn and soybeans using large farm equipment (tractors and combines). Perennial energy grasses, with their low fertilizer, pesticide, and planting requirements, were seen as a good option for troublesome rights-of-way areas.

To explore participants' perspectives on land suitability at a scale familiar to them, we zoomed in on a farm parcel in the study region (Figure 3). In viewing the parcel, participants immediately identified a "triangular area," located north of the stream. Cut off by the stream from the remainder of the farm, participants described this as a "nuisance area" that would be good for energy grass cultivation. They also described areas along treelines where it is difficult to maneuver large farming equipment and where crop yields are typically low because of shading by trees. At the farmstead scale, marginal lands suitable for energy grass cultivation were identified not just on the basis of environmental features but in relation to everyday farming practices and operations.

Social relations of farming also emerged as important. As tenant farmers, several participants described the contingency of cropping decisions and the fact that they would need to get the landowner's approval to plant energy grasses. Although landowners might be amenable to planting grasses on uncultivated land, taking land out of corn or soybean cultivation to plant energy grasses would be much more controversial because it goes against the grain of established farming practices and might place the owner in the position of being too progressive. Moreover, in deciding which crops to plant, many landowners engage in complex profitability calculations that include not just the market value and costs of crops produced but also factors such as subsidy payments, rental agreements, and conservation incentives. These play out in specific ways for particular land parcels, impacting energy grass cultivation decisions.

Throughout the discussions, participants defined lands suitable for energy grass production not purely in environmental terms but in relation to existing cropping patterns, farming operations, land parcel characteristics, and the social relations of farming. These

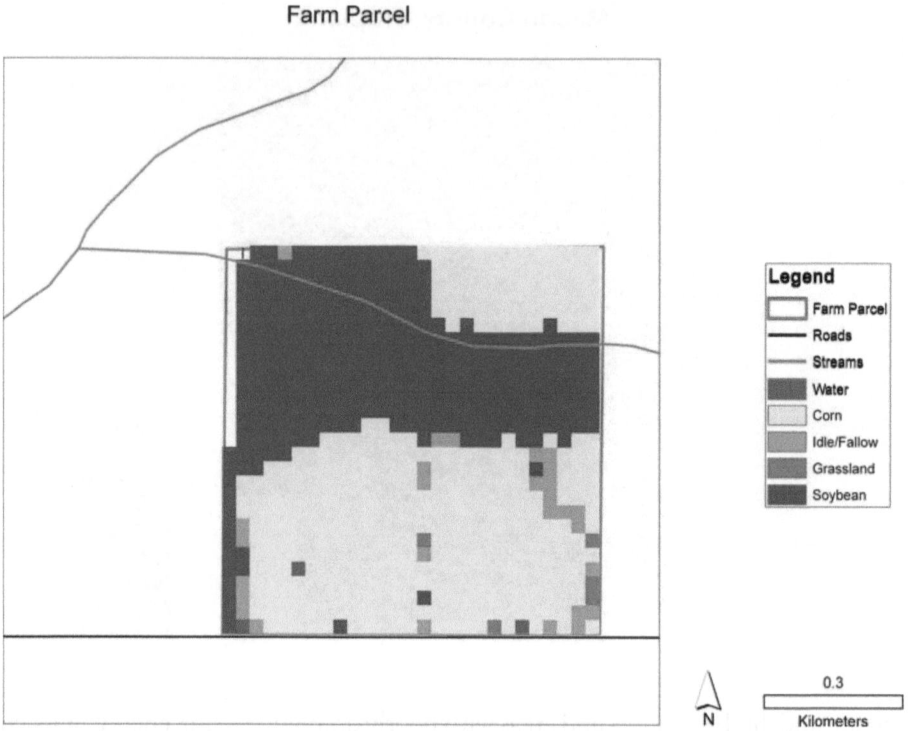

Figure 3. A sample farm parcel discussed by participants in the GIS-aided focus group.

marginal lands were defined at multiple scales from the farmstead scale to the regional and national scales. One participant commented that energy grasses are "a better option for places like Missouri" where the farmland is less productive. Although participants sometimes described specific criteria for land suitability, they also used relational reasoning, contrasting energy grasses with corn and soybeans: Land that is difficult or less productive to cultivate for the two dominant crops drew attention for energy grass planting. Research in geography and other disciplines argues that marginalization is a relational process: So-called marginal groups or places can only be defined in relation to a nonmarginal other (Halfacree 2001; Collins 2010). Marginalization reflects imbalances of power. In the farm landscapes of central Illinois, power is embedded in networks of infrastructure, equipment, agro-industries, and farming practices that support corn and soybean production. These networks served as a focal point in participants' conceptions of marginal land.

Conclusions

Federal, regional, and local initiatives to promote renewable energy and environmental sustainability by encouraging planting of perennial energy grasses hinge on farmers' willingness and ability to cultivate these crops. Farmer participation in community-based biofuels initiatives and farmers' local knowledge of planting opportunities and constraints are critically important to the success of these initiatives. The findings from this initial study suggest that such local knowledge is inherently spatial: Farmers' willingness to plant energy grasses is tied up with understandings of land suitability for planting at the farmstead and regional scales. The method of a GIS-aided focus group provided an essential tool for drawing out and illuminating these understandings, which are critical for community-based efforts to explore the potential for generating a local biofuels market.

Findings from both the survey and GIS-aided focus group highlight the economic, social, and geographical contingency of farmers' decision making about if and where to grow energy grasses. Farmers' perspectives on land suitability were broadly consistent with the goal of local conservation organizations to encourage planting of energy grasses on highly erodible land. Farmers also, however, described social barriers such as tenancy arrangements and pragmatic considerations about farm operations, market constraints, and transportation that are likely to limit the success of efforts to achieve environmental goals through grass cultivation. Another key component of federal and local policies is the effort to develop energy grass markets that rely on grasses cultivated by local farmers. Although limited by small sample size, our survey findings indicate that in Macon County, an area dominated by corn and soybean production, a fraction of farmers are willing to plant energy grasses on at least a limited scale if a local market exists. How this willingness extends across Illinois and other Midwestern regions should be evaluated in future studies.

This research also demonstrates the value of an innovative method, a GIS-aided focus group, in ascertaining farmers' local knowledge of energy grass cultivation. The familiar visual language of maps and the ability to shift between the farmstead and regional scales helped to reveal farmers' knowledge and attitudes in relation to the spaces and settings of daily life. These kinds of advantages have been highlighted by GIS researchers in other contexts (Kwan 2002; St. Martin and Hall-Arber 2008). Elwood (2006) described how community organizations use GIS to construct spatial narratives to support their objectives. Similarly, our farmer participants created narratives about the potential for energy grass cultivation in central Illinois that reflected their understandings of local farmland resources, their everyday farming practices, and their rootedness in current cropping patterns. In viewing GIS data layers, participants identified marginal lands—environmentally vulnerable lands, interstitial areas, and "nuisance" areas—as prime candidates for energy grass cultivation. Farmers' diverse conceptions of marginal land indicate that assessments of local capacity for energy grass production and the viability of local markets need to consider more complex notions of land suitability than those embedded in traditional concepts of marginal land.

Several factors limit the generality of these findings. The sample sizes for the mail survey and GIS-aided focus group were each small; results of this study should be viewed as suggestive, rather than definitive, and need to be evaluated through additional studies in more diverse geographic contexts based on more extensive information. The context for this study—the highly productive agricultural lands of central Illinois—strongly conditioned farmers' responses and influenced their attachment to corn and soybean production. Farmers in less productive agricultural areas might be more willing to engage in energy grass cultivation on a large scale. Despite these limitations, our results reveal the situatedness of farmers' decisions in social, economic, and geographic webs, a perspective that can inform efforts to understand and influence energy grass cultivation across the United States.

Acknowledgement

This research was supported by a Community Informatics Initiative grant from the University of Illinois, Urbana-Champaign. The authors would like to thank Steve John, Executive Director of the Agricultural Watershed Institute, for his support throughout the research project.

References

Agrawal, A., and C. Gibson. 1999. Enchantment and disenchantment: The role of the community in natural resource conservation. *World Development* 27:629–49.

Armitage, D. 2005. Adaptive capacity and community-based natural resource management. *Environmental Management* 35:703–15.

Atwell, R., L. Schulte, and L. Westphal. 2010. How to build multifunctional agricultural landscapes in the U.S. Cornbelt: Add perennials and partnerships. *Land Use Policy* 27:1082–90.

Bernard, T., and J. Young. 1997. *The ecology of hope: Communities collaborate for sustainability*. Gabriola Island, BC, Canada: New Society.

Biswas, A. 1979. Management of traditional resource systems in marginal areas. *Environmental Conservation* 6:257–64.

Born, S. M., and W. C. Sonzogni. 1995. Integrated environmental management: Strengthening the conceptualization. *Environmental Management* 19:167–81.

Breunig-Madsen, H., A. Reenberg, and K. Holst. 1990. Mapping potentially marginal land in Denmark. *Soil Use & Management* 6:114–20.

Castree, N. 2005. *Nature*. London and New York: Routledge.

Collins, T. 2010. Marginalization, facilitation and the production of unequal risk: The 2006 Paso del Norte floods. *Antipode* 42:258–88.

Cope, M., and S. Elwood. 2009. *Qualitative GIS: A mixed methods approach*. Los Angeles: Sage.

Dale, V., K. Kline, J. Wiens, and J. Fargione. 2010. Biofuels: Implications for land use and biodiversity. *Biofuels and Sustainability Reports* January.

Elwood, S. 2006. Beyond cooptation or resistance: Urban spatial politics, community organization and GIS-based spatial narratives. *Annals of the Association of American Geographers* 96:323–41.

———. 2010. Thinking outside the box: Engaging critical geographic information systems theory, practice and politics in human geography. *Geography Compass* 4:45–60.

Focht, W., and Z. Trachtenberg. 2005. A trust based guide to stakeholder participation. In *Swimming upstream: Collaborative approaches to watershed management*, ed. P. Sabatier, W. Focht, M. Lubell, Z. Trachtenberg, A. Vedlitz, and M. Matlock, 85–135. Cambridge, MA: MIT Press.

Greider, T., and L. Garkovich. 1994. Landscapes: The social construction of nature and the environment. *Rural Sociology* 59:1–24.

Gruber, J. 2010. Key principles of community based natural resource management: A synthesis and interpretation of identified effective approaches for managing the commons. *Environmental Management* 45:2–66.

Halfacree, K. 2001. Constructing the object: Taxonomic practices, "counterurbanisation" and positioning marginal rural settlement. *International Journal of Population Geography* 7:395–411.

Heaton, E., F. Dohleman, and S. Long. 2008. Meeting U.S. biofuel goals with less land: The potential of Miscanthus. *Global Change Biology* 14:2000–14.

Jensen, K., C. Clark, P. Ellis, B. English, J. Menard, M. Walsh, and D. de la Torre Ugarte. 2007. Farmer willingness to grow switchgrass for energy production. *Biomass and Bioenergy* 31:773–81.

Koontz, T. M., and C. W. Thomas. 2006. What do we know and need to know about the environmental outcomes of collaborative management? *Public Administration Review* 66:111–21.

Kwan, M. 2002. Feminist visualization: Re-envisioning GIS as a method in feminist geographic research. *Annals of the Association of American Geographers* 92:645–61.

Larson, K. L., D. D. White, P. Gober, S. Harlan, and A. Wutich. 2009. Divergent perspectives on water resource sustainability in a public-policy-science context. *Environmental Science & Policy* 12:1012–23.

Longhurst, R. 2010. Semi-structured interviews and focus groups. In *Key methods in geography*. 2nd ed., ed. N. Clifford, S. French, and G. Valentine, 103–15. Los Angeles: Sage.

Lurie, S., and M. Hibbard. 2008. Community-based natural resource management: Ideals and realities for Oregon watershed councils. *Society & Natural Resources* 21:430–40.

Margerum, R. D. 2008. A typology of collaboration efforts in environmental management. *Environmental Management* 41:487–500.

Mitchell, B. 2005. Participatory partnerships: Engaging and empowering to enhance environmental management and quality of life? *Social Indicators Research* 71:123–44.

Morgaine, K. C., H. M. Firth, G. P. Herbison, A. M. Feyer, and D. I. McBride.2005. Obtaining health information from farmers: Interviews *versus* postal questionnaires in a New Zealand case study. *Annals of Agricultural and Environmental Medicine* 12:223–28.

Napier, T. L., and M. Tucker. 2001. Use of soil and water protection practices among farmers in three Midwest watersheds. *Environmental Management* 27:269–79.

Nassauer, J. I. 1989. Agricultural policy and aesthetic objectives. *Journal of Soil and Water Conservation* 44:384–87.

Nyerges, T., P. Jankowski, D. Tuthill, and K. Ramsey. 2006. Collaborative water resource decision support: Results of a field experiment. *Annals of the Association of American Geographers* 96:699–725.

Organisation for Economic Co-operation and Development. 2001. Glossary of statistical terms: Marginal land. http://stats.oecd.org/glossary/detail.asp?ID=1591 (last accessed 29 December 2010).

Pennings, J. M. E., S. H. Irwin, and D. L. Good. 2002. Surveying farmers: A case study. *Review of Agricultural Economics* 24:266–77.

Plieninger, T., and M. Gaertner. 2011. Harnessing degraded lands for biodiversity conservation. *Journal for Nature Conservation* 19:18–23.

Prell, C., K. Hubacek, and M. Reed. 2009. Stakeholder analysis and social network analysis in natural resource management. *Society & Natural Resources* 22:501–18.

Reed, M. 2008. Stakeholder participation for environmental management: A literature review. *Biological Conservation* 141:2417–31.

Rhoads, B. L., D. Wilson, M. Urban, and E. E. Herricks. 1999. Interaction between scientists and nonscientists in community-based watershed management: Emergence of the concept of stream naturalization. *Environmental Management* 24:297–308.

Rodriguez, J., J. Molnar, R. Fazio, E. Sydnor, and M. Lowe. 2008. Barriers to adoption of sustainable agriculture practices: Change agent perspectives. *Renewable Agriculture and Food Systems* 24:60–71.

Royal Society. 2008. Sustainable biofuels: Prospects and challenges. Policy Document 01/08, The Royal Society, London.

Ryan, R., D. Erickson, and R. De Young. 2003. Farmers' motivations for adopting conservation practices along riparian zones in a Midwestern agricultural watershed. *Journal of Environmental Planning and Management* 46:19–37.

Schmer, M., K. Vogel, R. Mitchell, and R. Perrin. 2008. Net energy of cellulosic ethanol from switchgrass. *Proceedings of the National Academy of Sciences* 105:464–69.

St. Martin, K., and M. Hall-Arber. 2008. The missing layer: Geo-technologies, communities, and implications for marine spatial planning. *Marine Policy* 32:779–86.

Urban, M. A. 2005. Values and ethical beliefs regarding agricultural drainage in central Illinois, USA. *Society & Natural Resources* 18:173–89.

Urban, M. A., and B. L. Rhoads. 2003. Conceptions of nature: Implications for an integrated geography. In *Contemporary meanings in physical geography: From what to why?* ed. S. T. Trudgill and A. Roy, 211–31. London: Edward Arnold.

U.S. Census Bureau. 2005–2009 American Community Survey. http://www.factfinder.census.gov/home/saff/main.html?_lang=en (last accessed 10 January 2011).

U.S. Department of Agriculture. 2007. Census of agriculture: Macon County, Illinois. http://www.agcensus.usda.gov/Publications/2007/Online_Highlights/County_Profiles/Illinois/cp17115.pdf (last accessed 5 March 2010).

———. 2009a. Biomass Crop Assistance Program—CHST Matching Payments Program. http://www.fsa.usda.gov/FSA/newsReleases?area=newsroom&subject=landing&topic=pfs&newstype=prfactsheet&type=detail&item=pf_20090911_consv_en_bcap.html (last accessed 5 February 2010).

———. 2009b. Soil survey of Macon County, Illinois. http://soils.usda.gov/survey/online_surveys/illinois/ (last accessed 5 March 2010).

U.S. Department of Energy. 2008. Renewable energy as share of total primary energy consumption. http://www.eia.doe.gov/emeu/aer/pdf/pages/sec10.pdf (last accessed 5 February 2010).

Villamil, M., A. Silvis, and G. Bollero. 2008. Potential miscanthus' adoption in Illinois: Information needs and preferred information channels. *Biomass and Bioenergy* 32:1338–48.

Waller, T., 1995. Knowledge, power, and environmental policy: Expertise, the lay public, and water management in the western United States. *The Environmental Professional* 17:153–66.

Walter, G. 1997. Images of success: How Illinois farmers define the successful farmer. *Rural Sociology* 62 (1): 48–68.

Weber, E. 2000. A new vanguard for the environment: Grassroots ecosystem management as a new environmental movement. *Society & Natural Resources* 13:237–59.

White, S. S., J. C. Brown, J. Gibson, E. Hanley, and D. Earnhardt. 2009. Planting food or fuel: An interdisciplinary approach to understanding farmers' decision to grow second-generation biofuel feedstock crops. *Comparative Technology Transfer and Society* 7:287–302.

Wilson, D., M. Urban, M. Graves, and D. Morrison. 2003. Beyond the economic: Farmer practices and identities in central Illinois, USA. *The Great Lakes Geographer* 10:21–33.

Wilson, G. 1997. Factors influencing farmer participation in the environmentally sensitive areas scheme. *Journal of Environmental Management* 50:67–93.

———. 2001. From productivism to post-productivism and back again? Exploring the (un)changed natural and mental landscapes of European agriculture. *Transactions of the Institute of British Geographers* 26:77–102.

Wright, L. 2007. Historical perspective on how and why switchgrass was selected as a high-potential energy crop. Oak Ridge National Lab Report Series, ORNL/TM-2007/109, Oak Ridge, TN.

Zimmerer, K. S. 2011. New geographies of energy: Introduction to the special issue. *Annals of the Association of American Geographers* 101 (4): 705–11.

Renewable Energy and Human Rights Violations: Illustrative Cases from Indigenous Territories in Panama

Mary Finley-Brook and Curtis Thomas

Department of Geography and the Environment, University of Richmond

Local implementation of international climate policies is frequently obscure. The objective of our research is to unpack the "black box" of carbon offsetting as it is being conducted in Latin American indigenous territories. Our two case studies of renewable energy projects under construction in Naso and Ngöbe villages in western Panama show that carbon offsets in oppressive societies have the potential to cause social harm. Our cases illustrate processes of green authoritarianism, spatial control, and social restructuring. The private developers constructing the Chan 75 and Bonyic dams did not follow international standards for free, prior, and informed consent, and state agencies reinforced private rights with physical violence. As the hydro developers await decisions on their applications for verification under the Clean Development Mechanism (CDM), we recommend CDM procedural reforms to assure respect for human rights, including the special rights codified in the 2007 UN Declaration on the Rights of Indigenous Peoples. If not, project developers could use low-carbon objectives to justify social oppression. *Key Words: carbon offsets, hydroelectric dams, Indigenous peoples, Panama, renewable energy.*

国际气候政策在地方一级的实施经常是模糊不清的。我们的研究目标是要解开在拉丁美洲的固有领土所进行的碳补偿的“黑盒子”。我们对巴拿马西部的纳苏和恩哥贝农村正在建设的两个再生能源项目进行了个例分析，研究显示，社会压迫情况下的碳补偿有可能会造成社会危害。我们的案例展示了绿色独裁，空间控制和社会转型过程。建设 Chan 75 和 Bonyic 大坝的私人开发商为遵循国际标准而付出了代价，事先知情并且同意的国家机构用身体暴力来强化私有权利。随着水电开发商等待他们根据清洁发展机制（CDM）提交申请的核查决定，我们建议 CDM 进行程序改革以确保对人权的尊重，包括依据 2007 年联合国土著人民权利宣言中所定义的特殊权利。如果不这样，项目开发人员可以使用低碳目标作为社会压迫的理由。*关键词：碳补偿，水电坝，土著人民，巴拿马，可再生能源。*

La implementación local de políticas climáticas internacionales es frecuentemente oscura. El objetivo de nuestra investigación es destapar la "caja negra" de las compensaciones por carbono, como están siendo aplicadas en los territorios indígenas latinoamericanos. Nuestros dos estudios de caso sobre proyectos de energía renovable en contrucción en las aldeas Naso y Ngöbe, en el occidente de Panamá, indican que los bonos de carbono en sociedades opresivas potencialmente pueden causar daño social. Nuestros casos ilustran procesos de autoritarismo verde, control espacial y reestructuración social. Los empresarios que están construyendo las represas de Chan 75 y Bonyic no siguieron los estándares internacionales sobre consentimiento libre, anticipado y consciente, a la vez que entidades estatales reforzaban los derechos privados con violencia física. En tanto que los promotores del desarrollo hidroeléctrico esperan las decisiones sobre sus propuestas para verificación de acuerdo con el Mecanismo de Desarrollo Limpio (CDM), nosotros recomendamos reformas de procedimiento del CDM para apuntalar el respeto por los derechos humanos, incluyendo los derechos especiales codificados en la Declaración de las NU sobre Derechos de los Pueblos Indígenas. Si no se hace eso, quienes desarrollan este tipo de proyectos podrían usar objetivos orientados hacia la baja en emisiones de carbono para justificar opresión social. *Palabras clave: bonos de carbono, represas hidroeléctricas, pueblos indígenas, Panamá, energía renovable.*

Should low-carbon development fragment communities and disrupt Indigenous peoples' collective institutions and subsistence practices? Emerging from structural inequities and physical violence, the two hydro development projects we assess potentially extend Radcliffe's (2007) Latin American indigenous geographies of fear, racism, and unevenness to the clean energy sector.

The Clean Development Mechanism (CDM), an international framework to reduce greenhouse gas (GHG) emissions as industrial countries finance low-carbon projects in developing regions, has the potential to significantly influence global processes. There were US$6.5 billion dollars of project-based CDM finance transferred in 2008 alone (Capoor and Ambrosi 2009). By January of 2011, 2,700 validated projects existed in seventy countries. In spite of a bifurcated mandate to promote sustainable development along with GHG emissions reductions, the focus of CDM regulatory tools on measuring carbon (e.g., monitoring methodologies, additionality assessment,[1] etc.) limits attention to socioeconomic factors (Capoor and Ambrosi 2009; Gilbertson and Reyes 2009; Lövbrand, Rindefjäll, and Nordqvist 2009).[2]

In this article we assess whether carbon offsets can cause social harm. In 2008, the Indigenous Environmental Network (IEN) and Society for Threatened Peoples (2008) identified GHG mitigation projects they believe violate indigenous rights in acts of carbon colonialism, including several Panamanian dam sites. IEN's claims, and documents from Cultural Survival (e.g., Lutz 2007), spurred our 2009 fieldwork to analyze two projects in Naso and Ngöbe villages.[3] In western Panama, renewable energy projects create intense pressure for governance and livelihood transitions (Cordero et al. 2006; Paiement 2007; Jordán 2008; Finley-Brook and Thomas 2010).

The central objective of our research is to unpack the "black box" of carbon offsetting (Lovell and Liverman 2010). National and subnational implementation of the policies emerging from international climate institutions is often opaque (Lövbrand, Rindefjäll, and Nordqvist 2009; Bulkeley and Newell 2010). Since 2008, we have completed fieldwork on eleven carbon projects in four countries, including Nicaraguan bagasse cogeneration, forest carbon and hydroelectric dams in Costa Rica and Panama, and Dominican wind development. Research methods include semistructured interviews with project developers, state officials, nongovernmental organizations (NGOs), and impacted populations. The nineteen interviews contributing to this article drew from each of these sectors. Our case study approach documents how decision-making power and processes influence cost–benefit distribution among stakeholders (e.g., host communities, investors, government agencies, offset purchasers, etc.).[4]

Naso and Ngöbe community members charged the Panamanian government with human rights violations due to dam construction in their territories. As we sought to contextualize their claims within broader clean development trends, we uncovered an avoidance of issues pertaining to indigenous land and cultural rights in CDM project documents throughout Latin America, even in those located in Mexican *ejidos* (communal lands). Academic research addressing injustice in Latin American offsets tends to focus on biofuel or forestry projects (e.g., Boyd 2009; McAfee and Shapiro 2010; Hazlewood forthcoming), creating a research gap. In popular media (e.g., Dyer 2009), concerns about renewable energy CDM projects in indigenous territories are addressed.

Although we caution against generalizing from our illustrative cases, we recognize similar patterns in other indigenous territories (see, e.g., Hale 2002, 2004; Radcliffe 2007; Baldwin 2009). During hydro development, private and state partners often attempt to extinguish indigenous land claims (Paiement 2007; Jordán 2008). With the rise of offset markets, we argue that project developers might use low-carbon objectives to justify their demands for local sociocultural change. We highlight "stick and carrot" approaches that use carbon credits as positive incentives while employing physical force to assure project implementation. As global environmental change influences expectations for energy projects (Zimmerer 2011), we identify a resurgence of historical prejudices that categorize subsistence practices as inefficient and indigenous customs as inferior. Before presenting the details of our two case studies, we seek to contextualize events in Panama, where state agencies and private firms selectively define sustainable development in renewable energy projects in ways that allow them to pursue neoliberal agendas while further marginalizing indigenous communities.

Expanding Interests and Inequities in Hydro Power

There is a historic pattern of unequal distribution of costs and benefits in large-scale energy projects. *Hydrologic colonialism* is a spatial process imposed by big dams: high costs (e.g., ecological degradation, resettlement, loss of resource access, etc.) are felt in source landscapes and benefits (e.g., electricity, profit, etc.) are exported (Bakker 1999; Bonta 2004; Desbiens 2004; Sneddon and Fox 2008). State agencies have long justified large-scale energy projects as essential for national economic development, in spite of obvious spatial and social inequities.

We suggest that some large-scale CDMs might extend these inequities. Fair distribution and social consensus might be sidelined in CDM decision making because a major goal is to save money: Industrialized nations finance mitigation in developing countries because it is less expensive than cutting domestic emissions (Bumpus and Liverman 2008; Gilbertson and Reyes 2009; Bulkeley and Newell 2010). For host governments, renewable energy CDM projects are an opportunity to capture foreign investment for the expansion of infrastructure (Lokey 2009; Schreuder 2009), a pressing concern in countries like Panama with rapid growth in electrical demand.

Approximately 60 percent of CDM projects create renewable energy and the largest CDM sector in terms of total number of projects is the hydro sector.[5] As of January 2011, there were 800 CDM-verified hydro projects globally and 750 in the pipeline.[6] Our cases involve foreign-financed, large-scale (>15 megawatts) dams. Multinational firms sponsor a significant share of large hydro CDM projects (Haya 2007).

The hydro industry suggests large dams can be socially responsible as well as ecologically friendly (International Hydropower Association 2007). The project developers in our case studies are Colombia's state utility company (*Empresas Públicas de Medellín*, or EPM) and a Panamanian subsidiary of U.S.-based AES Corp.[7] Both firms claim to be industry leaders in corporate social responsibility (AES-Changuinola 2008; EPM 2009). Our case studies suggest these firms recast negative social consequences as economic development in ways that are both superficial and harrowingly profound.

Private–State Energy Sector Partnerships

Neoliberal economic reforms, such as privatization and deregulation, generally expand or reinforce the power of multinational firms, and this is certainly evident in the energy sector (Ahmed 2010). The Central American Electrical Interconnection System (SIEPAC), an integrated regional grid under construction across 2,000 kilometers, required extensive foreign private investment. Energy projects linked to SIEPAC, including the two Panamanian case studies, fit within broader private–state economic development strategies to exploit rural peripheries to fuel industrial and urban areas.

Involvement of the private sector in Latin American energy development signifies an important political-economic shift with broad implications for industry, trade, tax collection, and much more (Lokey 2009; Schreuder 2009). Impacts from neoliberal reforms are varied and remain tied to other international and domestic policies (Mansfield 2004; Radcliffe 2007). For example, the CDM transfers extensive authority to the private sector because it is a project-based framework often relying on implementation by firms, although the vast rule-based structure of the CDM maintains an instrumental role for governmental and quasi-governmental agencies (Giddens 2009; Bulkeley and Newell 2010). State agencies continue to influence the type, pace, and form of resource commodification tied to specific CDM projects (Lokey 2009; Schreuder 2009).

Processes in Panama reflect the hollowing out of the state as responsibility for watershed management, public education, and energy infrastructure shifts to the private sector. At the same time, authoritarianism seeps out of complex rearticulations of governance and regulation (*sensu* Swyngedouw 2000). The state and private sector join forces to constrict local resource access and disempower Indigenous peoples, in spite of Panama's recognition of semiautonomous indigenous territories (*comarcas*) over the past century.[8] Across national and institutional landscapes, support for racial and class-based privilege rooted in colonial and imperial histories remains clear (Swyngedouw 2000; Hale 2002, 2004; Radcliffe 2007; Jordán 2008).

When Latin American Indigenous peoples oppose development projects, they might become targets of state violence (Radcliffe 2007; Jordán 2008). Yet Hale (2002, 2004) identified concurrent politics of recognition where states legally codify ethnic rights in ways that seem progressive but might still be manipulated to limit access to land and resources. Tying race relations to the valorization of GHGs, Baldwin (2009) suggested that carbon markets can lead to the entrenchment of racial hierarchies and limit Indigenous peoples' economic and political options. In western Panama, the selective transfer of benefits (e.g., land payments, jobs, gifts, carbon credits, etc.) to local individuals willing to allow hydro development creates community division such that the organization of alternatives or unified opposition becomes unviable (Jordán 2008).

Although neither project we assess had been verified by the CDM at the time of writing, project developers had submitted proposals and, more important, justified the dams using global climate change arguments (e.g., AES-Changuinola 2008; Rodriguez 2010).[9] Bonyic developers suggested in their CDM application that a main motivation for the project was to share carbon market benefits with Naso communities (Rodriguez 2010).

Hydro Development in Bocas del Toro, Panama

After determining strong hydro potential in Bocas del Toro Province in the 1970s, Panama set aside Palo Seco Forest Reserve (Figure 1) in the 1980s to protect the watershed for energy production, but state attempts to build dams in the region were unsuccessful (Paiement 2007; Barber 2008; Jordán 2008). Following privatization of Panama's Institute of Hydrologic Resources and Electrification (IRHE) in 1998, there was vast investment leading to more than seventy new hydro concessions (Cordero et al. 2006). Eighty percent of Panama's two dozen proposed and verified CDM projects involve dams.

The Panamanian state plans to redefine Bocas del Toro Province. Paiement (2007, 126) described a vision a governmental official shared with him: "in the next ten years the rivers and forests of Bocas del Toro will be transformed by a series of dams, artificial lakes, access roads, and transmission lines."[10] In this context, state efforts to assure implementation of the first hydro projects in the province, described next, garner additional importance.

Green Authoritarianism in Naso-Tjërdi

> I didn't know that the development I was promised had people in uniforms militarizing our communities . . . every day more than ten or twelve police enter our community. . . . They are taking care of the machines. That is what they do. (Naso leader, conversation, 7 June 2009)

This Naso leader's allegation, supported by other interviews and media coverage, was that Panamanian officials protect dam construction equipment from sabotage in the face of local resistance. We label this process of state oppression to defend renewable energy sources and market-valorized ecological processes *green authoritarianism* (see Peluso 1992; Neumann 1998).

As an example of green authoritarianism, Bonyic dam obstructed progress toward legal recognition of the Naso homeland (*Naso-Tjërdi*; Paiement 2007; Jordán 2008; World Bank Inspection Panel 2009). The Naso were in the final stages of negotiating their comarca when the Colombian state utility company EPM (and two partner firms with minor holdings) purchased the dam concession (Paiement 2007; Jordán 2008).[11] Naso leaders believe demarcation of Naso-Tjërdi was stalled to assure dam construction, particularly after Naso began to vocalize opposition to the energy project (Anonymous, conversation, 7 June 2009).

The Naso are proud to have one of the few remaining monarchies in the Americas, but disagreement over the Bonyic dam split their kingdom (Paiement 2007). King Tito Santana signed a weak agreement in 2003 with the firm Hydroecológica Teribe (HET), of which EPM is the majority owner (Paiement 2007). Later, HET promised to transfer 25 percent of the project's carbon credits to the local community (Rodriguez 2010). Santana's cooperation with HET, however, led many Naso to accuse him of being corrupt, although he maintained support from a group of followers (Paiement 2007; Jordán 2008). The majority of Naso elected a new king, but the Panamanian state and private investors continue to recognize Tito Santana. Naso institutions were sufficiently disrupted that in 2004 the Inter-American Development Bank cancelled loans promised to HET for dam construction (Paiement 2004; Rodriguez 2010).

In June 2009, after HET decided to finance the dam with internal funding, the National Environmental Authority (ANAM) granted a 1,246-hectare hydro concession (Rodriguez 2010). Construction began in spite of Naso protests (Finley-Brook and Thomas 2010). In June 2010, Naso, seeking to halt Bonyic construction, filed a petition with the Inter-American Commission on Human Rights (IACHR). A series of IACHR petitions expose a pattern of Panamanian state oppression of Indigenous peoples (see Mayhew, Jordán, and Rolnick 2009; Finley-Brook and Thomas 2010).

In June 2010, the Bonyic project applied for CDM verification. According to the application, at the time of negotiation with HET, the Naso were "in the midst of a prolonged leadership crisis, caused by historical and political factors, as well as by family feuds and personal interests" (Rodriguez 2010, 16). This version of history removes blame from HET and the Panamanian state, although Paiement (2007) provides a detailed firsthand account of their involvement in Naso internal conflicts.

We have argued elsewhere that CDM application processes do not provide adequate, accessible opportunities for impacted communities to document concerns (Finley-Brook and Thomas 2010). In addition to a one-month public commenting period on the Internet,[12] developers are required to comply with national standards, as Bonyic did.[13] Describing the consultation process, the Bonyic CDM application vaguely mentions "a public discussion with local stakeholders" in 2005, and greater attention is drawn to "highly positive" comments generated in Panama City at a clean production symposium (Rodriguez 2010, 49). Although it is likely CDM verifiers will request additional details,[14] this example suggests applicants might attempt to omit essential information. In the next case study, the CDM application notes "ample support" from local populations (TÜV-SÜD 2008, 35), but when the UN Special Rapporteur on the situation of human rights and fundamental freedoms of Indigenous peoples visited the

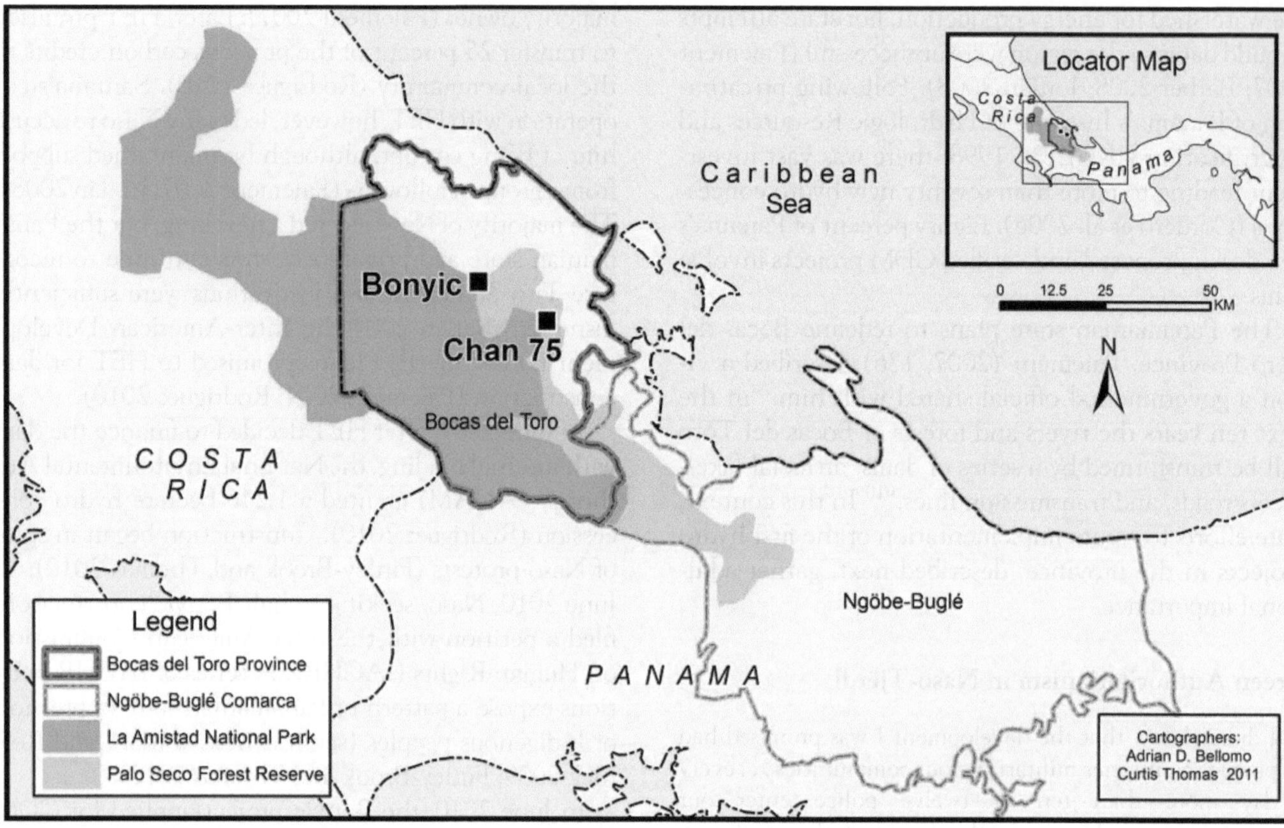

Figure 1. Bocas del Toro Province with dam locations.

project area, he found "significant discontent" (Anaya 2009, 10, authors' translation). Both statements are selectively true, as the dam created a social rift.

Spatial Control and Social Restructuring in Chan 75[15]

In 2007, Ngöbe villagers blocked Chan 75 dam construction for two weeks until national police beat and arrested protesters, including women and children (Barber 2008; Cultural Survival 2008; Jordán 2008). Since this protest, national police receiving salaries from AES-Changuinola screen movement to and from the zone. According to the UN Special Rapporteur on the rights of Indigenous peoples, the contractual relationship between state security forces and AES-Changuinola is concerning due to evidence of unequal power and pressure tactics (Anaya 2009).

Four Ngöbe villages (Changuinola Arriba, Charco de la Pava, Nance de Risco, and Valle del Rey) are being resettled without free, prior, and informed consent (Anaya 2009). AES-Changuinola notes compliance with World Bank standards for involuntary resettlement.[16] Based on the 2007 UN Declaration on the Rights of Indigenous Peoples, however, an international norm Panama supported, involuntary resettlement is prohibited in indigenous territories.

In 2007, Panama's National Assembly Resolution Number 1228 decreed that the four Ngöbe villages in the proposed flood zone would be permitted to stay within the Palo Seco Forest Reserve if they relocate to allow dam construction (Jordán 2008). Modern Ngöbe settlements in this area predate the reserve's creation (Anaya 2009), but inhabitants were not allowed to register their lands prior to the dam concession (Jordán 2008; Mayhew, Jordán, and Rolnick 2009). The World Bank loaned funds in 2001 to Panama's National Land Administration Program to title Ngöbe-Buglé Comarca (Figure 1) annex areas (World Bank Inspection Panel 2009), including Ngöbe and Naso villages in this study. A decade later these villages remain without land titles.

AES-Changuinola's 2007 contract with the state environmental agency ANAM privatized responsibility for 6,215 hectares of the Palo Seco Forest Reserve (Jordán 2008). Because watershed protection is necessary for hydroelectric production, the firm will restrict the clearing of agricultural fields: "[Resettled populations] are being trained on farming techniques and

Figure 2. Chan 75 dam in relation to a Ngöbe village (upper right). *Source:* Photo by Mary Finley-Brook.

efficient production. Moreover, they are receiving training on the sustainable environmental management the communities must adhere to" (AES-Changuinola 2008, 47).

Company employees designed new forms of subsistence for local populations as tree farmers and as artisans producing crafts for tourists expected to visit the artificial reservoir. Paternalistic social programs followed state violence to quell local opposition to the dam (Barber 2008; Cultural Survival 2008; Jordán 2008; Anaya 2009; Finley-Brook and Thomas 2010). Spatial control (i.e., fences, travel restrictions, loss of river transportation) and pressure tactics (i.e., unwarranted house-to-house searches, death threats, destruction of crops and property) promoted isolation, fear, and desperation (IACHR 2009a). Dynamite blasts were loud enough to force school closures in the nearby village of Charco de la Pava (Figure 2).

Chan 75 creates fundamental change in surrounding villages and positive social development could be possible. For example, AES-Changuinola (2008) has established education and health care programs. Although the firm suggests that these efforts are evidence of corporate social responsibility, the investments were integral to a broader partnership with state agencies and paved the way for AES-Changuinola to receive the hydro concession. Furthermore, AES-Changuinola's social programs, as well as the firm's commitment to share 20 percent of earnings from carbon offsets with the state environmental agency (Finley-Brook and Thomas 2010), allows the government of Panama to avoid paying for basic programs and services generally defined as state responsibilities. Social assistance can also be used to counteract allegations of harm caused by dam development or to mask other injustices.

AES-Changuinola's (2008) plans to improve Ngöbe living conditions create significant cultural change, such as splitting multigenerational households into nuclear families. Traditional houses were categorized as instituting "confinement" due to "inadequate use of construction materials" (AES-Changuinola 2008, 43). New cement structures are "dignified," in contrast to customary wood and palm structures that left inhabitants "exposed to rain and diseases" (AES-Changuinola 2008, 50).

AES-Changuinola representatives negotiated relocation accords and compensation for land use by household, creating upheaval within families because individuals signed resettlement agreements in representation of other family members even when they lacked this legal right (Lutz 2007; Jordán 2008). Discrete negotiations with each household unit were inappropriate because land is communally owned (Barber 2008; Jordán 2008). The firm's disbursement of gifts and money during negotiations also contributed to social tension (Jordán 2008).[17]

In 2009, the IACHR advised the Panamanian state to halt dam construction and consult with the Ngöbe in good faith. Negotiations occurred while construction advanced. Decisive meetings were held in Panama City, meaning select village representatives negotiated without community support and other villagers were unable to monitor events (Mayhew, Jordán, and Rolnick 2009). Displeased Ngöbe suggest state officials made sure that dam opponents were not allowed "at the table" (IACHR 2009a, audio recording, authors' translation). The accord signed between village and state representatives stated that both parties agreed to work together to assure timely dam completion.

The IACHR brought the Panamanian state to the Inter-American Human Rights Court (CIDH) in May 2010. The court decided to disregard the IACHR's recommendation to halt dam construction because judges were not convinced the matter was "extremely serious" or "urgent" (CIDH 2010, 9–12, authors' translation). Panamanian officials testified that by mid-2010 only a handful of families had not signed relocation agreements with AES-Changuinola (CIDH 2010). Judges highlighted acceptance of money and land settlements on the part of various members of the indigenous communities, while noting the existence of unclear dates for alleged death threats and pressure tactics.

The Chan 75 application has been in the CDM pipeline since 2008. The verdict remains unclear at the time of writing. The consultation process leading to the application was poor. ANAM's public meetings in reference to the dam in 2005 were held in a town located six hours away from the four communities to be resettled (Anaya 2009). A letter from members of the Charco de la Pava village opposing the project was sent to ANAM in 2007 during the prevalidation public consultation period for AES-Changuinola's concession, but state officials approved the concession nonetheless (Anaya 2009). Chan 75's CDM application made no mention of the land tenure conflict or social opposition (TÜV-SÜD 2008). No comments were received during the thirty-day online CDM public commenting period.

Neocolonial Carbon Projects and Indigenous Communities

Green authoritarianism and carbon colonialism are evident in the construction of both the Chan 75 and Bonyic dams. In these case studies, state agencies and private firms worked in partnership to dominate and oppress local populations. With support from the state, developers used physical force to assert claims to exploit or protect natural resources with market value. Working in partnership, state actors and private firms have obligated Naso and Ngöbe villages to experience what Radcliffe (2007) defined as Latin American Indigenous peoples' geographies of fear and inequitable development.

Our research suggests GHG reduction projects in indigenous territories can adversely affect local self-governance, land tenure, resource access, and subsistence practices (see also Lohmann 2006; Baldwin 2009; Lövbrand, Rindefjäll, and Nordqvist 2009; Mate and Ghosh 2009; Hazlewood forthcoming). The protection of cultural rights is lacking in CDM requirements and Indigenous peoples have insufficient influence over carbon offset decisions (see also Finley-Brook and Thomas 2010). For example, there are no CDM guidelines to protect cultural heritage. We are aware of two Panamanian projects in the CDM pipeline that destroyed or displaced ancient ancestral sites including cemeteries.

Hydro development and carbon markets involve spatial imbalance in terms of the distribution of costs and benefits. Although Latin American CDM projects are often linked to high-stake energy markets, the United Nations Framework Convention on Climate Change oversight processes focused on GHG emissions currently lack safeguards against exclusionary and harmful practices. Findings from our two cases suggest that the CDM might contribute to hydrologic neocolonialism in some instances; however, constraints to local participation are not limited to dam projects. Regardless of CDM project type, greater attention appears necessary to defend local populations from authoritarian spatial control linked to the imposition of externally defined institutional arrangements and neoliberal ecological practices.

We expect that solving social justice issues in international carbon offset and renewable energy projects will be complex. For example, stipulations such as prohibiting CDM projects in untitled indigenous territories could create perverse incentives for the privatization of communal property if broader sociopolitical injustices are not addressed first. Although the UN Declaration on the Rights of Indigenous Peoples provides important guidelines, its impact might remain limited if even the Inter-American Human Rights Court averts the Declaration's fundamental call for free, prior, and informed consent, as apparently occurred in the case of Chan 75. For low-carbon development to be considered truly "clean," it is necessary to eliminate pervasive social inequalities and oppression in addition to reducing GHG emissions.

Acknowledgments

We would like to thank Karl Zimmerer, Osvaldo Jordán, Greg Knapp, Ken Young, Stan Stevens, Karl

Offen, David Salisbury, Todd Lookingbill, and anonymous reviewers for helpful comments. We are indebted to anonymous Panamanian sources. Both authors received University of Richmond research grants. Curtis Thomas, Kimberley Klinker, and Adrian Bellomo produced the map.

Notes

1. Additionality suggests that a project would not have occurred without CDM finance. See Schneider (2009) for problems associated with assuring additionality.
2. Emissions markets separate GHGs from other elements in the development equation, such as ecosystem integrity and community resilience. Carbon calculations remain imperfect (e.g., exaggerated baselines, indirect emissions, externalities, exceptions for transportation, etc.; Lohmann 2006; Haya 2007; Gilbertson and Reyes 2009).
3. We also assessed older dams in Chiriquí and Panamá Provinces (see Finley-Brook and Thomas 2010).
4. Although we record material flows, we recognize limitations to this approach because value cannot be placed on key components such as cultural change.
5. Updates can be found at http://cdmpipeline.org/cdm-projects-type.htm (last accessed 7 April 2011).
6. A guide to the CDM project cycle can be found at http://www.undp.org/energy/docs/cdmchapter2.pdf (last accessed 7 April 2011).
7. Founded in 1981 as Applied Energy Systems, the name was changed to AES Corporation and later shortened.
8. See Paiement (2007) and Jordán (2008) for details about comarca recognition.
9. Both firms have implemented carbon offsets in other locations, including indigenous territories (Lokey 2009; Wittman and Caron 2009; Finley-Brook and Thomas 2010).
10. In this article we focus on the role of national-level state actors. Lutz (2007), Paiement (2007), Jordán (2008), and Mayhew, Jordán, and Rolnick (2009) documented how the actions of provincial officials generally coincide with central government efforts.
11. The Naso and Ngöbe in Bocas del Toro refused entry into the Ngöbe-Buglé Comarca (Figure 1) when it was formed in 1997 due to concerns over representation (Paiement 2007; Jordán 2008).
12. Fewer than 10 percent of verified Central American CDMs (2005–2008) received online comments (Finley-Brook and Thomas 2010).
13. See Paiement (2007) for details.
14. A CDM applicant is responsible for contracting a UN-designated operational entity (DOE) to carry out verification. The contractual basis of this relationship might create situations where DOEs are biased toward approval because rejecting projects could harm their ability to obtain future contracts.
15. This dam is sometimes referred to as Changuinola 1 or Chan 1.
16. A 2009 AES-Changuinola slideshow is available from the authors.
17. Details of financial settlements have not been publicly disclosed. Resettlement contracts transferring land holdings were signed with one person in each family, even though in some instances other family members were opposed to relocation (Jordán 2008). Firm representatives pressured illiterate individuals to "sign" (by thumbprint) resettlement agreements (Lutz 2007; Jordán 2008).

References

AES-Changuinola. 2008. *Sustainability and corporate performance report: Hydroelectric project Changuinola 1 (Chan 75)*. Panama City, Panama: AES-Changuinola.

Ahmed, W. 2010. Neoliberalism, corporations, and power: Enron in India. *Annals of the Association of American Geographers* 100 (3): 621–39.

Anaya, S. J. 2009. *Observaciones sobre la situación de la Comunidad Charco la Pava y otras comunidades afectadas por el Proyecto Hidroeléctrico Chan 75 (Panamá)* [Observations about the situation of Charco la Pava Community and other communities affected by Chan 75 Hydroelectric Project]. Panama City, Panama: United Nations.

Bakker, K. 1999. The politics of hydropower: Developing the Mekong. *Political Geography* 18:209–32.

Baldwin, A. 2009. Carbon nullius and racial rule: Race, nature and the cultural politics of forest carbon in Canada. *Antipode* 41 (2): 231–55.

Barber, J. 2008. *Paradigms and perceptions: A chronology and analysis of the event of the Chan 75 hydroelectric project and the role and relationships of participants, Bocas del Toro, Panama*. Panama City, Panama: School for International Training.

Bonta, M. 2004. Death toll one: An ethnography of hydro power and human rights violations in Honduras. *GeoJournal* 60 (1): 19–30.

Boyd, E. 2009. Governing the clean development mechanism: Global rhetoric versus local realities in carbon sequestration projects. *Environment and Planning A* 41:2380–95.

Bulkeley, H., and P. Newell. 2010. *Governing climate change*. London and New York: Routledge.

Bumpus, A., and D. Liverman. 2008. Accumulation by decarbonization and the governance of carbon offsets. *Economic Geography* 84 (2): 127–55.

Capoor, K., and P. Ambrosi. 2009. *State and trends of the carbon market 2009*. Washington, DC: World Bank.

Cordero, S., R. Montenegro, M. Mafla, I. Borgués, and J. Reid. 2006. *Análysis de costo beneficio de cuatro proyectos hidroélectricos en la cuenca Changuinola-Teribe* [Cost–benefit analysis of four hydroelectric projects in the Changuinola-Teribe watershed]. Panama City, Panamá: Alianza para la Conservación y el Desarrollo.

Corte Interamericana de Derechos Humanos (CIDH). 2010. *Medidas provisionales solicitadas por la Comisión Interamericana de Derechos Humanos Respecto de la República de Panamá* [Provisional measures solicited from the Republic of Panama by the Inter-American Human Rights Commission]. San José, Costa Rica: CIDH.

Cultural Survival. 2008. Panama dam construction steps up the pace. *Cultural Survival Quarterly* 32 (1): 4–5.

Desbiens, C. 2004. Producing North and South: A political geography of hydro development in Quebec. *The Canadian Geographer* 48 (2): 101–18.

Dyer, Z. 2009. Clean energy plays dirty in Mexico. *North American Congress on Latin America.* https://nacla.org/node/5638 (last accessed 18 November 2009).

Empresas Públicas de Medellín (EPM). 2009. *Informe de Sostenibilidad* [Sustainability report]. http://www.epm.com.co/epm/institucional/general/Balances/2009/Informe_Sostenibilidad_2009.pdf (last accessed 10 August 2010).

Finley-Brook, M., and C. Thomas. 2010. Treatment of displaced indigenous populations in two large hydro projects in Panama. *Water Alternatives* 3:269–90.

Giddens, A. 2009. *The politics of climate change*. Malden, MA: Polity.

Gilbertson, T., and O. Reyes. 2009. *Carbon trading: How it works and why it fails*. Uppsala, Sweden: Dag Hammarskjöld Foundation.

Hale, C. R. 2002. Does multiculturalism menace? Governance, cultural rights and the politics of identity in Guatemala. *Journal of Latin American Studies* 34:485–524.

———. 2004. Rethinking indigenous politics in the era of the "Indio Permitido." *NACLA Report on the Americas* 38 (2): 16–21.

Haya, B. 2007. *Failed mechanism: How the CDM is subsidizing hydro developers and harming the Kyoto Protocol.* Berkeley, CA: International Rivers.

Hazlewood, J. A. Forthcoming. C0210nialism and the "unintended consequences" of commoditizing climate change: Geographies of hope amid a sea of oil palms in the northwest Ecuadorian Pacific region. *Yale Journal of Sustainable Forestry* 30 (5–6).

Indigenous Environmental Network and Society for Threatened Peoples. 2008. Indigenous peoples' guide: False solutions to climate change. http://www.earthpeoples.org/CLIMATE_CHANGE/Indigenous_Peoples_Guide-E.pdf (last accessed 19 July 2009).

Inter-American Commission on Human Rights (IACHR). 2009a. Case 12.717/ precautionary measure PM 56/08—Indigenous communities Ngöbe and others, Panama participants: Cultural Survival, State of Panama. Panama City, Panama: IACHR.

———. 2009b. Report 58/09, Petition 12.354, admissibility, Kuna of Madungandí and Emberá of Bayano indigenous peoples and their members, Panama. Panama City, Panama: IACHR.

International Hydropower Association. 2007. Sustainability assessment protocol. http://www.hydropower.org/downloads/IHA_SAP.pdf (last accessed 4 August 2009).

Jordán, O. 2008. "I entered during the day, and came out during the night": Power, environment, and indigenous peoples in a globalizing Panama. *Tennessee Journal of Law and Policy* 4 (2): 467–505.

Lohmann, L. 2006. Carbon trading: A critical conversation on climate change, privatisation and power. Development dialogue No. 48. Uppsala, Sweden: The Dag Hammarskjöld Centre.

Lokey, E. 2009. *Renewable energy project development under the clean development mechanism: A guide for Latin America.* Sterling, VA: Earthscan.

Lövbrand, E., T. Rindefjäll, and J. Nordqvist. 2009. Closing the legitimacy gap in global environmental governance? Lessons from the emerging CDM market. *Global Environmental Politics* 9 (2): 74–100.

Lovell, H., and D. Liverman. 2010. Understanding carbon offset technologies. *New Political Economy* 15 (2): 255–73.

Lutz, E. 2007. Dam nation. *Cultural Survival Quarterly* 31 (4): 16–23.

Mansfield, B. 2004. Rules of privatization: Contradictions in neoliberal regulation of North Pacific fisheries. *Annals of the Association of American Geographers* 94 (3): 565–84.

Mate, N., and S. Ghosh. 2009. The CDM scam: Wind power projects in Karnataka. *Mausam* 1–2:30–34.

Mayhew, C., O. Jordán, and A. Rolnick. 2009. Panama is in breach of its obligations to indigenous peoples under the convention on the elimination of all forms of racial discrimination. Report submitted to the UN Committee on the Elimination of Racial Discrimination. Panama City: Alianza para la Conservación y el Desarrollo (Alliance for Conservation and Development).

McAfee, K., and E. Shapiro 2010. Payments for ecosystem services in Mexico: Nature, neoliberalism, social movements, and the state. *Annals of the Association of American Geographers* 100 (3): 579–99.

Neumann, R. 1998. *Imposing wilderness: Struggles over livelihood and nature preservation in Africa.* Berkeley: University of California Press.

Paiement, J. J. 2007. The tiger and the turbine: Indigenous rights and resource management in the Naso territory of Panama. Unpublished PhD dissertation, Department of Anthropology, McGill University, Montreal, Canada.

Peluso, N. L. 1992. *Rich forests, poor people: Resource control and resistance in Java.* Berkeley: University of California Press.

Radcliffe, S. A. 2007. Latin American indigenous geographies of fear: Living the shadow of racism, lack of development, and antiterror measures. *Annals of the Association of American Geographers* 97:385–97.

Rodriguez, R. 2010. *Bonyic hydroelectric plant.* Buenos Aires, Argentina: MGM Worldwide. http://cdm.unfccc.int/UserManagement/FileStorage/7GDQP6E1HZWMS540R829TXFCBK30NU (last accessed 25 August 2010).

Schneider, L. 2009. Assessing the additionality of CDM projects: Practical experiences and lessons learned. *Climate Policy* 9:242–54.

Schreuder, Y. 2009. *The corporate greenhouse: Climate change policy in a globalizing world.* New York: Zed Books.

Sneddon, C., and C. Fox. 2008. Struggles over dams as struggles for justice: The World Commission on Dams (WCD) and anti-dam campaigns in Thailand and Mozambique. *Society and Natural Resources* 21:625–40.

Swyngedouw, E. 2000. Authoritarian governance, power and the politics of rescaling. *Environment and Planning D: Society and Space* 18:63–76.

TÜV-SÜD. 2008. Changuinola I hydroelectric project in Panama. http://cdm.unfccc.int/Projects/Validation/DB/

TRB10UH4NZYNIGESERW4L7QGC126FR/view.html (last accessed 20 May 2009).

Wittman, H. K., and C. Caron. 2009. Carbon offsets and inequality: Social costs and co-benefits in Guatemala and Sri Lanka. *Society and Natural Resources* 22 (8): 710–26.

World Bank Inspection Panel. 2009. Report and recommendation. Panama: Land administration project. Report No. 49004-PA. Washington, DC: World Bank.

Zimmerer, K. S. 2011. New geographies of energy: Introduction to the special issue. *Annals of the Association of American Geographers* 101 (4): 705–11.

Downstream Effects of a Hybrid Forum: The Case of the Site C Hydroelectric Dam in British Columbia, Canada

Nichole Dusyk

Institute for Resources, Environment, and Sustainability, University of British Columbia

Attempting to scale up the deployment of renewable energy technology has come with considerable controversy and opposition. Research exploring this opposition has highlighted the importance of project control and decision-making structures, including public engagement and consultation. This article contributes to the discussion by considering how participatory processes might have varying effects across space and time. Combining the concept of hybrid forums with spatial theories of change, it explores participatory processes and how they can result in uneven change in sociotechnical networks. It applies this theoretical framing to one hydroelectric project in northeastern British Columbia, to show how lessons learned from this project, and, from the legacy of hydroelectricity more generally, are not consistent throughout the province and how attempts to manage these differences have led to further conflict and opposition. *Key Words: British Columbia, hybrid forums, hydroelectricity, public participation, renewable energy policy.*

试图扩大可再生能源技术的部署已经具有相当的争议和反对。研究探索这种对立突出了项目控制和决策结构，包括公众参与和协商的重要性。本文通过考虑参与过程如何产生跨越时空的不同的影响，而有助于对此问题的讨论。结合混合论坛概念与变化的空间理论，本文探索参与过程以及它们如何能导致在社会技术网络的不均匀变化。它把这一理论框架应用到不列颠哥伦比亚省东北部的一个水电项目，以显示从这个项目汲取的，和从水电遗留下来的更普遍的经验教训，在全省是不一致的，以及处理这些分歧的企图如何导致了进一步的冲突和对立。*关键词：不列颠哥伦比亚省，混合论坛，水电，公众参与，可再生能源政策。*

En el intento por ampliar su uso, la instalación de tecnología de energía renovable ha llegado rodeada de considerable controversia y oposición. La investigación dedicada a explorar esta oposición ha destacado la importancia del control del proyecto y de las estructuras de toma de decisiones, incluyendo el compromiso y consulta públicos. Este artículo contribuye en la discusión al considerar la manera como los procesos participativos podrían tener efectos variados a través del espacio y el tiempo. Combinando el concepto de foros híbridos con teorías espaciales de cambio, se exploran los procesos participativos y cómo estos pueden resultar en el cambio desigual en redes socio-técnicas. Se aplica este marco teórico a un proyecto hidroeléctrico del nordeste de la Colombia Británica, para mostrar cómo las lecciones aprendidas de este proyecto y, más generalmente, del legado de la hidroelectricidad, no son consistentes en toda la provincia y cómo los intentos para manejar estas diferencias han llevado a mayor conflicto y oposición. *Palabras clave: Colombia Británica, foros híbridos, hidroelectricidad, participación pública, política de energía renovable.*

Attempting to scale up the deployment of renewable energy technology has come with considerable controversy and opposition. Recent scholarship has explored the rationale and trajectory of opposition, complicating not-in-my-backyard (NIMBY) theories with more nuanced explanations. This research has highlighted the importance of project control and decision-making structures, including public engagement processes, ownership structures, community benefits, and the presentation of alternatives (see, for example, Bell, Gray, and Haggett 2005; Szarka 2006; Wolsink 2007).

What is less understood is how renewable energy controversies, and our collective means for overcoming them, produce varying effects through space and time. By combining the concept of hybrid forums with Doreen Massey's recent spatial theories, this article explores participatory processes and how they could result in uneven change in sociotechnical networks.

This theoretical framing is applied to a case in British Columbia, Canada, where a hydroelectric dam first proposed, and publicly debated, three decades ago has resurfaced as part of provincial climate policy.[1] The case supports the importance of democratic engagement in renewable energy deployment and the transformative potential of hybrid forums. At the same time, it also describes disjuncture that has occurred as a result of change at different points in the network. It demonstrates how lessons learned from this project, and from the legacy of hydroelectricity more generally, are not consistent throughout the province and how attempts to manage these differences have led to further conflict and opposition.

Transforming Sociotechnical Networks

Supporters of democratic engagement in environmental and technological decision making argue that, among other things, participatory processes have transformative potential. This includes the possibility of creating new, potentially more robust, forms of knowledge and helping to shape individual and collective identities (Fischer 2000; Healey 2006; Callon, Lacoumes, and Barthe 2009; Whatmore 2009). Callon, Lacoumes, and Barthe (2009) proposed the term *hybrid forums* to refer to the dialogical spaces that allow for the exploration and transformation of knowledge and identities. Hybrid forums are either formal or informal participatory spaces where collective learning takes place and possible futures are explored. They can range from advocacy coalitions to formalized citizens' juries with different formats allowing more or less room for collective learning. A key component of this conceptualization is its emphasis on the long-term sociotechnical transformation that can emerge from participatory processes (Joly 2007).

Energy networks are highly technical, far-reaching systems deeply integrated in social and political practices. As such, they can be usefully theorized using the "flat," relational model of interaction developed in actor-network theory (Callon 1991; Latour 2005). This model provides a powerful way to interpret change in energy networks because it repositions actors, whether human or not, as active and implicated in the collective regardless of their position in the network. This offers a way of theorizing how change might occur in energy production and use, even (or especially) from the ground up. Hybrid forums represent a dialogical moment when network reconfiguration and realignment may be initiated.

What remains as an empirical question is how local change or transformation in energy networks might vary across space and over time and what this might mean for the trajectory of the network. Hughes (1983) used the term *regional style* to describe how electrical networks are regionally configured based on geography, institutions, financing opportunities, and key actors. This concept explains regional variation in energy networks based on, for instance, the availability of resources and political motivations, but it does not have the resolution to explain how energy networks and policies (as well as the related controversies) might be differently experienced across the network. Similarly, it does not explain how members of the network might change in different ways or at different rates.

The concern here is that system changes are not evenly or simultaneously felt across the network; that reconfigured infrastructures result in a redistribution of resources and impacts (Guy, Graham, and Marvin 1997). It follows that regardless of the apparent evenness of the outcome, asymmetrical change, or disjuncture, is inevitable particularly in sociotechnical systems as far-reaching as electrical networks. All parts of the system cannot change simultaneously or in identical ways.

Massey (1984) has used the term *sedimentary layering* to describe how uneven landscapes develop over time and space. More recently she has emphasized the dynamic, relational, multiplicity of space such that layers themselves cannot be understood as homogenous or static events. Rather each layer is composed of "meetings" that vary across time and space. She theorized "layers as accretions of meetings. Thus something which might be called there and then is implicated in the here and now" (Massey 2005, 139).

Combining Massey's interpretation of space and time with the concept of hybrid forums leads us to consider how the effects of such meetings are implicated in sociotechnical change. Although hybrid forums, as "meetings," offer the opportunity for learning and transformation, this, too, will be differently experienced from different vantage points.

Basing an analysis on an essentialized notion of territorial scale (e.g., the local) is problematic in a number of ways (see, for instance, Jessop, Brenner, and Jones 2008; Moore 2008). The argument here, however, is that all points in the network can be viewed as local sites (Latour 2005) that, although connected, accumulate different histories and change at different rates. The heterogeneity of space and time creates uneven terrain—in everyday experience but also in and from formalized gatherings and forums. Therefore, even though collective learning might occur, there is no guarantee that this leads to resolution or consensus. Furthermore, these variations in space and time are embedded within power geometries. "The multiple layering is thus neither neutral nor value-free" (Graham and Healey 1999, 642) and attention should be paid to how disjuncture and conflict are negotiated and managed.

The remainder of this article explores these concepts through the lens of a proposed hydroelectric project in northeast British Columbia. It illustrates how accumulation and change have varied across space and time and how disjuncture, created by asymmetrical transformation, has contributed to conflict and misunderstanding between opponents and proponents of a hydroelectric project in the region.

Hydroelectricity in Northeastern British Columbia

In Canada, energy development falls under provincial jurisdiction. Therefore, the physical and institutional structure of electrical networks depends, to a large extent, on provincial energy resources and policies. In British Columbia, Canada's western-most province, 89 percent of electricity is generated by hydroelectricity (Nyboer and Nyboer 2009). The majority of the hydroelectric capacity was installed in the 1960s and 1970s along two river basins: the Columbia in the southeast of the province and the Peace in the northeast. This article is focused on hydroelectric development on the latter. Figure 1 is a map of British Columbia's Peace River region, including the local communities, the existing hydroelectric infrastructure, and, the topic of this article, the proposed Site C hydroelectric dam.[2]

In the mid-twentieth century, the Peace River was extensively mapped for its hydroelectric power potential as part of a plan to industrialize northern British Columbia (Loo 2007). In 1968, the 2,730-MW W.A.C. Bennett Dam, the first and largest dam on the Peace River, came online, creating the massive Williston Reservoir, 250 km in length. In 1974, only three years after the Williston Reservoir had completely filled, work began on a second hydroelectric dam. The 700-MW Peace Canyon Dam, 22 km downstream from the Bennett Dam, was completed in 1980.

The construction of the two hydroelectric dams profoundly changed the physical and cultural landscape along the Peace River. Accompanying the physical transformation (of submerged land, altered climate, and new flow and ice cover patterns) was a process of sociocultural transformation (Zimmerer 2011). At the time of construction, the dams were not considered controversial.[3] Yet, the sociospatial "meetings" that occurred as a result of these projects created permanent change. Firsthand experience on the worksite and on the land, the temporary boom in the community of Hudson's Hope, the treatment of displaced people, and the spotty record of promises regarding local employment and economic development was instructive to local residents (Pollon and Matheson 1989). These two projects resulted in new knowledge and new meanings regarding the river itself and mega-project construction in the region.

The hydroelectric development also created new lines of connection and expectation between the region and the rest of the province. The hydroelectric capacity on the Peace River forms a core component of the province's hydro-based electricity system. Together, these two dams produce approximately 29 percent of the provincial utility's annual electricity generation (British Columbia Hydro and Power Authority 2010), and the generating station at the Bennett Dam remains the single largest source of electricity in British Columbia. These material and nonmaterial lines of connection serve as an interface between events in the region and the province as a whole. As will be discussed in the next section, the existing infrastructure and its actual and perceived value have become rationales for further development on the Peace River.

Creating a Hybrid Forum: The Original Site C Proposal

The first two hydroelectric dams on the Peace River transformed the landscape and the communities in the region. New knowledge of the river and of the social and political forces at work resulted from technical studies and from firsthand experience (Pollon and Matheson 1989; Loo 2007). Yet, this was far from what could be called a hybrid forum. The projects were planned and built by the provincial utility, BC Hydro and Power Authority (BC Hydro), and with little to no public debate. It was not until a third hydroelectric dam was proposed in the mid-1970s[4] that a hybrid forum began to take shape.

At the time, growing awareness of the environmental impacts of hydroelectricity and new expectations regarding public input into decision making, coupled with questions regarding the expansion and financial

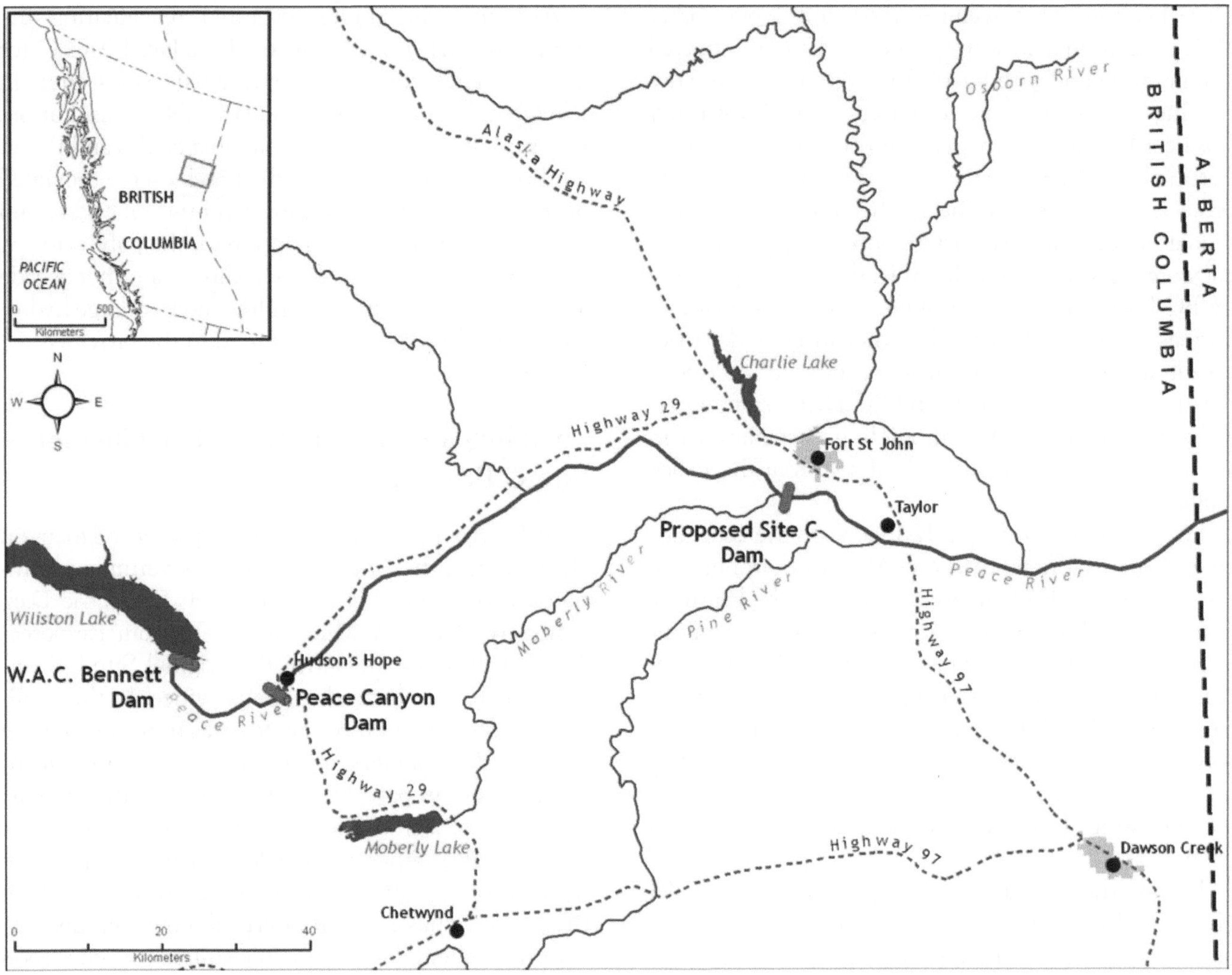

Figure 1. British Columbia's Peace River region.

integrity of provincial power utilities, were creating concern about the British Columbia energy planning regime (Smith 1988). In 1980, the provincial government passed the Utilities Commission Act (R.S.B.C 1996, c. 473), creating the British Columbia Utilities Commission (BCUC) with the mandate to review major energy projects and regulate public utilities and the petroleum industry in the province. The intent of the regulation was to "rein in" BC Hydro by increasing oversight and creating a mechanism for independent project assessment (Smith 1988).

As a result of the new regulatory regime, BC Hydro was required to apply for an energy project certificate for all new developments. The application for the Site C Hydroelectric Dam was submitted only weeks after the Utilities Commission Act came into effect. The application was for the construction of a 900-MW hydroelectric generating station and associated transmission facilities at Site C on the Peace River. The proposed facility would be 93 km downstream from the Peace Canyon Dam and 7 km southwest of the community of Fort St. John. Like the Peace Canyon Dam, Site C could make use of the energy potential stored upstream in the Williston Reservoir and therefore have a reservoir that was considerably smaller than Williston. Yet, Site C would still create a reservoir 85 km long and flood approximately 5,340 hectares between the communities of Fort St. John and Hudson's Hope (British Columbia Hydro and Power Authority 2007).

The application was referred to the newly created BCUC and in April 1981 a special panel was appointed to review the application. As part of the review process, the panel opted to hold both formal and informal hearings. In total, 116 days of formal hearings took place between 24 November 1981 and 2 November 1982. Additional meetings were held in six affected communities and with First Nations groups. During the hearings the professional testimony of BC Hydro staff was heard along with the testimony of consultants supporting both the proponents and the opponents of the project. The panel also heard the testimony of individuals, municipalities, First Nations groups, and advocacy organizations. In addition to the reports prepared by BC Hydro and its consultants, the panel reviewed various artifacts submitted by other interested parties and individuals including maps, copies of correspondence between landowners and BC Hydro staff, a partial transcript for a film chronicling First Nations in the region, and even four vials of soil from the Peace Valley.[5]

Local residents were clearly divided on the issue. Local opponents formed the Peace Valley Environmental Association (PVEA) and were in favor of the hearings, participating both as members of PVEA and as individuals. As one founding member of the PVEA put it, "First it was rumoured that they were going to build Site C, and then [BC] Hydro started having their little softening-up meetings. We said, 'Hold it, there's got to be a forum.' So the first thing we went after, some of us, was a hearing" (quoted in Pollon and Matheson 1989, 310).

Despite the limitations of the public hearing model (Innes and Booher 2004), the Site C hearings served as a kind of hybrid forum, providing a gathering place for heterogeneous actors and for multiple rationales for and against the project. The forum included the formal hearings and opportunities to challenge the formal process as exemplified by local supporters of the project, anxious for the promised employment, who picketed the hearings with signs reading "Stop the hearings, start the dam." In addition to gathering testimony regarding Site C, the hearings also served as a much-needed outlet for citizens, advocacy organizations, municipalities, First Nations communities, independent consultants, and even government ministries to voice an opinion on the direction of electricity planning in the province (Smith 1988).

On 3 May 1983 the panel issued its recommendation stating that, "The evidence does not demonstrate that construction must or should start immediately or that Site C is the only or best feasible source of supply . . . in the system plan" (British Columbia Utilities Commission 1983, 10). Following the recommendation, the provincial cabinet rejected the application for the energy project certificate.

Although the ruling suggested reevaluating the project in one year, it effectively delayed Site C for decades. The project was revisited six years later, in 1989, but despite another round of public consultation, it was not pursued (British Columbia Hydro and Power Authority 1991). Although the flood reserve remained on the Peace Valley, shaping land-use practices, and the project remained as an option for the public utility, it was not seriously considered again for another decade and a half. In fact, since the ruling, no new large hydroelectric facilities have been started in the province.

Avoiding a Hybrid Forum: The Current Site C Proposal

In 2007, as part of its new climate policy and focus on clean energy, the British Columbia government reinitiated investigation of the Site C Hydroelectric Dam (Ministry of Energy Mines and Petroleum Resources 2007). In the decades since the original Site C hearings, much had changed in the province. In the 1980s the provincial utility, BC Hydro, was transformed from a supply development organization to a management organization, in part due to the new regulatory regime and the lessons from the Site C hearings (Smith 1988; Kellow 1996). Since the last large hydroelectric facility came on line in the mid-1980s, most new supply in the province has come from private power producers.[6] As environmental regulation has shifted toward carbon control and climate change mitigation (While, Jonas, and Gibbs 2010), however, the province's existing supply of firm, dispatchable, and relatively low-carbon hydroelectricity has increased in value; thus the potential for increasing its capacity has become politically attractive. In its current incarnation, the $6 billion Site C project is promoted as a large block of firm, clean, and—over its lifetime—inexpensive energy. The project literature cites a domestic need for the energy although, significantly, the provincial government has recently formalized its intent to export clean electricity to U.S. jurisdictions. The project is also justified in part due to the existing infrastructure, including the long-distance transmission lines and the potential energy stored upstream in the Williston Reservoir that increases the ratio of energy output-to-reservoir area for Site C.

The intervening years have also witnessed the evolution of public consultation in environmental decision making such that formal consultation processes are mandated (and expected) even though, in Canada, there remains considerable discretion in the implementation of the existing environmental regulation (VanNijnatten 1999). Therefore, unlike the original proposal, a structured, five-stage development process has been established for the project. Stage 1, Review and Project Feasibility, was completed in 2007 and Stage 2, Consultation and Technical Review, was completed in 2009.

As in previous decades, there is considerable opposition to the project and much of the rationale against it remains the same. There is concern that flooding the valley for a single use—electrical generation—will destroy all the current uses that the valley serves. This includes submerging prime agricultural land, wildlife habitat, and important cultural sites; displacing residents; and lowering recreational and tourism values. There is also concern about the social and infrastructural impacts that the construction process will have on the community of Fort St. John and, as with the first round, the ultimate need and final destination of the electricity produced. More specifically, there is concern that the electricity is intended primarily for export markets and, in a somewhat contradictory effort to reduce greenhouse gas emissions, for electrification of the region's rapidly expanding oil and gas sector.

In the wake of the original Site C hearings, the current consultation process has become a focal point of opposition. There is widespread concern that the decision to go ahead with the project is a political one that had already been made before the consultation began. As a member of Fort St. John city council put it,

> My personal view, held long before I got into municipal government and totally unchanged by anything that's happened since, is that Site C is a sell job. The decision to build Site C was made already. (Participant #9)

In any case, many participants in this research felt that the consultation was not addressing the real issues (such as whether or not the Site C project should be built) and instead focused on insignificant details such as access roads and recreational use for the new reservoir. When asked what kind of process would be adequate for affected landowners, one landowner responded,

> There should be an independent review. And I guess one of the first questions should be do we want the dam in our valley? That's not even raised. One of the first things they bring up in the meetings that we've had is, what's their word for bribery . . . benefits. What benefits do you want to see from this? (Participant #4)

Convinced of the need for widespread debate, project opponents have sought alternative forums. In an attempt to get answers to questions that were not addressed or answered in the consultation meetings, some residents have submitted detailed questions to BC Hydro and to their elected representatives. Local advocates have appeared on provincial and national radio programs and written newspaper editorials to get the word out beyond the region. Although the First Nations communities in the region are conducting a separate consultation process, the PVEA has been invited to a number of communities to present their views and share information. In addition, the PVEA has sought the support of provincial environmental groups.

The Treaty 8 Tribal Association, representing some (but not all) affected First Nations communities, has officially expressed concerns about the consultation process and the Stage 2 Consultation Agreement they entered into with BC Hydro. In an appendix to the BC Hydro Stage 2 Summary Report, the Association asserts that their Stage 2 consultation has not been completed and the government cannot make a fair and informed decision to go ahead with Stage 3 until it has been completed (Treaty 8 Tribal Association 2009).

Nevertheless, on 19 April 2010 the British Columbia premier flew to the community of Hudson's Hope to announce that the government had reviewed the findings from Stages 1 and 2 and had decided to proceed with Stage 3, Environmental and Regulatory Review.

Shortly after, in June 2010, the provincial government passed the Clean Energy Act (S.B.C. 2010, c. 22). The Act officially states the government's intent to export "clean and renewable" electricity and, by way of legislating requirements for surplus "insurance" energy and limiting the use of a natural gas-fired electrical station, creates a clear need for the block of electricity that Site C would provide. More significantly, the Act circumscribes the role of the BCUC, transferring planning oversight to the provincial cabinet and, in the case of Site C, removing the requirement for regulatory review under the Utilities Commission Act. In effect, the Clean Energy Act further reduces the opportunities for independent scrutiny and public debate regarding Site C, virtually eliminating any possibility of repeating the 1981–1982 public hearings.

The form that the Site C public consultation has taken combined with the regulatory streamlining of the Clean Energy Act suggests that provincial decision makers are uninterested in publicly debating the project and might even be actively avoiding the type of hybrid forum that the original proposal generated.

An Uneven History

Even a hydroelectric project that does not get built can transform the social and physical landscape (Heasley 2005). This is indeed the case in the Peace Valley. Site C and the "meetings" that have accumulated as a result of it have changed the physical and sociocultural landscape in the region despite remaining at the project proposal stage for over three decades. Lessons drawn from the existing dams on the Peace River and from the original Site C forum continue to accumulate and to shape divergent meanings and responses to the project.

The original Site C hearings were a learning opportunity for opponents of the project. For instance, it created a contingent of local residents who familiarized themselves with the issues and with effective forms of intervention. As one landowner put it,

> [W]hen your house is about to be, or could potentially be flooded, it focuses your attention. . . . And I guess after that I just started keeping on top of it. (Participant #5)

This meant not only studying provincial energy policy, but also keeping informed of international politics and trends, particularly in key export markets like California.

Many of the original opponents, or their children, still live in the region. As a result, there has been a knowledge and cultural accumulation around Site C and the other two hydroelectric projects on the Peace River.

> The way the compensation was handled around the Bennett dam was really poor. It was typical of [BC] Hydro at the time and it happened not only here but also in the Columbia system. It left a sour taste in a lot of people's mouths. And a number of people that are dealing with Site C have some history related either through families or local knowledge or whatever. So the feeling is there. (Participant #11)

> And if you talk to people who were really involved in the battle in the Valley, you can go back and look at their testimony. And you may as well scratch off the word [19]80 and put now. It's still the same arguments. Not all, but a lot of it is. Some of the people have moved on or they're just tired. Some are still fighting. (Participant #3)

Although the accumulated experience does translate to intelligent and informed advocacy, it has also led to the kind of frustration displayed in the preceding quote. In contrast to the knowledge and experience of local residents, many participants in this study expressed frustration with the perceived lack of institutional memory in BC Hydro and its representatives' inability or lack of authorization to discuss the issues at sufficient depth. For local opponents, all the reasons not to build the dam remain basically the same as they were in 1980. Thus, there is the feeling (and frustration) that they have been through this before and are repeating history. As a Fort St. John city councilor put it, "We've been down that dam path before" (Participant #18).

Living with the legacy of hydroelectricity on the Peace River, including the dams that have been built and those that have not, has led many local residents to the conclusion that large-scale hydroelectricity is an outdated mode of development. The original hearings provided a legitimate public forum to share their experiences and opinions. The outcome, although officially only delaying the project, helped to validate their concerns and to build collective rationale against large-scale hydroelectricity in the province. The sense is that if another inclusive forum took place the project would again prove inappropriate; hence the effort by opponents to open up dialogue and to explore alternative interpretations.

In May 2010, on a national radio program, two residents of the Peace Valley mused that between the two of them, they had seventy years of experience fighting Site C.[7] Significantly, the cultural accumulation around the project has been quite different at the provincial level. This has included both change and inertia. The original Site C hearings were a significant learning opportunity for BC Hydro, contributing to a reorientation and restructuring of the public utility as well as a lesson for the government of the day regarding the potential impact of its new regulatory structure and the need to reassert some political influence on the BCUC (Smith 1988). More generally, since the project hearings in the early 1980s, there have been seven provincial elections and three different governing parties. Yet this is not a case of static, historical places and dynamic, ahistorical spaces. There is inertia at the provincial level that, despite changing staff, politics, and rationale, leads them back to the same project and even the same exclusive, politically driven decision-making processes that produced the existing hydroelectric infrastructure half a century ago. Space limitations do not allow a full account of the political, economic, and environmental agendas that are propelling British Columbia energy policy; however, it is clear that there is considerable technological momentum (Hughes 1983) behind the province's hydro-based electricity system. Thus, although the climate change imperative is new and there has been an institutional shift in BC Hydro (and in effect the proponent of Site C is now the provincial government instead of BC Hydro), the political and economic vision of hydroelectric development is strikingly similar to three decades ago.

The provincial framing of large hydroelectricity as a solution to the problem of climate change is no doubt influenced by the inertia and perceived value (financial and otherwise) of existing hydroelectric infrastructure. At the same time, the lessons from the Site C hearings appear to be more of a cautionary tale for the provincial government, serving as a reminder that extensive public debate could again derail the project. The 2010 Clean Energy Act is a step away from public deliberation; what Callon, Lacoumes, and Barthe (2009) might describe as a retreat away from hybrid forums and toward the structure of delegative democracy where elected representatives assume the authority to decide on behalf of the collective.

Thus, even though the original Site C hearings served as a hybrid forum, the geographical, political, and institutional terrain has resulted in the emergence of divergent meanings. In the end, the "accretion of meetings" around Site C has accumulated differently across the provincial energy network, creating disjuncture and contributing to conflict. At the provincial level, Site C has remained the first and best option for public power generation. This is despite political and institutional change at a broad level and evolving project justifications. Within the region, the existing hydroelectric dams and the original Site C hearings have left a different legacy—one that appears to be increasingly at odds with both the government's energy planning objectives and processes.

Opponents have negotiated this disjuncture by appealing to alternative forums where public debate might be possible. On the other hand, the provincial government has sought to avoid conflict and challenge by limiting the opportunities for meaningful public deliberation in the formal consultation process and pursuing regulatory reform that limits other potential forums.

Conclusion

The case of the Site C hydroelectric dam is an argument for the transformative power of hybrid forums. In this case, landscape transformation and collective learning resulted from the forum that occurred during the Site C hearings in 1981–1982. The official outcome was a decision to delay the project. More broadly, the forum led to new understandings of large-scale hydroelectricity in the province and the potential for public engagement in energy planning. Accumulation and change have been uneven across space and time, however, and the legacy of hydroelectricity in British Columbia has come to mean different things across the network. Rather than providing a forum to work through these differences, it appears that, for the time being, power has trumped rationality (Flyvbjerg 1998) and that managing conflict has led to less, rather than more, democratic participation in the province.

This is cause for both optimism and caution in anticipating the outcomes of participatory forums, particularly when they emerge around large, centralized networks such as electrical grids. Although these forums do have the potential for collective learning and network transformation, disjuncture and conflict are likely to remain. Thus, using participatory forums purely as instrumental tools to achieve public buy-in or consensus on the deployment of renewable energy technologies might prove ineffective. On the other hand, the potential for hybrid forums to facilitate collective learning and sociotechnical transformation remains.

Acknowledgments

I would like to thank Melissa Nones for producing the map included in this article. This research received funding from the Social Sciences and Humanities Research Council and from the British Columbia Hydro and Power Authority.

Notes

1. The analysis is based on document analysis and interviews conducted in two communities in northeast British Columbia between April 2009 and January 2010. The fieldwork included thirty-seven interviews with forty participants including elected officials, municipal staff, and community members engaged in energy-related issues.
2. Multiple potential sites for hydroelectricity were mapped on the Peace River and, until completed and formally named, they are referred to by generic titles Sites A, B, C, and so on.
3. This is based on the historical account of Pollon and Matheson (1989), who described local concerns and protests around specific aspects of the projects but suggested that the projects in their entirety were not highly contested.
4. Early discussions with landowners began in 1975 (British Columbia Hydro and Power Authority 1991).
5. In total 442 exhibits were catalogued. See the British Columbia Utilities Commissions Resource Library

(http://www.bcuc.com/SearchLibrary.aspx) for a complete list.

6. BC Hydro was originally created in 1961 to publicly finance hydroelectric development in British Columbia. The economics of large-scale hydroelectricity remain such that it is only feasible as a public investment.
7. This discussion occurred on 21 May 2010 on the radio program *The Current*, aired by the Canadian Broadcasting Corporation.

References

Bell, D., T. Gray, and C. Haggett. 2005. The "social gap" in wind farm siting decision: Explanations and policy responses. *Environmental Politics* 14 (4): 460–77.

British Columbia Hydro and Power Authority. 1991. *Peace Site C summary status report*. Vancouver, BC, Canada: British Columbia Hydro and Power Authority.

———. 2007. Peace River Site C hydro project: An option to help close B.C.'s growing electricity gap. Summary of Stage 1 Review of Project Feasibility. Vancouver, BC, Canada: British Columbia Hydro and Power Authority.

———. 2010. Our system 2009. http://www.bchydro.com/about/our_system/generation.html (last accessed 15 February 2010).

British Columbia Utilities Commission. 1983. Site C Report: Report and recommendations to the Lieutenant Governor-in-Council. Vancouver, BC, Canada: British Columbia Utilities Commission.

Callon, M. 1991. Techno-economic networks and irreversibility. In *A sociology of monsters: Essays on power, technology, and domination*, ed. J. Law, 132–61. London and New York: Routledge.

Callon, M., P. Lacoumes, and Y. Barthe. 2009. *Acting in an uncertain world: An essay on technical democracy*, trans. G. Burchell. Cambridge, MA: The MIT Press.

Fischer, F. 2000. *Citizens, experts, and the environment: The politics of local knowledge*. Durham, NC: Duke University Press.

Flyvbjerg, B. 1998. *Rationality and power: Democracy in practice*. Chicago: University of Chicago Press.

Graham, S., and P. Healey. 1999. Relational concepts of space and place: Issues for planning theory and practices. *European Planning Studies* 7 (5): 623–46.

Guy, S., S. Graham, and S. Marvin. 1997. Splintering networks: Cities and technical networks in 1990s Britain. *Urban Studies* 34 (2): 191–216.

Healey, P. 2006. *Collaborative planning: Shaping places in fragmented societies*. 2nd ed. New York: Palgrave Macmillan.

Heasley, L. 2005. *A thousand pieces of paradise: Landscape and property in the Kickapoo Valley*. Madison: University of Wisconsin Press.

Hughes, T. P. 1983. *Networks of power: Electrification in western society, 1880–1930*. Baltimore: The Johns Hopkins University Press.

Innes, J. E., and D. E. Booher. 2004. Reframing public participation: Strategies for the 21st century. *Planning Theory and Practice* 5 (4): 419–36.

Jessop, B., N. Brenner, and M. Jones. 2008. Theorizing sociospatial relations. *Environment and Planning D: Society and Space* 26:389–401.

Joly, P.-B. 2007. Scientific expertise in public arenas: Lessons from the French experience. *Journal of Risk Research* 10 (7): 905–24.

Kellow, A. J. 1996. *Transforming power: The politics of electricity planning*. Cambridge, UK: Cambridge University Press.

Latour, B. 2005. *Reassembling the social: An introduction to actor-network-theory*. Oxford, UK: Oxford University Press.

Loo, T. 2007. Disturbing the Peace: Environmental change and the scales of justice on a northern river. *Environmental History* 12:895–919.

Massey, D. 1984. *Spatial divisions of labour: Social structures and geography of production*. London: Macmillan.

———. 2005. *For space*. London: Sage.

Ministry of Energy Mines and Petroleum Resources. 2007. *The BC Energy Plan: A vision for clean energy leadership*, ed. P. R. Ministry of Energy Mines. Victoria, BC, Canada: Province of British Columbia.

Moore, A. 2008. Rethinking scale as a geographical category: From analysis to practice. *Progress in Human Geography* 32 (2): 203–25.

Nyboer, J., and E. Nyboer. 2009. *Review of energy supply, consumption and GHG emissions in British Columbia, 1990 to 2007*. Burnaby, BC, Canada: Canadian Industrial Energy End-use Data and Analysis Centre.

Pollon, E. K., and S. S. Matheson. 1989. *This was our valley*. Calgary, AB, Canada: Detselig Enterprises.

Smith, L. G. 1988. Taming B.C. Hydro: Site C and the implementation of the B.C. Utilities Commission Act. *Environmental Management* 12 (4): 429–43.

Szarka, J. 2006. Wind power, policy learning and paradigm change. *Energy Policy* 34:3041–48.

Treaty 8 Tribal Association. 2009. Treaty 8 First Nations' report on Stage 2 consultation. Fort St. John, BC, Canada: Treaty 8 Tribal Association.

VanNijnatten, D. 1999. Participation and environmental policy in Canada and the United States: Trends over time. *Policy Studies Journal* 27 (2): 267–87.

Whatmore, S. J. 2009. Mapping knowledge controversies: Science, democracy and the redistribution of expertise. *Progress in Human Geography* 33 (5): 587–98.

While, A., A. Jonas, and D. Gibbs. 2010. From sustainable development to carbon control: Eco-state restructuring and the politics of urban and regional development. *Transactions of the Institute of British Geographers* 35:76–93.

Wolsink, M. 2007. Wind power implementation: The nature of public attitudes: Equity and fairness instead of "backyard motives." *Renewable and Sustainable Energy Reviews* 11:1188–1207.

Zimmerer, K. S. 2011. New geographies of energy: Introduction to the special issue. *Annals of the Association of American Geographers* 101 (4): 705–11.

A Study of the Emerging Renewable Energy Sector Within Iowa

Peter Kedron and Sharmistha Bagchi-Sen

Department of Geography, University at Buffalo, State University of New York

This article offers an understanding of the evolution of Iowa's ethanol industry landscape. A conceptual framework based on a techno-economic paradigm of networked organizations and associated regional innovation systems is used to understand linkages among organizations involved in the ethanol production value-chain. Iowa is an adapter region—federal and state policies conducive to renewable energy and ethanol production encouraged new entrants to the sector. Farmers became business owners through cooperatives and outsourced refinery design and sometimes management. The state also attracted integrated large firms, including agri-processors and an oil refiner, which leveraged additional support from the local government and indirectly affected federal policy through their position in industry associations. University research, extension services, government laboratories, and foreign companies got involved in the process, directly and indirectly, through research and development (R&D), farmers' support, and other related services. The ethanol industry's future depends on progress in R&D (cellulosic ethanol), innovation, shifts in embedding environmental concerns in economic processes, and policy. Subsidies and mandates across the value-chain and innovative shifts in products and processes will continue to influence the restructuring of how the industry is organized. *Key Words: ethanol, policy, regional system, renewable energy.*

本文有助于了解美国爱荷华州乙醇工业格局的演变。针对网络化组织和相关的区域创新系统，本研究采用了一个基于技术经济范式的概念性框架，用于了解涉及乙醇生产价值链内不同组织之间的联系。爱荷华州是一个转接器类型的区域，一方面联邦和州的政策有利于可再生能源，同时乙醇生产鼓励了该行业的新加入者。农民通过合作、外包的炼油厂设计和部分管理业务而成为业主。爱荷华州还吸引了大型综合性公司，包括农业处理公司和一个炼油厂，从而得以从当地政府得到更多的支持，并通过他们在行业协会的地位间接地影响了联邦政策。直接或间接地通过研究与开发（R&D），对农民的支持，以及其他相关服务，大学的研究，推广服务，政府实验室，和外国公司都介入到这一过程中。乙醇行业的未来取决于研发（纤维素乙醇）的进展和创新，在经济进程中嵌入环境问题的变化，以及政策。在整个价值链中的补贴和指令，以及产品和工艺创新的转变，将继续影响该行业的组织结构的重组。*关键词：乙醇，政策，区域系统，可再生能源。*

Este artículo ofrece un entendimiento sobre la evolución del paisaje de la industria del etanol en Iowa. Se utiliza un marco conceptual basado en un paradigma tecno-económico de cadenas de organizaciones y sistemas de innovación regional asociados, para entender los lazos entre aquellas organizaciones que están implicadas en la cadena económica de producción de etanol. Iowa es una región adaptadora—donde las políticas federales y estatales orientadas hacia la energía renovable y la producción de etanol estimularos la entrada de nuevos participantes en el sector. Los agricultores se convirtieron en propietarios de negocios a través de cooperativas y por diseños de refinería subcontratados, y a veces por actividades de manejo. El estado también atrajo grandes firmas integradas, incluyendo procesadores de productos agrícolas y un refinador de petróleo, que aseguraron el apoyo adicional del gobierno local e indirectamente afectaron la política federal a través de sus posiciones en las asociaciones industriales. La investigación universitaria, servicios de extensión, laboratorios del gobierno y compañías extranjeras se involucraron en el proceso, directa e indirectamente, a través de investigación y desarrollo (I&D), ayuda a los agricultores y otros servicios relacionados. El futuro de la industria del etanol depende del progreso que se logre en I&D (etanol celulósico), innovación, cambios en la inclusión de asuntos ambientales en los procesos económicos, y de las políticas que se adopten. Los subsidios y mandatos por encima de la cadena de valor y las transformaciones innovadoras en productos y procesos, seguirán determinando la reestructuración sobre cómo organizar la industria. *Palabras clave: etanol, políticas, sistema regional, energía renovable.*

Renewable energy constitutes 11 percent of the U.S. energy supply; biomass accounts for 50 percent of that portion (U.S. Energy Information Administration 2011). From 2001 to 2007, the effect of policies, market incentives, and technologies raised annual biofuel production from 1 billion to 7 billion gallons—of which 6 billion gallons came from corn, making the United States a leader in corn-based biofuel production (Boyle et al. 2008). Industrial ethanol, a type of biofuel, drew investment to the corn-growing states (e.g., Iowa) through a complex value-chain of national agri-processors, refineries, local farmer cooperatives, specialized engineering firms, national laboratories, universities, and industry associations. Federal and state legislation supporting ethanol created new market opportunities for corn producers and users. The purpose of this article is to examine the factors shaping the evolution of the ethanol-based renewable energy landscape in Iowa, the leading state in corn-based ethanol production. This article contributes to geographic research through an understanding of the changing spatial interdependencies of economic, environmental, and political agents and their respective processes in the formation of landscapes of renewable energy in a region without a strong industrial legacy (Zimmerer 2011).

Research Context

Sterzinger (2008, 3) stated: "Renewable energy is conceived as basic science, created in labs and universities, commercialized by developers, and then manufactured as component parts assembled into final products. This process chain indicates renewable energy is a manufactured energy industry and is driven by technology innovation." Industrial ethanol, a renewable energy, in the United States is driven not only by technology innovation (e.g., advanced process control, genetically modified corn) but also by government policy (e.g., subsidies), environmental concerns (e.g., emissions reduction), and regional advantages (e.g., corn, agri-processing, agricultural lobbying). Industrial ethanol production brings together consideration of the environment and the economy to produce landscapes of renewable energy. Bridge (2008, 79–80) indicated three lines of inquiry to examine the difference environmental concerns make to economic processes: the effect of environmental processes on economic decision making, the effect of economic processes on environmental conditions and livelihoods, and "how forms of 'environmental governance' relate to the restructuring of economic relations." All of these have uneven geographic outcomes. Discussing the third line of inquiry, Bridge went on to state

> The rapid uptake of biofuel production by some firms and farmers in the U.S., for example, illustrates how coupling environmental and energy-security issues and framing them in terms of domestic supply can entrench existing relations of political patronage and economic subsidy (in the U.S. case, around arable agriculture on the Great Plains). In short, a focus on modes of environmental governance . . . also produces particular political-economic relations. (80)

The evolution of the energy landscape rests within a larger framework of techno-economic paradigm (TEP) shifts, which are influenced by and have influenced government policy (Hayter and LeHeron 2002; Hayter, Barnes, and Bradshaw 2003; Hayter 2008; Stormer 2008).[1] "Each TEP is connected . . . and economic evolution is geographically differentiated by place-based institutions, notably with respect to innovation systems" (Hayter 2008, 832). As opposed to the Fordist system of vertically integrated firms, the current TEP, facilitated by information and communication technologies and strategic partnerships (Freeman and Perez 1988; Freeman 1992; Hayter 2008), involves a network of producers and users, large and small firms, and other organizations based on "close cooperation in technology, quality control, training, investment planning and production planning" (Dicken 2007, 76). More specifically, technological, market, and organizational factors characterize TEPs and innovation systems, which have dominant industries and modes of industrial organization; overall policies distinguish one system from another (Freeman and Louca 2001). In geographic terms, agglomeration or clustering of industry over space and time is evidence of TEP-based regional shifts (Boschma 2004; Martin and Sunley 2006). According to Oinas and Malecki (1999, 2002), national, regional, and sectoral systems of innovation are at work in producing distinct regions: Innovator regions create completely new technologies and lead the establishment of best practices (e.g., Silicon Valley); adapter regions build on technologies of innovators by providing an environment conducive to incremental innovation (e.g., ethanol in Iowa); and regions that are technological latecomers and provide little innovation are adopters (Cheshire and Malecki 2003; Iammarino 2005; Fromhold-Eisebith 2007). Bridge, Cooke, and Hayter argued that future TEP will integrate environmental concerns within the value-chain (research and development [R&D], production, distribution) and the question of renewable energy will be important in every transaction (Bridge

2002; Hayter and LeHeron 2002; Cooke 2002, 2004; Hayter 2004). The need for resources will continue to tie renewable energy industries to specific locations, even as the need for emerging innovation and markets will expand the importance of extraterritorial linkages.

Cooke (2009) identified the importance of place-specific ties in the emergence of renewable energy production clusters worldwide (e.g., wind, solar, biomass; Table 1), in particular he finds incentives for collaboration (eco-industrial parks) and production (tax incentives, production capacity standards) to be important. Denmark capitalized on local competencies in marine and agricultural engineering to develop its North Central Jutland wind turbine cluster. In Austria's Murau region, wood biomass was developed in response to the informal Murau Energy Objectives 2015 (Spath and Rohracher 2010). Studying the United Kingdom, Smith (2007) demonstrated how national government mandates negatively impacted regional renewable energy industry development because no clear coordinating mechanism was put in place. In contrast, national policies encouraging renewable energy adoption (solar) and the scaling up of local innovation (self-built solar technology) resulted in Austria exporting 80 percent of its solar collector production (Lund 2009). Similarly, Norwegian government policies focused on R&D support to energy firms, resulting in the production of 25 percent of the world's solar wafers (components in solar cells) in 2006. In Wales, the Institute for Grassland and Environmental Research developed sugargrass, a feedstock with high sugar content. Now, a government-supported experimental biorefinery is testing the viability of converting sugargrass to bioenergy on an industrial scale (Cooke 2009).

Related research focuses on the environmental impact of ethanol by examining overall benefits (e.g., emissions reductions; Pimentel, Patzek, and Cecil 2007; Searchinger et al. 2008; Yacobucci 2008; Capehart 2009) and net energy balance (NEB).[2] Research on NEB demands that the ethanol value-chain (1) produce more energy than it requires, (2) produce environmental benefits (reduced emissions, water conservation, habitat conservation), (3) be economically competitive, and (4) make minimal impact on food supplies (Hill et al. 2006; Koh and Ghazoul 2008). Some studies of NEB include cost of feedstock production, transportation and processing, coproduct production, facility building, farm machinery production, and chemical release (Pimentel 2003; Pimentel and Patzek 2005; Hammerschlag 2006; Hill et al. 2006; Koh and Ghazoul 2008). The method used to assess NEB is openly debated (Hammerschlag 2006) and the findings are varied depending on assumptions made. Biofuel impact analysis is beyond the scope of this study.

Iowa's ethanol industry is part of a larger first-generation U.S. ethanol industry dependent on numerous technologies and policies (Rajagopal et al. 2009). Studies specifically examine the location and impact of ethanol refineries. Haddad, Taylor, and Owusu (2010) identify regional factors important to ethanol plant location decisions in four midwestern states. Location decisions are made in a two-stage process, considering first if a region has necessary feedstock abundance (corn) and second a series of other location factors (population, rail access). Low and Isserman (2009) test how economic impacts of refineries in Nebraska and Illinois depend on value-chain assumptions. The authors concluded that employment creation effects of plant location increase with the complexity of the local economy, estimating secondary employment (cattle industry, corn production) and economic (job creation, rent to landowners) impacts to be smaller than those directly related to ethanol production. This study examines the role of major players in the evolution of the ethanol industry landscape in Iowa with a goal toward understanding how "highly varied functional interdependencies are organized within and among populations of firms and shaped by corporate strategies, structures and government policies" (Hayter 2008, 844).

Methodology and Data

Iowa leads the United States in ethanol production and total refineries (Table 2). Therefore, the selection of Iowa represents purposive sampling with the goal of analytical generalization (Curtis et al. 2000; Eisenhardt and Graebner 2007). Review of trade publications; company, government, and trade association reports; and federal and state government records dating back to the early 1980s, with a specific focus on the period of 2000 to 2009, are used to collect qualitative and quantitative information about functional interdependencies within the biofuel industry.

Studying Iowa's biofuel landscape helps us understand renewable energy industries. Renewable energy industries (e.g., ethanol, wind, solar) depend on both innovation and resources. Resources tie industries to specific geographies. Wind and solar energy require consistent wind and sunlight, just as ethanol requires feedstock abundance. Unlike other forms of renewable energy, the main resource of U.S. ethanol production, corn, is an agricultural product that receives subsidy

Table 1. Selected literature on renewable energy and ethanol

Authors	Year	Summary
Haddad, Taylor, and Owusu	2010	Using probit regression analysis, the authors identify location conditions important to ethanol location decisions in four Midwestern states. Refinery location decisions are a two-stage process. First a potential site must have an abundance of corn for production. When corn is abundant throughout a state (e.g., Iowa, Illinois), the location decision proceeds to secondary considerations (rail access, population factors).
Spath and Rohracher	2010	Collaboration between government officials, renewable energy activists, and industry groups led to creation of an informal plan (Murau Energy Objectives for 2015) for renewable industry development in Murau, Austria. Informal institutional connections gave credibility to the objectives leading to their incorporation as formal guiding principles for industrial and government partnership and policy.
Cooke	2009	Government innovation funding for a regional research laboratory targeted feedstock development for the livestock industry. The resulting basic innovation, sugargrass, is also an excellent feedstock for bioenergy production. To capitalize on sugargrass, local and national interests combined to create policies supporting industrial development (biorefinery funding) and feedstock improvement (R&D funding).
Lund	2009	In Austria, national policies encouraging renewable energy adoption (solar) and local policies encouraging the scaling up of local innovation (self-built solar technology) resulted in export of 80 percent of national solar collector production. Norwegian research assistance policies for high-performing firms resulted in production of 25 percent of the world's solar wafers in 2006.
Low and Isserman	2009	Authors examine the location, policy support, and future potential of the U.S. corn ethanol industry. Developing four case study scenarios in Illinois and Nebraska, the authors attempt to quantify the impact of ethanol refineries under different location conditions. The authors found that economic impacts (job creation, income generation) increase when selected locations have more diversified and complex economies.
Rajagopal et al.	2009	Each segment of the U.S. value-chain has environmental sustainability and economic viability problems. Development of next-generation biofuels (cellulosic) through research partnerships involving universities, industry, and government is needed. Dissemination of basic knowledge (by universities) and its incorporation into industrial process (via industry partnership) is likewise critical to development.
Zhang and Cooke	2009	The authors identified renewable energy production clusters (e.g., wind, solar, biomass). Cluster development is supported by incentives for collaboration (eco-industrial parks) and production (tax incentives, production capacity standards). Eco-industrial parks in particular establish networks among firms that create economic and environmental gains by reducing waste. The by-products of one firm become the inputs of another and firms work in collaboration to achieve economic gains.
Capehart	2009	In a policy issues report prepared for the U.S. Congress, the author examined the state of the U.S. ethanol industry. Highlighted is the extensive support provided by the federal government to ethanol producers and the ongoing debate over the necessity of that support as well as the form it should take in the future. Interconnections among multiple interest groups (farmers, biorefiners, food producers, research laboratories) and their position in the debate frame the analysis of industry direction. Net energy balance studies reviewed by the author demonstrate generally positive energy balances ranging from 1.25 and 1.67 return on inputs.
Yacobucci	2008	Summarizing federal incentives for the U.S. Congress the author traces connections between the government and U.S. ethanol producers. Detailed are responsibilities of five federal agencies in the management of twenty-four programs supporting the U.S. ethanol industry.
Smith	2007	Examined policy support for renewable energy industries within the United Kingdom and analyzed what factors lead to successful regional renewable energy industry development. The author found failures in coordination among national and regional objectives hinder development.
Hammerschlag	2006	Reviewed six studies of the net energy balance of corn ethanol and found claims that ethanol returned between 0.84 and 1.65 of energy inputs. Variation in claims depended greatly on assumptions of what production processes were included in the model and how they operated.
Hill et al.	2006	Including numerous upstream production processes (e.g., fertilizer production), the authors found that corn ethanol production yielded more energy than production required. Most energy benefits derived from the production of distillers' grains and not ethanol.
Pimentel and Patzek	2005	Using an inclusive definition of production processes (e.g., fertilizer and machinery production), the authors found the corn ethanol only returned 84 percent of the energy required to make it. The authors assumed electricity-intensive production process raising energy inputs.

Table 2. Geographic distribution of biorefineries in 2010

Region[a]	Capacity (mgy)	Operating production (mgy)	Biorefineries	
			Completed	Under construction
State				
Iowa	3, 439	3, 439	39	0
Nebraska	1, 744	1, 719	22	1
Illinois	1, 226	1, 226	14	2
Minnesota	1, 136	1, 094	20	0
South Dakota	1, 016	1, 016	15	0
Indiana	908	816	13	2
Ohio	538	424	9	0
Wisconsin	498	498	10	0
Kansas	491	436	13	1
Total	10, 996	10, 668	159	6
National				
National total	13, 519	12, 829	187	15

Source: Compiled by the authors from Renewable Fuels Association and Nebraska Energy Office.
[a]State capacity > 450 mgy.

through extensive lobbying. Therefore, policies affecting corn are instrumental in the evolution of Iowa's ethanol sector. To demonstrate empirically the evolution of this sector, interlinkages among firms, government, and nongovernmental organizations (NGOs) are examined. Iowa's ethanol industry is first placed within the larger context of U.S. energy and environmental policies. Next, major players, their functions, and linkages are discussed using quantitative and qualitative data from published sources (e.g., industry association reports, company reports, newspaper archives). The current approach can be replicated when studying the evolution of the ethanol industry elsewhere in the United States, especially the Corn Belt states.

The Evolution of Ethanol Production in Iowa

Table 3 provides a framework to illustrate the evolution of Iowa's ethanol industry—the framework is based on the TEP connecting environment and economy (Hayter 2008, 2009). Iowa's ethanol industry began in response to the OPEC oil crisis (Table 3). Ethanol production in Iowa combines high- and low-technology process knowledge. Grain distillation and fermentation have been understood for centuries and Iowa companies have long specialized in grain processing. Ethanol production, however, also depends on designer enzymes, bacteriological materials, and specialized production equipment to produce high ethanol yields and distillers' grains (animal feed). Corn used in Iowa's ethanol production is often a high-tech resource, genetically modified for disease resistance and high yields. In 2010, 90 percent of all corn harvested in Iowa was grown from genetically modified seed (U.S. Department of Agriculture 2010).[3]

Policy Support

Building on energy security, national security, and environmental issues, a system of subsidization for ethanol was created alongside existing support for corn. Responding to the OPEC oil crisis and political lobbying from Archer Daniels Midland (ADM), an agriprocessor, the federal government selected ethanol as a potential replacement for foreign oil (Table 3). In support of ethanol (Table 4), Congress established the Federal Excise Tax Exemption of 1978 that reduced taxes on blended fuels, tariff protection through the Omnibus Reconciliation Act of 1980 (continued today in the form of a 2.5 percent tariff and $0.54 per gallon most-favored-nation duty on imported ethanol), and the Consolidated Farm and Rural Development Act of 1972, which provided a source of project financing for ethanol refineries.

In the 1990s (Tables 3 and 4), environmental concerns led to expansion of government support and increased ethanol production. Interest groups positioned ethanol as a way to preempt environmental degradation, increasing public and government support.

Table 3. Factors affecting Iowa's biofuel landscape

Policy Support		Firms and industry associations[b]	Iowa biofuel landscape
Policy priorities (Scale)	Policy[a] (Year)		
1970–1990			
Energy security and independence[c] (Federal)	Energy Tax Act (1978)	Archer Daniels Midland Corn Lobby	OPEC oil crisis motivates a search for domestic energy independence and early funding for biofuel R&D.
Industry protection[d] (Federal)	Omnibus Reconciliation Act (1980)	Archer Daniels Midland POET	Archer Daniels Midland a national agri-processor, establishes Iowa's first ethanol plant. POET, a national biorefiner, begins operations.
1990–2000			
Clean-up[e] (Federal)	Clean Air Act (1992)	—	Concerns about emissions and air pollution levels lead to establishment of environmental policies that create incentives for Iowa biofuel production.
Preempt environmental impacts[f] (Federal)	Energy Policy Act (1992)	—	Vehicle regulations are introduced that aim to reduce emissions through the use of alternative fuels.
Expand alternatives[g] (Federal)	Biomass Research and Development Act (2000)	Archer Daniels Midland POET	Federal government increases the number of funding programs and dollar amounts available for ethanol research projects. POET and ADM gradually expand throughout the corn belt.
Expand alternatives (State)	Iowa Code 266.39C (2000)	Corn Lobby	Iowa state government establishes Alternative Energy Revolving Loan Program, the state Office of Renewable Fuels, and makes available grant money for biofuel research.
2000–2010			
Expand alternatives (Federal)	Energy Policy Act (2005)	POET	Poet enters into international partnerships to develop Broin Fractation and other assets necessary for industrial cellulosic ethanol production.
Expand alternatives (State)	Iowa House File 918 (2007)	Cooperative Biorefiners	Biorefineries of all sizes emerge across Iowa (Golden Grain, Tall Corn Ethanol, Hawkeye Renewables), some fail during economic downturn (VeraSun).
Preempt environmental impacts (Federal)	Food, Conservation, and Energy Act (2007)	Cooperative Biorefiners	Iowa refineries continue to expand corn-based biofuel production, more resources are directed to developing efficient cellulosic alternatives.
Energy security & independence (Federal)	Energy Policy Act (2005)	—	The 25×'25 agenda calling for 25 percent of U.S. energy supply to be sourced from renewable by 2025 receives widespread support in Iowa.
National security[h] (Federal)	Energy Independence and Security Act (2007)	Archer Daniels Midland POET Renewable Fuels Association	Renewable Fuels Association, Iowa politicians, and industry leaders couple national and energy security issues with environmental issues increasing biofuel R&D funding and production incentives in Iowa.

[a]See Table 5 and 6 for details on each federal and state policy.
[b]Engaged in ethanol (corn-based) production and related industries.
[c]Limit U.S. need of foreign energy sources, primarily oil and gas.
[d]Laws limit competition facing an industry and directly assist in development.
[e]Reduction of current environmental impacts.
[f]Limit or eliminate future damage to the environment.
[g]Locate and develop nonconventional energy sources to replace existing sources.
[h]Secure national borders and limit threats to the United States.

Table 4. Examples of federal ethanol policies

Policy	Date	Impacts	Iowa biorefineries[a]
Consolidated Farm and Rural Development Act (Pub. Law 92-385)	1972	Initiated the Business and Industry Guaranteed Loan Program, frequently used by biorefineries to secure project financing.	0
Energy Tax Act (Pub. Law 95-618)	1978	Gasoline blended with ethanol is first exempted from federal excise tax.	1
Omnibus Reconciliation Act (Pub. Law 96-499)	1980	Established an ad valorem tariff of 2.5 percent and a most-favored-nation duty of $0.54 per gallon of ethanol; later extended through 2010.	1
Clean Air Act (Pub. Law 101-549)	1990	Required the use of oxygenated gasoline in areas with high levels of air pollution. Led to increased use of MTBE and ethanol.	1
Omnibus Budget Reconciliation Act (Pub. Law 101-508)	1990	Introduced the Small Ethanol Producer Tax Credit, giving producers up to $1.5 million per year at refineries with a capacity of under 30 million gallons per year.	1
Energy Policy Act (Pub. Law 102-486)	1992	Provides renewable energy incentives, regulations regarding vehicle alternative fuel use, and tax incentives to promote alternative fuel use.	2
Biomass Research and Development Act (Pub. Law 106-224)	2000	Aimed at the promotion of biomass research through research grants and a federal biomass initiative. Authorized at $200 million per year.	4
American Jobs Creation Act (Pub. Law 108-357)	2004	Authorized through 2010 the "blenders credit," providing refiners a $0.51 tax credit per gallon on ethanol, incentivizing its use.	13
Energy Policy Act (Pub. Law 109-058)	2005	Doubled the renewable fuels standard to 7.5 billion gallons per year by 2012. Extends tax credits and R&D grants. Expanded the Small Ethanol Producer Tax Credit to 60 mgy capacity.	19
Energy Independence and Security Act (Pub. Law 110-140)	2007	Expanded the Energy Policy Act, raising renewable fuel use requirements to 36 bgy by 2022.	34
Food, Conservation, and Energy Act (Pub. Law 110-234)	2008	Over $1 billion allocated to renewable energy programs. Specific targets included $250 million in loan guarantees for renewable energy investments, $300 million to expand cellulosic biofuel production.	37
American Recovery and Reinvestment Act (Pub. Law 111-005)	2009	Created and extended investment tax credits associated with biorefining. Authorized an additional $16.8 billion for the Energy Efficiency and Renewable Energy Program focused on biofuel R&D.	39

Source: Compiled by authors from Congressional Research Service (2009), Congressional Hearings, U.S. Government Accountability Office (2009), and industry Web sites (10 January 2010).
[a]Cumulative number of Iowa biorefineries that remain in operation.

Responding to air pollution issues in major U.S. cities, the Clean Air Act of 1990 created emission standards requiring use of oxygenated gasoline, which is blended with additives (ethanol or Methyl Tert-Butyl Ether [MTBE]) to increase oxygen release during burning. In the past decade, ethanol use increased following the banning of MTBE in 2004 by California and New York in response to groundwater contamination concerns. The Energy Policy Act of 2005 expanded ethanol markets by establishing a renewable fuel standard (RFS) that required production of 7.5 billion gallons per year (bgy) of renewable fuel by 2012, a target later expanded to 36 bgy by 2022 in the Energy Independence and Security Act of 2007. The elimination of MTBE and the establishment of the RFS created a dedicated market for ethanol.

Tax credits encourage the production of ethanol (Table 4). The largest current ethanol tax credit is the blender's credit established in the 2004 American Jobs Creation Act. The blender's credit provides to refiners $0.45 per gallon of ethanol blended fuel. Partial benefits of the blender's credit are often passed to ethanol refiners. Other federal policies directly benefit ethanol refiners by providing investment, infrastructure, and production subsidies. The small ethanol producers' tax credit expanded in the Energy Policy Act of 2005 provides $0.10 per gallon to ethanol refiners producing 60 million gallons per day (mgy) or less. Growth in ethanol production is consistent with policy objectives put forward by interest groups and government officials. One such initiative is 25 × 25, a broadly supported (U.S. Congressional members, farmers, ethanol

Table 5. Examples of Iowa state ethanol policies

Policy	Date	Impacts
Iowa Code § 476.46	1996	Establishment of the Alternative Energy Revolving Loan Program that funds renewable energy production facilities in the state of Iowa.
Iowa Code § 159 A3	1997	Established Office of Renewable Fuels and Co-Products within the State Department of Agriculture.
Iowa Code § 266.39C	2000	Makes available energy research grants to universities, nonprofits, and foundations for renewable and energy-efficiency research. Past successful grant applications include biofuel projects.
Iowa Code § 422.11C	2002	Ethanol blended tax credit for state gasoline refiners. Amended and extended in January 2009 (Iowa Code § 422.11N).
Iowa Code § 8A.362	2003	Mandates state motor vehicles use of ethanol-blended gasoline.
Iowa Code § 15.333	2003	Provides investment tax credits for ethanol projects.
Iowa Code § 15G.203	2006	Infrastructure grants for retailers and distributors upgrading their facilities to handle ethanol sale and distribution.
Iowa Code § 422.33	2006	Retailer incentives for ethanol blend gasoline sales.
Iowa Acts 2006 § 2241	2006	Establishes an Iowa renewable fuels standard to replace 25 percent of gasoline in state with biofuel and ethanol by 2020.
House File 918	2007	Establishment of Iowa Office of Energy Independence (IOEO) and appropriated $100 million to Iowa Power Fund to support renewable energy research.
House File 927	2007	Extension of House File 918, appropriating $25 million to Iowa Power Fund 2008–2011.
Executive Order 3	2007	By June 2009, a minimum of 60 percent of state-owned flexible fuel vehicles must use an E85 ethanol to gasoline mixture.

Source: Compiled by authors from Iowa General Assembly, Iowa Climate Change Advisory Council, and Iowa Clean Cities Coalition.

refiners, university members) agenda calling for 25 percent of the U.S. energy supply to come from renewable sources by 2025 (25 × '25 Alliance 2010).

Recognizing the economic opportunity that a corn-based ethanol industry provided, the Iowa state government expanded its policy support (Tables 3 and 5). Offices, dedicated to the coordination of industry, university, and government R&D efforts, administer grants for industry development (IC §159 A3) and encourage ethanol research through project funding (IC §266.39C). One of Iowa's largest research grants, established through House File 918, provided $100 million in funding distributed over four years to renewable energy projects (extended by another $25 million in 2007). The state's Alternative Energy Revolving Loan Program funds ethanol refineries directly by providing low-interest guaranteed loans. State investment tax credits and infrastructure improvements (IC §15G.203) for biorefineries directly facilitate development. Numerous state tax incentives (IC §422.11C) are available to ethanol distributors and wholesalers. State tax code applies differential tax rates to blended and nonblended gasoline favoring ethanol. Furthermore, usage mandates for state-owned vehicles and a state RFS that aims to replace 25 percent of state gasoline use with biofuel by 2020 (IA 2006 §2241) create a dedicated market niche for ethanol.

Producer Firms and Industry Associations

Until recently, ethanol production was dominated by a small number of firms. ADM and POET (a biorefinery) began ethanol production in their home states of Illinois and South Dakota, respectively, and expanded into Iowa locations by the mid-1980s. Iowa's abundance of corn is important, but until mandates and subsidies encouraging ethanol use were passed, capacity was limited. Of the thirty-nine ethanol producers in Iowa, only six were operating prior to 2004. Production now occurs throughout the state, at refineries with capacities ranging from 20 mgy to 275 mgy. Large nonlocal ethanol refiners operate in multiple locations, provide value-chain services, and undertake basic R&D. Research projects at those firms, funded by the state and federal governments, are completed in collaboration networks consisting of multiple firms and NGOs. A few large nonlocal refiners dominate production and R&D within the Iowa ethanol landscape (e.g., POET, ADM, Hawkeye Renewables, Valero). The four largest nonlocal refiners produce approximately 50 percent (1,666 mgy) of Iowa's ethanol, but the majority of local and nonlocal biorefiners are small, with capacities below 100 mgy (Table 6).

Local producers with smaller operations undertake less R&D and are involved in fewer segments of the

Table 6. Characteristics of Iowa biorefineries

Company	Location	Capacity (mgy)
Local ownership		
Golden Grain Energy, LLC	Mason City	115
Absolute Energy, LLC	St. Ansgar	110
Platinum Ethanol, LLC	Arthur	110
Southwest Iowa Renewable Energy	Council Bluffs	110
Big River Resources, LLC	West Burlington	100
Little Sioux Corn Processors, LP	Marcus	92
Corn, LP	Goldfield	60
Siouxland Energy and Livestock Coop	Sioux Center	60
Amaizing Energy, LLC	Denison	55
Lincolnway Energy, LLC	Neveda	55
Tall Corn Ethanol	Coon Rapids[a]	52
Plymouth Ethanol, LLC	Merrill	50
Quad-County Corn Processors	Galva	30
Total locally owned production capacity	—	1,019
Nonlocal ownership		
Archer Daniels Midland (2)	Clinton; Cedar Rapids	550
Hawkeye Renewables (4)	Multiple[b]	420
POET Biorefining (7)	Multiple[c]	366
Valero Renewable Fuels (4)	Multiple[d]	330
Big River United Energy	Dyersville	110
Green Plains Renewable Energy (2)	Shenandoah; Superior	110
Tate and Lyle	Fort Dodge	105
Homeland Energy	New Hampton	100
Louis Dreyfus Commodities	Grand Junction	100
Global Ethanol/Midwest Grain Processors	Lakota	98
Penford Products	Cedar Rapids	45
Cargill, Inc.	Eddyville	35
Pine Lake Corn Processors, LLC	Steamboat Rock	31
Grain Processing Corp	Muscatine	20
Total nonlocally owned production capacity	—	2,420
All ownership types		
Total capacity of Iowa biorefineries	—	3,439

Source: Compiled by the authors from Renewable Fuels Association industry statistics.
[a]Tall Corn Ethanol operated by POET Biorefining was initially developed by a 450-member cooperative of investors.
[b]Valero locations (mgy): Albert City (110), Fort Dodge (110), Hartley (110).
[c]Hawkeye locations (mgy): Fairbank (110), Iowa Falls (90), Menlo (110), Shell Rock (110).
[d]POET locations (mgy), partially owned or operated: Ashton (56), Coon Rapids (54), Corning (65), Emmetsburg (55), Gowrie (69), Haniontown (56), Jewell (69).

industry's value-chain. As a result, these producers are more apt to contract either nonlocal refiners or specialty firms located outside of Iowa for industrial services. Ethanol production by local refiners is 1,019 mgy compared to 2,420 mgy by nonlocal refiners. Only thirteen of the state's forty refineries have significant local ownership (Table 6). Local ownership usually is cooperative—self-organized groups of farmers pool financial assets to fund refinery development. Purchasing cooperative shares often requires investors supply a set amount of corn. Many locally owned entrants built ethanol refineries as value-added extensions of existing corn businesses. Being smaller than subsidiaries of large refineries, local biorefiners capitalize on specific policy opportunities. Two of the earliest local entrants present in the industry, Golden Grain Energy and Quad-County Corn Processors, qualified for the federal small producers' tax credit when first built. Golden Grain Energy was founded as a private company, with membership shares initially sold to the local public. More than 400 investors, many local farmers, founded Quad-County Corn Processors in 2000 as an alternative market for their corn crop.

Following a period of rapid expansion between 2004 and 2007, Iowa refinery expansion peaked near the end of 2007. Abundant low-priced corn coupled with high oil prices and government subsidy created large profit margins for ethanol refiners. Despite government support, industrial ethanol remains a commodities industry dependent on market conditions. Production expansions and declining oil prices reduced profit margins, putting pressure on ethanol refiners. In response, many refineries temporarily suspended production and some exited the industry. Bankruptcy filings by major refiners like VeraSun led to industry consolidation and the entry of new corporate interests. Valero, a multinational oil refiner, integrated four of VeraSun's Iowa ethanol refineries through its subsidiary Valero Renewable Fuels. The following section provides a framework to show how different organizations and functions came together to create this emerging sector in a state dominated by agriculture.

Value-Chain Connections Within the Iowa Ethanol Industry

Figure 1 captures the linkages of biorefineries (local and nonlocal), research laboratories, government, and industry associations within the state's ethanol industry. Within the value-chain, a single producer or user might perform multiple functions. For example, oil refiner Valero Energy, through its acquisition of eight VeraSun biorefineries (total capacity of 640 mgy, 330 mgy in Iowa), is now both a fuel and ethanol refiner. Linkages might not exist for every refinery and might vary in strength. For example, ADM has a historically strong direct lobbying connection with the federal government that is absent for other biorefineries.

Nonlocal Biorefiners' Functions and Linkages

Nonlocal biorefiners (e.g., POET and ADM) account for approximately 60 percent of Iowa's ethanol production capacity and coordinate industry development. Nonlocal biorefiners produce ethanol and distillers' grains sold to oil refiners and livestock producers. Production is dependent on innovations developed in collaboration with universities, national laboratories, and agri-biotech firms and the purchase of inputs from corn producers and engineering firms (production equipment, plant services). The nonlocal biorefinery value-chain depends on direct lobbying connections with government and indirect membership relationships with industry associations, which have lobbying services. The benefit of lobbying is subsidization. Individual national biorefiners differ in industry connection and role. POET and ADM are considered because they representative almost all possible value-chain relationships of biorefiners.

Production

POET Biorefining (POET) closely matches the value-chain of Figure 1. The company produces more than 1 billion gallons of ethanol annually in twenty-six refineries (seven in Iowa) located in seven states and sells distillers' grains to livestock producers. POET also sells plant services to local biorefiners with divisions dedicated to design, construction, and management of biorefineries. For instance, the company manages and partially owns Iowa cooperative refineries Siouxland Energy and Livestock Coop and Tall Corn Ethanol.

R&D

POET's value-chain includes an international network of R&D operations. Through collaboration, POET developed two significant innovations, Broin Fractation (Bfrac) and BPX technology. Bfrac and BPX technologies increase ethanol production efficiency and yields while simultaneously lowering costs. Federally funded development partners located throughout the United States include the National Renewable Energy Laboratory (NREL) and South Dakota State University. Projects were funded through Energy Policy Act of 2005 and the Iowa Power Fund. Collaboration with Danish agri-biotech firm Novozymes and Japanese fractation company Satake was instrumental in the development of designer enzymes essential to Bfrac. Developed in Colorado, South Dakota, Denmark, and Japan, POET applied these technologies in its Emmetsburg, Iowa, facility in 2007, the first combined corn-to-ethanol and cellulose-to-ethanol refinery. The Emmetsburg refinery also capitalizes on POET's connection with Dupont. For the refinery, Dupont provided its Integrated Corn-Based BioRefinery (ICBR) technology. ICBR uses proprietary microorganisms to convert corn stalks to ethanol and was developed in partnership with Diversa, NREL, Michigan State University, and Deere & Company.

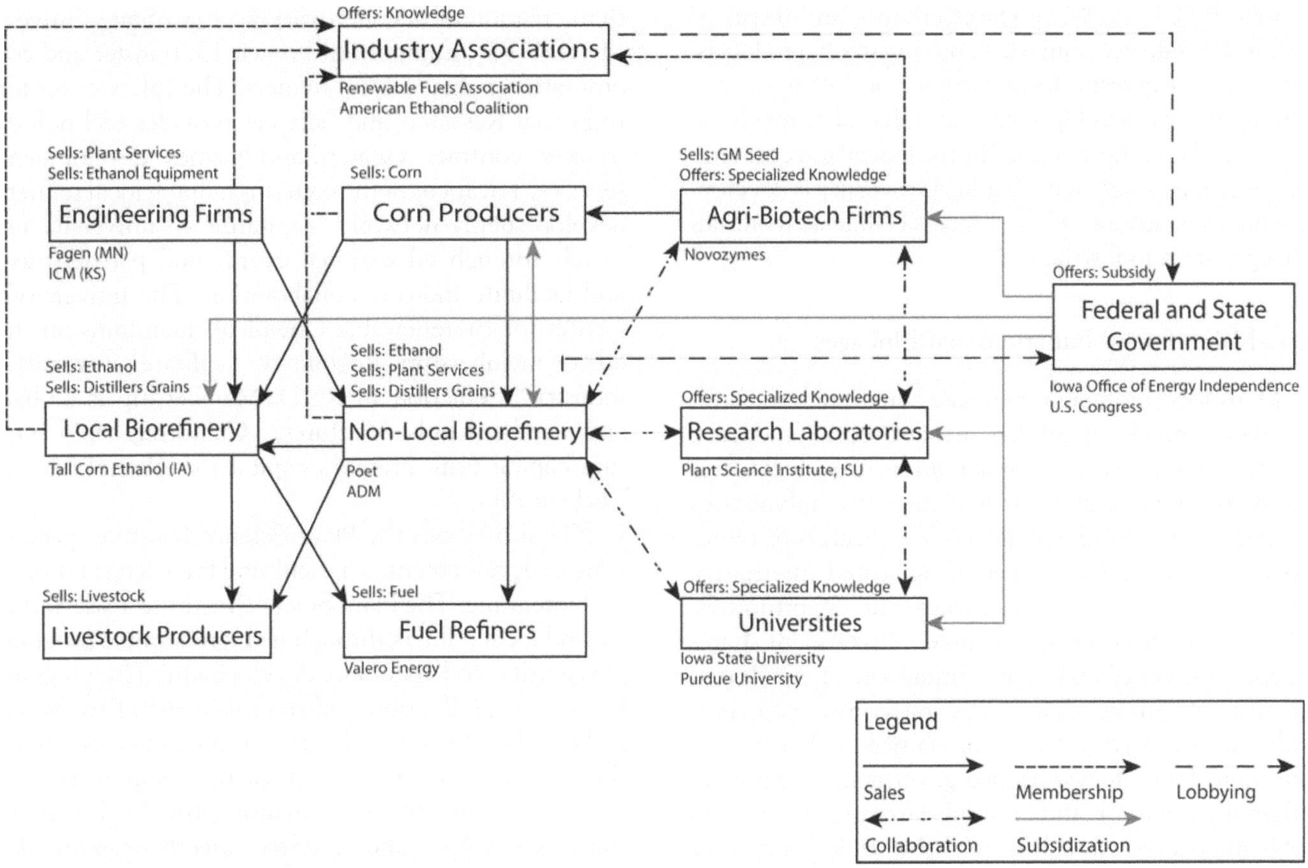

Figure 1. Value-chain connections within the Iowa ethanol industry.

Networks of Support

In 2007, lobbying efforts secured $3.9 billion in federal renewable energy tax credits for ethanol production. Under current federal policies ethanol tax credits will reach $6.75 billion per year by 2015 (Environmental Working Group 2009). POET's research is supported directly by government grants and indirectly through subsidization of ethanol production. POET is a founding member of the Ethanol Promotion and Information Council (EPIC), currently merged with industry association Growth Energy. In 2008, the Renewable Fuels Association (RFA) made specific suggestions about government support for ethanol directly to President Obama's transition team. The association recommended credit and loan assistance to ethanol refiners and a requirement for auto manufacturers to produce vehicles able to run on 85 percent ethanol if they wish to receive federal aid (Carnevale 2008).

Whereas POET actively lobbies the government, ADM leads coordination of federal support for industrial ethanol production. ADM was instrumental in creating early federal support (excise tax, tariff). Since 2006, ADM has spent more than $4.9 million on political lobbying (Center for Responsive Politics 2010) including contributions to congressional members from Corn Belt states holding important subcommittee positions. The company's long relationship with Senator Bob Dole of Kansas is well documented in the press (Bovard 1995), and Senator Dole introduced several early pieces of ethanol legislation during his tenure. The company is a leading voice in both the RFA and Corn Refiners Association. The RFA, American Ethanol Coalition, and Growth Energy bring support in the form of tax credits, production incentives, and research grants. Growth Energy's board of directors includes executives from POET, ICM, Siouxland Energy, and former Iowa Congressman Jim Nussle. Executives appear as witnesses at congressional hearings on renewable energy and ethanol policy, and state and federal reports use industry statistics provided by the RFA and the American Ethanol Coalition. Executives and associations consistently use energy security, national security, environmental, and economic development issues to secure support.

Like POET, ADM produces ethanol and distillers' grains for sale to refineries and livestock producers. ADM has a current Iowa capacity of 550 mgy; it is attempting to develop cellulosic ethanol through research collaborations funded by the federal government. An ongoing project with Purdue University is developing microorganisms able to process cellulosic materials cheaply at an industrial scale.

Local Biorefiners' Functions and Linkages

Many local biorefiners purchase corn from Iowa farmers, but some local refiners are also corn producers. Cooperatively owned ethanol refiners (e.g., Tall Corn, Golden Grain Energy, Siouxland Energy) combine corn production and local biorefiner roles. Local co-op biorefiners provide farmer-owners a dedicated alternative market for their crop and investment opportunities. The two are often explicitly linked: Investment shares in cooperatives often require annual sale of an allotted quantity of corn per share. Cooperative ownership provides an incentive for local biorefiners and corn producers to lobby for continued government support of ethanol production and corn subsidization. Most often lobbying occurs through membership in industry associations. The 40 percent of ethanol capacity owned by local biorefiners and cooperatives makes ethanol policy as much an economic development issue as an environmental issue.

Local biorefiners often outsource specific production activities. Local biorefiners are purchasers of plant services from national biorefiners and engineering firms. Iowa's leading sellers of production equipment and plant services to local biorefiners are Minnesota-based refinery design firm Fagen and Kansas engineering firm ICM. Fagen has completed seventeen projects in Iowa. Of Iowa's thirteen local refineries, seven were built by Fagen (Absolute Energy, Amaizing Energy, Big River Resources, Corn LP, Golden Grain, Lincolnway Energy, Little Sioux). ICM provided design services and production equipment to seven local biorefiners and has completed twenty-one Iowa ethanol refineries in total. Like the national biorefiner POET, Fagen and ICM are drawn to Iowa for market opportunities created by government incentives. National biorefiners provide plant services in the form of refinery operation, as with the case of Tall Corn Ethanol, and management of ethanol and distillers' grain sales.

Connections between local biorefiners and universities focus on transfer of specialized knowledge rather than creation of new innovations. Iowa State University (ISU) is an important knowledge transfer and coordination point to local refiners. The ISU Center for Industrial Research and Services provides technology transfer, contract research, and business development services. The focus of the center is helping local refiners develop supply networks, capitalize on university research through educational events and partnerships, and facilitate industry collaboration. The university's Center for Biorenewable Chemicals maintains an industry membership program to facilitate university–industry interaction in R&D. Membership is global, and member Kholsa Ventures, a California-based venture capital firm, provides capital for ethanol-related R&D to ISU.

ISU also houses the Plant Science Institute specializing in development of agricultural knowledge tailored for biorefining. The Plant Science Institute assists farmers and local refiners through its biorenewable program of scientific and economic development. The program focuses on production and improvement of feedstock dedicated for biofuel and creates tools for assessment of the socioeconomic impacts of the biofuel industry.

Innovation is critical to ethanol. Other R&D centers assisting Iowa's ethanol industry, directly or indirectly, include (1) the National Corn-to-Ethanol Research Center (NCERC) conducting research in analytics and fermentation through work with the Iowa Department of Agriculture and Land Stewardship, which specializes in the development of distillers' grains; (2) NREL and ISU, which are part of a series of research laboratories working to develop biofuels compatible with current national energy infrastructure[4] funded through the American Recovery and Reinvestment Act at $33.8 million; and (3) the Iowa Power Fund of the Iowa Office of Energy Independence, which coordinates government support with industry, community, and institutional agencies. The Power Fund provides government funding to renewable energy projects. The largest approved project ($14.7 million) is conversion of POET's Emmetsburg refinery to a combined corn-kernel and cellulosic ethanol refinery. Other projects focus on improving existing ethanol production technologies in an effort to raise efficiency and reduce energy use during production. Cellencor Inc. received funding to complete R&D on the replacement of coal-fired dryers with microwave dryers. Smaller projects such as I-Renew EXPO create important industry connections and raise public involvement in the ethanol industry.

Conclusions

This article analyzes the evolution of the Iowa ethanol industry from its initial development in the 1970s to the present. Environmental priorities created policies that led to the production of corn-based industrial ethanol in the state of Iowa. With the OPEC oil crisis and the emergence of the environmental movement as a backdrop, lobbying efforts positioned corn-based ethanol as an alternative to foreign oil reserves. In response, federal policies encouraged ethanol R&D and production (Energy Tax Act 1978) and protectionist measures (Omnibus Reconciliation Act 1980) ensured development of the market. Areas abundant in corn became the center of ethanol production. Iowa's abundant corn, early role within the industry, and leading role in the agricultural lobby made the state a primary location of ethanol investment.

As Iowa's ethanol value-chain expanded to include a greater number of biorefiners (ADM, POET, Valero) and a greater diversity of interests (small biorefiners, farmer cooperatives, research laboratories, agri-biotech firms), the industry lobbied for additional government support. Iowa's expanding ethanol industry includes local and nonlocal companies. Nonlocal companies led by POET benefited from federal government support such as tariffs on foreign ethanol (Omnibus Reconciliation Act 1980), renewable fuels standard (Energy Policy Act 2005), R&D funding (Biomass Research and Development Act 2000), and state government support, such as renewable fuels standards (Iowa Acts 2006 2241) and tax credits (Iowa Code 15.333). Local biorefiners benefited from those same policies as well as the small ethanol producer tax credit (Omnibus Budget Reconciliation Act 1990, Energy Policy Act 2005) and state incentives.

Investment within Iowa's ethanol industry took many forms. Opportunistic investment started (e.g., VeraSun's entry in 2004) but failed during the economic downturn and resulted in the entry of the first oil refinery (Valero Renewable Fuels) in 2009. As the industry expanded, so did supporting services and industries. R&D services from the NREL, National Corn-to-Ethanol Research Center, and the Biomass Conversion Research Laboratory receiving federal (Energy Policy Act 2005) and state (House File 918) funding expanded their R&D. Necessary engineering services entered the industry through collaborations and purchases from Fagen and ICM, which later became investors in local refineries. Lobbying efforts of Iowa's ethanol industry expanded and were accompanied by an increase in lobby efforts by agricultural lobbying groups.

Iowa's ethanol industry demonstrates how environmental concerns can affect economic processes through government actions and firm-level strategic response. The evolution also shows limitations of growth if subsidies are critical to the creation of markets. Uncertainty remains whether the industry really produces claimed environmental benefits. To address environmental concerns, innovation is critical to ethanol production and if the focus of industrial production and subsidies shifts to cellulosic ethanol, the renewable energy landscape in Iowa will experience further restructuring, possibly involving consolidation and restricted entry to companies with access to R&D as well as new process knowledge.

Notes

1. TEPs include Industrial Revolution (1760–1920), Fordism (1920–1970), ICT (1970–2020), and Green (2020–). The present ICT paradigm can be characterized by its network structures emphasizing heterogeneity, diversity, and adaptability with a focus on knowledge as capital. Industry organization emphasizes cooperation facilitated by instant global and local contact and often takes the form of clusters (Freeman 1987; Hayter 2004).
2. Net energy balance is a comparison of the amount of energy used to produce a certain amount of biofuel related to the amount of energy that biofuel will produce when used (Hammerschlag 2006; Hill et al. 2006).
3. The percentage Iowa's total corn production derived from genetically modified seed rose sharply from 2000 to 2010 (2000, 30 percent; 2004, 54 percent; 2007, 78 percent; 2010, 90 percent; U.S. Department of Agriculture 2010). Although genetically modified corn is not a necessary resource for ethanol production, in Iowa over half the corn crop has been planted from genetically modified seed since 2004 when the number of ethanol refineries started to rise sharply
4. Ethanol cannot currently be shipped through oil and gas pipelines. This substantially raises not only its production cost but also production emissions and environmental impacts. Ethanol must be shipped by rail or truck to market.

References

25 × '25 Alliance. 2010. Lutherville, MD: 25 × '25 Alliance. http://www.25x25.org/ (last accessed 12 September 2010).

Boschma, R. A. 2004. Competitiveness of regions from an evolutionary perspective. *Regional Studies* 38 (9): 1001–14.

Bovard, J. 1995. Archer Daniels Midland: A case study in corporate welfare. CATO Policy Analysis No. 241. Washington, DC: CATO Institute.

Boyle, R., C. Greenwood, A. Hohler, M. Liebreich, V. Sonntag-O'Brien, A. Tyne, and E. Usher. 2008. *Global trends in sustainable energy investment*. Paris: United Nations Energy Program.

Bridge, G. 2002. Grounding globalization: The prospects and perils of linking economic processes of globalization to environmental outcomes. *Economic Geography* 78:361–86.

———. 2008. Environmental economic geography: A sympathetic critique. *Geoforum* 39:76–81.

Capehart, T. 2009. Ethanol: Economic and policy issues. Congressional Research Service R40488.

Carnevale, M. L. 2008. Ethanol industry lines up for Washington help. *Washington Wire*. http://blogs.wsj.com/washwire/2008/12/16/ethanol-industry-lines-up-for-washington-help/ (last accessed 20 February 2010).

Center for Responsive Politics. 2010. Washington, DC: The Center for Responsive Politics. http://www.opensecrets.org/lobby/index.php (last accessed 20 February 2010).

Cheshire, P. C., and E. J. Malecki. 2003. Growth, development, and innovation: A look back and forward. *Papers in Regional Science* 83 (1): 249–67.

Cooke, P. 2002. Regional innovation systems: General findings and some evidence from biotechnology clusters. *Journal of Technology Transfer* 27:133–45.

———. 2004. The regional innovation system in Wales. In *Regional innovation systems: The role of governance in a globalized world* 2nd ed., ed. P. Cooke, M. Heidenreich, and H. C. Braczyk, 214–33. London and New York: Routledge.

———. 2009. Jacobian cluster emergence: Wider insights from "green innovation" convergence on a Shumpeterian "failure." In *Emerging clusters: Theoretical, empirical, and political perspectives on the initial stage of cluster evolution*, ed. D. Fornahl, S. Henn, and M. P. Menzel, 17–43. Cheltenham, UK: Edward Elgar.

Curtis, S., W. Gesler, G. Smith, and S. Washburn. 2000. Approaches to sampling and case selection in qualitative research: Examples in the geography of health. *Social Science and Medicine* 50:1001–14.

Dicken, P. 2007. *Global shift*. New York: Guilford.

Eisenhardt, K., and M. Graebner. 2007. Theory building from cases: Opportunities and challenges. *Academy of Management Journal* 50 (1): 25–32.

Environmental Working Group. 2009. Ethanol's federal subsidy grab leaves little for solar, wind and geothermal energy. Washington, DC: Environmental Working Group. http://www.ewg.org/reports/Ethanols-Federal-Subsidy - Grab - Leaves - Little - For - SolarWind - And-Geothermal-Energy (last accessed 20 February 2010).

Freeman, C. 1987. *Technology and economic performance*. London: Pinter.

———. 1992. A green techno-economic paradigm for the world economy. In *The economics of hope*, ed C. Freeman, 190–211. London: Pinter.

Freeman, C., and F. Louca. 2001. *As time goes by*. Oxford, UK: Oxford University Press.

Freeman, C., and C. Perez. 1988. Structural crisis of adjustment, business cycles and investment behavior. In *Technical change and economic theory*, ed. G. Dosi, C. Freeman, R. Nelson, G. Silverberg, and L. Soete, 38–66. London: Frances Pinter.

Fromhold-Eisebith, M. 2007. Bridging scales in innovation policies: How to link regional, national, and international innovation systems. *European Planning Studies* 15 (2): 217–33.

Haddad, M., G. Taylor, and F. Owusu. 2010. Location choices of the ethanol industry in the Midwest Corn Belt. *Economic Development Quarterly* 24 (1): 74–86.

Hammerschlag, R. 2006. Ethanol's energy return on investment: A survey of the literature 1990–present. *Environment, Science, and Technology* 40:1744–50.

Hayter, R. 2004. Economic geography as dissenting institutionalism: The embeddedness, evolution, and differentiation of regions. *Geografiska Annaler B* 86:1–21.

———. 2008. Environmental economic geography. *Geography Compass* 2 (3): 831–50.

———. 2009. Teaching and learning guide for: Environmental economic geography. *Geography Compass* 3 (4): 1596–1601.

Hayter, R., T. Barnes, and M. J. Bradshaw. 2003. Relocating resource peripheries to the core of economic geography's theorizing: Rationale and agenda. *Area* 35:15–23.

Hayter, R., and R. B. LeHeron. 2002. Industrialization, techno-economic paradigms and the environment. In *Knowledge, industry, and the environment: Institutions and innovation in territorial perspectve*, ed. R. Hayter and R. B. LeHeron, 11–30. London: Ashgate.

Hill, J., E. Nelson, D. Tilman, S. Polasky, and D. Tiffany. 2006. Environmental, economic, and energetic costs and benefits of biodiesel and ethanol biofuels. *Proceedings of the National Academy of Sciences* 103 (30): 11206–10.

Iammarino, S. 2005. An evolutionary integrated view of regional systems of innovation: Concepts, measures and historical perspectives. *European Planning Studies* 13 (4): 497–519.

Koh, L. P., and J. Ghazoul. 2008. Biofuels, biodiversity, and people: Understanding the conflicts and finding opportunities. *Biological Conservation* 141:2450–60.

Low, S., and A. Isserman. 2009. Ethanol and the local economy: Industry trends, location factors, economic impacts, and risks. *Economic Development Quarterly* 23 (1): 71–88.

Lund, P. D. 2009. Effects of energy policies on industry expansion in renewable energy. *Renewable Energy* 34:53–64.

Martin, R., and P. Sunley. 2006. Path dependence and regional economic evolution. *Journal of Economic Geography* 6 (4): 395–437.

Oinas, P., and E. J. Malecki. 1999. Spatial innovation systems. In *Making connections: Technological learning and regional economic change*, ed. P. Oinas and E. J. Malecki, 7–33. Aldershot, UK: Ashgate.

———. 2002. The evolution of technologies in time and space: From national and regional to spatial innovation systems. *International Regional Science Review* 25 (1): 102–31.

Pimentel, D. 2003. Ethanol fuels: Energy balance, economics, and environmental impacts are negative. *Natural Resources Research* 12 (2): 127–34.

Pimentel, D., and T. Patzek. 2005. Ethanol production using corn, switchgrass, and wood; biodiesel production using soybeans and sunflower. *Natural Resources Research* 14 (1): 65–76.

Pimentel, D., T. Patzek, and G. Cecil. 2007. Ethanol production: Energy, economic, and environmental losses. *Review of Environmental Contamination and Toxicology* 189:25–41.

Rajagopal, D., S. Sexton, G. Hochman, and D. Zilberman. 2009. Recent developments in renewable technologies: R&D investments in advanced biofuels. *Annual Review of Resource Economics* 1:621–44.

Searchinger, T., R. Heimlich, R. A. Houghton, F. Dong, A. Elobeid, A. Fabiosa, S. Tokgoz, D. Hayes, and T. H. Yu. 2008. Use of US croplands for biofuels increases greenhouse gases through emissions from land-use change. *Science* 319:1238–40.

Smith, A. 2007. Emerging in between: The multi-level governance of renewable energy in the English regions. *Energy Policy* 35:6266–80.

Spath, P., and H. Rohracher. 2010. "Energy regions": The transformative power of regional discourses on socio-technical futures. *Research Policy* 39:449–58.

Sterzinger, G. 2008. Energizing prosperity: Renewable energy and reindustrialization. Briefing paper no. 205. Washington, DC: Economic Policy Institute.

Stormer, E. 2008. Greening as strategic development in industrial change—Why companies participate in eco-networks. *Geoforum* 39:32–47.

U.S. Department of Agriculture. 2010. Adoption of genetically engineered crops in the US: Corn varieties. Washington, DC: U.S. Department of Agriculture. http://www.ers.usda.gov/Data/BiotechCrops/ExtentofAdoptionTable1.htm (last accessed 12 September 2010).

U.S. Energy Information Administration. 2011. *Monthly energy review—March 2011*. DOE/EIA-0035 (2011/03). Washington, DC: U.S. Energy Information Administration. http://www.eia.doe.gov/ (last accessed 3 April 2011).

Yacobucci, B. 2008. Biofuels incentives: A summary of federal programs. Congressional Research Service R40110. Washington, DC: Congressional Research Service.

Zhang, F., and P. Cooke. 2009. Global and regional development of renewable energy. Working paper for Dynamics of Institutions and Markets in Europe (DIME). http://www.dime-eu.org/working-papers/sal3-green/ (last accessed 3 April 2011).

Zimmerer, K. S. 2011. New geographies of energy: Introduction to the special issue. *Annals of the Association of American Geographers* 101 (4): 705–11.

A Regional Evaluation of Potential Bioenergy Production Pathways in Eastern Ontario, Canada

Warren E. Mabee and Jaconette Mirck

Department of Geography, Queen's University

The potential for bioenergy and biofuel development is considered for five counties in the eastern Ontario region. The downward trend in Ontario's forest sector has resulted in the idling or closure of many forest products mills, creating an opportunity for bioenergy development supported by the Ontario Green Energy and Green Economy Act. In eastern Ontario, additional opportunities are afforded by the presence of a large agricultural base, which could be the source of residual as well as purpose-grown biomass feedstocks. The biological potentials of the region have been mapped and quantified. A range of scenarios for bioenergy and biofuel production based on these resources are considered, and impacts of this potential development are considered in terms of energy demand within the region. Both county- and regional-level approaches are compared to similar analyses at the provincial scale to highlight potential advantages to developing energy policy at local or regional levels. Biomass-sourced processes might be able to meet varying levels of energy demand across three major categories, including home and industrial heating, electricity generation, and production of fuels for transportation. In terms of conventional bioenergy options, the combined heat and power option can provide significant amounts of energy to the eastern region of the province while meeting most of the policy goals laid out by the Ontario government. *Key Words: biomass supply, combined heat and power, Green Energy and Green Economy Act, liquid biofuels for transport, wood pellets*.

安大略省东部地区的五个县正在考虑生物能源和生物燃料的发展潜力。安大略省林业产业下降的趋势已导致许多林业产品工厂的闲置或关闭，安大略绿色能源和绿色经济法案支持创造一个发展生物能源的机遇。在安大略东部地区，大型农业基地的存在提供了额外的机遇，这些农业基地可以是农业生产剩余以及专门种植的生物质原料的来源。该区域的生物潜力已经被绘制成地图和量化分析过。文章对基于这些资源的生物能源和生物燃料生产的条件范围进行了分析，从该区域内的能源需求方面，对这一潜在发展的影响也进行了探讨。我们对县和区域水平的方法进行了比较，在省级规模的类似分析突出了在地方或区域水平上能源开发政策的潜在优势。生物质原料的处理也许能够满足横跨三大类的不同级别的能源需求，包括家庭和工业供热，发电，以及运输燃料的生产。在常规生物能源选项方面，热电联产的选项可以提供大量的能量到该省的东部地区，同时也能满足安大略省政府制定的大部分政策目标。*关键词：生物质供应，热电联产，绿色能源和绿色经济法案，交通运输所用的液体生物燃料，木材颗粒*。

Se considera en este estudio el potencial de desarrollo en bioenergía y biocombustibles de cinco condados de la región de Ontario Oriental. La tendencia hacia el descenso en el sector forestal de Ontario ha dado lugar a la inactividad o cierre de muchas factorías de productos forestales, lo cual ha creado una oportunidad para desarrollo bioenergético con el apooyo de la Ley de Energía y Economía Verdes de Ontario. En la parte oriental de esta provincia, se presentan oportunidades adicionales por la existencia de un gran basamento agrícola, que podría ser fuente de biomasa residual o deliberadamente cosechada para obtener materias primas. Los potenciales biológicos de la región se cartografiaron y cuantificaron. Se consideró un campo de escenarios para la producción de bioenergía y biocombustibles a partir de estos recursos, lo mismo que el impacto de este desarrollo potencial en términos de la demanda de energía dentro de la región. También se compararon los enfoques tanto a nivel de condado como regional, con análisis similares a escala provincial para relievar las ventajas potenciales de las políticas de desarrollo energético a niveles local o regional. Los procesos originados en biomasa podrían tener la capacidad de atender varios niveles de demanda de energía en tres categorías principales que incluyen calefacción doméstica e industrial, generación de electricidad y producción de combustibles para el transporte. En términos de opciones bioenergéticas convencionales, la opción combinada de calor y energía puede suministrar cantidades significativas de energía para la región Oriental de la provincia, a tiempo que puede satisfacer la mayoría de las metas formuladas en las políticas del gobierno de Ontario. *Palabras clave: oferta de biomasa, calor y energía combinados, Ley de Energía y Economía Verdes, biocombustibles líquidos para el transporte*, pellets *de madera*.

Of the sixteen pulp and paper facilities that were present in Ontario in 2005, only five are still operating on a regular schedule; three are on constrained production schedules, one has been sold, three have been demolished, and four have been idled indefinitely (Canadian Imperial Bank of Commerce 2005; Canadian Parks and Wilderness Society 2007). With the loss of this production capacity, there has been an increase in unused chips and residues from sawmill operations within the province, which in turn negatively impacts the profitability of primary forest products production—the loss of a profit stream has at least in part contributed to the intermittent idling of significant sawmill capacity across Ontario. Like other parts of the province, the forest industry in eastern Ontario has been hit hard: Losing the Domtar pulp and paper facility in Cornwall directly impacted the workers in this facility, as well as the employment of those involved in forest harvesting and sawmilling operations in the region, due to the interconnected nature of the forest sector.

A partial solution to the forest industry's problems is the development of bioenergy and biofuel production to use these excess forest feedstocks. *Bioenergy* generally refers to the energy derived from biomass, both as heat and as electricity; biofuels, on the other hand, are a carrier of bioenergy that can be utilized for downstream energy production (i.e., via combustion in the engine of a vehicle). In 2009, the government of Ontario introduced the Green Energy and Green Economy Act (Ontario 2009), which creates incentives for the renewable energy sector within the province as well as support for an increased focus on energy conservation across the province. Within the Act, a range of technologies are covered, including solar, wind, hydro, and biomass-to-electricity. The primary incentive tool included in the Act is a Feed-in Tariff (FIT), which provides guaranteed pricing structures for electricity production (note that this does not include heat and transport fuels). The FIT is designed to help boost investor confidence and increase access to financing for projects of varying sizes. FIT levels vary extensively by the technology at hand, and currently range from 10.3 cents per kilowatt hour for electricity from landfill gas (greater than 10 megawatts in scale) to 80.2 cents per kilowatt hour for electricity from rooftop solar panels (less than 10 kilowatts in scale; Ontario Power Authority 2010). Agricultural and forest biomass-to-electricity options are offered between 13.0 and 13.8 cents per kilowatt hour, considerably less than solar, wind, and even most biogas options but competitively priced with hydropower technologies; this reflects the perceived maturity of both biomass and hydro-to-electricity options. The Act is in part a response to the phased shutdown of coal-fired electricity generation in the province, which accounts for about 19 percent of total generation capacity (6,400 MW in 2007) and about 15 percent of the province's total greenhouse gas emissions (28 million metric tons of CO_2 equivalent; Zhang et al. 2010).

Electricity generation is only one of the areas of energy use in the province. In 2003, Ontario used 901,751 TJ of electricity. Home, institutional, and industrial heating within the province consumed 1,619,719 TJ, and transportation required 787,667 TJ of energy in 2003 (Statistics Canada 2004). Although bioenergy technology options exist to deliver energy in each of these categories, it is possible that the FIT program will encourage biomass-to-energy projects that focus on electricity, perhaps taking away from heat and potential biofuel generation. A strategic approach to deploying these technologies is to consider the potential contribution that biomass-based energy could make toward consumer demand in each of these categories and to identify locations where facilities for the production of different forms of biomass-to-energy could be best located.

Regional- and county-level assessments will be used to identify likely pathways and locations for bioenergy development in eastern Ontario, and the efficacy of a regional approach to bioenergy planning will be explored. A contiguous area of five census regions, made up of four counties (Lanark, Leeds & Grenville, Prescott & Russell, and Stormont, Dundas & Glengarry) and one census division (Ottawa) in eastern Ontario, was selected for the subprovincial assessment. These jurisdictions have been chosen because they illustrate the balance between productive forest and agricultural land common in southern Ontario and also display the urban–rural split characteristic of the Canadian landscape. Development of regional assessments for energy has been explored in the literature extensively. A recent paper by Berberi, Thodhorjani, and Aleti (2009) analyzed development of renewable energy to supply a remote coastal region comparing various alternatives including hydro, solar, and wind, and optimizing energy production to support planning policies. Models of energy deployment at the regional scale available in the literature (i.e., Li et al. 2010) have informed the analysis in this article. Regional analysis has been shown to be useful in assessing the sustainability of energy systems within the wider context of globalization, local resource use, and regional energy development (Pulselli et al. 2009). This article examines the potential of bioenergy options at the regional and provincial scales to inform debate and assess the relative scale of opportunity in each of the census regions selected.

Biomass Resources

The annual increment (i.e., the growth measured in the forest) for the southern Ontario forest region (which includes the study region) and the entire province of Ontario were reported in the *Forest Resources of Ontario: State of the Forest Report 2006* (Ontario Ministry of Natural Resources [OMNR] 2006). The mean annual increment (MAI, averaged growth over the life span of the forest) in the southern Ontario forest region was estimated to be 3.3 m^3/ha/year; the current annual increment (CAI, the growth recorded in the year of observation) was reported as 2.5 m^3/ha/year in 2006 (OMNR 2006). The same document reported MAI for the entire province of Ontario at 0.8 m^3/ha/year and CAI at 0.6 m^3/ha/year in 2006. In this study, CAI is used to calculate conservative estimates of wood production at the county, regional, and provincial scales. This is likely an underestimate of total wood growth but anticipates restrictions on commercially available wood for bioenergy due to environmental or social considerations. Cumulative wood production in the eastern Ontario region is estimated at just over 400,000 metric tons per year, and provincial production is estimated at 13.6 million metric tons per year, given average wood densities and moisture contents of common Ontario tree species (Sjöström 1993). Of this production, only a small fraction is currently utilized in eastern Ontario. In 1998 (the last year for which detailed information is available), about 124,000 m^3 was removed for forest products across the five counties under consideration, or about 10 percent of the estimate of regional forest growth (Johnson and Heaven 1999). Since this inventory was taken, a major pulp mill in the region (the Domtar plant in Cornwall, Ontario) and several sawmill installations have been permanently closed, and thus the low level of demand for wood fiber observed in 1998 has likely declined further since. Over the same period, total forest harvests in Ontario have been consistently less than the estimated sustainable wood supply. In 2004 (the last year for which data are available), the total harvest was approximately 23 million m^3, or about 70 percent of the total sustainable wood supply estimated to be available under the annual allowable cut (OMNR 2006).

Table 1. Estimated biomass productivity for eastern Ontario by census division

County	Estimated production (thousands of tons per year)				
	Wood	Stover	Straw	Hay	Tons per capita
Lanark	149	60	8	75	4.7
Leeds and Grenville	113	147	10	169	4.5
Prescott and Russell	38	334	29	209	8.0
Ottawa (Census division)	56	262	20	172	0.7
Stormont, Dundas and Glengarry	57	584	59	286	9.0
Eastern Ontario	412	1,387	126	911	2.5
Ontario	13,555	11,666	4,167	6,078	3.1

Source: Johnson and Heaven (1999); Statistics Canada (2007); Agriculture and Agri-Food Canada (2007).

The other major source of biomass in the eastern Ontario region is agricultural residues. These are reported in the most recent Census of Agriculture (Agriculture and Agri-Food Canada [AAFC] 2007) and presented in concise form in Table 1. Graham et al. (2007) indicate that current corn varieties provide about one ton of stover (residue) for every ton of corn produced. For other cereal crops, residue generation rates of 36, 23, and 18 kilograms per bushel of wheat, barley, and oats, respectively, have been reported (Shanahan et al. 1999). These figures can be converted using standard crop densities to show that total straw production is on the order of 1.3, 1.0, and 1.2 tons per ton of wheat, barley, and oats, respectively (Bowyer and Stockmann 2001). Based on these criteria and farm data reported in the Canadian Census of Agriculture (AAFC 2007), the production of corn stover in eastern Ontario is almost 1.4 million tons per year, and other cereal crops could provide about 126,000 additional tons per year. Farms in this region also produce about 900,000 tons per year of hay, a grass that could be used as an energy crop, although it is currently used for other purposes within the agricultural food chain.

In addition to forest and agricultural biomass, significant amounts of biomass material can be found in residues from industrial operations as well as residential waste. These are not considered in this article due to space constraints but should be considered important in overall biomass supply. It is also important to realize that the actual amount of biomass that might be available for heat, electricity, or transportation will be a subset of the total amount of biomass estimated. There are environmental constraints on the recovery of this material; for example, in agricultural systems soil conservation requirements could account for 50 percent or more of the total residues (Lindstrom et al. 1979; Shanahan et al. 1999). In addition, there are economic constraints on the use of this material; for example, hay is widely used in existing agricultural systems as fodder for livestock. We make the assumption in this article that biomass might be available for use but note that true availability will be constrained by these issues. Accounting for the factors of soil conservation, livestock feed, and season variation, Bowyer and Stockmann (2001) suggested that available residues might be as low as 15 percent of total production, and Graham et al. (2007) estimated that residue recovery rates as high as 30 percent might be achieved in a sustainable and cost-effective manner. This report utilizes the 30 percent figure in estimating available agricultural residue from corn and other cereal crop production. The assumption is also made that 30 percent of the hay harvest might be made available to bioenergy options under ideal circumstances. Finally, we assume that 30 percent of the annual increment in forest growth might be available for bioenergy purposes; this is the amount that is currently projected to be sustainably available but unused by the Ontario forest industry across the province (OMNR 2006). In eastern Ontario, 2.8 million dry tons of feedstock are estimated to be available, with the majority of these feedstocks coming from the agricultural sector. Across all of Ontario, 35.5 million dry tons of feedstock are estimated to be available annually, with slightly more agricultural residue than forest material. These estimates are summarized in Table 1.

Bioenergy Technologies

Residential Heating

Heat can be recovered from biomass in a number of different ways. Wood stoves and pellet stoves are useful options for residential applications, providing significantly better energy recovery than open flame. Wood pellets are made from wood fibers, often in the form of ground wood. Once dried, the wood can be run through a pellet mill, which extrudes wood under pressure to create high-density pellets. Wood pellets can be used in residential furnaces that resemble wood stoves, which offer improved combustion and minimal air emissions (Karlsson and Gustavsson 2003). Wood pellets are also used in district heating applications in Europe; the Canadian wood pellet industry has grown steadily over the past ten years due to strong demand in Europe driven by climate-related incentives (Junginger et al. 2008). There is no good uptake mechanism for wood pellets in Canada; thus, most wood pellets continue to be shipped overseas, leaving the domestic availability of this feedstock uncertain. It should also be noted that the wood pellet industry depends greatly on both inexpensive feedstocks and incentive support for consumption and that interruptions on either account could impact the industry negatively.

Electricity Generation and Cogeneration of Heat and Power

The largest existing dedicated biomass power facility in Canada is in Williams Lake, British Columbia, a 60-MW electricity generating plant that has been operational since 1993. This facility consumes about 600,000 tons of wood residues including bark, chips, and sawdust annually. The wood waste fuel is provided by five surrounding sawmill operations (McCloy 1999). The Williams Lake facility is used for electricity generation only, but process steam from these types of facilities could supply other industrial processes or support district heating grids for residential heating; this is called *combined heat and power* (CHP). Modern CHP facilities deliver about 65 percent electricity to 35 percent heat (based on U.K. figures; Institution of Engineering and Technology 2008); this ratio can be adjusted if the heat and power users have different requirements. It should be noted that the success of CHP depends greatly on the proximity of power generation to useful applications for steam generated; thus, isolated facilities might not be able to utilize steam and thus would lose overall efficiency.

Thermal power is also generated in Kraft pulp mills, where recovery boilers are used to recycle black liquor and recover pulping chemicals, as well as to produce steam that drives the pulping process. The design of recovery boilers has improved significantly over the last fifty years, and they can now produce a surplus of electricity without firing mill bark and wood residues

(Vakkilainen, Kankkonen, and Suutela 2008). The Canadian government's Pulp and Paper Green Transformation Program (announced summer 2009) will provide up to $1 billion to increase energy, paying $0.16 for every liter of black liquor burned.

Energy for the Transport Sector

There are essentially two technology platforms for the production of liquid fuels that can be used in the transport sector. The first of these, the bioconversion platform, uses biological agents to convert the sugars within biomass to ethanol. It is estimated that ethanol yields from the bioconversion of lignocellulosic feedstocks range between 110 and 300 l/ton dry matter (Mabee et al. 2006). Given the energy content of biomass is around 20 GJ/dry ton, process conversion efficiency can range between 2.3 and 2.6 GJ/dry ton at the low end of the scale to as high as 6.4 GJ/ton, which reflects the theoretical maximum conversion efficiency possible based on a lignocellulosic material containing 70 percent carbohydrates and virtually complete conversion of carbohydrate to ethanol (i.e., 51 percent efficiency; Sims et al. 2010).

The bioconversion process is not yet commercial and further development must reduce product prices (Wooley et al. 1999). Options for reducing cost include mechanical and chemical processes (e.g., Mais et al. 2002; Mabee et al. 2006); selection of these options will often be dictated by feedstock characteristics (e.g., Boussaid et al. 1999; Mosier et al. 2005; Chandra et al. 2007). Liberation of sugars from biomass must also be achieved in a cost-effective manner; costs have been reduced significantly in recent years (Mabee et al. 2006). Process efficiency could be improved further through the addition of specialty chemicals, although there is a trade-off between efficiency and cost that must be optimized (Tengborg, Galbe, and Zacchi 2001). Finally, new yeast strains able to process all of the sugars present within forest or agricultural biomass at high efficiency are being developed (Ragauskas et al. 2006). This might also improve overall process efficiency, reducing costs. Bioconversion facilities are likely to process approximately 650,000 to 700,000 tons of biomass per year, producing approximately 200 million liters of ethanol product (Mabee et al. 2006). At this scale, existing wood handling and transport options would be well able to meet supply requirements for the facility.

A second option for the production of liquid fuels from biomass is the thermochemical platform. Thermochemical conversion uses heat to reduce wood to its basic constituents. At lower temperatures, ranging from 450°C to 600°C, combustion in the absence of oxygen can produce a liquid condensate (commonly known as *bio-oil*) and very little gas, as well as a char component (Garcia et al. 2000). Bio-oil is a strong acid and thereby necessitates expensive storage and handling equipment (Faaij 2006). At higher temperatures (between 700°C and 1200°C), biomass can be further converted to gases (carbon monoxide, carbon dioxide, hydrogen, and other trace gases; Garcia et al. 2000; Cetin et al. 2005). These gases can be cleaned and passed through specialized processes such as the Fischer-Tropsch (FT) process to create a range of liquid fuels and chemicals (Putsche 1999; Sims et al. 2010).

The thermochemical process is not yet commercial; challenges include reactor design, which strongly influences the quality of gas produced. Clean gas production is the most costly component in the process (Spath and Dayton 2003), and the presence of impurities in the gas can inhibit the production of end products (Sims et al. 2010). Sometimes these impurities can also be turned to advantage, and used as a feedstock for value-added chemicals (as done by companies such as Choren, Ensyn, and Enerkem; Branca and Di Blasi 2006). A final consideration is scale; it could be assumed that *n*th-scale plants would be in the range of 0.5 billion to 1 billion liters in capacity, and at this scale, existing wood handling and transport options would be stretched to meet supply requirements for the facility.

Discussion

The five jurisdictions considered in eastern Ontario have more forest area than agricultural area, but their agricultural zones are currently far more productive than the forests, as shown in Table 1. The presence of Ottawa, with its large population, means that at the regional scale, productivity is actually lower in eastern Ontario than when averaged across the province as a whole (Table 1). This difference can be observed in Tables 2 and 3, where total Ontario figures are compared to the eastern Ontario region under consideration.

The efficiency of different biomass-to-energy platforms, and the contribution that these platforms might make to heat and electricity energy requirements for all of Ontario and for the eastern region, are shown in Table 2 (using data from Karlsson and Gustavsson 2003; Statistics Canada 2004; Mabee et al. 2006). The

Table 2. Estimated contribution of biomass to heat and electricity pathways, Ontario

Technology	Average conversion efficiency (η_c)	Average energy yields (GJ/bdt)	Estimated contribution to heat and electricity requirements (%)[a] Eastern Ontario: Heat	Eastern Ontario: Electricity	Ontario total: Heat	Ontario total: Electricity
Traditional wood stoves	0.36	7.2	3.9	—	4.7	—
Wood pellet stoves, low	0.78	15.6	8.4	—	10.2	—
Wood pellet stoves, high	0.81	16.2	8.7	—	10.6	—
Steam-turbine power boiler	0.40	8	—	7.7	—	9.4
Gas-turbine generator	0.47	9.4	—	9.0	—	11.1
Combined heat and power, low	0.70	14	2.6	8.8	3.2	10.7
Combined heat and power, high	0.80	16	3.0	10.0	3.7	12.3

Sources: Karlsson and Gustavsson (2003); Statistics Canada (2004, 2007); Mabee et al. (2006).
[a]Based on 2004 figures of 1,619,719 TJ (residential and industrial heat) and 901,751 TJ (electricity; Statistics Canada 2004).

most effective option seems to be high-efficiency CHP. In eastern Ontario, these facilities could provide up to 10 percent of regional electricity requirements and about 3 percent of regional need for heat energy. It is likely that total biomass consumption by a single 60-MW facility would not exceed 400,000 tons per year; in eastern Ontario, as many as seven of these facilities could be sourced with local biomass feedstocks. These facilities would have to be sited carefully to ensure a use for the heat outputs of the plant and attain maximum efficiency. With large local populations within the region, however, this consideration would be easier to meet than in more remote locations. This option would best meet the clean energy generation goal of Ontario's Green Energy Act, which focuses on the production of clean electricity from sources including hydro, solar, wind, and biomass.

Wood pellet stoves used for residential heating also represent an interesting opportunity. If available biomass in eastern Ontario were directed to this option, about 8.7 percent of heat requirements could be met, providing about 43,000 TJ of energy. It has been calculated that Canadian homes built since World War II require between 70 and 80 GJ of heat energy per year.[1] Therefore, locally grown biomass could meet the requirements of about 580,000 homes in the five jurisdictions under consideration. Achieving this goal would require multiple installations of modern pellet stoves, as well as installation of feed systems and supply of pellets. As an employment opportunity, this option would probably provide the most significant return for the region and might best serve the jobs creation goal of the Green Energy Act, which is meant to facilitate the creation of up to 50,000 jobs in the province by 2013 (Ontario 2009). This option would require policy to ensure a constant wood pellet supply for residential users.

The efficiency of different biomass-to-biofuel platforms and the contribution that these emerging biofuel platforms might make to Ontario's transportation energy requirement are shown in Table 3 (using data from Putsche 1999; Mabee et al. 2006). Two options seem best: high-efficiency bioconversion to ethanol and high-efficiency thermochemical conversion to FT diesel. Based on the available feedstock, up to 7 percent of total annual transport energy requirements for eastern Ontario could be met. As ethanol is blended with gasoline, most of this energy would be used in light-duty vehicles—passenger cars and light trucks. The biomass supply in eastern Ontario could probably support three or four 200-million-liter per year facilities, based on conversion efficiencies as shown in Table 3. Enough FT diesel fuel, another biofuel option, could be generated to meet about 7.4 percent of total annual transport energy needs in the region. This biofuel would be used in heavy-duty applications (buses and trucks). Thermochemical biofuel facilities are likely to be very large (in the 1-billion-liter per year range), however; it is likely that the eastern Ontario region could support such a facility with locally grown biomass, as this scale would require almost double the biomass currently estimated to be available. It is worth repeating that these technologies are not yet commercially available.

Table 3. Estimated contribution of biomass to liquid fuel pathways, Ontario

Technology	Average conversion efficiency (liters/bdt[a])	Average energy yields (GJ/bdt[a])	Estimated contribution to Ontario's transportation requirements (%)[b]	
			Eastern Ontario	Ontario total
Bioconversion to ethanol, low	125	2.6	2.9	3.5
Bioconversion to ethanol, high	300	6.4	7.0	8.6
Syngas-to-FT diesel, low	75	2.6	2.9	3.5
Syngas-to-FT diesel, high	200	6.7	7.4	9.1
Syngas-to-ethanol, low	120	2.5	2.8	3.4
Syngas-to-ethanol, high	160	3.4	3.7	4.6

Note: FT = Fischer-Tropsch.
Sources: Putsche (1999); Statistics Canada (2004, 2007); Mabee et al. (2006).
[a]Efficiency and yield reported per bone-dry ton of biomass input.
[b]Based on 2004 figure of 787,667 TJ (Statistics Canada 2004).

In Figure 1, basic statistics for the five jurisdictions under consideration are provided, as are the potential biomass supplies (green circles) and the estimated bioenergy contribution that each county might make to its local heat, electricity, or transport energy requirements. The county-level analysis shows that Stormont, Dundas, and Glengarry could meet up to 31 percent of its local requirement for residential and industrial heat (H) with locally grown agriculture and forest biomass. With the same biomass the jurisdiction could produce up to 36 percent of its requirement for electricity (E), or up to 26 percent of its need for transportation energy

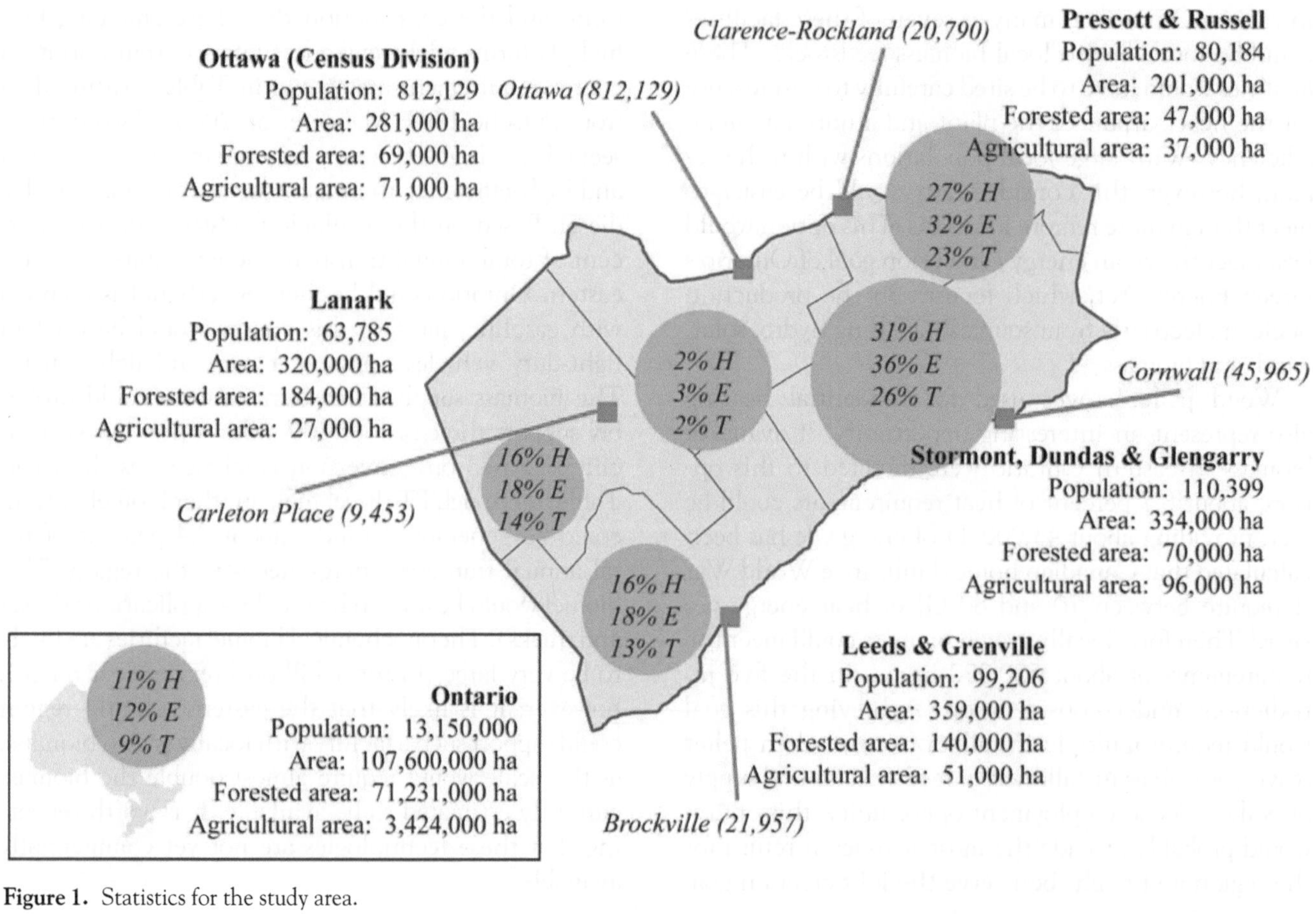

Figure 1. Statistics for the study area.

(T). These figures are significantly higher than the regional averages, due to the lower population and greater biomass resources of this county. In contrast, the large population present in Ottawa means that locally grown biomass could only meet 2 percent of heat, 3 percent of electricity, or 2 percent of transport requirements.

Conclusions

This article has reviewed the opportunities for bioenergy development in Ontario, based on the availability of biomass that can be sustainably sourced from forest and agricultural operations in the province. In terms of conventional bioenergy options, the CHP option can provide significant amounts of energy to the province.

For eastern Ontario, it is estimated that combined heat and power facilities could provide 8.8 percent of electricity requirements as well as 2.6 percent of heat energy needs. As many as seven facilities in the 60 MW range could be sourced with locally grown biomass supplies. The analysis suggests that the two easternmost counties of this region are the most promising for initial facilities; local population centers (Cornwall and Clarence-Rockland) could supply uses for heat outputs and easy connections to the power grid. Development of CHP facilities in these rural centers would keep employment opportunities local and reduce transport distances required for the biomass itself, which would increase the sustainability of this option.

Available biomass could also be used to provide residential heat through wood pellet stoves, serving up to 580,000 households in the region. The large population of the city of Ottawa creates the most obvious opportunity, but this would dictate that most biomass be shipped from surrounding areas to supply the city population. This might reduce opportunities for rural employment, but maximize total employment across the region.

The production of biofuels might also be an interesting future option for the biomass grown in eastern Ontario, but at this time both the bioconversion and thermochemical pathway remain at the late research and development stage, and the ability of biofuels to contribute to transport energy needs is lower proportionally than heat or electricity potentials. If no other renewable options exist for sustainable transport energy generation, however, biofuels might remain an attractive option.

This article explores the trade-offs between different bioenergies. Based on the existing biomass supply, no technology can provide a very large proportion of eastern Ontario's energy requirements. A localized approach could help optimize job creation. Supporting this development requires engineers and tradespeople to build and maintain facilities; it will require people with the financial acumen to fund the work, the computer networking skills to tie the plants together, and the ongoing expertise to run that system. Many challenges remain, but significant opportunity exists for bioenergy in eastern Ontario.

The methodology of regional analysis for specific renewable energy options, as explored in this article, might be applicable in a variety of settings. The region around much of the Great Lakes of North America is characterized by similar trends in land use toward forestry and agriculture, interspersed with urban development, as seen in eastern Ontario, and it might be assumed that bioenergy assessments within these regions would be defined by similar biomass productivity, efficiency of conversion, and energy requirements to those observed in the Ontario case study. Similar methodologies could be used to explore other renewable energy alternatives. Policy drivers in different jurisdictions, however, can be significantly different, and thus optimal renewable energy choices are likely to be very different, particularly when crossing provincial, state, or international borders.

Note

1. Natural Resources Canada—Office of Energy Efficiency, Table 32: Gross Output Thermal Requirements per Household by Building Type and Vintage. Available online at http://www.oee.nrcan.gc.ca/corporate/statistics/neud/dpa/tablestrends2/res_on_32_e_3.cfm?attr=0.

References

Agriculture and Agri-Food Canada (AAFC). 2007. *Canadian census of agriculture 2006*. Ottawa, ON, Canada: AAFC.

Berberi, P., S. Thodhorjani, and R. Aleti. 2009. Integration and optimization of alternative sources of energy in a remote region. In *8th symposium on Advanced Electromechanical Motion Systems (Electromotion 2009)*, IEEE, 41–44. doi: 10.1109/ELECTROMOTION.2009.5259146.

Boussaid, A., J. Robinson, Y. Cai, D. J. Gregg, and J. N. Saddler. 1999. Fermentability of the hemicellulose-derived sugars from steam-exploded softwood (Douglas-fir). *Biotechnology and Bioengineering* 64 (3): 284–89.

Bowyer, J. L., and V. E. Stockmann. 2001. Agricultural residues: An exciting bio-based raw material for the global panels industry. *Forest Products Journal* 51 (1): 10–21.

Branca, C., and C. Di Blasi. 2006. Multistep mechanism for the devolatization of biomass fast pyrolysis oils. *Industrial and Engineering Chemistry Research* 45 (17): 5891–99.

Canadian Imperial Bank of Commerce. 2005. Ontario government's package for the forest products industry: Equity Research Industry Update, 29 September 2005. Toronto: CIBC World Markets.

Canadian Parks and Wilderness Society. 2007. Ontario's forest economy . . . adrift in a global marketplace. Special Report, June 2007. Toronto: CPAWS Wildlands League. http://www.wildlandsleague.org/attachments/Adrift.Ontario.mill.closures.2007.pdf (last accessed 5 March 2010).

Cetin, E., B. Moghtaderi, R. Gupta, and T. F. Wall. 2005. Biomass gasification kinetics: Influences of pressure and char structure. *Combustion Science and Technology* 177 (4): 765–91.

Chandra, R., R. Bura, W. E. Mabee, A. Berlin, X. Pan, and J. N. Saddler. 2007. Substrate pretreatment: The key to effective enzymatic hydrolysis of lignocellulosics? *Advances in Biochemical Engineering/Biotechnology—Biofuels* 108:67–93.

Faaij, A. P. C. 2006. Bio-energy in Europe: Changing technology choices. *Energy Policy* 34 (3): 322–42.

Garcia, L., R. French, S. Czernik, and E. Chornet. 2000. Catalytic steam reforming of bio-oils for the production of hydrogen: Effects of catalyst composition. *Applied Catalysis A: General* 201 (2): 225–39.

Graham, R. L., R. Nelson, J. Sheehan, R. D. Perlack, and L. L. Wright. 2007. Current and potential US corn stover supplies. *Agronomy Journal* 99:1–11.

Institution of Engineering and Technology. 2008. Combined heat and power (CHP). Factfile. London: The Institution of Engineering and Technology.

Johnson, L., and P. Heaven. 1999. *The eastern Ontario model forest's 1998–1999 state of the forest report*. Kemptville, ON, Canada: Eastern Ontario Model Forest.

Junginger, M., T. Bolkesjo, D. Bradley, P. Dolzan, A. Faaij, J. Heinimo, B. Hektor et al. 2008. Developments in international bioenergy trade. *Biomass and Bioenergy* 32 (8): 717–29.

Karlsson, A., and L. Gustavsson. 2003. External costs and taxes in heat supply systems. *Energy Policy* 31 (14): 1541–60.

Li, Y. F., G. H. Huang, Y. P. Li, Y. Xu, and W. T. Chen. 2010. Regional-scale electric power system planning under uncertainty: A multistage interval-stochastic integer linear programming approach. *Energy Policy* 38 (1): 475–90.

Lindstrom, M., E. Skidmore, S. Gupta, and C. Onstad. 1979. Soil conservation limitations on removal of crop residues for energy-production. *Journal of Environmental Quality* 8 (4): 533–37.

Mabee, W. E., D. J. Gregg, C. Arato, A. Berlin, R. Bura, N. Gilkes, O. Mirochnik, X. Pan, E. K. Pye, and J. N. Saddler. 2006. Updates on softwood-to-ethanol process development. *Applied Biochemistry and Biotechnology* 129:55–70.

Mais, U., A. R. Esteghlalian, J. N. Saddler, and S. D. Mansfield. 2002. Enhancing the enzymatic hydrolysis of cellulosic materials using simultaneous ball milling. *Applied Biochemistry and Biotechnology* 98–100 (1–9): 815–32.

McCloy, B. W. 1999. *Opportunities for increased woodwaste cogeneration in the Canadian pulp and paper industry*. Vancouver, BC, Canada: BW McCloy & Associates Inc.

Mosier, N., C. E. Wyman, B. Dale, R. Elander, Y. Y. Lee, M. Holtzapple, and M. Ladisch. 2005. Features of promising technologies for pretreatment of lignocellulosic biomass. *Bioresource Technology* 96 (6): 673–86.

Ontario. 2009. Green Energy and Green Economy Act. Bill 150, Government of Ontario, Toronto, ON Canada.

Ontario Ministry of Natural Resources (OMNR). 2006. *Forest resources of Ontario: State of the forest report 2006*. Ottawa, ON, Canada: Forests Division, Ministry of Natural Resources, Government of Ontario.

Ontario Power Authority. 2010. Feed-in tariff prices for renewable energy projects in Ontario—13 August 2010. Ottawa, ON, Canada: Ontario Power Authority. http://fit.powerauthority.on.ca/Storage/102/11128_FIT_Price_Schedule_August_13_2010.pdf (last accessed 22 August 2010).

Pulselli, R. M., P. Romano, D. Bogunovich, and F. M. Pulselli. 2009. Integrating human and natural systems sustainably: Energy evaluation and visualization of the Abruzzo region. *WIT Transactions on Ecology and the Environment* 122:273–78.

Putsche, V. 1999. Complete process and economic model of syngas fermentation to ethanol. C Milestone Completion Report, National Renewable Energy Laboratory, Golden CO.

Ragauskas, A. J., C. K. Williams, B. H. Davison, G. Britovsek, J. Cairney, C. A. Eckert, W. J. Frederick Jr., et al. 2006. The path forward for biofuels and biomaterials. *Science* 311 (5760): 484–89.

Shanahan, J., D. Smith, T. Stanton, and B. Horn. 1999. *Crop residues for livestock feed*. Fort Collins: Colorado State University Cooperative Extension. http://www.ext.colostate.edu/pubs/livestk/00551.html (last accessed 5 February 2010).

Sims, R. E. H., W. E. Mabee, J. N. Saddler, and M. Taylor. 2010. An overview of 2nd-generation biofuel technologies. *Bioresource Technology* 101 (6): 1570–80.

Sjöström, E. 1993. *Wood chemistry, fundamentals and application*. 2nd ed. New York: Academic.

Spath, P., and D. Dayton. 2003. Preliminary screening—Technical and economic assessment of synthesis gas to fuels and chemicals with emphasis on the potential for biomass-derived syngas. National Renewable Energy Laboratory report NREL/TP-510–34929, U.S. Department of Energy, Washington, DC.

Statistics Canada. 2004. Report on energy supply-demand in Canada. Catalogue No. 57-003-X, 1990 and 2003. Ottawa, ON: Statistics Canada.

———. 2007. Population and dwelling counts, for Canada, provinces and territories, and census subdivisions (municipalities), 2006 and 2001 censuses. In *Census of Canada 2006*. Ottawa, ON: Statistics Canada.

Tengborg, C., M. Galbe, and G. Zacchi. 2001. Reduced inhibition of enzymatic hydrolysis of steam-pretreated softwood. *Enzyme and Microbial Technology* 28:835–44.

Vakkilainen, E. K., S. Kankkonen, and J. Suutela. 2008. Advanced efficiency options: Increasing electricity generating potential from pulp mills. *Pulp and Paper Canada* 109 (4): 14–18.

Wooley, R., M. Ruth, D. Glassner, and J. Sheehan. 1999. Process design and costing of bioethanol technology: A tool for determining the status and direction of research and development. *Biotechnology Progress* 15:794–803.

Zhang, Y. M., J. McKechnie, D. Cormier, R. Lyng, W. E. Mabee, A. Ogino, and H. L. MacLean. 2010. Life cycle emissions and cost of producing electricity from coal, natural gas, and wood pellets in Ontario, Canada. *Environmental Science and Technology* 44:538–44.

Opposing Wind Energy Landscapes: A Search for Common Cause

Martin J. Pasqualetti

School of Geographical Sciences and Urban Planning, Arizona State University

Although wind power is local, sustainable, affordable, and carbon free, mounting public opposition to the landscape changes it produces threatens its expansion. In an era when many countries are looking to renewable energy as an answer to questions about national security and the risks of climate change, it is important to explain the sources of this reaction. This article looks for similarities in public resistance to wind developments in four diverse settings: Palm Springs, California; Cape Cod, Massachusetts; the Isle of Lewis, Scotland; and Oaxaca State, Mexico. Despite the natural and cultural diversity among these places, there are five common threads in the opposition that has been experienced: immobility, the site specificity of the resource; immutability, an expectation of landscape permanence; solidarity, the close relationship between people and the land; imposition, a sense of marginalization; and place identity, a loss of security. Considering more deeply the relationship between land and life, in advance of the development of renewable energy resources, will help smooth the otherwise bumpy road toward a more sustainable future. *Key Words: landscapes, place, wind energy.*

虽然风力发电是本地的、可持续的、价格适中并且无碳的，公众反对风能机组的安装所造成的景观改变威胁着风力发电的进一步扩张。目前的时代，可再生能源是许多国家都在寻找的应对国家安全和气候变化风险问题的答案，解释上述反应的根源是非常重要的。本文着眼于四个不同背景下公众对风能发电进行抵制的相似之处：加利福尼亚州的棕榈泉，马萨诸塞州的科德角，苏格兰的刘易斯岛，和墨西哥的瓦哈卡州。尽管这些地方存在自然和文化的多样性，其已经历过的抵制中有五个共同的思路：固定性，资源位点的特异性；不变性，对景观持久性的期望；团结性，人与土地的紧密关系；强加性，被边缘化的感觉；及地方认同性，一种安全感的丧失。在开发可再生能源资源之前更深入地考虑土地和生活之间的关系，将有助于顺利走向更加可持续发展的未来，否则将会是一条不平坦的道路。*关键词：景观，地点，风能。*

Aunque la energía eólica es local, sustentable, barata y libre de carbono, la creciente oposición pública que se le hace por los cambios que aquélla produce en el paisaje está amenazando su expansión. En un tiempo en el que muchos países se encuentran buscando fuentes de energía renovable como respuesta a cuestiones de seguridad nacional y a los riesgos del cambio climático, es importante explicar las fuentes de esa reacción. Este artículo busca similitudes en la resistencia pública por desarrollos eólicos en cuatro escenarios diferentes: Palm Springs, California; Cabo Cod, Massachusetts; la Isla de Lewis, Escocia; y el Estado de Oaxaca, México. A pesar de las diferencias naturales y culturales entre estos lugares, existen cinco hilos comunes en la oposición que se ha experimentado: inmovilidad, la especificidad de sitio del recurso; inmutabilidad, una expectativa de permanencia del paisaje; solidaridad, la cercana relación entre la gente y la tierra; imposición, un sentido de marginación; e identidad de lugar, una pérdida de seguridad. Si se considerara más profundamente la relación entre tierra y vida, antes del desarrollo de recursos energéticos naturales renovables, ayudaría a limar lo que de otro modo sería camino lleno de asperezas hacia un futuro más sostenible. *Palabras clave: paisajes, lugar, energía eólica.*

Most visions of a sustainable future foresee a turn toward renewable energy, but it is a change that will not come without a fight. Wind energy is a case in point; in many places, reservations are mounting about how wind turbines change landscapes and our relationship with them. It is not a concern of small import; wind power is expanding faster than any other renewable energy resource, and it already has a significant presence in dozens of countries from Denmark to China. Growing at a yearly rate of 38 percent, by June 2010 wind installations reached a global capacity of about 175,000 megawatts (MW). These installations generate in excess of 340 terrawatt-hours (TWh) of electricity annually, about as much as forty-five large nuclear power plants (World Wind Energy Association 2010). Given its leadership position, the study of wind power might yield some clues as to what is in store for other renewable energy prospects as they begin to expand (Maloney 2008). In the case of many, as with wind, the central issues tend to revolve around land use conflicts, and geographers have been diligently trying to sort it all out (Pasqualetti 2001a, 2001b; Wolsink 2007a; van der Horst and Toke 2009; Devine-Wright and Howes 2010; Swofford and Slattery 2010; Warren and McFadyen 2010).

Wind's attraction stems from its wealth of advantages. It generates electricity carbon free and with no long-term wastes, cooling water is unneeded, turbines are simple to install on a variety of terrain, and the successful operations produce electricity both reliably and profitably. Despite these attributes, plans to increase wind power's share of the energy portfolio are encountering unexpected opposition. Consternation over the energy landscapes from hydro, nuclear, and fossil fuels have persisted for decades (centuries for coal), and now such misgivings are spreading to substitute fuels as they grow in popularity and importance. Taking wind power, for example, we find hundreds of antiwind groups of various stripes and intensities, turbines being burned in effigy in Scotland, rock-throwing resistance in Mexico, and people being killed during protests over wind projects in China (Ang 2005; Davies 2007; Hawley 2009; Bohn and Lant 2009; Penicuik Environment Protection Association 2010).

Responding to public disquiet over wind projects can take several forms, but several inexpensive and simple modifications are already common. For example, turbines are now routinely painted to blend in and be less noticeable in their environments, busy lattice towers have been giving way to sleeker monopoles, and various technical adjustments have reduced noise. These changes, however, will never appease everyone for a simple reason inherent in the resource: Whatever we do to make the wind turbines less conspicuous, we can do nothing to make them invisible. That, in a nutshell, is the problem. People see them, hear them, and even feel them, and in response they often reject them, a reaction that has become more common with their proliferation and their increasing size (Pasqualetti, Gipe, and Righter 2002; Agterbosch, Meertens, and Vermeulen 2007; Aitken 2010). Moreover, these reactions do not change with location, culture, economy, history, geography, or jurisdictional boundaries (Hinshelwood 2001; Pasqualetti 2001b; Pasqualetti, Gipe, and Righter 2002; Ang 2005; Szarka, 2007; Wolsink 2007a, 2007b; Wüstenhagen, Wolsink, and Bürer 2007; Eltham, Harrison, and Allen 2008; Graham, Stephenson, and Smith 2009; Moses 2009; Phadke 2010; van der Horst and Toke 2010; Wolsink 2010).

What causes such reactions? Motivations are not always clear, but sophisticated methods are being developed to identify them (Devine-Wright 2005a, 2005b; Ellis, Barry, and Robinson 2007; Torres-Sibille et al. 2009; Graham, Stephenson, and Smith 2009). The most popular explanation is NIMBY (not in my backyard), although this is increasingly considered too simplistic (Wolsink 2000; Devine-Wright 2005a; Ek 2005; van der Horst 2007). In some places, the focus of attention has been the impacts of wind turbines on birds and bats (Johnson et al. 2004; Saito 2004; Blum 2005; De Lucas, Janss, and Ferrer 2007; Kunz et al. 2007; National Research Council 2007; Lilley and Firestone 2008; National Wind Coordinating Collaborative 2008). Other times the concern has centered on potential interference with visual aesthetics, radar operations, property values, tourist attractions, and a sense of serenity (Moller 2006; Whitcomb and Williams, 2007; Ciardi and Crum 2009; Hoen et al. 2009; Lilley, Firestone, and Kempton 2010). Resistance in many communities accompanies the perception that wind projects are being imposed on them by outsiders (Hinshelwood 2001; Wolsink 2007a, 2007b).

This article seeks to identify principal explanations for public resistance to wind power developments by melding a review of the literature with brief case studies from four diverse settings. Of the four, the first two are relatively familiar: the California desert adjacent to Palm Springs, a two-hour drive east of Los Angeles; and the shallow waters between Cape Cod and Nantucket Island, Massachusetts, sixty miles southeast of Boston. The third study highlights a wind project proposed for peat-rich Isle of Lewis, Scotland, remotely located about 200 miles northwest of Glasgow. The fourth and last study is taken from the agricultural lands of coastal Oaxaca in southern Mexico, on the Pacific side of the Isthmus of Tehuantepec. Considered collectively, they should hold clues to the common causes for public resistance to wind energy landscapes.

Wind Turbines in Four Settings

A large part of the appeal of wind power is that it is in many ways environmentally benign, especially compared with its nonrenewable associates. This attribute often makes wind power attractive, particularly in countries that have committed themselves to reduce greenhouse gases. Add to these motivations that wind power offers other bonuses such as reliability and profitability and the attraction to the technology is no mystery.

Despite environmental and business advantages, however, public resistance to wind energy continues. The pattern began in California when large clusters of turbines were first installed there in the 1980s. Among

Figure 1. Wind turbines interfering with the view of Mt. San Jacinto, the western boundary of Palm Springs, California. Photo by the author, March 2006.

such installations were those in San Gorgonio Pass, immediately north of the resort city of Palm Springs. Once in place, the nearby machines were impossible to miss or abide (Pasqualetti 2001b). With hundreds of them concentrated along the principal routes into the city, business owners and public officials fretted that they would be considered eyesores by visitors seeking the calm, restful, and sophisticated lifestyles that their desert oasis tries to sell. Media attention rose quickly, lawsuits ensued, and research studies were commissioned (Pasqualetti and Butler 1987). All this attention came as a bit of a shock to wind developers, who considered the windy sites at the east end of San Gorgonio Pass so prone to sandstorms as to be inhospitable. Most presumed that no one would complain when the turbines went up.

That presumption might have been valid in other locations but not near Palm Springs. Sheltered from winds by the mass of Mt. San Jacinto to the west, the city is generally immune to blustery conditions. It is a place with beds of flowers lining the streets and snow-capped peaks serving as backdrop to over 100 nearby golf courses that extend eastward down the Coachella Valley. There are spas and healing centers, oases of palm trees, clear-water streams, world-class museums, and celebrities in number. From November to April, throngs arrive ready to relax, rejuvenate, and escape the cold that molds their lives back home. Given this setting, the last thing community leaders wanted was an industrial landscape that could interfere with the enjoyment of the visitors who were the backbone of the local economy (Pasqualetti and Butler 1987; Pasqualetti 2001b; Figure 1).

Opposition to wind landscapes near Palm Springs helped establish a pattern of public response that would show up elsewhere, most notably at the equally affluent recreational areas of the Eastern seaboard near Cape Cod. A 420-MW project called Cape Wind would install 130 turbines in the waters between the Cape and Nantucket Island. Indefatigable opposition surfaced early, and it has persisted much longer, over nine years at last count. The controversy has prompted innumerable meetings, protests, broadcasts, reports, articles, and books that chronicle the long tug-of-war between developers and preservationists (Kempton et al. 2005; Firestone and Kempton 2007; Whitcomb and Williams 2007; Firestone, Kempton, and Krueger 2009).

The primary objection to Cape Wind continues to be the visual change it would produce and the impact of such changes on the local economy. Save Our Sound (SOS), the leading opposition organization, has summarized its objections this way: "Occupying 25 square miles, an area the size of Manhattan, the Cape Wind project would be highly visible both day and night from Cape Cod and from the islands of Nantucket and Martha's Vineyard. The plant would dramatically alter the natural landscape" (SOS 2010; Figure 2).

Figure 2. Protest banner aimed at the planned installation of wind turbines in Nantucket Sound. Used with permission.

The location of the Cape Wind project near popular and prosperous communities has stimulated heightened attention, attention that at times has come from unexpected directions. For example, even Robert F. Kennedy, Jr., an ardent environmentalist, has expressed displeasure with the project (as did his late uncle, U.S. Senator Edward Kennedy), arguing that developers are "trying to privatize the commons" (Kennedy 2005). Despite the predominant concern about visual impacts, many other arguments have also been thrown into the mix, including threats to the health of marine life and birds, navigational safety, water quality, and infringement on ancient Indian burial sites. Indeed, just about any possible impact is accepted if it helps slow or defeat the project. Eventually, after long debate, the decision came across the desk of U.S. Interior Secretary Ken Salazar, who, in April 2010, ruled in favor of Cape Wind (Corcoran 2010).

The saga of Cape Wind has been tracked closely by both adherents and opponents of wind development (Agterbosch, Glasbergen, and Vermeulen 2007; Snyder and Kaiser 2009; Gee 2010; Meyerhoff, Ohl, and Hartje 2010). SOS, among other groups, is determined to fight on, hoping for a reversal. Part of their motivation is the fear that a successful Cape Wind project might open the floodgates to similar developments along the entire Atlantic Coast and the Gulf of Mexico (Wind Energy Systems Technology 2010). It could be the first of many similar projects, not just in the United States but abroad, where wind developments are also running into difficulties.

The United Kingdom can take pride in an attractive wind energy resource, among the best in Europe. A single modern turbine constructed at a reasonable site in the United Kingdom can generate 6.5 million kWh of electricity per year, enough to meet the needs of 1,400 households (or as the British Wind Energy Association [2010] phrases it, "run a computer for 2,250 years . . . or make 230 million cups of tea"). Among the many windy places in the United Kingdom, some of the best are in Scotland (Moran and Sherington 2007), and particularly in the remote Western Isles (the Outer Hebrides).

One such place is the Isle of Lewis. Wind development would seem an ideal use for Lewis and not just because of its ample resource. Despite 8,000 years of habitation, it has a sparse population of only about 25,000, giving developers a lot of open land to work with and not many people to please. It is often rainy and cold, the growing season is short, and it is a long way out in the Atlantic Ocean, as close to southeast Iceland as to southeast England. Given these isolated conditions, one might justifiably assume that a proposal for wind projects there would attract little attention. As the title of the protest group Mòinteach gun Mhuileann (Moorland without Turbines) suggests, however, that is not the case (Lewis Wind 2010).

One reason prospective wind projects on Lewis draw public ire is the worry that they will bring about a weakening of the cultural roots and conservative lifestyles that people have established there. It remains a simple place where livestock until recently commonly slept in the same house as their owners; where residents practice a fundamentalist form of Presbyterianism; and where Gaelic continues in use, alongside English. As with those following similar lifestyles elsewhere, residents of Lewis tend to instinctively resist change—especially blatant change—to the appearance of the land.

Resistance to the wind proposals on Lewis is tied in part to its large scale. The initial project would build 181 turbines with a total capacity of 651.6 MW, making it the largest wind installation in Europe, roughly twice the size of the mature wind complex near Palm Springs. Each 3.8-MW turbine intended for Lewis would reach a total height of 140 m (460 ft) and a rotor diameter of 107 m (358 ft), roughly equivalent to the wingspan of a 747 jetliner. In addition, the installation would require 200 transmission pylons and conductors, several new roads, and many construction platforms (Figure 3). As a reflection of its size, the project would generate up to 7 percent of Scotland's energy, enough to meet the average needs of 1 million people. In addition, it would satisfy 36 percent of Scotland's original 2010 renewable energy target of 18 percent electricity from renewable sources (Lewis Wind 2010). This, plus the profit motive,

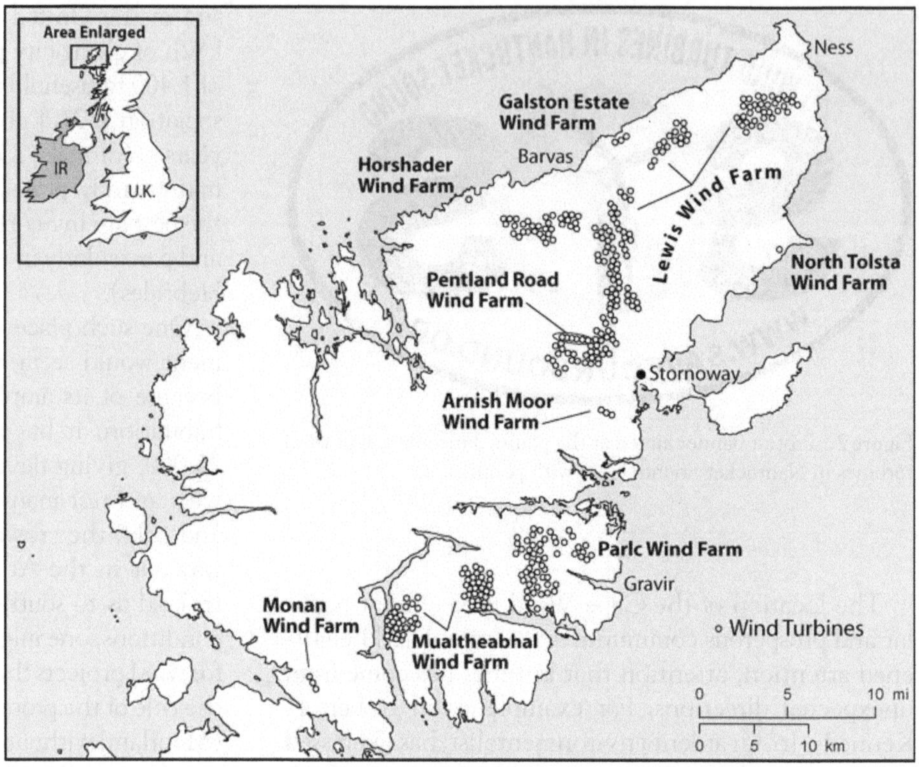

Figure 3. Distribution of proposed wind turbines on the Isle of Lewis, Scotland.

provides ample motivation to see the project through to full operation.

The controversy over the wind farms on Lewis followed hard on the heels of its initial announcement, and it continues to take many forms, just as it has in Cape Cod and Palm Springs. In addition to visual changes, opponents point out that it would have negative impacts on the economy, ruin the peat bogs, and threaten the integrity of some of the most impressive megaliths in Europe, the Callanish Standing Stones (Gray 2009; Figure 4). The other objection to the project, perhaps an overriding one, is that local residents hold little stake in its success (Ittmann 2005; Warren et al. 2005; Vidal 2006; Fisher and Brown 2009; Warren and McFadyen 2010). To those on the island and to many people elsewhere in the United Kingdom, the entire project seems inappropriate.

By April 2008, the Scottish political authority charged with making the final decision announced that they had received 10,924 letters of opposition and 98 letters of support. This overwhelming sentiment influenced the decision to deny the petition, although the official reason was the serious damage the project would cause to the Lewis Peatlands Special Protection Area, an area that is designated under the European Commission Birds Directive and protected under the EC Habitats Directive ("Lewis Wind Farm Refused" 2008; Fisher and Brown 2009). For those who reviled the intrusion of wind energy on their landscape, the rejection was more a reprieve than a victory, however; less than two years later, an amended—somewhat more modest plan—was submitted for regulatory approval (Lewis Wind 2010). No decision has yet been handed down. Resistance continues.

Although public resistance to wind energy landscapes has been attracting wide attention in many locations, similar resistance in the Pacific Lowlands of Oaxaca has been getting little notice (Figure 5). Such obscurity will not last. Planned development there would create the largest concentration of wind turbines in the world. More than 5,000 hectares of land have been reserved already in the windy municipalities of Juchitán de Zaragoza, Union Hidalgo, El Espinal, and San Dionisio del Mar. By 2012 the generating capacity proposed for these locations would total 2,500 MW (Secretaría de Energía 2009). This is roughly equivalent to the entire installed capacity in California, a mark that required twenty-five years to reach.

The Isthmus of Tehuantepec creates perfect conditions for wind projects. After crossing the open waters of the Gulf of Mexico, the winds come onshore and concentrate their power as they funnel through the narrowing topography on their move southward. Near

Figure 4. The Standing Stones of Callanish on the Isle of Lewis, Scotland.

places like La Venta, they spread out across a broad area of farmlands that are ideal for the erection of turbines (Figure 6).

The largest city in the area is Juchitán. Its history informs some of the reactions to the wind projects. Founded in 1486, it is now home to approximately 75,000, mostly Zapotecs and Huaves. It is also the seat of the Coalition of Workers, Peasants, and Students of the Isthmus, an influential popular movement that matured in the 1970s to meld into a single group of local socialists, peasants, students, and indigenous people. Favorable conditions for an agricultural economy and the relative autonomy that its location offers from the political influences of Mexico City have contributed to the formation of a long and close relationship between the people and the land (O'Connor and Kroefges 2008).

Oaxaca has a history of political unrest and activism. A revolt took place there in 1834. Life was again disrupted by the Mexican-American War in 1847. Less than twenty years on, the people of Juchitán defeated the French. When Porfirio Díaz was vying for a leadership position in Mexico, he populated his army mostly with citizens of Juchitán. In 1910, natives of the town organized in support of the revolutionaries Villa and Zapata. By 1980 the area again gained attention by electing a left-wing, prosocialist municipal government, the first Mexican community to do so in the twentieth century. In February 2001, Juchitán welcomed a caravan of the Zapatista Army of National Liberation.

Given their activist predilection, it is little surprise that those working the local communal farms (*ejidos*) do not welcome the proposals for large-scale wind development there. Initially, the nearby population held little understanding of what the projects would bring, but once the first phase of installations was completed, complaints started. Rather quickly, the penchant for activism evolved into clashes that have become increasingly frequent between locals and the plans to seed the fields with whirling machines of fiberglass and steel.

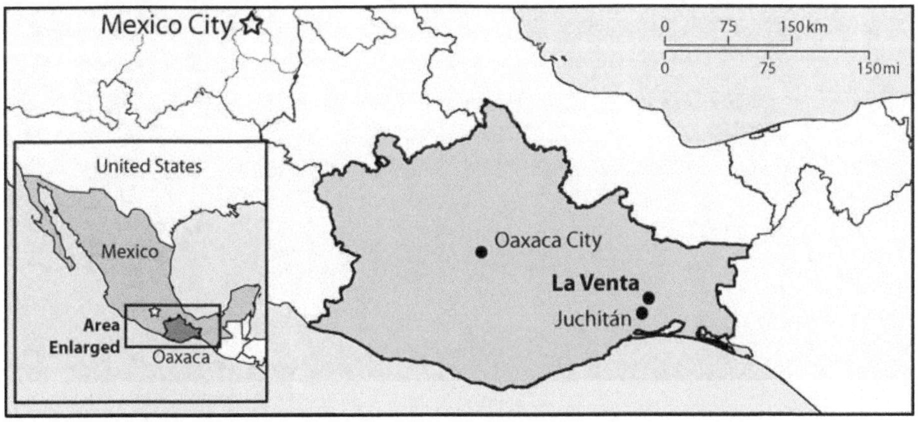

Figure 5. Location of wind turbines in lowland Oaxaca, Mexico.

Table 1. Summary of wind sites and their areas of concern

Location & Primary Characteristics	Visible Impacts	Focus of Economic Vulnerability	Lasting Economic Benefits	Significant Challenges to Cultural Values	Sense of Victimization
Palm Springs, California—Desert winter resort, affluent present use < 100 years	Yes, desert scenery	Recreation & tourism	Small	No	No
Cape Cod, Massachusetts—Shoreline summer resort, affluent, present use < 100 years	Yes, shoreline scenery	Recreation & tourism	Small	No	No
Isle of Lewis, Scotland—Remote, quiet lifestyle with long occupancy, present use > 500 years	Yes, cultural artifacts	Tourism (minor)	Small	Yes	Yes
Lowlands of Oaxaca, Mexico—Agrarian, long occupancy, political activism, present use > 500 years	Yes, non-specific	Farming	Small	Yes	Yes

Note: The two remote and more traditional sites (Lewis and Oaxaca) have greatest vulnerability to a sense of intrusion from wind power than the two more urban ones.

As with many energy projects, local development does not mean local benefit. Among the grievances is that people who live and work in the fields will receive meager lease payments from the projects, perhaps $125 per hectare per year for a single turbine (Sanchez 2007). Other estimates of compensation have been lower, from $98 to $117 per hectare (Hawley 2009). In comparison, U.S. wind turbines typically return $3,000 to $5,000 per year. Such perceived inequities helped prompt the formation of opposition organizations such as the *Grupo Solidario de la Venta*.

Now alert, local farmers and others are posing sharp questions to developers, regional politicians, national government officials, and representatives of the *Comisión Federal de Electricidad* (CFE), the federal electricity provider. At a public meeting in Juchitán in late May 2009, they asked why they are paid so little.[1] A similar question was asked by a reporter from *USA Today* at

Figure 6. Wind turbines at La Venta development within the preexisting agricultural fields. Photo by the author, June 2009.

Figure 7. Banner protesting the further installation of wind turbines in the Isthmus of Tehuantepec near La Venta. The banner reads: "If they plant these today, what will we be harvesting tomorrow?" Used with permission.

a public presentation at the Benjamin Franklin Library in Mexico City less than a week later (Hawley 2009). In both places, there was an evident awareness that residents in the area of La Venta were being expected to accept, with little compensation, historic changes to their familiar natural and cultural surroundings.

National Wind Watch, a public advocacy organization, explains the reservations of the local residents by suggesting that "the growing resistance to wind farm construction in southern Oaxaca . . . is based on local landowners' negative negotiating experiences with the CFE, discomfort with the broad freedoms seemingly granted to multinational corporations and an increasing concern about the possible environmental consequences of the wind farms themselves" (Sanchez 2007, 1). Judging by the complaints so far expressed, tensions have resulted, at least in part, from insufficient consultation between developers and the communities, an oversight that—as in other locations such as those described earlier—often leads to conflict (Walker et al. 2010).

Reactions have been more than words. Protestors have barricaded roads leading to the wind sites and have displayed antiwind banners (Figure 7). There have been incidents resulting in minor injuries. A local leftist farm group known as the Assembly in Defense of Land has complained about the treatment the *campesinos* have been receiving from wind power promoters, saying, "They promise progress and jobs, and talk about millions in investment in clean energy from the winds that blow through our region, but the investments will only benefit businessmen; all the technology will be imported . . . and the power won't be for local inhabitants" (Associated Press 2009). The group has called on its supporters to "defend the land we inherited from our ancestors" and say "no to the wind energy megaprojects in the isthmus that desecrate our lands and cultural heritage" (Sanchez 2007, 1).

Conclusions

This article has summarized public opposition to wind energy developments in four diverse settings (Table 1). Although in the aggregate the list of complaints against wind projects can be long, relatively few issues stand out consistently. The case studies presented here identified five.

The first core issue is immobility. Wind energy is site specific and must adjust to existing natural, cultural, and social conditions within a very narrow range of spatial options. The siting of turbines cannot be adjusted to less contentious sites without sacrificing productivity.

The second core issue is immutability. It is part of the human condition to believe that the landscapes with which we are most familiar, those that provide both our livelihoods and our greatest comfort, will not change over time. Such faith in "landscape permanence" is common in all cultures, as Jackson (1994) often reminded us, but few energy projects change a landscape as quickly and as fundamentally as a large collection of wind turbines.

The third core issue is solidarity. Knowing the intensity of the landscape changes that wind projects produce, development planning should integrate deeper understanding of the ties between land and life. The landscapes themselves can help tell the story if people stop long enough to "read" them. This suggests that those proposing to change landscapes long tilled and held dear should incorporate the advice of Mitchell (1994), who proposed that we think of them as more than just unoccupied swaths of nature. Rather, we should consider landscapes in terms of what they can tell us about the human condition. We should "be shifting the notion of landscape from an object to be seen or a text to be read to a process by which social and subjective identities are formed" (1).

The fourth core issue is imposition. It stems from the belief that such wind projects are someone else's idea, for someone else's benefit, and for someone else's profit. To one degree or another, local residences from desert to coastline, from Scotland to Mexico, were asked to bear costs for the production of something that would not flow to them directly and would not be in their best interest to support. In the instances of the Isle of Lewis and lowland Oaxaca, where life is simpler, the price being levied is greater still because of the wide gap between traditional ways of the past and unknown ways of the future. This gap broadens into a chasm when, as in Oaxaca, residents have no stake in the project planning or its success, and where they perceive that the entire intrusion is a continuation of their business-as-usual marginalization.

The fifth core issue is place. Wind energy projects, more than most others, are considered threats to place identity, something that is not only apparent in the four case studies presented here but is being identified elsewhere as well (Devine-Wright and Howes 2010). Ruptures in the composure of places of wind power development originate in an interference with place attachment, an attachment that originates from the accumulated affection and comfort people feel in maintaining their investment in the way the land is but not in the way it might become.

When we look at these four wind energy projects, they remind us of an uninvited guest who plants himself in our favorite easy chair; not only are we put out, but we feel a loss of balance. Yet the disaffection we might experience is more than a new example of a "machine in the garden," to use Marx's (1964) phrase. It is also more than a reaction just to the landscapes that wind turbines reshape. It is a response to the threat they pose to the way we fashion how we live. Although conflicts over wind development and renewable energy projects will not cease, considering more deeply the relationship between landscapes and the people who occupy and value them, in advance, will help smooth the otherwise bumpy road toward a more sustainable future.

Acknowledgement

I wish to thank Richard Crowell and Jacob Sowers for help on earlier drafts of this article.

Note

1. This meeting in Juchitán in May 2009 was among local residents, government and utility-company officials, representatives of the U.S. Embassy, and the author.

References

Agterbosch, S., P. Glasbergen, and W. J. V. Vermeulen. 2007. Social barriers in wind power implementation in The Netherlands: Perceptions of wind power entrepreneurs and local civil servants of institutional and social conditions in realizing wind power projects. *Renewable and Sustainable Energy Reviews* 11 (6): 1025–55.

Agterbosch, S., R. M. Meertens, and W. J. V. Vermeulen. 2007. The relative importance of social and institutional conditions in the planning of wind power projects. *Renewable and Sustainable Energy Reviews* 13 (2): 393–405.

Aitken, M. 2010. Why we still don't understand the social aspects of wind power: A critique of key assumptions within the literature. *Energy Policy* 38 (4): 1834–41.

Ang, A. 2005. Chinese village surrounded after shootings. http://www.activeboard.com/forum.spark?aBID=33011&p=3&topicID=5024635 (last accessed 25 February 2009).

Associated Press. 2009. Mexico looks beyond oil and toward the wind. http://www.msnbc.msn.com/id/28798851/ (last accessed 6 September 2010).

Blum, J. 2005. Researchers alarmed by bat deaths from wind turbines. *Washington Post* 1 January: A01. http://www.washingtonpost.com/wp-dyn/articles/A39941-2004Dec31.html (last accessed 4 April 2011).

Bohn, C., and C. Lant. 2009. Welcoming the wind? Determinants of wind power development among U.S. states. *The Professional Geographer* 61 (1): 87–100.

British Wind Energy Association. 2010. Onshore wind. http://www.bwea.com/onshore/index.html (last accessed 5 September 2010).

Ciardi, E. J., and T. Crum. 2009. Wind farm—Weather radar issues, current initiatives, and research results. Paper prepared for the State of the Art in Wind Siting conference, Washington, DC. http://www.nationalwind.org/assets/blog/Ciardi_NWCC_WashDC_Oct21_Final.pdf (last accessed 21 February 2010).

Corcoran, S. 2010. Fate of offshore wind farm in government's hands. http://www.wbur.org/npr/124010199 (last accessed 27 February 2010).

Davies, N. 2007. Organized chaos in Oaxaca: PFP evicts farmers to construct wind park on the Isthmus of Tehuantepec. http://www.narconews.com/Issue45/article2611.html (last accessed 4 September 2010).

De Lucas, M., G. F. E. Janss, and M. Ferrer, eds. 2007. *Birds and wind farms: Risk assessment and mitigation*. London: Quercus Books.

Devine-Wright, P. 2005a. Beyond NIMBYism: Towards an integrated framework for understanding public perceptions of wind energy. *Wind Energy* 8:125–39.

———. 2005b. Local aspects of UK renewable energy development: Exploring public beliefs and policy implications. *Local Environment* 10 (1): 57–69.

Devine-Wright, P., H. Devine-Wright, and F. Sherry-Brennan. 2010. Visible technologies, invisible organisations: An empirical study of public beliefs about electricity supply networks. *Energy Policy* 38 (8): 4127–34.

Devine-Wright, P., and Y. Howes. 2010. Disruption to place attachment and the protection of restorative environments: A wind energy case study. *Journal of Environmental Psychology* 30 (3): 271–80.

Ek, K. 2005. Public and private attitudes towards "green" electricity: The case of Swedish wind power. *Energy Policy* 33:1677–89.

Ellis, G., J. Barry, and C. Robinson. 2007. Many ways to say "No," different ways to say "Yes": Applying Q-methodology to understand public acceptance of wind farm proposals. *Journal of Environmental Planning and Management* 50 (4): 517–51.

Eltham, D. C., G. P. Harrison, and S. J. Allen. 2008. Change in public attitudes towards a Cornish wind farm: Implications for planning. *Energy Policy* 36 (1): 23–33.

Firestone, J., and W. Kempton. 2007. Public opinion about large offshore wind power: Underlying factors. *Energy Policy* 35:1584–98.

Firestone, J., W. Kempton, and A. Krueger. 2009. Public acceptance of offshore wind power projects in the United States. *Wind Energy* 12 (2): 183–202.

Fisher, J., and K. Brown. 2009. Wind energy on the Isle of Lewis: Implications for deliberative planning. *Environment and Planning* A 41:2516–36.

Gee, K. 2010. Offshore wind development as affected by seascape values on the German North Sea Coast. *Land Use Policy* 27:185–94.

Graham, J. B., J. R. Stephenson, and I. J. Smith. 2009. Public perceptions of wind energy developments: Case studies from New Zealand. *Energy Policy* 37:3348–57.

Gray, F. 2009. Wind farm threat looms over ancient Lewis stones. http://business.scotsman.com/alternativeenergysources/Wind-farm-threat-looms-over.5142997.jp (last accessed 26 May 2009).

Hawley, C. 2009. Clean-energy windmills a "dirty business" for farmers in Mexico. *USA Today* 17 June.http://www.usatoday.com/money/industries/energy/environment/2009–06–16-mexico-wind-power_N.htm?POE=click-refer (last accessed 22 July 2009).

Hinshelwood, E. 2001. Power to the people: Community-led wind energy-obstacles and opportunities in a South Wales valley. *Community Development Journal* 36 (2): 95–110.

Hoen, B., R. Wiser, P. Cappers, M. Thayer, and G. Sethi. 2009. The impact of wind power projects on residential property values in the United States: A multi-site hedonic analysis. Report No. LBNL-2829E. Ernest Orlando Lawrence Berkeley National Laboratory, Berkeley, CA.

Ittmann, N. 2005. Urgent petition to save Isle of Lewis from massive windfarm development. http://www.gopetition.com/online/7650.html (last accessed 25 February 2010).

Jackson, J. B. 1994. *A sense of place, a sense of time*. New Haven, CT: Yale University Press.

Johnson, G. D., M. K. Perlik, W. P. Erickson, and M. D. Strickland. 2004. Bat activity, composition, and collision mortality at a large wind plant in Minnesota. *Wildlife Society Bulletin* 32 (4): 1278–88.

Kempton, W., J. Firestone, J. Lilley, R. Tracey, and P. Whitaker. 2005. The offshore wind power debate: Views from Cape Cod. *Coastal Management* 33: 119–49.

Kennedy, R. F., Jr. 2005. An ill wind off Cape Cod. *New York Times* 16 December. http://www.nytimes.com/2005/12/16/opinion/16kennedy.html (last accessed 6 September 2010).

Kunz, T. H., E. B. Arnett, W. P. Erickson, A. R. Hoar, G. D. Johnson, R. P. Larkin, M. D. Strickland, R. W. Thresher, and M. D. Tuttle. 2007. Ecological impacts of wind energy development on bats: Questions, research needs, and hypotheses. *Frontiers in Ecology and the Environment* 5 (6): 315–24.

Lewis Wind. 2010. Moorland without turbines. http://www.mwtlewis.org.uk/ (last accessed 5 September 2010).

Lewis Wind farm refused. 2008. *Stornoway Gazette* 21 April. http://www.stornowaygazette.co.uk/news/Lewis-Wind-Farm-refused.4002752.jp (last accessed 25 August 2009).

Lilley, M. B., and J. Firestone. 2008. Wind power, wildlife, and the Migratory Bird Treaty Act: A way forward. *Environmental Law* 38 (4): 1167–1214.

Lilley, M. B., J. Firestone, and W. Kempton. 2010. The effect of wind power installations on coastal tourism. *Energies* 3:1–22.

Maloney, P. 2008. Solar projects draw new opposition. *New York Times* 28 September. http://www.nytimes.com/2008/09/24/business/businessspecia12/24shrike.html (last accessed 5 January 2010).

Marx, L. 1964. *The machine in the garden: Technology and the pastoral ideal in America*. Oxford, UK: Oxford University Press.

Meyerhoff, J., C. Ohl, and V. Hartje. 2010. Landscape externalities from onshore wind power. *Energy Policy* 38 (1): 82–92.

Mitchell, W. J. T. 1994. *Landscape and power*. Chicago: University of Chicago Press.

Moller, B. 2006. Changing wind-power landscapes: Regional assessment of visual impact on land use and population in Northern Jutland, Denmark. *Applied Energy* 83:477–94.

Moran, D., and C. Sherington. 2007. An economic assessment of wind farm power generation in Scotland including externalities. *Energy Policy* 35:2811–25.

Moses, S. 2009. Wind energy bad for West Virginia, Allegheny Front Alliance claims. *Cumberland Times-News* 15 July. http://times-news.com/archive/x1048579937 (last accessed 24 July 2009).

National Research Council. 2007. *Environmental impacts of wind-energy projects*. Washington, DC: National Academies Press.

National Wind Coordinating Collaborative. 2008. Wind and wildlife: Key research topics. http://www.nationalwind.org/assets/publications/NWCC_ResearchPriorities.pdf (last accessed 21 February 2010).

Oakes, B., and B. Swasey. 2010. Salazar's cape wind decision is difficult, for a consensus builder. http://www.wbur.org/2010/02/22/cape-wind-two-way (last accessed 27 February 2010).

O'Connor, L., and P. C. Kroefges. 2008. The land remembers: landscape terms and place names in Lowland Chontal of Oaxaca, Mexico. *Language Sciences* 30:291–315.

Pasqualetti, M. J. 2001a. Morality, space, and the power of wind-energy landscapes. *The Geographical Review* 90 (3): 381–94.

———. 2001b. Wind energy landscapes: Society and technology in the California desert. *Society and Natural Resources* 14 (8): 689–99.

Pasqualetti, M. J., and E. Butler. 1987. Public reaction to wind development in California. *International Journal of Ambient Energy* 8 (3): 83–90.

Pasqualetti, M. J., P. Gipe, and R. Righter. 2002. *Wind power in view: Energy landscapes in a crowded world*. San Diego, CA: Academic.

Penicuik Environment Protection Association. 2010. Auchencorth wind farm appeal rejected. http://www.auchencorth.org.uk/ (last accessed 4 September 2010).

Phadke, R. 2010. Steel forests and smoke stacks: The politics of visualization in the Cape Wind controversy. *Environmental Politics* 19 (1): 1–20.

Saito, Y. 2004. Machines in the ocean: The aesthetics of wind farms. *Contemporary Aesthetics* 2. http://www.contempaesthetics.org/newvolume/pages/article.php?articleID=247 (last accessed 22 February 2010).

Sanchez, S. 2007. Grassroots resistance: Contesting wind mill construction in Oaxaca. *National Wind Watch* 8 November. http://www.wind-watch.org/newsarchive/2007/11/08/grassroots-resistance-contesting-wind-mill-construction-in-oaxaca/ (last accessed 4 September 2010).

Save our Sound (SOS). 2010. Cape Wind threats: View. http://www.saveoursound.org (last accessed 1 September 2010).

Secretaría de Energía. 2009. Programa Especial para el Aprovechamiento de Energías Renovables [Special Program for the Development of Renewable Energy]. http://www.sener.gob.mx/webSener/res/0/Programa%20Energias%20Renovables.pdf (last accessed 5 January 2010).

Snyder, B., and M. J. Kaiser. 2009. A comparison of offshore wind power development in Europe and the US: Patterns and drivers of development. *Applied Energy* 86:1845–56.

Swofford, J., and M. Slattery. 2010. Public attitudes of wind energy in Texas: Local communities in close proximity to wind farms and their effect on decision-making. *Energy Policy* 38 (5): 2508–19.

Szarka, J. 2007. *Wind power in Europe: Negotiating political and social acceptance*. New York: Palgrave Macmillan.

Torres-Sibille, A., V.-A. Cloquell-Ballester, V-A. Cloquell-Ballester, and R. Darton. 2009. Development and validation of a multicriteria indicator for the assessment of objective aesthetic impact of wind farms. *Renewable and Sustainable Energy Reviews* 13 (1): 40–66.

van der Horst, D. 2007. NIMBY or not? Exploring the relevance of location and the politics of voiced opinions in renewable energy siting controversies. *Energy Policy* 35: 2705–14.

van der Horst, D., and D. Toke. 2010. Exploring the landscape of wind farm developments; local area characteristics and planning process outcomes in rural England. *Land Use Policy* 27:214–21.

Vidal, J. 2006. Take money or safeguard the land: Plans for world's biggest windfarm divide Lewis. *The Guardian* 20 July. http://www.guardian.co.uk/environment/2006/jul/20/energy.renewableenergy (accessed 4 September 2010).

Walker, G., P. Devine-Wright, S. Hunter, H. High, and B. Evans. 2010. Trust and community: Exploring the meanings, contexts and dynamics of community renewable energy. *Energy Policy* 38:2655–63.

Warren, C., C. Lumsden, S. O'Down, and R. Birnie. 2005. "Green on Green": Public perceptions of wind power in Scotland and Ireland. *Journal of Environmental Planning and Management* 48 (6): 853–75.

Warren, C. R., and M. McFadyen. 2010. Does community ownership affect public attitudes to wind energy? A case study from south-west Scotland. *Land Use Policy* 27:204–13.

Whitcomb, R., and W. Williams. 2007. *Cape Wind: Money, celebrity, energy, class, politics, and the battle for our energy future*. New York: Public Affairs.

Wind Energy Systems Technology. 2010. http://www.windenergypartners.biz/home.html (last accessed 24 February 2010).

Wind Watch. 2007. Grassroots resistance: Contesting wind mill construction in Oaxaca. http://www.wind-watch.org/news/2007/11/08/grassroots-resistance-contesting-wind-mill-construction-in-oaxaca/ (last accessed 25 August 2009).

Wolsink, M. 2000. Wind power and the NIMBY-myth: Institutional capacity and the limited significance of public support. *Renewable Energy* 21:49–64.

———. 2007a. Planning of renewables schemes: Deliberative and fair decision-making on landscape issues instead of reproachful accusations of non- cooperation. *Energy Policy* 35:2692–2704.

———. 2007b. Wind power implementation: The nature of public attitudes: Equity and fairness instead of "backyard motives." *Renewable & Sustainable Energy Reviews* 11:1188–1207.

———. 2010. Near-shore wind power—Protected seascapes, environmentalists' attitudes, and the technocratic planning perspective. *Land Use Policy* 27:195–203.

World Wind Energy Association. 2010. *World wind energy report 2009 executive summary*. http://www.wwindea.org/home/index.php (last accessed 5 September 2010).

Wüstenhagen, R., M. Wolsink, and M. J. Bürer. 2007. Social acceptance of renewable energy innovation: An introduction to the concept. *Energy Policy* 35:2683–91.

Burning for Sustainability: Biomass Energy, International Migration, and the Move to Cleaner Fuels and Cookstoves in Guatemala

Matthew J. Taylor,* Michelle J. Moran-Taylor,* Edwin J. Castellanos,† and Silvel Elías‡

**Department of Geography, University of Denver*
†Centro de Estudios Ambientales, Universidad del Valle de Guatemala
‡Facultad de Agronomía, Universidad de San Carlos

More than a century after the introduction of electric power transmission, almost 3 billion people still rely on biomass fuels to meet their energy needs. Use of this renewable fuel in unvented cooking stoves results in disastrous consequences for human health and global warming. These negative outcomes have led governmental and nongovernmental organizations (NGOs) to push for improved wood-burning stoves and cleaner burning, but nonrenewable, alternatives like liquefied petroleum gas (LPG). The move up the energy ladder to cleaner fuels and improved stoves is thought to be associated with rising income and increased levels of urbanization. Increased income in developing countries often comes in the form of remittances from millions of migrants working abroad. Thus, migrants and their money could arguably be agents of change in the transition to cleaner fuels or the more efficient use of existing renewable energy sources. This article examines the case of Guatemala, where 88 percent of rural households use firewood for cooking, and where almost 15 percent of the country's 14 million population migrates to the United States. A continued preference for firewood, despite increased income, can be explained as a rational decision based on cost, experience, and cooking methods. Additionally, through an analysis of forest cover in firewood source areas, we demonstrate that this energy source is, for the most part, used in a fashion that makes it renewable. Recognizing these patterns of, and reasons for, this resource use permits us to make realistic recommendations for sustainable livelihoods and use of this renewable energy source. *Key Words: cookstoves, energy, firewood, Guatemala, migration.*

电力传输技术的引入已经超过一个世纪，大约 30 亿人仍然依靠生物燃料来满足其能源需求。使用这种可再生燃料的无排气管烹饪炉具对人类健康和全球变暖产生灾难性的后果。这些消极的结果导致政府和非政府组织（NGO）致力于推动改善烧木柴的炉子和清洁燃烧，使用不可再生的能源替代品，例如液化石油气（LPG）。向上移动到更清洁的燃料和改良炉灶的能量阶梯被认为是与收入的增加和城市化水平的提高相关的。发展中国家收入的增加通常具有下述形式：数百万在海外工作的移民汇款。因此，移民和他们的钱也许可以说是过渡到更清洁燃料或者更有效地利用现有再生能源这一改革的推动者。本文考察了危地马拉案例，该国百分之 88 的农村家庭使用木柴做饭，全国 1400 万人口里迁移到美国的几乎有百分之 15。尽管收入增加，对木柴的继续偏好可以被解释为在成本、经验和烹调方法基础上的合理决策。此外，通过对木柴源区森林覆盖的分析，我们证明了这种能量来源在大部分情况下是以可再生的方式使用的。认识到这些资源利用的模式以及这些模式的产生原因，可以使我们对可持续生计和可再生能源利用提出现实的建议。*关键词：锅灶，能源，木柴，危地马拉，迁移。*

Más de un siglo después de que se introdujera la transmisión de energía eléctrica, casi 3 mil millones de personas aun dependen de combustibles de biomasa para satisfacer sus necesidades energéticas. El uso de este combustible renovable en estufas para cocinar sin desfogue termina en desastrosas consecuencias para la salud humana y en calentamiento global. Estos resultados negativos han llevado a organizaciones gubernamentales y no gubernamentales (ONGs) a presionar por el uso de estufas mejoradas a base de la quema de madera y por alternativas de combustión más limpia, aunque no renovables, como el gas de petróleo licuado. El ascenso en la escala de la energía hacia combustibles más limpios y estufas mejoradas se toma como asociada con mejores ingresos y niveles incrementados de urbanización. El aumento del ingreso en los países en desarrollo viene a menudo en la forma de remesas de millones de migrantes que trabajan en el extranjero. Así pues, los migrantes y su dinero podrían ser considerados como agentes de cambio en la transición hacia combustibles más limpios o al uso más eficiente de las fuentes existentes de energía renovable. Este artículo examina el caso de Guatemala, donde el 88 por ciento de los hogares rurales usan leña para cocinar y donde casi el 15 por ciento de los 14 millones de habitantes del país migra a los Estados Unidos. Una preferencia continuada por la leña, a pesar de la mejora del ingreso, puede explicarse como una decisión racional basada en costo, experiencia y métodos de cocinar. Adicionalmente, a través de un análisis de la cubierta de bosque en las áreas de donde procede la leña, demostramos que esta fuente de energía es, en su mayor parte, usada de una manera que la hace renovable. Reconociendo estos patrones del uso de este recurso, y las razones para hacerlo, nos permite hacer recomendaciones realistas de medios de vida sustentables y el uso de esta fuente de energía renovable. *Palabras clave: estufas para cocinar, energía, leña, Guatemala, migración.*

Almost 3 billion people in the developing world rely on biomass fuels to meet their household energy needs, accounting for 10 percent of all human energy use and 78 percent of the global supply of renewable energy. Biomass fuels include firewood, charcoal, dung, and crop residues (Granderson et al. 2009; Jetter and Kariher 2009). Even though many biomass users will transition to fuels like kerosene, liquefied petroleum gas (LPG), and electricity, most analyses predict that the total number of biomass users will increase over the next four decades (Barnes et al. 1994; International Energy Agency 2006; Legros et al. 2009). Biomass burning for cooking results in high levels of indoor and outdoor pollution. These emissions have implications for a number of important and interrelated aspects of development, including human health, natural resource use, climate change, and household economy. We touch on some of these aspects to frame our study.

Results from longitudinal epidemiological studies exploring the relationship between biomass burning and human health demonstrate the causal link between biomass smoke and the deaths of more than 1.6 million people annually and the development of chronic obstructive pulmonary disease, which is responsible for 2 percent of the global burden of disease (Smith-Sivertsen et al. 2009; Northcross et al. 2010). In addition to health impacts, recent studies link biomass cookstoves to global warming. Biomass cookstoves not only emit carbon dioxide (CO_2) but also other products of incomplete combustion, such as carbon monoxide, methane, nonmethane hydrocarbons, nitrous oxide, oxides of nitrogen, particulate matter, organic carbon, organic matter, and black carbon (MacCarty et al. 2008). Black carbon (BC), recent research reveals, is the number two contributor to rising global temperatures after CO_2. Researchers believe that BC from various sources is responsible for 18 percent of global warming, whereas CO_2 is responsible for 40 percent of global warming. CO_2 has a lifetime in the atmosphere on the centuries to millennial timescale, whereas BC's lifetime is less than a few weeks. Thus, because of their shorter life span, reduction in the output of BC and other short-lived climate forcers (SLCF) has the potential to offset CO_2-induced warming for several decades, giving policymakers time to develop and implement effective measures to reduce CO_2 emissions (Bond, Venkataraman, and Masera 2004; Ramanathan and Carmichael 2008; Gustafsson et al. 2009).

The widespread use of biomass and knowledge about its effects on human health and the environment illustrate why the study of biomass is important and also why many institutions around the world attempt to provide biomass users with fuel-efficient and cleaner burning biomass stoves (Ezzati and Kammen 2002; Berrueta, Edwards, and Masera 2008). Simply, more efficient and chimney-vented stoves use less fuel, reduce the amount of time and money spent on fuel acquisition, save lives because the harmful products of combustion are vented outside, and reduce the emission of SLCFs like BC. The scale at which biomass fuel is used around the world also explains why so much research focuses on the factors that determine household fuel and stove choice. Researchers suggest that the ascent of the energy ladder to cleaner fuels and improved stoves is associated with rising income and increased levels of urbanization (Heltberg 2004, 2005; Edwards and Langpap 2005; Madubansi and Shackleton 2007). Increased income often comes from millions of migrants who live and work abroad and inject billions of dollars into their home countries. Thus, migrants and their money could arguably be agents of change in the transition to cleaner fuels or the more efficient use of existing renewable energy sources. This article examines the relationship between fuel use and income in Guatemala, where about 15 percent of the population migrates to the United States and where biomass makes up 52 percent of the national energy budget.

In this article we investigate how migration influences cooking fuel choice in Guatemala. Do migrant households transition to other cooking fuels like LPG, use various fuels, or continue using firewood? Also, we explore the landscapes that result from the annual increase in the use of firewood (although firewood's contribution to Guatemala's total energy budget declines every year, the amount of firewood used continues to increase). Are these energy landscapes renewable (Zimmerer 2011), or is the increased demand for firewood driving deforestation as suggested in the "fuelwood crisis" literature of the 1970s and 1980s (e.g., Eckholm 1975; Openshaw 1978; Dewess 1989)?

We offer a case study from a Maya community in Guatemala's Western Highlands. Our discussion begins with a review of the literature surrounding the fuel and income ladder theory, the impacts of firewood consumption on forests, and the relationships between migration and the environment. We also reveal the importance of firewood to Guatemala's energy budget and environment. Then, we provide outcomes of the research, and finally discuss the ramifications of the results in terms of renewable energy use in the developing world.

The Move to Cleaner Cooking Fuels and Improved Biomass Stoves

Modern cooking fuels like LPG and improved biomass stoves provide significant health, environmental, and productivity benefits (Boy et al. 2000; MacCarty et al. 2008). Yet understanding the social, cultural, and economic behaviors involved with users making the fuel and stove switches is more complicated. One theory suggests an "energy ladder" where each of the three rungs corresponds with income levels, and the energy rungs rise as income rises. Biomass sits at the bottom; transitional fuels such as kerosene, charcoal, and coal occupy the middle rung; and fuels like LPG and electricity characterize the top rung. The energy ladder implies that moves are made from inferior to superior fuels. Moreover, this theory uncritically places firewood—the most widely used renewable resource in the world—at the bottom of the ladder (Arnold, Kohlin, and Persson 2006). The reality with many households in the developing world, however, is far more complex because they often use multiple fuels at the same time. This is referred to as the *household fuel mix* or *portfolio*. Fuel portfolios depend on multiple factors, including culture, household size and age structure, price, opportunity cost, fuel availability, precipitation regimes, and variations in household economy and labor (Heltberg 2004; Moran-Taylor and Taylor 2010).

Many governmental and nongovernmental organizations (NGOs) recently realized that there is no silver bullet to create a fast transition to modern fuels and thus focused their efforts on the fuel that billions of households already use—biomass. Take the case of India. In 2009, recognizing that a renewable resource was already part of the lifestyles of its citizens, the Indian government launched the National Biomass Cookstove Initiative, which aims to provide an affordable and reliable clean cooking energy option for the poorest households that rely on biomass. The new options revolve around stoves that burn biomass (Venkataraman et al. 2010). At the global scale, the Global Alliance for Clean Cookstoves (GACC), launched in September 2010, intends to provide 100 million households with cleaner biomass-burning stoves by 2020.

Firewood and Forests: Degradation or Sustainable Use of a Renewable Resource?

Firewood use in the developing world has often been linked to deforestation. Early research predicted massive energy crises as a result of increased firewood consumption (e.g., Eckholm 1975). More recent research shows that demand for woodfuel is unlikely to result in large-scale deforestation. Case studies and models demonstrate that tropical deforestation and changes in forest cover in general have multiple, rather than single causes (Allen and Barnes 1985; Geist and Lambin 2002; Arnold, Kohlin, and Persson 2006; Hecht and Saatchi 2007). Local demand for firewood, however, can result in changes (degradation and improvement) in local forest composition and extent (McCrary, Walsh, and Hammett 2005; Madubansi and Shackelton 2007). In short, in some cases firewood use results in local or distant forest degradation (depending on local control of forests and the network of improved roads to tap more distant forests) or it can result in the increase of forested areas, especially in the form of energy forests, which are tree plantations specifically for energy purposes (Arnold, Kohlin, and Persson 2006).

International Migration, Development, and the Environment

Around the world, economic remittances (the monies that migrants send to their countries of origin) transform household and national economies. For example, in 2007 the amount of cash remittances reaching developing countries dwarfed official development assistance and were about half as large as both net inflows of foreign direct investment and private debt (Gabriel 2008). In Guatemala, remittances amounted to nearly US$4 billion in 2010. This amount is equivalent to one fifth of Guatemala's gross domestic product (Banco de Guatemala 2010). Clearly, such funds are key for the economies of sending countries and much scholarly and policy work examines their role in development (e.g., Rhoades 1978; Georges 1990; Grasmuck and Pessar 1991; Rubenstein 1992; Cohen 2001; Jokisch 2002; Taylor, Moran-Taylor, and Rodman 2006; Moran-Taylor 2008; de Haas 2010). Research that examines the linkages between migration and the environment is now gaining momentum. Hugo (1996) reviewed scholarship on the environmental factors (both proximate and distant) that force human migration. Other recent research reveals how droughts, landslides, and hurricanes contribute to outmigration and create environmental refugees (Organización Internacional para las Migraciones 2008). But, we need to pay more attention to how migration feeds back into the environment and land use in the countries of origin of migrants. Aptly, the National Research Council (1999) states:

> There is some agreement that, in the future migration, rather than changes in human fertility and mortality, will be the key demographic link between the two dynamic processes of land use and land cover changes. Causation and feedback will probably move in both directions: environmental changes will likely cause migration, and migration will likely change the environment. . . . Data on migration and other social variables must be linked with biophysical data from remote and land-based sources on soils, climate, and other biophysical factors. (349)

Research examining the relationships between migration and the environment in Latin America shows how return migrants and cash remittances play a pivotal role in sending countries. Bilsborrow (1992) illustrates how return migrants in Ecuador can be linked to lower levels of deforestation. In the Caribbean islands some migrants bring back new notions about ecosystem services and commitments toward environment preservation and invest in the formation of nonprofit organizations for that purpose (Conway and Lorah 1995). Recent studies show how international migration can lead to forest recovery (Rudel, Perez-Lugo, and Zichal 2000; Hecht and Saatchi 2007; Schmook and Radel 2008). The interaction of migration and the environment in sending countries merits continued attention by social and natural scientists because migration makes up, and will continue to make up, an integral part of culture and economies in both the developed and developing world.

Cooking, Fuel Consumption, and Forest Change in Guatemala

Guatemala, with 45 percent of its population living in urban areas, is one of the least urbanized countries in Latin America. Urbanization is on the increase in Guatemala, and recent estimates place the rate of change at 3.4 percent per year (Brunn et al. 2008). Guatemala also has the highest natural rate of increase in population in Latin America at 2.8 percent per year (Population Reference Bureau 2011). Biomass makes up 52 percent of the national energy budget, and 75 percent of households in Guatemala use this renewable fuel as part of their energy portfolio. In rural areas, about 88 percent of households exclusively use wood for cooking and space heating. Moreover, the total amount of firewood consumed every year continues to grow (Elías et al. 1997; Taylor 2005; Programa de Las Naciones Unidas para el Desarollo 2008). LPG is available in Guatemala. From a unit of energy perspective, LPG is cheaper than firewood at current prices. Despite this cost disadvantage, most rural households, and even 46 percent of urban households, continue to use firewood. The continued reliance on firewood can be explained, in part, by the high startup costs (US$110 in 2005) for stoves and cylinders involved in a transition to LPG (Edwards and Langpap 2005). Despite the importance of firewood to the national energy budget and to the households who burn wood, it is not taken into account in national budgets because of the informal nature of the firewood economy. In fact, the true size of the firewood economy is not known in Guatemala. The Guatemalan National Institute of Forestry (INAB) estimated the value of the firewood economy at US$1.5 billion for 2006 (Martínez 2009).

In Guatemala, firewood consumption is on the increase and forested areas are decreasing. Early research by the Food and Agricultural Organization placed forest cover at 53 percent of the total area of the country in 1988. A later study conducted by the Ministry of Agriculture in 1999 placed forest cover at 40 percent. In 2001, a group of institutions produced a forest-change map comparing images of three years (1991, 1996, and 2001). They reported a deforestation rate of 1.4 percent, equivalent to 73,000 hectares of forest lost per year. Two thirds of that deforestation happened in Petén (the northern one third of the country) where there is a rapid expansion of the agricultural frontier and the establishment of large cattle ranches (Universidad del Valle Guatemala [UVG], Instituto Nacional de Bosques [INAB], and Consejo Nacional de Areas Protegidas

[CONAP] 2006). If, in countries like Guatemala, we are at the incipient stages of understanding the value and impacts of the fuel used for cooking by three quarters of households in the country, it is important to interrogate how other critical issues, like migration and subsequent changes in household income, have the potential to alter firewood consumption patterns.

Study Site: San Cristóbal Totonicapán

The municipality of San Cristóbal Totonicapán (hereinafter San Cristóbal), with a population of almost 30,000 inhabitants, lies in the heartland of the Maya K'iche' indigenous culture in Guatemala's Western Highlands (Figure 1). San Cristóbal is located in the department of Totonicapán, which is recognized for its forested areas and for local control of those forested areas (Veblen 1978; Elías et al. 1997). In the last three decades, the municipality gradually moved away from subsistence agriculture, local furniture manufacturing, and weaving of textiles to a more diverse economy that includes intensive vegetable production for local and international markets (Moran-Taylor and Taylor 2010). Like many other rural towns in Guatemala, at

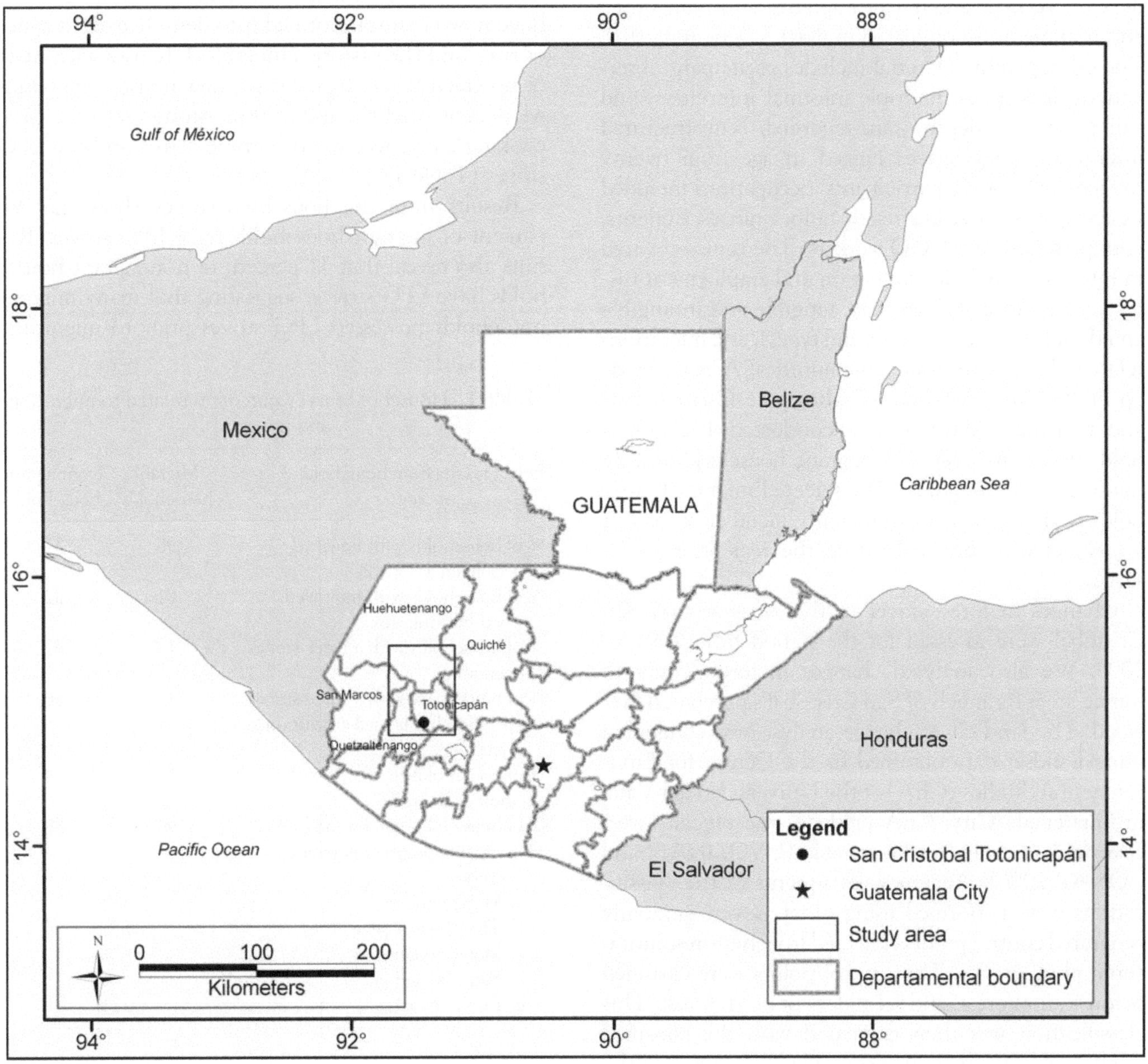

Figure 1. Location of the study area in Guatemala.

least 15 percent of San Cristobal's residents reside and work in the Unites States. Most migrants remit part of their earnings back home on a regular basis (US$200–$300 per month), and many return home after several years of work abroad. Migrants' achievements are clearly evident in their building of large houses (Moran-Taylor 2008, 2009), yet these monuments to success do not tell us anything about the fuel used in migrant kitchens.

Methods

This article relies on ethnographic work and survey data among male and female migrants, return migrants, and nonmigrants conducted in 2001, 2006, and 2010. The ethnographic material includes participant observation, field notes, multiple informal interviews, and thirty-seven in-depth, tape-recorded, semistructured interviews. Interviewees ranged in age from twenty to eight-two. Study participants' occupations included weavers, artisans, seamstresses, tailors, nurses, students, entrepreneurs, and NGO workers. The topics covered in interviews included migration and employment history, transnational flows (e.g., tangible and intangible and their frequency, density, and types), and migration-related changes in home communities. A team made up of the authors (three of whom are Guatemalan) and two local Maya women conducted 102 household surveys in 2006 (77 migrant household and 25 nonmigrant households). We selected migrant households that had at least one family member abroad or a member who returned within the year prior to the survey.

Changes in forest cover in the areas around San Cristóbal were assessed for the period from 1991 to 2006. We also analyzed changes in forest cover in source areas for much of San Cristobal's purchased firewood. The land cover change analysis was conducted using Landsat data obtained by the Center for Environmental Studies (CEA) at the Universidad del Valle in Guatemala City. CEA produces the official forest cover change maps for Guatemala (UVG, INAB, and CONAP 2006). Accuracy assessments of the classifications were performed using ninety-seven randomly generated control points collected from high-resolution aerial photographs. The control points were classified as forest, nonforest, and secondary or brush forest. This classification was then compared with the classified satellite image. The points coincided 85 percent of the time. The points classified as forest and nonforest coincided 95 percent and 100 percent of the time, respectively. Secondary brush or forest had a lower coincidence (60 percent), which lowered the overall accuracy.

Results and Discussion: Migration and Firewood Use in Guatemala

San Cristobaleños report that migrants use remittances to build new homes, buy more land, send their children to better schools, start businesses, buy vehicles, and invest in agriculture. This behavior is common in Guatemala (e.g., Camus 2007; Falla 2008). Whole neighborhoods that in the past were one-level adobe homes interspersed with forest patches and maize plots have now been transformed into densely packed zones of two- and three-story, cinderblock houses with little or no green space. Inside these new house compounds we documented the use of three-stone fires (open-fire cooking), improved wood-burning stoves, and new LPG stoves (Table 1).

Results from the household survey show that 98 percent of migrant households have LPG stoves. Results also reveal that 31 percent of nonmigrant households have LPG stoves, suggesting that many migrant households possessed LPG stoves prior to migrating.

Table 1. Household survey outcomes related to migration and firewood

Select results from household survey[a]	Migrant households[b]	Nonmigrant households[c]
% of households with liquefied petroleum gas stoves	98	31
% of households with improved wood-burning stoves	81	22
% of households with a three-stone open fire	12	67
% of households with a three-stone open fire that is used at least once a month	12	36
% of households that cook with wood "most of the time"	77	94
% of households purchasing wood	94	88
Municipality origin of purchased wood		
a. Malacatancito	35%	
b. The coastal coffee farms	21%	
c. Momostenango	3%	
d. San Carlos Sija	5%	
e. Don't know	27%	

[a] $n = 102$.
[b] $n = 77$.
[c] $n = 25$.

Although almost all migrant households have LPG stoves, 77 percent of migrant households continue to do most of their cooking with wood (Table 1). Why? In 2009 and 2010 a simple three-burner LPG stovetop and 25-lb. gas cylinder sold for US$110. Filling a 25-lb. tank cost $13 and the cost to fill a 100-lb. tank was $52. Most households purchase 25-lb. tanks. A *tarea* of wood (one tarea of firewood measures 1 m high × 5 m long × 35 cm wide) costs between US$25 and $30 depending on the type of wood and delivery options. Although LPG is a more economical choice on paper and in terms of energy efficiency per unit of energy than wood (Edwards and Langpap 2005), most migrant households continue to use wood as their primary fuel. Our survey results indicate that more migrant households purchase firewood (94 percent vs. 88 percent) versus collecting their own wood, which is an onerous task. Overall, more so than nonmigrant households, migrant households use multiple fuels and cooking methods. This points to fuel stacking or mixing, rather than a clean transition from one fuel to another. We also found that 81 percent of migrant households build and use improved wood-burning stoves (Table 1). These stoves are better for human health because they channel smoke out of the cooking area using a chimney. If used properly, these stoves are also more fuel efficient and reduce emissions (Granderson et al. 2009). Thus, even if migrants are not making a quick transition to LPG, their use of improved wood-burning stoves bodes well for human and environmental health. The responses we received about the types of energy use and cooking methods mask the complexity of the energy mix that households employ. The amount of each fuel used depends on household factors that include economy, time, and cooking preference.

The greater ease with which migrant households can purchase firewood does not explain why they do not make a complete switch to LPG. Like Heltberg (2005), we realized that we needed to look beyond cost factors to understand why so many people continue to use firewood. Indeed, culture often proves a wonderful confounding variable. But, good anthropological field work and years of participant observation often help solve the conundrum of what seems like nonrational behavior from a purely economic perspective (i.e., examining the cost per calorie of different fuels). San Cristobaleños explain that they continue to cook with wood for several reasons. Many offer comments such as, "The food tastes better when cooked on firewood and for longer periods slowly," or, "You can always keep hot water boiling/warm" when using firewood. Additionally, because Guatemala's two staple foods, beans and corn, require many hours of cooking, gas becomes more expensive (e.g., corn is cooked twice: once for boiling and again after taking it to the mill to make tortillas or tamales). Many locals also say that using propane gas for these long processes is too expensive, so they prefer to rely on firewood. Moreover, the burners on conventional gas stoves cannot generate enough power to handle the quantity of food cooked and the size of the pots used for cooking corn, beans, and tortillas. Also, interviews reveal that the rising price of LPG makes wood a much more attractive alternative as a domestic fuel choice.

In our surveys we also asked residents to name the source of their firewood. Large trucks trundle through San Cristóbal on a weekly basis selling firewood from various nearby municipalities and distant coffee and rubber plantations (Table 1). Foresters in Totonicapán (which surrounds San Cristóbal) are famous for their management of their high pine, fir, and mixed pine and oak forests. Indeed, while the rest of Guatemala undergoes rapid deforestation, the forests of Totonicapán survive because local laws and indigenous forest managers closely regulate use of these forests (Veblen 1978; Elías and Wittman 2005). Firewood is collected from the local forests, but it is done within the norms, rules, supervision, and sanctions established by the users themselves and the local government. Thus, families can only collect firewood if they are members of that community, if they have completed their community service and forest maintenance obligations, and if the firewood is not for sale and is used for domestic purposes only. Added to this regulation of forests, which varies in strictness from community to community, there is also an increase in conservation discourse among forest managers. This has led community leaders to ban firewood collection from older forests in protected areas and establish some forest areas for carbon fixation projects. This new regulation has impacted the poorest of families who relied on the now-protected forest for the collection of dry wood from the forest floor (Elías and Wittman 2005; Elías 2011).

Because use of these forests is restricted, residents of San Cristóbal and other Totonicapán communities must purchase their fuel from more distant or private forest sources (Table 1). For example, about 35 percent of respondents say that the firewood they buy comes from Malacatancito, about 50 km to the north of San Cristóbal via the Pan-American Highway. The purchase of fuelwood, studies from around the world state, often reflects local fuelwood scarcities (Madubansi and Shackleton 2007). Analysis of satellite data from 1991 to 2006 reveals that in all studied municipalities, forest

Table 2. Forest cover for the years 1991, 2001, and 2006 in hectares for seven municipalities of the Western Highlands in Guatemala

Municipality	Area in hectares (ha.)	Forest 1991 ha.	Forest 2001 ha.	Forest 2006 ha.
Huehuetenango	18,953	8,170	8,359	8,135
Malacatancito	41,193	19,529	18,721	18,267
Momostenango	35,915	17,366	17,062	17,164
San Andrés Xecul	1,649	320	338	337
San Bartolo Ag Ca	5,636	3,468	3,301	3,366
San Carlos Sija	22,651	10,402	10,689	10,624
San Cristóbal Toto	4,425	737	792	783

cover is roughly 50 percent of the municipal land except for the two smallest municipalities of San Andrés Xecul and San Cristóbal Totonicapán, where forest cover is less than 20 percent (Table 2). The national average for forest cover is around 40 percent, so these two municipalities are considerably more deforested than the rest of the country. Malacatancito is the only municipality that shows a steady decline in its forest cover over the fifteen years that the data span (Table 2). The analysis also revealed that most of the change within Malacatancito is found in the northern part of the municipality on the border with Huehuetenango (Figure 1). The rest of the municipalities show a smaller change in direction, increase, or decrease of forest cover for the period studied; but in many cases, the changes are relatively small and within our range of error, indicating a fairly stable forest cover. This stability in forest areas outside of the municipality of Malacatancito can be attributed to the fact that these forests do not serve as principal firewood source areas.

Firewood for the stoves of San Cristóbal must come from somewhere. Much of the firewood comes from the forests of Malacatancito. Malacatancito is a community that has a long tradition of harvesting trees for firewood sale, thus the decline in forest coverage around Malacatancito is not a direct result of increased regulation of Totonicapán forests. Our data, however, suggest that some of the decline in the forest area of Malacatancito can be attributed to the fact that San Cristobaleños do not get their firewood from forests in their immediate vicinity. This conclusion suggests that the forests of Malacatancito are not managed in a sustainable manner. We must note, though, that in our analysis of forest cover through time, secondary growth and brush demonstrated the lowest coincidence levels in the error analysis. This might mean that the analysis is not capable of capturing regrowth of forest in its early stages in the Malacatancito area.

About one fifth of households in San Cristóbal report that they purchase firewood from coffee plantations on the Pacific Slope. This practice is becoming more common as more Guatemalans move into towns and small cities where they no longer have direct access to free firewood. Purchase of wood from any source in Guatemala has increased dramatically as forest area decreases and competition over resources increases. Indeed, owners of large coffee, cattle, and rubber estates that we have interviewed in the course of our field work have stated that theft of firewood from their land has increased in the last decade. Large coffee plantations trim their shade trees every year to ensure optimal shade for the growth of the coffee bushes. Likewise, wood from coffee bushes makes its way into the energy market when the bushes are drastically cut back every ten to fifteen years. If coffee continues as a valuable cash crop and it is grown under shade trees, these plantations could prove to be sustainable sources of firewood.

Conclusions

Migrant households in San Cristóbal, for the most part, continue to use firewood for cooking. The results of this study add evidence to the growing number of studies that call for a nuanced fuel and income ladder theory that includes variables beyond income (Arnold, Kohlin, and Persson 2006; Maconachie, Tanko, and Zakariya 2009). Instead of moving up the ladder, migrants with increased income make rational decisions about their cooking fuel purchases and use a mix of fuels. Also, although migrants can afford LPG and indeed purchase LPG stoves, they rarely use these stoves because the adoption of this new fuel also requires changes in food preparation traditions. In current and pending discussions about the adoption of improved cookstoves by entities like the GACC, it is important to note that adoption of improved wood-burning stoves by migrants illustrates how increases in income and access to credit can lead to the adoption of more efficient stoves. This is an important revelation as these institutions move ahead with their plans to provide 100 million households with improved biomass cookstoves.

In this study we hesitate to make a determination on the sustainability of this renewable resource in Guatemala because wood is sourced from many areas in both sustainable and nonsustainable ways. Our analysis does show, however, that forest area has declined in an area that is cited as a significant firewood source area.

Future studies on firewood use must take into account the recent global recession. Cash remittances to Guatemala fell by 10 percent in 2010. In an economy that is cash-strapped, will even more people turn to "free" or more affordable cooking fuels like firewood and will more people begin to collect their own wood and put pressure on local forests? Based on this study in Guatemala, we also recommend in-depth studies of household energy use around the world. These studies must include sufficient detail and cultural considerations that can be incorporated into the design of more efficient biomass stoves. Detailed studies that move away from simplistic versions of the fuel and income ladder theory will ensure that new stoves are widely disseminated and adopted.

Importantly, and this bears repeating, the number of households and the amount of firewood used each year in the developing world increases. Thus, because billions of people around the world already employ a form of renewable energy—biomass—scientists and policymakers must examine the use of this energy with local integrated production systems in mind. Simply, woody biomass, if managed in a renewable way and if efficient conversion technologies are employed, is a major existing renewable energy source. If we are serious in our attempts to meet the Millennium Development Goals, we must, as others observe (Bailis, Ezzati, and Kammen 2008; Cherian 2009), address biomass burning in cookstoves on a massive scale because its use is directly tied to several of the goals, including health, gender equity, and environmental sustainability.

Acknowledgments

We express our sincere gratitude to the residents of San Cristóbal, the reviewers, and the team at the Centro de Estudios Ambientales at the Universidad del Valle, Guatemala. *Gracias!* Funding for research was provided by the Fulbright-Hays Program, AAUW, the Wenner Gren Foundation, and the Offices of Internationalization and Public Good at the University of Denver.

References

Allen, J. C., and D. Barnes. 1985. The causes of deforestation in developing countries. *Annals of the Association of American Geographers* 75 (2): 163–84.

Arnold, J. E. M., G. Kohlin, and R. Persson. 2006. Woodfuels, livelihoods, and policy interventions: Changing perspectives. *World Development* 34 (3): 596–611.

Bailis, R., M. Ezzati, and D. M. Kammen. 2005. Mortality and greenhouse gas impacts of biomass and petroleum energy futures in Africa. *Science* 308 (5718): 98–103.

Banco de Guatemala. 2010. Estadísticas de remesas familiares [Statistics on family remittances]. http://www.banguat.gob.gt/inc/ver.asp?id=/estaeco/remesas/remfam2008.htm&e=51802 (last accessed 3 January 2011).

Barnes, D. F., K. Openshaw, K. R. Smith, and R. van de Plas. 1994. What makes people cook with improved biomass stoves: A comparative international review of stove programs. World Bank Technical Paper 242, Energy Series, World Bank, Washington, DC.

Berrueta, V. M., R. D. Edwards, and O. R. Masera. 2008. Energy performance of wood-burning cookstoves in Michoacán, Mexico. *Renewable Energy* 33:859–70.

Bilsborrow, R. 1992. *Rural poverty, migration and the environment in developing countries: Three case studies*. Washington, DC: World Bank.

Bond, T. C., C. Venkataraman, and O. Masera. 2004. Global atmospheric impacts of residential fuels. *Energy for Sustainable Development* 8 (3): 54–66.

Boy, E., N. Bruce, K. R. Smith, and R. Hernandez. 2000. Fuel efficiency of an improved wood-burning stove in rural Guatemala: Implications for health, environment and development. *Energy for Sustainable Development* 4 (2): 23–31.

Brunn, S. D., M. Hays-Mitchell, D. J. Zeigler, and E. R. White. 2008. *Cities of the world: World regional urban development*. Plymouth, UK: Rowman and Littlefield.

Camus, M. 2007. *Comunidades en Movimiento: La Migración Internacional en el Norte de Huehuetenango* [Communities on the move: International migration from Northern Huehuetenango]. Guatemala City: INCEDES.

Cherian, A. 2009. Bridging the divide between poverty reduction and climate change through sustainable and innovative energy technologies: Scaling up sustainable energy innovations that can address climate change concerns and poverty reduction needs. Expert paper, United Nations Development Program, New York.

Cohen, J. 2001. Transnational migration in rural Oaxaca, Mexico: Dependency, development and the household. *American Anthropologist* 103 (4): 954–67.

Conway, D., and P. Lorah. 1995. Environmental protection policies in Caribbean small islands: Some St. Lucian examples. *Caribbean Geography* 6 (1): 16–27.

de Haas, H. 2010. Migration and development: A theoretical perspective. *International Migration Review* 44 (1): 227–64.

Dewess, P. A. 1989. The woodfuel crisis reconsidered: Observations on the dynamics of abundance and scarcity. *World Development* 17:1159–72.

Eckholm, E. 1975. The other energy crisis: Firewood. World Watch Institute Paper # 1, Washington, DC.

Edwards, J. H. Y., and C. Langpap. 2005. Startup costs and the decision to switch from firewood to gas fuel. *Land Economics* 81(4): 570–86.

Elías, S. 2011. From communal forests to protected areas: The implications of tenure changes in natural resource management in Guatemala. Unpublished manuscript.

Elías, S., G. Gellert, E. Pape, and E. Reyes. 1997. *Evaluación de la Sostenibilidad en Guatemala* (Evaluation of sustainability in Guatemala). Guatemala City, Guatemala: FLACSO.

Elías, S., and H. Wittman. 2005. State, forest and community: Decentralization of forest administration in Guatemala. In *The politics of descentralization*, ed. C. J. Pierce Colfer and D. Capistrano, 282–96. London: Earthscan.

Ezzati, M., and D. Kammen. 2002. Household energy, indoor air pollution, and health in developing countries: Knowledge base for effective interventions. *Annual Review of Energy and the Environment* 27:233–70.

Falla, R. 2008. *Migración Transnacional Retornada* [Return international migration]. Guatemala City, Guatemala: Editorial Universitaria, Universisdad de San Carlos de Guatemala.

Gabriel, I. 2008. The political economy of remittances: What do we know? What do we need to know? Working Paper Series No. 184, Political Economy Research Institute, University of Massachusetts, Amherst.

Geist, H. J., and E. L. Lambin. 2002. Proximate causes and underlying driving forces of tropical deforestation. *Bioscience* 52 (2): 143–50.

Georges, E. 1990. *The making of a transnational community: Migration, development, and cultural change in the Dominican Republic*. New York: Columbia University Press.

Granderson, J., J. S. Sandhu, D. Vasquez, E. Ramirez, and K. R. Smith. 2009. Fuel use and design analysis of improved woodburning cookstoves in the Guatemalan Highlands. *Biomass and Bioenergy* 33 (2): 306–15.

Grasmuck, S., and P. Pessar. 1991. *Between two islands: Dominican international migration*. Berkeley: University of California Press.

Gustafsson, O., M. Krusa, Z. Zencak, R. J. Sheesley, L. Grennat, E. Engström, P. S. Praveen, P. S. P. Rao, C. Leck, and H. Rodhe. 2009. Brown clouds over South Asia: Biomass or fossil fuel combustion? *Science* 323:495–98.

Hecht, S. B., and S. S. Saatchi. 2007. Globalization and forest resurgence: Changes in forest cover in El Salvador. *BioScience* 57 (8): 663–72.

Heltberg, R. 2004. Fuel switching: Evidence from eight developing countries. *Energy Economics* 26:869–87.

———. 2005. Factors determining household fuel use in Guatemala. *Environment and Development Economics* 10:2337–61.

Hugo, G. 1996. Environmental concerns and international migration. *International Migration Review* 30 (1): 105–31.

International Energy Agency. 2006. Energy for cooking in developing countries. In *World energy outlook 2006*, 419–45. Paris: International Energy Agency.

Jetter, J. J., and P. Kariher. 2009. Solid-fuel household cook stoves: Characterization of performance and emissions. *Biomass and Bioenergy* 33:294–305.

Jokisch, B. D. 2002. Migration and agricultural change: The case of smallholder agriculture in highland Ecuador. *Human Ecology* 30 (4): 523–50.

Legros, G., I. Havet, N. Bruce, and S. Bonjour. 2009. *The energy access situation in developing countries*. New York: United Nations Development Programme.

MacCarty, N., D. Ogle, D. Still, T. Bond, and C. Roden. 2008. A laboratory comparison of the warming impact of five major types of biomass cooking stoves. *Energy for Sustainable Development* 12 (2): 5–14.

Maconachie, R., A. Tanko, and M. Zakariya. 2009. Descending the energy ladder? Oil price shocks and domestic fuel choices in Kaon, Nigeria. *Land Use Policy* 26 (4): 1090–99.

Madubansi, M., and C. M. Shackleton. 2007. Changes in fuelwood use and selection following electrification in the Bushbuckridge lowveld, South Africa. *Journal of Environmental Management* 83:416–26.

Martínez, F. M. 2009. Fuente de Energía [A source of energy]. *Prensa Libre* 11 January:D-19.

McCrary, J. K., B. Walsh, and A. L. Hammett. 2005. Species, sources, seasonality, and sustainability of fuelwood commercialization in Masaya, Nicaragua. *Forest Ecology and Management* 205:299–309.

Moran-Taylor, M. J. 2008. Guatemala's Ladino and Maya Migra landscapes: The tangible and intangible outcomes of migration. *Human Organization* 67 (2): 111–24.

———. 2009. Going north, coming south: Guatemalan migratory flows. *Migration Letters* 6 (2): 175–82.

Moran-Taylor, M. J., and M. J. Taylor. 2010. Land and Leña: Linking transnational migration, natural resources and the environment. *Population and Environment* 32:198–215.

National Research Council. 1999. *Global environmental change: Research pathways for the next decade*. New York: National Academy Press.

Northcross, A., Z. Chowdhury, J. McCraken, E. Canuz, and K. R. Smith. 2010. Estimating personal PM2.5 exposures using CO measurements in Guatemalan households cooking with wood fuel. *Journal of Environmental Monitoring* 12:873–78.

Openshaw, K. 1978. Wood fuel—A time of re-assessment. *Natural Resources Forum* 13:35–51.

Organización Internacional par alas Migraciones. 2008. Encuesta sobre remesas y medio ambiente [Survey on remittances and the environment]. *Cuadernos de Trabajo sobre Migración* 26. http://www.oim.org.gt (last accessed 15 December 2010).

Population Reference Bureau. 2011. Rate of natural increase. Guatemala summary: Demographic highlights. http://www.prb.org/Datafinder/Topic/Bar.aspx?sort=v&order=d&variable=87 (last accessed 10 January 2011).

Programa de Las Naciones Unidas para el Desarollo. 2008. *Guatemala: Una Economía al Servicio del Desarollo Humano? Informe Nacional del desarollo Humano 2007–2008* [Guatemala: An economy at the service of human development? National report on human development 2007–2008]. Guatemala City, Guatemala: Programa de Las Naciones Unidas para el Desarollo.

Ramanathan, V., and G. Carmichael. 2008. Global and regional climate changes due to black carbon. *Nature Geoscience* 1:221–22.

Rhoades, R. 1978. Intra-European return migration and rural development: Lessons from the Spanish case. *Human Organization* 37:136–47.

Rubenstein, H. 1992. Migration, development and remittances in rural Mexico. *International Migration* 30:127–53.

Rudel, T., M. Perez-Lugo, and H. Zichal. 2000. When fields revert to forest: Development and spontaneous reforestation in post-war Puerto Rico. *The Professional Geographer* 52 (3): 386–97.

Schmook, B., and C. Radel. 2008. International labor migration from a tropical development frontier: Globalizing households and an incipient forest transition. *Human Ecology* 36 (6): 891–908.

Smith-Sivertsen, T., E. Díaz, D. Pope, R. T. Lie, A. Díaz, J. McCracken, P. Bakke, B. Arana, K. R. Smith, and N. G. Bruce. 2009. Effect of reducing indoor air pollution on women's respiratory symptoms and lung function: The RESPIRE randomized trial, Guatemala. *American Journal of Epidemiology* 170 (2): 211–20.

Taylor, M. J. 2005. Electrifying rural Guatemala: Central policy and local reality. *Environment and Planning C* 23 (2): 173–89.

Taylor, M. J., M. Moran-Taylor, and D. R. Ruiz. 2006. Land, ethnic, and gender change: Transnational migration and its effects on Guatemalan lives and landscapes. *Geoforum* 37 (1): 41–61.

Universidad del Valle Guatemala (UVG), Instituto Nacional de Bosques (INAB), and Consejo Nacional de Areas Protegidas (CONAP). 2006. *Dinámica de la Cobertura Forestal de Guatemala durante los Años 1991, 1996 y 2001 y Mapa de Cobertura Forestal 2001* [Dynamics of forest cover in Guatemala in 1991, 1996, and 2001, and map of forest cover in 2001]. Guatemala City, Guatemala: Ediciones Superiores.

Veblen, T. 1978. Forest preservation in the western highlands of Guatemala. *Geographical Review* 68 (4): 417–34.

Venkataraman, C., A. D. Sagar, G. Habib, N. Lam, and K. R. Smith. 2010. The Indian National Initiative for Advanced Biomass Cookstoves: The benefits of clean combustion. *Energy for Sustainable Development* 14:63–72.

Zimmerer, K. S. 2011. New geographies of energy: Introduction to the special issue. *Annals of the Association of American Geographers* 101 (4): 705–11.

The Impact of Brazilian Biofuel Production on Amazônia

Robert Walker

Department of Geography, Michigan State University

Global energy demand will increase through the twenty-first century. Competition for energy resources has already revealed geopolitical fault lines, and the dependence of industrial economies on fossil fuel promises to keep nations on edge. A widespread consensus has emerged that societies must transition to a new energy basis, given that fossil fuel is nonrenewable and its combustion leads to global warming. Although alternatives like nuclear energy and hydropower provide important electrical supplies locally, the search goes on, and recently much attention has focused on biofuel. Although biofuel represents a renewable and "green" energy, there is also a downside. This article considers one potential problem, namely, the impact of growing international biofuel demand on Amazônia. The article focuses on Brazil, given the explosive growth of Brazilian agriculture, and notable effects on forests within its national borders. The article seeks to answer this question: How will global demand for Brazil's land-based commodities, including biofuel, impact its tropical forest in the Amazon basin? In attempting to answer this question, the article describes recent agricultural expansion in Brazil and its emergent landscape of renewable energy. Using an adaptation of rent theory, it frames a concept of landscape cascade and shows how Brazil's expanding landscape of renewable energy is impacting forest areas at a great distance. The article then considers recent projections of demand for Amazonian land out to 2020, given growth of Brazilian biofuel production and cattle herds. The projections indicate that more Amazonian land will be demanded than has been made available by Brazilian environmental policy. With this result in mind, the article discusses the discursive dismemberment of Amazônia and how this articulates with efforts by Brazilian politicians to increase the region's land supply. The article points out that agricultural intensification holds the key to meeting global demand without degrading the Amazonian forest, a landscape unique in the world for its ecological and cultural riches. *Key Words: Amazônia, biofuel, Brazil, deforestation*.

整个二十一世纪都将面临全球能源需求的增加。对能源资源的竞争已经被证实为地缘政治的断层线，工业经济对化石燃料的依赖使得国家常常处于边缘境界。由于化石燃料不可再生而其燃烧导致全球变暖，一个普遍的共识得以出现，即人类社会必须过渡到一个新的能源基础。尽管替代性能源，如核能和水电给当地提供了重要的电能，对新能源的搜索仍在持续，最新的研究焦点是生物燃料。尽管生物燃料是可再生和"绿色"能源，它也有缺点。本文探讨了一个潜在的问题，即日益增长的国际生物燃料需求对亚马孙流域的影响。文章着重对巴西进行了研究，由于巴西农业爆炸性的增长，对其国家边界地区的森林产生了显著的影响。本文试图回答这样一个问题：全球对巴西土地为基础的商品需求，包括生物燃料，是如何影响亚马逊河流域热带雨林的？为了试图回答这个问题，文章介绍了最近在巴西的农业扩张和新兴的可再生能源格局。使用改编的地租理论，本文建立了景观阶梯这一概念，并显示了巴西再生能源景观的扩展是如何影响到遥远的森林地区的。鉴于巴西的生物燃料生产和牛群的增长，文章接着考虑了直到 2020 年对亚马逊流域土地需求的最新预测。预测表明，相比由巴西环境政策已经提供的，将会有更多的亚马逊土地被需求。考虑到这一结果，文章讨论了亚马逊流域推论式的解体，并阐述了巴西政客如何努力提高本地区的土地供应。文章指出，亚马逊森林因其生态和文化财富，是世界上独一无二的景观，在不造成亚马逊森林退化的前提下，农业集约化是满足全球需求的关键。*关键词：亚马逊河流域，生物燃料，巴西，砍伐森林。*

La demanda global de energía se incrementará durante el siglo XXI. La competencia por los recursos energéticos ya revela líneas de falla geopolíticas, y la dependencia de las economías industriales de combustibles fósiles promete mantener algunas naciones en vilo. Ha emergido un consenso generalizado de que las sociedades deben hacer la transición hacia nuevas bases de energía, en consideración a que el combustible fósil no es renovable y a que su combustión lleva al calentamiento global. Aunque alternativas como la energía nuclear y la energía hidráulica proveen localmente una importante oferta de electricidad, la búsqueda sigue adelante y recientemente mucha de la atención se ha concentrado en los biocombustibles. Aunque el biocombustible representa una energía renovable y "verde", también tiene su lado malo. Este artículo considera un problema potencial, es decir, el impacto de la creciente demanda internacional de biocombustibles sobre la Amazonia. Este artículo está enfocado en Brasil, dado el explosivo crecimiento de su agricultura y los notables efectos que eso tiene sobre las selvas ubicadas dentro de sus fronteras. El artículo busca la respuesta a esta pregunta: ¿Cómo será el impacto de la demanda global por los productos brasileños basados en tierra, incluyendo los biocombustibles, sobre sus bosques tropicales en la cuenca amazónica? Intentando resolver esta cuestión, el artículo describe la reciente expansión agrícola del Brasil y su emergente paisaje de energía renovable. Haciendo uso de una adaptación de la teoría de la renta, esta enmarca un concepto de cascada de paisajes y muestra cómo el paisaje de de la energía renovable en expansión está impactando las áreas de bosque a grandes distancias. El artículo considera luego recientes proyecciones de la demanda de tierra amazónica hasta el 2020, dado el crecimiento de la producción brasileña en biocombustibles y la expansión del hato ganadero. Las proyecciones indican que se demandará más tierra amazónica de la que pueda ponerse a disposición mediante la política ambiental brasileña. El artículo destaca el hecho de que la intensificación en la agricultura es la clave para satisfacer la demanda global, sin que se degrade la selva amazónica, un paisaje único en el mundo por su riqueza ecológica y cultural. *Palabras clave: Amazonia, biocombustibles, Brasil, deforestación.*

Energy demand will increase through the twenty-first century, particularly with economic growth in countries like China and India. This has sparked widespread concern, given societal dependence on nonrenewable fossil fuels, the exploitation of which enables industrialization but also makes the world dangerous. Now, global warming from greenhouse gas buildup joins violence in the Middle East as a grim consequence of our reliance on an energy resource we always knew would have to be replaced (Odum 1971). Although nuclear energy and hydropower provide important electrical supplies locally, the search goes on for a viable alternative, and much attention presently focuses on biofuel. Biofuel appears to provide a green panacea to the energy problem. It also raises its own issues (Zimmerer 2011), and this article considers one of them: the likely effects of growing biofuel demand on Amazônia.

Recently, Brazilian agriculture has boomed, and new demands for land have sparked speculative frenzies in real estate markets on the margins of Amazônia (Almeida 2009). The article's goal is to consider how expanding Brazilian agriculture, driven partly by biofuel markets, will impact this ecologically rich region. It poses a question: To what extent does Brazil's expanding landscape of renewable energy production impose an ecological compensation, by reducing in equal measure the ancient forests of Amazonia? As the watershed containing the river that bears its name, Amazônia, covers more than 7 million km^2 and includes parts of Ecuador, Bolivia, Brazil, Colombia, Peru, and Venezuela. Although certain of these countries contribute Amazonian landscapes to global markets for agricultural commodities, this article focuses on the Brazilian portion alone, given its disproportionate magnitude and the dynamism of Brazilian agriculture.

The article begins answering its central question in the next two sections, which consider Amazonian development efforts and the institutional forces that have stimulated Brazil's biofuel sector. It then conceptualizes landscape implications for Amazônia by appeal to rent theory, showing that more Amazonian land might soon be demanded than Brazilian conservation policy has made available. Having demonstrated the possibility of excess demands, the article next addresses the political campaign underway to increase Amazonian land supplies, including efforts to discursively reinvent the region. It concludes by pointing out how Brazil's expanding landscapes of renewable energy could exert unexpected impacts on Amazonian forests.

Brazilian Amazônia Today

Definitional Considerations

A hydrologic basin of continental magnitude, Amazônia attained juridical status in Brazil during the presidency of Getúlio Vargas, with the creation of *Amazônia Legal* (AML), a planning entity, in 1953. Presently, AML covers the Brazilian states of Acre, Amapá, Amazonas, Mato Grosso, Pará, Rondônia, Roraima, Tocantins, and part of Maranhão. More recently, Brazilian agencies have provided an ecological definition with the "Amazonian Biome," a topic discussed in the sequel. This bureaucratic designation follows the floristic fact that the Amazon basin includes both closed moist forest and drier savannas, or *cerrados*, on its southern and eastern margins. Thus, agricultural development in AML need not always occur at the direct expense of forest. As for biofuel, this term covers a wide range of plant-based energy forms, two of which are significant for Brazil: ethanol from sugarcane and diesel from soy. These two feedstocks also provide a wide range of nonfuel commodities, but their botanical chemistry yields combustible hydrocarbons in concentrated forms. The United States also produces ethanol, with corn as a feedstock, and presently there is much discussion of cellulosic conversion using perennials like switchgrass (James, Swinton, and Thelen 2010). The discussion here recognizes this long-run possibility but focuses on sugarcane and soy given the immediacy of their impacts for the Brazilian case (cf. Lapola et al. 2010).

Amazonian Agriculture and Ranching

Mechanized agriculture and ranching have expanded dramatically in AML, which now produces a wide variety of crops, notably soy, and supports a large cattle herd. Agricultural development began in earnest with policies implemented during the military regime (Moran 1981; Hecht 1985). The restoration of democracy in the mid-1980s redirected some attention to the environment and indigenous rights in the region, but successive administrations sustained infrastructure investments, and monetary reform has greatly facilitated export (Simmons 2002; Brandão, Rezende, and Marquest 2005). The advance of soy into Amazônia, especially along the basin's drier *cerrado* margins, has also been stimulated by genetic improvements that enable Amazonian farmers to produce about 3 ton-ha^{-1}, 30 percent more than the national average (Walker, DeFries, et al. 2009). Soy provides between 35 and 40 percent of gross farm revenues in AML, which translates into a third of Brazil's entire harvest (IBGE 2010). As a propulsive sector, soy partly accounts for the importance of other crops, like maize, with which it grows in rotation. For its part, cattle ranching has expanded in both *cerrado* and forested parts of AML with the control of foot and mouth disease and the development of forages from African grasses (Walker, Browder, et al. 2009). Modern ranching techniques, low land prices, transportation cost reductions, and abundant sunshine make Amazonian ranching profitable, and AML's $\sim$70 million animals account for about 35 percent of the national herd (Arima, Barreto, and Brito 2005). Biofuel expansion in Brazil cannot avoid impacting Amazônia by virtue of its soy production base and landscape links to the cattle economy, to be discussed later. Sugarcane, mostly grown in non-Amazonian locations due to humidity constraints that lower sugar content, also possesses potential impact via these same landscape mechanisms (Companhia Nacional de Abastecimento 2008; Lapola et al. 2010).

The Emerging Bioeconomy

The appeal of biofuel as an alternative to fossil fuel has stirred many governments into action (International Energy Agency 2004). The United States, for example, is attempting to steer its economy toward renewable fuels through legislative initiative, such as the Energy Policy Act (2005), which created renewable fuels standards. This was recently intensified by the Energy Independence and Security Act (2007), stipulating that 36 billion gallons of ethanol be used as motor fuel by 2022 (Low and Isserman 2009). Similar efforts are underway in Brazil, long a leader in biofuel development, an outcome of the ProAlcool program initiated in the wake of the oil embargo of the 1970s (International Energy Agency 2006). Currently, flex fuel vehicles capable of using any ratio of ethanol-to-gasoline blend account for 72 percent of Brazil's automotive production (ANFAVEA 2010). As part of its drive to bioenergize, Brazil has stimulated ethanol output and now commands 50 percent of the global market. Demands for ethanol, together with more traditional sugar production, place 80,000 km^2 under cane, mostly concentrated in São Paulo State (International Energy Agency 2006; IBGE 2010). Brazil has also diversified its biofuel sector with the ProDiesel program, aimed at increasing the production of diesel from soy (International Energy Agency 2006). Presently, soy-based "biodiesel" represents a fraction of the marketable use of the crop, a situation that could change as a function of future demands (Amaral 2010; Wilkinson and Herrera 2010).

Despite significant potential, biofuel has its downside. Shifting fields to fuel production raises food prices, with consequences for the poor, who pay for more expensive farm goods as they lose lands for subsistence agriculture (Naylor et al. 2007). Other problems are environmental. Biofuel production causes pollution and stresses renewable resources like water (International Energy Agency 2004). An additional concern arises from the demand for land at the global scale, likely to be provided by Africa and Latin America, particularly Brazil (Gurgel, Reilly, and Paltsev 2007). Because Brazil enjoys relatively good infrastructure and efficient commodity chains, it is well positioned to funnel global biofuel demands to its own producers (Pingali, Raney, and Wiebe 2008). Aside from displacing smallholders unable to exploit emerging market opportunities, this creates a "biofuel carbon debt" if Amazonian forest biomass is cleared and carbon released to the atmosphere as a consequence of expanding biofuel production (Ragauskas et al. 2006; Fargione et al. 2008; Lapola et al. 2010). Resulting greenhouse gas benefits might then be offset to the future, and with environmental costs (Fargione et al. 2008; Gibbs et al. 2008). An important question follows: How much carbon is at stake? Or, in landscape terms, how much forest?

Landscape Cascades

This article addresses the latter question while recognizing that carbon from vegetative biomass is related to forest extent, both above and below ground. Geography provides a convenient heuristic to address the forest extent question, the theory of rent, which posits that agriculture and ranching occur when rents are positive and that rents are determined by transportation costs and market conditions. Thus, agricultural landscapes organize spatially, with implications for the forested landscapes with which they form frontiers. The rational decision making of unitary agents implied by rent theory is highly conditioned by social processes, institutions, and history. For the Amazonian case, a great deal of research has called attention to the political economy of land use, as well as the impacts of place, household structure, and social movements on land managers (e.g., Moran 1981; Hecht 1985; McCracken et al. 1999; Simmons et al. 2007; Pacheco 2009). That said, rent theory enables the generalized description of landscape patterns at regional scale and is therefore useful to the purposes of the article (cf. Cronon 1991). Recently, it has been embedded in political economy to explain the advance of ranching into Amazônia, and how this is linked to soy production (Walker, Browder, et al. 2009).

Rents and Landscape Displacement

The rent theory adaptation focuses on ranching and mechanized agriculture, with ranching the extensive land use, and mechanized agriculture, the intensive one. Ranching forms a frontier with primary forest, beyond which rents vanish, although mineral extraction and predatory logging often occur. "Behind" ranching relative to the economic strongholds of Brazil to the south comes mechanized farming (Jepson, Brannstrom, and Filippi 2010). Deforestation takes place as land use, typically for pastures, encroaches on forest, although direct conversion to soy fields does occur (Brown et al. 2005; Morton et al. 2006). The conceptual task is to identify the land use change mechanisms that underlie the infringement of forested landscapes. With one commodity (e.g., beef), forests become pastures when rents increase for ranching due to rising meat prices, decreasing transportation costs, or both. With two commodities, the situation grows more complex. If only beef prices rise, the situation is as just stated, with an advancing cattle frontier. If only the crop price rises, it advances on pasture until only forest remains, after which direct encroachments take place.

A complication of interest to the analysis that follows involves strong market conditions for both crops and cattle, in which case crop expansion consumes pastures, which do not disappear but are instead displaced to the forest frontier via indirect land use change (ILUC; Walker, Browder, et al. 2009; Lapola et al. 2010). ILUC has emerged as a key consideration in assessing biofuel expansion in Brazil, with soy playing a key role given its use as a biofuel feedstock, and given the dramatic buildup of soy fields at the expense of Amazonian pastures, particularly in Mato Grosso (IBGE 2008; Sawyer 2008). For these reasons, the discussion now focuses on soy. Direct encroachments of Amazonian forests by soy agriculture have been observed in Mato Grosso and Rondônia, but its primary impact on forested landscapes could be due to ILUC (Brown et al. 2005; Morton et al. 2006; Lapola et al. 2010).

ILUC and Amazônia

Although anticipated by the dynamic landscape articulation of rent theory and widely hypothesized, the magnitude of ILUC remains an empirical question. The analysis now embeds this question within a context of globalization, assessing two cases bracketing the range of beef price impacts associated with pastures converting to soy fields. For the first case, let Amazônia supply only a small market share, in which case its regional output has no effect on price (*elastic* demand). If soy fields are placed in old pastures, prices for beef stay the same and the extensive margin remains in place without deforestation. Alternatively, if Amazonian supplies are globally significant (*inelastic* demand), pasture conversion to soy production reduces beef supplies, which raises beef prices and rents, stimulating ILUC on the forest frontier. Table 1 formalizes these statements with results from a technical appendix available from the author. Table 1 presents terms showing loss of forest due to expansions of the

Table 1. Indirect land use change and forest frontier

Perfectly elastic	Elastic	Perfectly inelastic
$\frac{dt_e}{dp_m} = 0$	$\frac{dt_e}{dp_m} = \frac{\theta}{\left[1 - \varphi\varepsilon\frac{Q_b}{p_b}\right]}\left\{\frac{1}{f_b}\right\}$	$\frac{dt_e}{dp_m} = \theta\frac{1}{f_b}$

Note: Derivative, or incremental change in margin (dt_e) with respect to incremental change in price (dp_m)

$\varphi = \frac{f_b(f_m q_m - f_b q_b)}{f_m q_m q_b}$

$\theta \equiv \frac{f_b}{f_m}$

p_m= price of mechanized crop; p_b= price of beef; t_e = extensive margin, or forest frontier; f_m = unit transportation cost of mechanized crop; f_b = unit transportation cost of beef; q_b = productivity, ranching; Q_b = demand for beef; ε = price elasticity of beef demand.

Table 2. Projected deforestation increment, by 2020 with biofuel expansion

Expanding land use (km^2)	Lapola et al. (2010)[a]	Walker, DeFries, et al. (2009)[b]
Soy + Sugarcane	86,117	121,332
Pasture	371,294	314,400
Total	457,411	435,732

[a]Lapola et al. (2010) projection period, 2003–2020; normalized to 2008–2020.

[b]Walker, DeFries, et al. (2009) projection period, 2005–2020; normalized to 2008–2020. Sugarcane (União de Industria de Cana-de-Açucar 2010) adds 60,000 km^2 to projections originally based on soy and pasture.

extensive margin (or forest frontier), t_e, as a function of changes in mechanized crop price, p_m (e.g., soy). Demand elasticity refers to the beef market and ranges from perfectly elastic on the left to perfectly inelastic on the right. As for actual market conditions, Brazil is the world's largest exporter of beef and possesses the world's largest commercial herd, with 18 percent of global stocks, over a third of them on Amazonian pastures (Herlihy 2008). These market conditions are consistent with some degree of ILUC. Quite apart from the indirect effects of biofuel expansion, Amazonian ranching will expand of its own accord, with growing demands for beef worldwide and widespread land constraints.

Figure 1 depicts Brazil's landscape cascade of recent years. Although the discussion has focused on soy and cattle given their preponderance in AML, Brazil's biofuel issue involves other parts of the country, and a second crop, sugarcane. As shown, cattle herds have migrated north between 1980 and 2006, a reallocation more dramatic over longer periods, given that Brazilian production was initially localized in São Paulo State. Movement into AML is consistent with ILUC, as can be seen in the middle panel, where soy also jumps north. Sugarcane shows landscape dynamics likely to account for displacements of soy, with diffusion away from historic coastal locations in the northeast, although São Paulo presently maintains extreme concentration. That said, expanding sugarcane must go somewhere, and indications are north (Lapola et al. 2010).

The Demand for Land

Table 2 gives two recent projections of Amazonian forest loss out to 2020, as a result of expanding landscapes of renewable energy (sugarcane, soy fields) and beef production that, when added together assuming ILUC, yield total deforestation (Lapola et al. 2010). The projections do not consider the land sparing effects of agricultural intensification; further, their magnitudes imply annual rates of forest loss in excess of the historic record, even though deforestation has diminished recently, probably due to some combination of global recession and new resolve in Brazil's enforcement of environmental law (Alves 2002; Nepstad et al. 2009). Such numbers nevertheless provide a benchmark for assessing the adequacy of Brazilian policies directed at the conservation of forested landscapes, the next topic to be addressed.

The Green Redoubt

Efforts to protect Amazônia primarily involve setting aside lands under federal and state control and restrictions on private holdings. Public lands dedicated to this purpose are referred to as *protected areas* (PAs) and comprise a large fraction of Amazônia. Alternatively, restrictions on land managers involve the creation of forest reserves on individual holdings, bans on the destruction or exploitation of particular tree species such as Brazil nut and mahogany, specifications of source areas for tropical hardwoods, and regulation of the use of fire. The article considers the two most important relative to landscape dimension, the PA program and laws defining "forest reserve," derivative of the forest code. Once it evaluates the forest expanses they are intended to sustain, the article reconsiders the land needed to satisfy the global economy's growing hunger for Brazilian biofuel and beef.

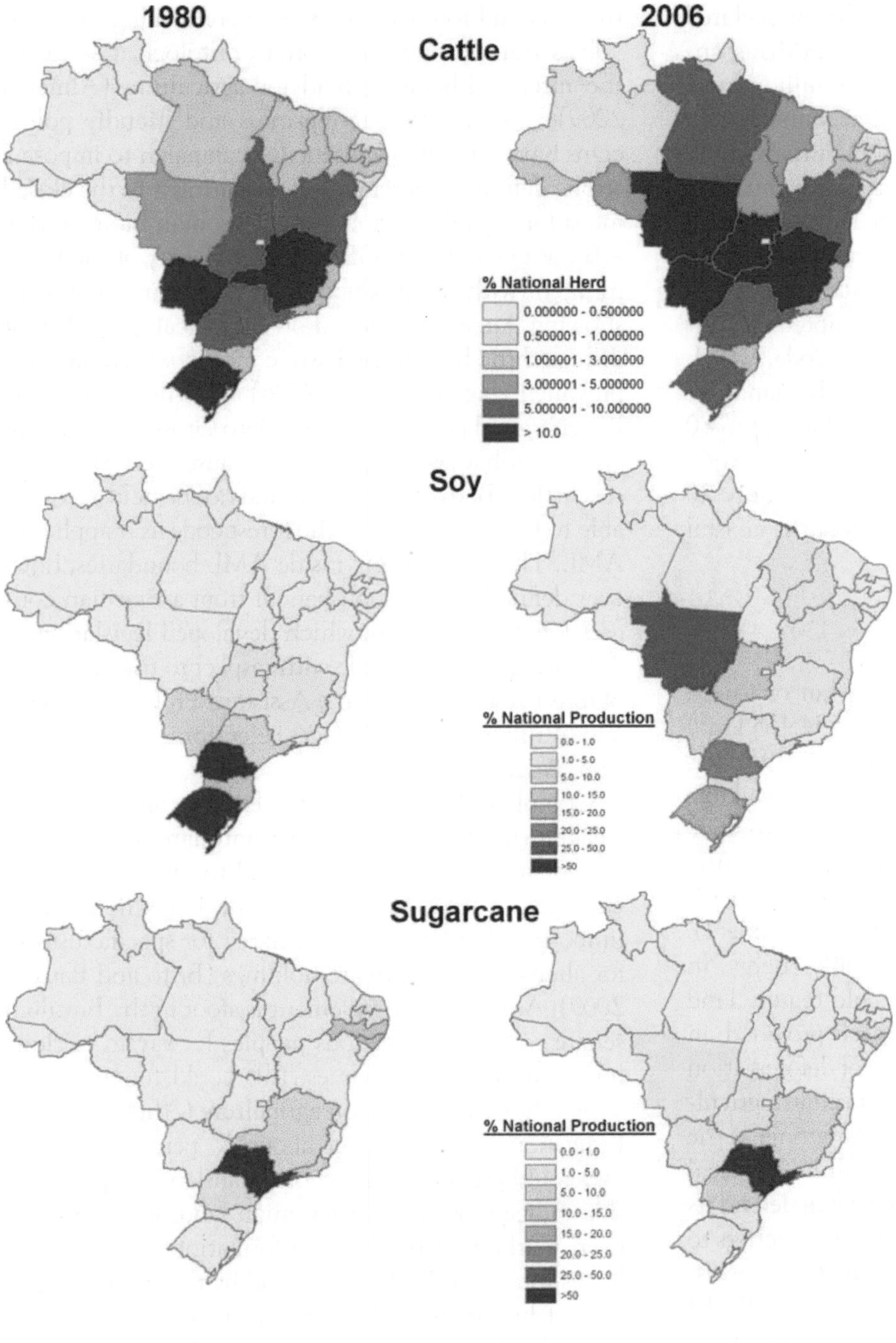

Figure 1. Spatial dynamics of Brazilian agriculture.

Protected Areas and the Forest Code

Although Brazil's environmental legislation dates to the 1930s (Machado 1995), the creation of conservation areas in Amazônia follows democratic reform in the 1980s (Simmons 2002). By 2000, about 10 percent of AML was declared for conservation under the Brazilian National System of Nature Conservation Units (SNUC; Law 9985 18 July 2000; Decree No. 4340, 22 August 2002). Since then, areas protected by federal and state governments have grown to more than 1.25 million km^2, about a quarter of AML. Critical to the system of PAs are Amazônia's indigenous reserves, guaranteed by Brazil's 1988 Constitution. Convention 169 of the International Labor Organization recognizes indigenous rights to natural resource use, although an expectation of environmental stewardship can be found in Agenda 21 from the Rio Summit. Despite pressures on PAs throughout AML, indigenous peoples often defend their lands, even near settlement frontiers (Euler et al. 2008). Many reserves (375) have been declared in recent years, and now about a fifth of AML is under indigenous control, spread over 1.06 million km^2. The PA system with indigenous land accounts for 43 percent of the forested part of AML (2.3 million km^2), with 37 percent permitting minimal to no disturbance (Walker, Moore et al. 2009). These federal, state, and indigenous lands contribute to the maintenance of the Amazonian forest. Nevertheless, a significant portion remains in private hands. Brazil has attempted to limit deforestation here as well via its Forest Code, which has long defined amounts of legally clearable land, set at 50 percent in AML in 1965. This was changed to 80 percent in 2001 by administrative decree (MP 2166–67). Outside AML, the forest code mandates only 20 percent, even in the Atlantic rainforests of the coastal states.

Adequate Supply?

AML originally possessed 4,196,943 $km^2$of forest; thus, with 17.5 percent gone as of 2008 (734,465 km^2) and 37 percent protected (1,281,116 km^2; Walker, Moore, et al. 2009), 2,181,362 km^2 remains available for occupation. The forest code requirement that 80 percent be maintained on private holdings evidently allows 432,272 km^2 for agricultural use. This number is less than both demand projections for 2020 (Table 2). Brazil's expanding landscapes of renewable energy, in concert with a growing cattle herd, could require land in excess of what conservation policy has provided, in little over a decade. Although the pace of deforestation has recently dropped, demands for Brazilian agricultural commodities will put pressure on Amazônia as the world economy regains its vigor. This will be a critical moment, as it remains to be seen if Brazil will defend its PAs and forest reserves, forcing farmers and ranchers to intensify production, sparing land in a manner consistent with the Borlaug hypothesis (Borlaug 2007; Rudel et al. 2009). Brazil has long enjoyed income levels high enough to reduce rates of land clearance, and possibly spark an Amazonian forest transition (Perz and Skole 2003; Walker 2004). Thus, the next decade looms as a moment of reckoning for Amazônia, given the globalization of Brazilian agriculture and the intensification of Brazilian efforts to conserve its Amazonian heritage.

Amazônia Under Erasure

The environmentalist response to landscape cascades into Amazônia is obvious: Defend the green redoubt of the PAs and forest reserves. A more insidious one involves transforming Amazonia's ontological status in the interest liberating land for agriculture (Almeida 2009). To this end, agribusiness and friendly politicians have launched a discursive campaign to impose a strong definition of Amazônia, equating it to the closed moist forest of the region, the Amazonian Biome (AB). AB was created by IBGE and the Ministry of the Environment with the production and distribution of maps showing Amazônia defined on ecological grounds that disregard the hydrologic basin concept, which has long prevailed (Figure 2; IBGE 2006). This new Amazônia has energized politicians from Mato Grosso, Tocantins, and Maranhão, who argue that because their states possess little AB, their citizens should not be held accountable to the restrictions of the forest code as it applies to AML. Those remaining inside AML boundaries, however defined, might soon benefit from a Brazilian congressional commission, which developed legislation in 2010 weakening the code with respect to the definition of forest reserve (E. Arima, Assistant Professor, Hobart and William Smith Colleges, electronic communication, August 2010).

Complementing discursive erasure are efforts to weaken the PA system, to facilitate land transactions by redefining property rights, and to sanction the occupation of *terras devolutas*, public lands in juridical limbo that have not been declared for specific uses or for alienation into private holdings (Brito and Barreto 2009). As for PAs, a movement is afoot in the Brazilian senate to allow indigenous peoples leeway in exploiting their mineral resources; this could function as a Trojan horse for follow-on agriculture (Almeida 2009). Indigenous reserves cover about 20 percent of AML and could greatly augment Brazilian land supply. Outside indigenous areas, Amazonian lands have recently been marketized by the transformation of *aforamento* leases into transactable titles and by the legal recognition of hundreds of thousands of private holdings on *terras devolutas*, ranging up to 1,500 ha in size (State of Pará 2009; Brito and Barreto 2009). This later maneuver provides incentives to occupy remaining *terras devolutas*, which might otherwise be declared for environmental protection, or turned over to indigenous claimants.

A Landscape of Salvation or Ruin?

The world must end its dependency on nonrenewable energy, and biofuel presents a viable alternative.

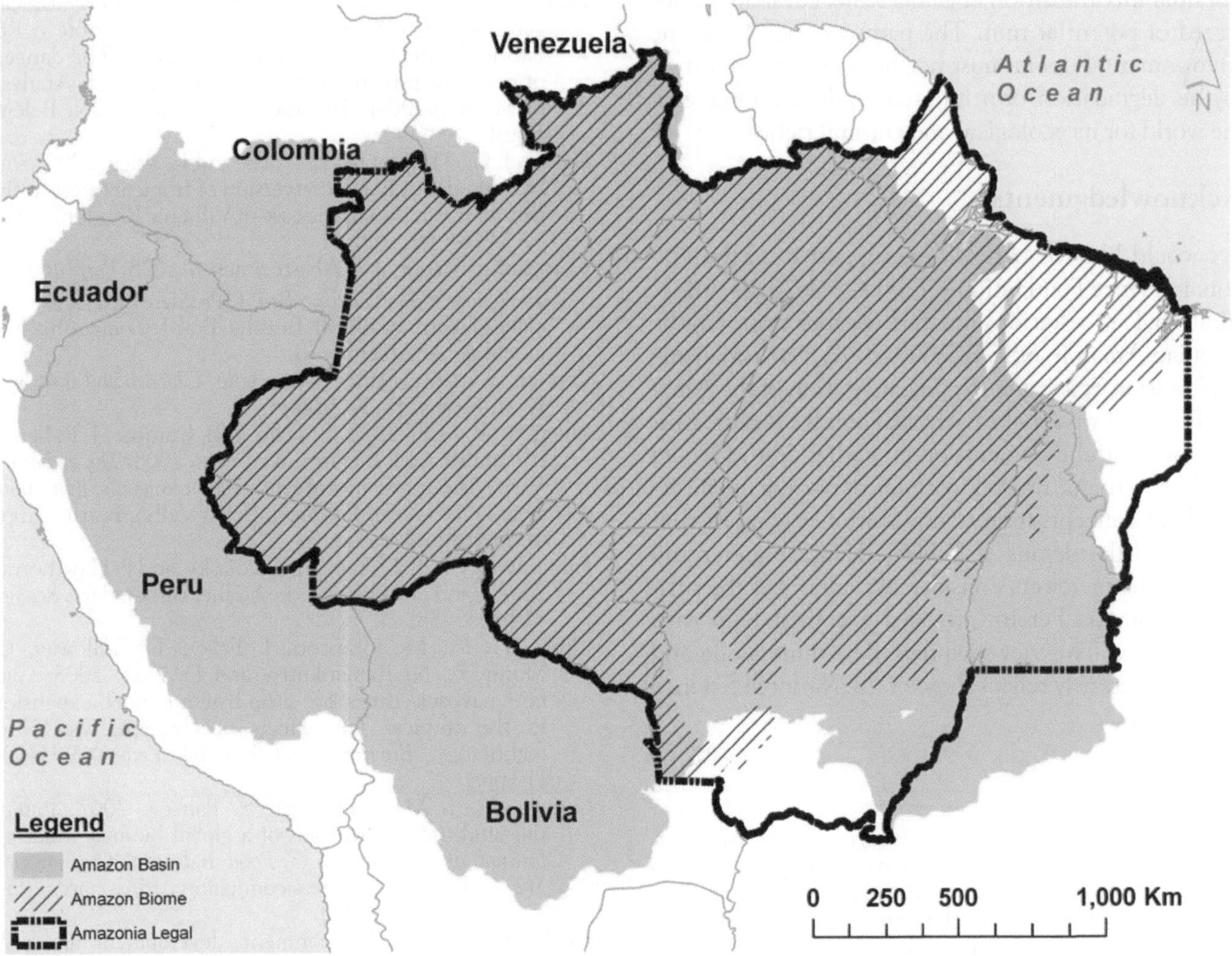

Figure 2. Defining Amazônia.

But at what cost? Although this article does not address food security, it calls attention to an environmental impact of considerable magnitude, Amazonian forest loss. To the extent the world relies on Brazil for new sources of energy, it does so at the possible expense of Amazônia's primary forest, or at least large parts of it. If agriculture and ranching significantly advance here, partly from biofuel-driven ILUC, greenhouse gases will reach an atmospheric equilibrium but only in the long run, after deforestation releases an initially large flux of carbon. At the same time, Amazônia's irreplaceable stocks of biodiversity will go up in smoke. Brazil has taken decisive steps to avoid this outcome by setting aside protected areas and also by announcing a deforestation reduction target in 2008 (Nepstad et al. 2009). The test will come in the near future, however, as the world economy recovers its appetite for Brazilian agricultural commodities.

One cause for concern is Amazônia's discursive dismemberment. If we no longer have an Amazônia of the mind, a distinct place requiring distinct protections, it becomes that much more difficult to provide them. Of course, Amazônia is a sum of vastly different parts, ecologically, culturally, and historically. But its imaginary has held in the face of change, until now. The United States also possessed a forest of continental scope, forming a closed canopy from Atlantic tidewater to the Mississippi. This forest sustained its indigenous inhabitants and provided habitat for the Carolina parakeet and Eastern wood bison, both extinct now (Mershon 1907; Rostlund 1960; Sauer 1971). How do we think of this wonder that has vanished, this silence that condemns us? In the same way we might soon think of Amazônia, as a forest and a way of life erased, a ghost of development history. The Brazilian landscape of renewable energy will prove crucial to sustainability transition at global scale, but it also holds a seed of potential ruin. The pathway back from one environmental doom must not help us down another, to the degradation of a forested landscape unique in the world for its ecological and cultural riches.

Acknowledgments

I would like to acknowledge support from the National Science Foundation (NSF-BCS-0620384, Globalization, Deforestation, and the Livestock Sector in the Brazilian Amazon; NSF-BCS-0822597, Territorializing Exploitation Space and the Fragmentation of the Amazon Forest). I would like to thank Peter Richards for his many insightful comments on an earlier draft. I would also like to thank Bruce Pigozzi for refreshing my memory about concepts from transportation geography that lie behind the details of Table 1. Karl Zimmerer and anonymous reviewers considerably improved the article. Ritaumaria Pereira provided excellent help with the graphics. The views expressed are mine alone and do not necessarily reflect those of the National Science Foundation.

References

Almeida, A. W. B. 2009. *Agroestrategias e desterritorializacao: Os direitos territoriais e etnicos na mira dos estrategistas dos agronegocios* [Agrostrategies and deterritorialization: Territorial rights and ethnicities in the gunsights of agribusiness]. In *O plano IIRSA: Na visao da sociedade civil Pan-Amazonica*, ed. A. W. B. Almeida and G. Carvalho, 57–105. Belem, Brazil: Actionaid.

Alves, D. 2002. Space-time dynamics of the deforestation in Brazilian Amazônia, *International Journal of Remote Sensing* 23 (14): 2903–08.

Amaral, D. F. 2010. Biodiesel no Brasil: Conjuntura atual e perspectivas [Biodiesel in Brazil: The current situation and perspectives]. Seminar presented at the symposium on the national biodiesel production program. Piracicapa, ESALQ, Universidade de São Paulo, Brazil.

ANFAVEA. 2010. Associação Nacional dos Fabricantes de Veículos Automotores [National Association of Automobile Manufacturers]. http://www.anfavea.com.br/tabelas/autoveiculos/tabela10_producao.pdf (last accessed 5 January 2011).

Arima, E. Y., P. Barreto, and M. Brito. 2005. Pecurária na Amazônia: Tendências e implicações para a conservação [Ranching in Amazonia: Tendencies and implications for conservation]. IMAZON Technical Report 75,IMAZON, Belém, Brazil.

Borlaug, N. 2007. Feeding a hungry world. *Science* 318:359.

Brandão, A., G. Rezende, and R. Marquest. 2005. Crescimento Agrícola no Brasil no Período 1999–2004: Explosão da Soja e da Pecuária Bovina e seu Impacto sobre o Meio Ambiente [Agricultural growth in Brazil between 1999 and 2004: The explosion of soy and beef and its environmental impact]. Working Paper 1103, IPEA, Rio de Janeiro, Brazil.

Brito, B., and P. Barreto. 2009. Os perigos da privatização generosa de terras na Amazônia: Análise sobre o Relatório da Medida Provisória n° 458/2009 [The dangers of generous privatization of Amazonian land: Analysis of MP 458/2009]. Technical Note, IMAZON, Belém, Brazil.

Brown, J. C., M. Koeppe, B. Coles, and K. Price. 2005. Soybean production and conversion of tropical forest in the Brazilian Amazon in the case of Vilhena, Rondônia. *Ambio* 34 (6): 462–69.

Companhia Nacional de Abastecimento. 2008. *Perfil do Setor do Açúcar e do Álcool no Brasil* [Profile of the sugar and alcohol sector of Brazil]. Brasilia, Brazil: Companhia Nacional de Abastecimento.

Cronon, W. 1991. *Nature's metropolis: Chicago and the Great West*. New York: Norton.

Euler, A., B. Millikan, E. B. Brito, I. B. Cardozo, J. P. Leroy, L. Caminha, M. I. Hargreaves, et al. 2008. *The end of the forest? The devastation of conservation units and indigenous lands in the state of Rondônia*. Porto Velho, Brazil: Grupo de Trabalho Amazônica.

Fargione, J., J. Hill, D. Tilman, S. Polasky, and P. Hawthorne. 2008. Land clearing and the biofuel carbon debt. *Science* 319:1235–38.

Gibbs, H. K., M. Johnston, J. Foley, T. Holloway, C. Monfreda, N. Ramankutty, and D. Zaks. 2008. Carbon payback times for crop-based biofuel expansion in the tropics: The effects of changing yield and technology. *Environmental Research Letters* 3:034001 (10 pp.).

Gurgel, A., J. M. Reilly, and S. Paltsev. 2007. Potential land use implications of a global biofuels industry. *Journal of Agricultural & Food Industrial Organization* 5(2). http://www.bepress.com/jafio/v015/iss2/art9 (last accessed 6 April 2011).

Hecht, S. B. 1985. Environment, development and politics: Capital accumulation and the livestock sector in eastern Amazonia. *World Development* 13:663–84.

Herlihy, J. 2008. USMEF beef export forecast. *Angus Beef Bulletin* March:74–75.

Instituto Brasileiro de Geografia e Estatística (IBGE). 2006. IBGE lança o Mapa de Biomas do Brasil e o Mapa de Vegetação do Brasil, em comemoração ao Dia Mundial da Biodiversidade [IBGE launches biome and vegetation maps in commemoration of the Day of Biodiversity]. Rio de Janeiro: IBGE. http://www.ibge.gov.br/home/presidencia/noticias/noticia_visualiza.php?id_noticia=169 (last accessed 2 February 2010).

———. 2008. *Produção da Pecuária Municipal 2007* [Production of municipal cattle ranching 2007]. http://www.ibge.gov.br/home/presidencia/noticias/noticia_impressao.php?id_noticia=1269 (last accessed 23 February 2010).

———. 2010. *2010 Censo Agropecuário; Produção Agrícola Municipal; Pesquisa Pecuária Municipal* [2010 agricultural census: Municipal agricultural production; municipal cattle report]. http://www.sidra.ibge.gov.br/bda/tabela/listabl.asp?c=1731&z=p&o=2&i=P, http://www.sidra.ibge.gov.br/bda/tabela/listabl.asp?z=t&o=11&i=P&c=1612, http://www.sidra.ibge.gov.br/bda/tabela/listabl.asp?z=t&o=23&i=P&c=73 (last accessed 16 February 2010).

International Energy Agency. 2004. *Biofuels for transport: An international perspective*. Paris: International Energy Agency.

———. 2006. *World energy outlook 2006*. Paris: International Energy Agency.

James, L., S. Swinton, and K. Thelen. 2010. Profitability analysis of cellulosic energy crops compared with corn. *Agronomy Journal* 102:675–87.

Jepson, W., C. Brannstrom, and A. Filippi. 2010. Access regimes and regional land change in the Brazilian cerrado, 1972–2002. *Annals of the Association of American Geographers* 100 (1): 87–111.

Lapola, D. M., R. Schaldach, J. Alcamo, A. Bondeau, J. Koch, C. Koelking, and J. A. Priess. 2010. Indirect land-use changes can overcome carbon savings from biofuels in Brazil, *Proceedings of the National Academy of Sciences* 107 (8): 3388–93.

Low, S. A., and A. Isserman. 2009. A. M. ethanol and the local economy. *Economic Development Quarterly* 23 (1): 71–88.

Machado, P. A. L. 1995. *Brazilian environmental law*. 5th ed. São Paulo, Brazil: Malheiros Editores Ltda.

McCracken, S. D., E. S. Brondizio, D. Nelson, E. F. Moran, A. D. Siqueira, and C. Rodrigues-Pedraza. 1999. Remote sensing and GIS at farm property level: Demography and deforestation in the Brazilian Amazon. *Photogrammetric Engineering and Remote Sensing* 65:1311–20.

Mershon, W. B. 1907. *The passenger pigeon*. New York: The Outing Publishing Company.

Moran, E. 1981. *Developing the Amazon*. Bloomington: Indiana University Press.

Morton, D., D. DeFries, Y. Shimabukuro, L. Anderson, E. Arai, F. Espirito-Santo, R. Freitas, and J. Morisette. 2006. Cropland expansion changes deforestation dynamics in the southern Brazilian Amazon. *Proceedings of the National Academy of Sciences* 103 (39): 14637–41.

Naylor, R., A. J. Liska, M. P. Burke, W. P. Falcon, J. C. Gaskell, S. D. Rozelle, and K. G. Cassman. 2007. The ripple effect: Biofuels, food security, and the environment. *Environment* 49 (9): 31–43.

Nepstad, D., B. S. Soares-Filho, F. Merry, A. Lima, P. Moutinho, J. Carter, M. Bowman, et al. 2009. The end of deforestation in the Brazilian Amazon. *Science* 326:1350–51.

Odum, H. T. 1971. *Environment, power, and society*. New York: Wiley-Interscience.

Pacheco, P. 2009. Agrarian reform in the Brazilian Amazon: Implications for land distribution and deforestation. *World Development* 37 (8): 1337–47.

Perz, S. G., and D. L. Skole. 2003. Secondary forest expansion in the Brazilian Amazon and the refinement of forest transition theory. *Society and Natural Resources* 16 (4): 277–94.

Pingali, P., T. Raney, and K. Wiebe. 2008. Biofuels and food security: Missing the point. *Review of Agricultural Economics* 30 (3): 506–16.

Ragauskas, A. J., C. K. Williams, B. H. Davison, G. Britovsek, J. Cairney, C. A. Eckert, W. J. Fredrick Jr., et al. 2006. The path forward for biofuels and materials. *Science* 311:484–89.

Rostlund, E. 1960. The geographic range of the historic bison in the Southeast. *Annals of the Association of American Geographers* 50 (4): 395–407.

Rudel, T. K, L. Schneider, M. Uriarte, B. L. Turner II, R. DeFries, D. Lawrence, J. Geoghegan, et al. 2009. Agricultural intensification and changes in cultivated areas, 1970–2005. *Proceedings of the National Academy of Sciences* 106 (49): 20675–80.

Sauer, C. O. 1971. *Sixteenth century North America: The land and the people as seen by the Europeans*. Berkeley: University of California Press.

Sawyer, D. 2008. Climate change, biofuels and eco-social impacts in the Brazilian Amazon and cerrado. *Philosophical Transactions of the Royal Society* B 363:1747–52.

Simmons, C. S. 2002. Development spaces: The local articulation of conflicting development, Amerindian rights, and environmental policy in Eastern Amazônia. *The Professional Geographer* 54 (2): 241–58.

Simmons, C. S., R. Walker, E. Arima, S. Aldrich, and M. Caldas. 2007. The Amazon land war in the south of Pará. *Annals of the Association of American Geographers* 97 (3): 567–92.

State of Pará. 2009. Decreto No. 1.805, de 21 de Julho de 2009 [Decree No. 1,805 of 21 July 2009]. Governo do Estado do Pará. Belem, Brazil: Para State Government.

União de Industria de Cana-de-Açucar. 2010. FAQ. http://english.unica.com.br/FAQ/ (last accessed 16 February 2010).

Walker, R. 2004. Theorizing land cover and land use change: The case of tropical deforestation. *International Regional Science Review* 27 (3): 247–70.

Walker, R., J. Browder, E. Arima, C. Simmons, R. Pereira, M. Caldas, R. Shirota, and S. Zen. 2009. Ranching and the new global range: Amazônia in the 21st century. *Geoforum* 40 (5): 732–45.

Walker, R., R. DeFries, M. de Carmen Vera-Diaz, Y. Shimabukuro, and A. Venturieri. 2009. The expansion of intensive agriculture and ranching in Brazilian Amazonia. In *Amazonia and global change*, ed. M. Keller, M. Bustamante, J. Gash, and P. Dias, 61–81. Washington, DC: American Geophysical Union.

Walker, R., N. Moore, E. Arima, S. Perz, C. Simmons, M. Caldas, D. Vergara, and C. Bohrer. 2009. Protecting Amazônia with protected areas. *Proceedings of the National Academy of Sciences* 106 (26): 10582–86.

Wilkinson, J., and S. Herrera. 2010. Biofuels in Brazil: Debates and impacts. *Journal of Peasant Studies* 37 (4): 749–68.

Zimmerer, K. S. 2011. New geographies of energy: Introduction to the special issue. *Annals of the Association of American Geographers* 101 (4): 705–11.

Shifting Networks of Power in Nicaragua: Relational Materialisms in the Consumption of Privatized Electricity

Julie Cupples

Department of Geography, University of Canterbury

Drawing primarily on actor-network theory, this article explores the aftermath of electricity privatization in Nicaragua. Privatization has not gone well for most low-income consumers in Nicaragua, who have faced rising and unaffordable tariffs and frequent power cuts that have been economically and psychologically devastating. Scholarship on privatization has focused on the social injustices and exclusions that privatization often engenders, but very little attention has been paid to how privatization is enacted materially. To implement privatization successfully across space, privatizers have to delegate some of their action to nonhumans that they anticipate will function as black-boxed intermediaries. In landscapes of economic hardship and popular suspicion, however, these intermediaries sometimes turn into mediators or technologies with political effects. Electricity consumers have resorted to a range of tactics to subvert the strategies of Spanish multinational Unión Fenosa, which took over the distribution of electricity, and these tactics have involved the creative and opportunistic enrollment of nonhumans. By tracing the shifting associations between the heterogeneous actors that make up the electricity (actor-) network, I seek to illuminate the relational materialisms in the consumption of privatized electricity and their potential for political transformation. An actor-network theory approach enables us to observe, amidst the entanglement of neoliberalizing maneuvers and disabling effects, material practices and translations in the network that are not always disempowering for ordinary people. It also reveals the contingency of multinational power and the (un)making of the global political economy in the spaces of everyday life. *Key Words: actor-network theory, consumer activism, electricity, Nicaragua, privatization.*

本文主要借鉴演员网络理论，探讨了尼加拉瓜电力私有化的后果。私有化对尼加拉瓜的大多数低收入消费者并不顺利，其所面临的高涨和负担不起的关税和频繁的断电已经给他们造成了经济上和心理上的摧残。有关私有化的学术研究焦点以往着重于社会的不公正和私有化常滋生的闭关政策，但私有化是如何被物质地立法却很少受到重视。为了成功地实施跨越空间的私有化，私有化主必须把其一些行动委托到作为黑装箱的非人文中介。然而，在经济困难和民众怀疑的境况，这些中介有时会变成调解员或有政治影响的技术。电力用户已使出了一系列的战术来颠覆接管电力分配的西班牙跨国电力公司 Uni'on Fenosa，这些策略已涉及到人类以外的创造性和机会主义的注册。通过追踪构成电力（演员）网络之间的异构行为者的关系转变，我寻求说明被私有化的电力中的关系唯物主义，和它们的政治变革的潜力。演员网络理论方法使我们能够观察，在新解放主义的回旋和致残的影响的缠结之中，物质实践和网络翻译并非总是剥夺普通百姓权利。它也揭示了日常生活空间中跨国力量和全球政治经济的意外事件。*关键词：演员网络理论，消费者的积极性，电力，尼加拉瓜，私有化。*

Primariamente con base en la teoría del actor-red, este artículo explora lo que ocurre en Nicaragua después de la privatización de la electricidad. La privatización no ha resultado buena para la mayoría de los consumidores de bajos ingresos en ese país, que han tenido que enfrentar tarifas cada vez más caras e impagables, lo mismo que frecuentes cortes de fluido que han sido devastadores económica y psicológicamente. Los estudios sobre la privatización se han concentrado en las injusticias sociales y exclusiones que ésta a menudo conlleva, pero muy poca atención se ha prestado a la manera como la privatización materialmente es puesta en ejecución. Para implementar con éxito la privatización a través del territorio, quienes están a cargo tienen que delegar algunas de sus acciones en instrumentos que con anticipación se sabe harán su trabajo de intermediación como cajas negras. En los paisajes de dificultad económica y desconfianza popular, sin embargo, a veces estos intermediarios se tornan en mediadores o tecnologías de efectos políticos. Los consumidores de electricidad han acudido a una serie de tácticas para subvertir las estrategias de la multinacional española Unión Fenosa, que adquirió la función de distribuidora de la electricidad, y estas tácticas han incluido la creativa y oportunista incorporación de colaboradores no humanos. Siguiéndole la pista a las cambiantes asociaciones entre los agentes heterogéneos (actor-) que constituyen la red de la electricidad, busco iluminar los materialismos relacionales en el consumo de electricidad privatizada y su potencial para la transformación política. Un enfoque de la teoría actor-red nos permite observar, en medio del embrollo de maniobras neoliberalizadoras y efectos incapacitadores, prácticas materiales y traslaciones en la red que no son siempre desempoderadoras para la gente ordinaria. Esto también revela la contingencia del poder multinacional y el desmonte de la economía política global en los espacios de la vida cotidiana. *Palabras clave: teoría de la relación actor-red, activismo consumista, electricidad, Nicaragua, privatización.*

The privatization of public utilities became a key component of International Monetary Fund (IMF), World Bank, and Inter-American Development Bank conditionality in many third world countries in the 1990s, based on a belief that privatization and deregulation of electricity promote "reliability and efficiency . . . and increase affordable access to modern energy systems for the poor" (Byer, Crousillat, and Dussan 2009, xxi). Accordingly, international financial institutions (IFIs) have dramatically reduced lending for public utilities, lending instead to private investors in the sector (McGuigan 2007). Cash-strapped governments in countries with inadequate public electricity systems have frequently created concessions for private investors in the hope that they will make the investments that the sector needs. In Central America, such energy conditionalities have exacerbated fossil fuel dependence (Solá Montserrat 2008) and dramatically increased spending on oil imports from US$47 million in 1990 to US$444 million in 2002 (CEPAL 2003).

The privatization of Nicaragua's electricity system was initiated under the government of Violeta Chamorro (1990–1996) and intensified under the government of Arnold Alemán (1996–2001). Under Alemán, Nicaragua was admitted into the Heavily Indebted Poor Countries Initiative (HIPC) and the IMF and the World Bank made privatization of electricity a condition for receiving debt relief under the HIPC. In 1998, the state-run Nicaraguan Electricity Company (ENEL) was unbundled into seven separate companies: two distribution companies (Disnorte and Dissur), one transmission company (ENATREL), and four generating companies (Gemosa, Hidrogesa, Geosa, and Gecsa). The most controversial aspect of the privatization process was the secretive sale in 2000 of Disnorte and Dissur to Spanish multinational Unión Fenosa, the only bidder, for US$115 million in a thirty-year concession under very favorable legal arrangements.

Drawing primarily on actor-network theory (ANT), this article explores the aftermath of electricity privatization in Nicaragua. It documents struggles for more sustainable energy futures in Latin America and analyzes the ways in which electricity consumers (especially the urban poor) are responding to the energy challenges facing the country. It seeks to make a contribution to a number of ongoing debates within the geographic literature. First, it contributes to the literature that is exploring the political dynamics of socio-technical and socio-natural assemblages. This work has demonstrated how nonhuman technological artifacts and environmental resources mediate social and political struggles (Marvin, Chappells, and Guy 1999; Law and Mol 2001; Barry 2002; Mitchell 2002; Kirsch and Mitchell 2004) and can come to embody or contest capitalist relations (Kaika and Swyngedouw 2000; Keil 2003; Loftus 2006; Loftus and Lumsden 2008). The meanings of resources and technologies are not therefore self-evident but emerge relationally through interactions with other nonhumans, human bodies, institutions, emotions, discourses, and ideas and through the overlapping of different networks.

Second, it joins the growing number of scholarly voices that assert that ANT analysis can illuminate the political contestation of neoliberalism (Kendall 2004; Larner and Walters 2004; Schwegler 2008; Holifield 2009), ongoing concerns notwithstanding (Kirsch and Mitchell 2004). Although a number of scholars believe that ANT lacks political or critical content (see, for example, Rudy 2005; Whittle and Spicer 2008), it enables us to conceptualize neoliberalism not as a completed set of social relations administered in a technical top-down fashion but as a performative work in progress that involves multiple and situated forms of improvisation, provisionality, and contingency (see Law and Mol 2001; Castree 2002; Larner and Walters 2004; Schwegler 2008; Larner and Laurie 2010). ANT analysis might therefore enable us to address questions less easily accessed through more conventional modes of social inquiry. As Holifield (2009, 639) wrote, "instead of explaining inequalities by contextualizing and situating them, actor-network approaches turn our attention to the forms and standards that make it possible to circulate new associations of entities, to generalize social orders, and to situate actors within a social context."

ANT can shed new light on the encounters generated by neoliberal economic policies and their potential not only for disempowerment and exclusion but for political action. In other words, it can reveal how neoliberalism and the global political economy are constructed in the spaces of everyday life (Zimmerer 2011).

Finally, it also contributes to recent scholarly debates on what Loftus (2006, 1032) called the "citizen-consumer nexus" to explore how consumption, because it is "good for thinking" (García Canclini 2001, 37), sometimes enables expanded and more inclusive forms of political and cultural citizenship. This is evident not just in the work focused on ethical consumption (Micheletti 2003; Barnett et al. 2005; Mansvelt 2008) or on entertainment media (Miller 2007; Dahlgren 2009) but also with respect to the consumption of privatized utilities, as the growing literature on opposition to the privatization of water in Bolivia and elsewhere in the global South has clearly demonstrated (see Shiva 2002; Nickson and Vargas 2002; Morgan 2005; Perreault 2005). It is apparent that the emergence of consumer citizenship around the question of electricity privatization and delivery is one of the important modes through which Nicaraguans come to know and understand the politics of neoliberalism and its geographic particularities. By tracing the shifting associations between the heterogeneous actors that make up the electricity (actor-)network, I seek to illuminate the relational materialisms (see White 2009, 11) of privatization by exploring "how things are 'stitched together' across divisions and distinctions" (Murdoch 1997, 322) and the potential for political transformation of these interactions.

Privatizing Electricity

Although many scholars writing on privatization processes in the global South have focused on the social injustices and exclusions privatization engenders (Shiva 2002; Green 2003; Peet 2003), little attention has been paid to how privatization is enacted materially, as well as discursively and ideologically. To implement privatization successfully across space, exert control at a distance, and stabilize the network, privatizers have to delegate some of their action to nonhumans and hope that this delegation translates into a set of what Kendall (2004, 68) referred to as "routinised solutions." Unión Fenosa anticipates that the nonhumans to which action is delegated will function as black-boxed intermediaries. Often, however, such delegation has political and unanticipated effects and the intermediaries turn into mediators.[1] Following such processes not only provides insight into why a network fails to establish itself successfully but also helps us to understand in greater depth, in the language of de Certeau (1984), how the privatization strategies of the powerful (in this case the IFIs, Unión Fenosa, and elite sectors within the Nicaraguan government) are transformed by the tactics of the socially weak (in this case low-income Nicaraguan consumers), which have involved the creative and opportunistic enrollment of nonhumans. We should therefore not only focus on the macrolevel dimensions of privatization, on privatization as a molar entity, but also explore how the material enactment of privatization in the spaces of everyday life causes it to break apart into molecular multiplicities (see Deleuze and Guattari 1987). Such an approach is valuable because of its refusal to presume the existence of a stable structure and to "determine the outcome of particular configurations as given by prefigured power relations" (Schwegler 2008, 688). In other words, it helps us to understand how and why neoliberal policy becomes (or indeed fails to become) effective in specific places and spaces.

The privatization process in Nicaragua was conducted amidst high levels of secrecy and political opposition. The distributing companies and some of the generating companies were sold off in highly secret deals at absurdly low prices to national elite entrepreneurs or foreign companies (Herrera Montoya 2005a). Not many investors were interested in Nicaragua's virtually obsolete electricity systems so the agreements contained incentivizing clauses, such as fixed capacity payments, which are paid whether or not the electricity is actually produced. The secrecy and lack of transparency that surrounded these deals represent an attempt at black boxing and could be seen, drawing on Barry's (2002) work, as a political action designed to have antipolitical effects, an action that might end up having deeply political and costly consequences. As Guy and Marvin's (1998) work in the United Kingdom has demonstrated, privatization can sometimes reconfigure electricity consumption in unanticipated ways, producing more dynamic modes of interaction between producers and consumers, with implications for the profitability of private electricity providers. The fact that privatization deals were conducted with very little transparency clearly fueled a landscape of popular suspicion in a country like Nicaragua in which citizens have frequently encountered and challenged the abuse of elite power. This popular suspicion in turn fuels new demands for democratic participation and has contributed to a failure to fully privatize the electricity network. Although the Nicaraguan government had agreed to IMF and World Bank conditions, confirming that in return for debt relief under the HIPC the state generating companies would be privatized, in the 2004 completion point document it was stated that they had not been able to privatize either Gecsa, because they were unable to find a buyer, or Hidrogesa, the hydroelectric power plant, because of organized political opposition. Similarly, the transmission company (ENATREL) has remained in state hands.

Most Nicaraguans would agree that the privatization of the electricity network has not gone well. Although the public systems that preceded privatization were highly inadequate,[2] today Nicaragua has the lowest levels of installed capacity and of electricity generated, the highest energy losses,[3] and the lowest level of coverage in Central America (see Table 1). In addition, 80 percent of Nicaragua's electricity continues to be generated by oil (Nitlapán-Envío 2007).[4]

In addition, electricity spending consumes a significant portion of household income, with many people spending one third of household income on electricity. Until 2008, the country was suffering from frequent blackouts and erratic supply, which has had devastating economic and psychological impacts on livelihoods, workplaces, hospitals, schools, home life, and security. During fieldwork in Nicaragua, I have lived through many power cuts, experiencing firsthand alongside Nicaraguan citizens having food spoil in the refrigerator, having to postpone working or studying, feeling fear when walking through the city at night in total darkness, dealing with sweaty nights without fans or air conditioners, attending meetings over the sound of noisy generators, and reorganizing activities around rationing. Nicaragua's ubiquitous home businesses, which sell baked goods, beans, milk, curd, homemade popsicles, hairdressing services, dressmaking, and the rental of time on a Playstation or an Internet-enabled computer, were devastated. Contributing factors to the crisis include rising oil prices, drought and low hydro lake levels, overexploitation of geothermal wells, government corruption, and the behavior of the privatized electricity companies. The private generating companies have invested little in upgrading their equipment and many plants produce below their installed capacity (Acevedo 2005). Unión Fenosa has also repeatedly failed to make the investment in maintenance and upgrading that it was contracted to do and has successfully pressed for both tariff increases and government subsidies (Wood 2005; Gómez 2006b).[5]

Consumers have mobilized in response to this situation in a diverse range of ways, including the formation of a vibrant and assertive Red de Defensa de los Consumidores (Consumer Defense Network). Most Nicaraguans direct their frustration not at the private generators but at the distributor, Unión Fenosa, the entity responsible for delivering electricity to the consumer and charging for it. The next sections explore some of the tactical enrollments of nonhumans that have emerged in response to Nicaragua's energy challenges. They highlight the ways in which new constitutive relations between humans and nonhumans, such as the electricity bill, the electricity meter, and media spaces, have been performed in response to the crisis. By refusing an analytical distinction between the human and the nonhuman, and the micro and the macro, an ANT approach enables us to observe, amidst the entanglement of neoliberalizing maneuvers and disabling effects, translations in the network that are not always disempowering for ordinary people.

Table 1. The Central American electricity sector

Countries of Central America (pop. in millions)	Installed capacity (MW)		Net electricity generated (GWh)		Cost to consumer per kWh (US$)		Losses (%)		Electricity coverage (% of population)	
	1990	2008	1990	2008	1998	2008	1985	2008	1985	2008
Costa Rica (4.53)	888.6	2446.6	3543.0	9412.9	0.067	0.097	8.5	10.6	80.5	98.8
El Salvador (6.13)	650.4	1441.3	2164.3	5916.2	0.098	0.157	12.8	12.8	41.3	85.8
Guatemala (13.68)	810.9	2250.9	2318.4	7903.7	0.078	0.179	16.6	17.1	29.8	83.8
Honduras (7.24)	532.6	1581.4	2273.6	6800.7	0.085	0.105	13	23.5	30.9	77.0
Panama (3.39)	883.4	1623.5	2624.8	6265.0	0.111	0.21	19.2	11.8	53.8	88.9
Nicaragua (5.68)	363.4	879.7	1251.0	3100.2	0.116	0.149	13.0	27.3	46.8	64.5

Source: CEPAL (2003, 2009)

The Electricity Bill and the Electricity Meter

Things often become black-boxed when they are too complex to decipher (Latour 1987). Indeed, keeping the diffuse spatiality of Nicaragua's electricity network in view is difficult in both academic research and consumer activism. Taking on the electricity network as a whole is far beyond the scope of both this article and the capacities of individual consumers, just as taking on the vast and seemingly faceless neoliberal apparatus enacted by too many robust institutional and discursive actors is daunting and too difficult. Consumers can, however, interact with particular nonhuman actors in the network in a way that undermines their black boxing. In Nicaragua, electricity meters and bills designed to function as intermediaries by Unión Fenosa instead became important mediators of political action.

One of Unión Fenosa's investments in the distribution network was to install new meters in people's homes to measure their electricity consumption. In Nicaragua, as has been documented elsewhere (Guy and Marvin 1998; Loftus 2006), the installation of new utility metering technologies sometimes enables providers to further disadvantage low-income consumers, and in most cases, people's bills doubled following the installation of the new meters (Herrera Montoya 2005b). In Hinchliffe's (1996, 666) study of the electricity network in Denmark, he described, drawing on Ackrich (1992), how the electricity meter functioned as an "unequivocal spokesperson" or as a "reassuring actor" for both the consumer and the electricity company. In black-boxed systems, meters act to stabilize the network and force consumers through obligatory points of passage (Latour 1987). In Nicaragua, the meters did not gain such coherence and many people put them to new uses, by bypassing them and hooking up directly to the line in the street or by covering them with black paint so that they could not be read. At times, Unión Fenosa employees assisted residents in altering the meter so that it provided a lower reading.

Latour (1987) suggested that we think of the difference between an unproblematic black box and a set of controversies in terms of a two-faced Janus, where one is already made and the other is in the making. If we take this approach into the realm of the electricity meter, we can see that it is not simply a matter of Unión Fenosa creating and installing a meter that works and, as a result of this action, the people will be convinced. Rather, as Latour's dictum (1987, 10) indicates, the meter "will work when all the relevant people are convinced," which they clearly were not. Because meters are "complex socio-technical systems" (Marvin, Chappells, and Guy 1999, 109), in situations of economic deprivation, they often extend their capacities beyond that of technical measurement, taking on social and political attributes. As Furlong (2007) has noted, the interface between consumers and meters is highly variable depending on how meters are managed and what their design aims to achieve. Loftus (2006, 1024) showed how low-income residents in Durban, South Africa, worked to "shatter the dominance of water-meters in their households" and a focus on such (inter)actions reveals "potentials for political changes and emancipation within what appears to be a bleak, reified drama of water delivery" (1029).

Similarly, in a stabilized electricity network, the electricity bill usually functions as an intermediary or unequivocal spokesperson. In Nicaragua, however, the bill has been subjected to a high degree of scrutiny and deconstruction by consumers and consumer activists, who have drawn attention to a whole series of both illegal and unfair charges that Unión Fenosa has included on bills. In addition to the kilowatt hours of electricity consumed, bills have included an illegal 1 percent tax for the municipality, a charge called *commercialization* (a charge to print the bill and deliver it to the consumer's house), a charge to hire the meter, a charge for public street lighting (which sometimes is not provided), and a charge for unregistered energy. After the installation of a new meter, which results in dramatic increases in recorded electricity consumption and much higher electricity bills for consumers, Unión Fenosa then alleges that people were not paying enough for their electricity prior to this, so they now need to be charged for this previous "unregistered energy." Such experiences are typical throughout the country and have had a mobilizing effect on consumers. The Red has raised a lot of awareness, encouraging consumers to take their complaints to both INE, the state regulator, and the offices of Unión Fenosa. National newspapers have provided readers with information on how to read their electricity bills (see Gómez 2006a). By 2008, INE was receiving 3,000 and Unión Fenosa 4,500 formal complaints a month. The Red has also successfully used Article 105 of the Nicaraguan Constitution, which establishes that the state must guarantee that a public service is accessible to the whole population, to challenge the way the Nicaraguan government has failed to enforce the conditions of Unión Fenosa's concession. The charges for nonexistent street lighting have also had a mobilizing effect on those who do not have it but are expected to pay for it. In 2005, one of the newly elected municipal councilors of the rural municipality of Achuapa told me he had decided to run for office as a result of being charged for street lighting that he and his neighbors did not receive.

So the properties of meters and bills are not fixed in advance of their use and become repositories of meaning (Pepperell and Punt 2000) and new allies in the struggle against neoliberalism. Just as Unión Fenosa delegates some of the work of measuring and charging for electricity to the meter and the bill, consumers can delegate some of their consumer activism to these entities. The blacked-out or bypassed meter and the deconstructed bill are used in ways not intended by the multinational distributor. In the confrontation between the projected user and the actual user, they undergo a process of translation or deinscription (Ackrich 1992) and take on new ontological statuses, as do the consumer-citizens who enroll them and put them to new uses. Amidst erratic supply, high prices, and the loss of income, the meter and bill could not hold consumption and consumers together in the way that was intended and began to mediate action differently.

In the neoliberal energyscape, electricity meters and bills become things rather than objects (Ingold 2010). For Ingold (2010, 4), whereas an object is like a fait accompli, with clearly defined edges and surfaces, a thing is "a certain gathering together of the threads of life" that becomes "caught with other threads in other knots." Meters and bills in the homes of low-income people, for whom daily life is a constant struggle to gain access to affordable and reliable electricity, are transformed by their interaction with the emotional, embodied, and experiential knowledges of such struggles, their properties rendered unpredictable. They have no choice but to "join in the processes of formation" (Ingold 2010, 6), which include the political struggles that have been part of the landscapes of electricity consumption. Like Deleuze and Guattari's book (1987, 22–23), the bill and the meter are assemblages "with the outside" that "in [their] multiplicity, necessarily act[s] on semiotic flows, material flows and social flows simultaneously."

Urban Communication and Mediated Activism

The interactions with bills and meters are only part of the complex and hybrid geographies of resistance enacted through the enrollment of things and technologies. To counter widespread consumer opposition and shift the blame for the crisis onto other stakeholders, Unión Fenosa engaged in a vigorous public relations campaign. In 2006, it ran prime-time television advertisements in which it tried to blame the generators for the power cuts by showing the lights going out in an Unión Fenosa office and stating, "We also suffer power cuts because we don't generate electricity, we only distribute it." It also posted billboards throughout Managua that attempted to blame the poor for the company's own failings. These read, "Stealing energy is no joke. It could leave us all in the dark." The cluttered complexity of Nicaragua's urban visual landscape means, however, that Unión Fenosa's billboards appeared alongside others advertising UHT milk with the slogan "Blackout proof" (*A prueba de apagones*), as well as alongside anti–Unión Fenosa graffiti scrawled on empty walls throughout Nicaragua's cities.

Unión Fenosa clearly has the financial resources to enact a public relations campaign to try to improve public perceptions, but such resources cannot automatically be equated with effective power as the campaign might not have the desired effect (see Allen 2003), particularly as such actions necessarily involve the mobilization of yet more mediators (Latour 2005). Indeed, to exert power across space, such advertisements would have to successfully dissociate the associations made by consumer activists and people's embodied experiences of electricity failure. In other words, what the Red firmly associates, Unión Fenosa would like to dissociate (see Callon and Latour 1981).

Diverse forms of mediated activism also make it difficult for Unión Fenosa to dissociate what the Red associates. Consumer-citizens have prioritized gaining media spaces, especially radio, in which to denounce Unión Fenosa and inform people of their rights as consumers and citizens. Critiques of Unión Fenosa proliferate across diverse media spaces, enrolling and being enrolled by new actants. The Internet abounds with information about Unión Fenosa's abuses, including an international campaign in which Unión Fenosa is renamed Unión Penosa (Shameful Union). In 2006 and 2007, consumers posted comments about their experiences dealing with the multinational on the killfenosa.org Web site (Zamora 2006). Online newspaper articles on the electricity crisis attract large numbers of comments and a video on YouTube shows Unión Fenosa as Darth Vader, who represents the dark side, appearing in a Managua barrio and being defeated by a local man armed with a machete.[6] The mediated resistance connects firmly with people's embodied experiences of electricity failure—the inability to make a living, the smell of rotten food, accidents, difficulty sleeping—and electricity consumers fail to be inscribed as easy sources of profit for a foreign company. It was somewhat neocolonial of Unión Fenosa to assume that they would be.

Conclusions

Even though the poor performance of privatized electricity has been economically and psychologically devastating for ordinary people, the modes of resistance in operation demand our scholarly attention. In a recent article, Marston, Woodward, and Jones (2007) advanced a flat materialist ontology of globalization to open up a greater sense of political alternatives and they draw on the ideas of Subcomandante Marcos of the Zapatistas. In response to the notion that changing the world is difficult because the world is so big, Marcos (2006, cited in Marston, Woodward, and Jones 2007, 58–59) outlined a "politics grounded . . . in material practice." He said that from above we see only injustice and despair in the world, but from below the world becomes so spacious that all kinds of things become possible. Seen from above, Nicaragua's electricity system is vast and complex; it is not easy to "untangle all the maneuvers carried out by the generating companies, Unión Fenosa and the INE against users" (Herrera Montoya 2005a, 54, author's translation) but at ground level, in the spaces of everyday life, there are some materials to work with. The privatization of electricity gives rise to new forms of connectivity (in a political sense as well as in a technical one) and new kinds of everyday material practice as people encounter privatization through the nonhuman artifacts through which electricity is provided, delivered, and billed.

Individual electricity consumers dealing with food that they have to throw away, an unpayable bill, a burned-out electrical appliance, or an inability to study for the next day's exam, the microactors in the network, have begun to connect their struggles, in part by enrolling nonhumans in new ways, and consequently have begun to grow to macro size (see Callon and Latour 1981). For example, the actions of Unión Fenosa were taken up in 2007 in a ruling by the Permanent People's Tribunal, where they have been conceptualized as "a violation of the human rights of the majority of Nicaraguans" (TNI 2007). As one founder of the Red told me in an interview in November 2006, "It is not just about electricity, it's also a process of people gaining more dignity" (*proceso de dignificación*; Santos Amador Mairena, personal communication, November 2006).

If neoliberal economic policy is partly about control at a distance exerted not through direct government but through technologies of governance, then it requires nonhuman intermediaries to which actions are delegated. In climates of popular suspicion and growing consumer activism, however, and because intermediaries are devoid of essential properties, there is no guarantee that these intermediaries will not begin to mediate action in ways that complicate the effective control of people at a distance and disrupt the exertion of power and implementation of privatization across space. Such an approach points to the contingency of multinational power and prevents us from developing an overly dichotomized analysis that pits an abusive dominating power (Unión Fenosa) on the one hand against a resistant subordinated group of people (low-income electricity consumers) on the other. As Allen (2003) has indicated, such an approach does not come to terms with how it is that power is exerted across space. Because power can only be understood as a relational network effect, a multinational company is not inherently powerful. It can only exert power for as long as it is able to "reproduce its capabilities or develop others" and its ability to "achieve certain objectives will be enhanced, modified or deflected by the ensemble of relationships in which it is a part" (Allen 2003, 24). We must not, therefore, decide in advance who is big and who is small, who is a microactor and who is a macroactor, because size depends on the ability to assimilate other elements or alternatively to dissociate elements enrolled by other actors as quickly as possible (Callon and Latour 1981). The economic success of Unión Fenosa is not therefore directly proportional to its ability to manipulate or rip off Nicaraguan consumers, because many of the actors in the network, including consumers, meters, and bills, might not be assimilated or translated in the way that Unión Fenosa anticipated. Therefore, we must not only focus on the behaviors enacted by Unión Fenosa in the pursuit of particular objectives, deplorable though these might be, but also look at their interaction with other actors in the network of which they are a part. Privatization is negotiated through distributed kinds of agency that we might fail to see if we only look at privatization from above, as a macrolevel system, dispensed by powerful actors.

The privatization of electricity has transformed everyday life in ways that are mostly negative for low-income consumers, but the creative and tactical enrollment of nonhumans, which is part of coping with and resisting privatization, reveals in a theoretically productive way the (un)making of neoliberalism and the (de)centering of the global political economy in the spaces of everyday life (see Aitken 2006, 80). One important lesson we can learn from Nicaragua's struggles for electricity is that we cannot destroy neoliberal capitalism (just yet), but we can tamper with its meters.

Notes

1. Latour (2005) made a conceptual distinction between intermediaries and mediators. Whereas an intermediary "transports meaning or force without transformation," holding its shape and coherence as it moves through space, mediators "transform, translate, distort and modify the meaning of the elements they are supposed to carry" (39).
2. Although there was some investment during the twentieth century in both hydroelectric and geothermal power generation, the public system remained heavily dependent on imported oil. During the Sandinista Revolution, electricity prices were kept low, but new connections did not keep up with population growth (CEPAL 2003) and blackouts were common. At the time of the Sandinista electoral defeat in 1990, the Nicaraguan electricity system urgently required investment, and only 49 percent of the population had access to electricity.
3. Electricity losses are divided between "technical losses," energy that is lost from the system during transmission because of infrastructure problems, and energy stolen by either large consumers or shanty town dwellers who establish their own illegal connections.
4. Whereas Costa Rica and Honduras generate 80 percent of their electricity from autochthonous renewable sources, and Guatemala, El Salvador, and Panama manage 50 percent, Nicaragua only reaches 15 percent of electricity generated in this category (CEPAL 2003). This percentage is likely to improve in the future owing to the creation of a wind farm in Rivas and Brazilian investment in hydroelectric power.
5. Although the complexities of the political landscape in Nicaragua are beyond the scope of this article, the current Nicaraguan president and leader of the Sandinista Front for National Liberation (FSLN), Daniel Ortega, was very critical of Unión Fenosa as leader of the opposition (AP 2007). The FSLN returned to power in 2007 and has since accommodated and purchased shares in the Spanish multinational. Concerns about the lack of government transparency continue particularly as Venezuelan power plants, oil, and aid, which have helped to alleviate the electricity crisis, are excluded from the national budget. Unión Fenosa remains in Nicaragua, but in a new attempt at ordering and gaining legitimacy now operates under a new name (Gas Natural). Although repeated calls to cancel their contract have not succeeded, it is clear from the multiple forms of ongoing contestation that, in the words of de Certeau (1984, 16) "this relationship of forces" has not "become any more acceptable."
6. See http://www.youtube.com/watch?v=c3qzSsHuZgE.

References

Acevedo, A. 2005. The energy crisis explained. *Envío* 287. http://www.envio.org.ni/articulo/2973 (last accessed 12 January 2010).

Ackrich, M. 1992. The description of technical objects. In *Shaping technology/building society: Studies in sociotechnological change*, ed. W. Bijker and J. Law, 259–64. Cambridge, MA: MIT Press.

Aitken, R. 2006. Performativity, popular finance and security in the global political economy. In *International political economy and poststructural politics*, ed. M. de Goede, 77–96. Basingstoke, UK: Palgrave Macmillan.

Allen, J. 2003. *Lost geographies of power*. Oxford, UK: Blackwell.

AP. 2007. Ortega: Unión Fenosa perjudica a Nicaragua [Ortega: Union Fenoso harms Nicaragua]. *La Prensa* 12 October. http://archivo.laprensa.com.ni/archivo/2007/octubre/12/noticias/ultimahora/220891.shtml (last accessed 12 January 2010).

Barnett, C., P. Cloke, N. Clarke, and A. Malpass. 2005. Consuming ethics: Articulating the subjects and spaces of ethical consumption. *Antipode* 37:23–45.

Barry, A. 2002. The anti-political economy. *Economy and Society* 31 (2): 268–84.

Byer, T., E. Crousillat, and M. Dussan. 2009. *Latin America and the Caribbean region energy sector—Retrospective view and challenges*. Washington, DC: The World Bank.

Callon, M., and B. Latour. 1981. Unscrewing the big Leviathan: How actors macro-structure reality and how sociologists help them to do so. In *Advances in social theory and methodology: Towards an integration of micro and macro-sociology*, ed. K. Knorr-Cetina and A. V. Cicouvel, 277–303. London and New York: Routledge.

Castree, N. 2002. False antitheses? Marxism, nature and actor-networks. *Antipode* 34 (1): 111–46.

CEPAL. 2003. *Evaluación de diez años de reforma en la industria eléctrica del istmo centroamericano* [Evaluation of ten years of reform in the Central American electricity industry]. México City, Mexico: CEPAL.

———. 2009. *Istmo centroamericano: Estadísticas del subsector eléctrico: Informe preliminar del segmento de la producción de electricidad* [Central America: Statistics of the electric subsector: Preliminary report on electricity production]. México City, Mexico: CEPAL.

Dahlgren, P. 2009. *Media and political engagement: Citizens, communication, and democracy*. Cambridge, UK: Cambridge University Press.

de Certeau, M. 1984. *The practice of everyday life*, trans. S. Rendall. Berkeley: University of California Press.

Deleuze, G., and F. Guattari. 1987. *A thousand plateaus: Capitalism and schizophrenia*. London: Athlone Press.

Furlong, K. 2007. Municipal water supply governance in Ontario: Neoliberalization, utility restructuring, and infrastructure management. Unpublished PhD thesis, University of British Columbia, Vancouver.

García Canclini, N. 2001. *Consumers and citizens: Globalization and multicultural conflicts*. Minneapolis: University of Minnesota Press.

Gómez, O. 2006a. ¿Cómo leer la factura de Fenosa? [How to read the bill from Union Fenoso]. *El Nuevo Diario*. 23 January. http://impreso.elnuevodiario.com.ni/2006/01/23/contactoend/10912 (last accessed 12 January 2010).

———. 2006b. Frank Kelly, acusado en la CGR [Frank Kelly, accused by the comptroller general]. *El Nuevo Diario*. 8 October:1A, 11A.

Green, D. 2003. *Silent revolution: The rise and crisis of market economics in Latin America*. London: Latin America Bureau.

Guy, S., and S. Marvin. 1998. Electricity in the marketplace: Reconfiguring the consumption of essential resources. *Local Environment* 3 (3): 313–31.

Herrera Montoya, R. S. 2005a. *Crisis del sector energético ¿Nicaragua apagándose?* [The electricity sector crisis: Nicaragua disconnected?]. Managua, Nicaragua: Red Nacional de Defensa de los Consumidores.

———. 2005b. Our electricity system is one of our political class' great failures. *Envío* 291. http://www.envio.org.ni/articulo/3075 (last accessed 12 January 2010).

Hinchliffe, S. 1996. Technology, power and space. *Environment and Planning D: Society and Space* 14:659–82.

Holifield, R. 2009. Actor-network theory as a critical approach to environmental justice: A case against synthesis with urban political ecology. *Antipode* 41 (4): 637–58.

Ingold, T. 2010. Bringing things to life: Creative entanglements in a world of materials. NCRM Working Paper 15. http://eprints.ncrm.ac.uk/1306/1/0510_creative_entanglements.pdf (last accessed 21 August 2010).

Kaika, M., and E. Swyngedouw. 2000. The environment of the city or . . . the urbanisation of nature. In *A companion to the city*, ed. G. Bridge and S. Watson, 567–80. Oxford, UK: Blackwell.

Keil, R. 2003. Urban political ecology. *Urban Geography* 24 (8): 723–38.

Kendall, G. 2004. Global networks, international networks, actor networks. In *Global governmentality: Governing international spaces*, ed. W. Larner and W. Walters, 59–75. London and New York: Routledge.

Kirsch, S., and D. Mitchell. 2004. The nature of things: Dead labour, nonhuman actors and the persistence of Marxism. *Antipode* 36:687–706.

Larner, W., and N. Laurie. 2010. Travelling technocrats, embodied knowledges: Globalising privatisation in telecoms and water. *Geoforum* 41:218–26.

Larner, W., and W. Walters. 2004. Globalization as governmentality. *Alternatives* 29:495–514.

Latour, B. 1987. *Science in action: How to follow scientists and engineers through society*. Cambridge, MA: Harvard University Press.

———. 2005. *Reassembling the social: An introduction to actor-network theory*. Oxford, UK: Oxford University Press.

Law, J., and A. Mol. 2001. Situating technoscience: An inquiry into spatialities. *Environment and Planning D: Society and Space* 19:609–21.

Loftus, A. 2006. Reification and the dictatorship of the water meter. *Antipode* 38:1023–45.

Loftus, A., and F. Lumsden. 2008. Reworking hegemony in the urban waterscape. *Transactions of the Institute of British Geographers* 33 (1): 109–26.

Mansvelt, J. 2008. Geographies of consumption: Citizenship, space and practice. *Progress in Human Geography* 32 (1): 105–17.

Marston, S. A., K. Woodward, and J. P. Jones. 2007. Flattening ontologies of globalization: The Nollywood case. *Globalizations* 4 (1): 45–63.

Marvin, S., H. Chappells, and S. Guy. 1999. Pathways of smart metering development: Shaping environmental innovation. *Computers, Environment and Urban Systems* 23:109–26.

McGuigan, C. 2007. *The impact of World Bank and IMF conditionality: An investigation into electricity privatization in Nicaragua*. London: Christian Aid.

Micheletti, M. 2003. *Political virtue and shopping: Individuals, consumerism, and collective action*. New York: Palgrave.

Miller, T. 2007. *Cultural citizenship: Cosmopolitanism, consumerism, and television in a neoliberal age*. Philadelphia: Temple University Press

Mitchell, T. 2002. *Rule of experts: Egypt, technopolitics, modernity*. Berkeley: University of California Press.

Morgan, B. 2005. Social protest against privatization of water: Forging cosmopolitan citizenship? In *Sustainable justice: Reconciling international economic, environmental and social law*, ed. M. C. Cordonier- Seggier and J. Weeramantry, 339–54. Leiden, The Netherlands: Martinus Nijhoff.

Murdoch, J. 1997. Towards a geography of heterogeneous associations. *Progress in Human Geography* 21 (3): 321–37.

Nickson, A., and C. Vargas. 2002. The limitations of water regulation: The failure of the Cochabamba concession in Bolivia. *Bulletin of Latin American Research* 21 (1): 99–120.

Nitlapán-Envío. 2007. The FSLN government pieces together its new international policy puzzle. *Envío* 314. http://www.envio.org.ni/articulo/3643 (last accessed 12 January 2010).

Peet, R. 2003. *Unholy trinity: The IMF, World Bank and WTO*. London: Zed Books.

Pepperell, R., and M. Punt. 2000. *The postdigital membrane: Imagination, technology and desire*. Bristol, UK: Intellect Books.

Perreault, T. 2005. State restructuring and the scale politics of rural water governance in Bolivia. *Environment and Planning A* 37 (2): 263–84.

Rudy, A. P. 2005. On ANT and relational materialisms. *Capitalism Nature Socialism* 16 (4): 109–25.

Schwegler, T. A. 2008. Take it from the top (down)? Rethinking neoliberalism and political hierarchy in Mexico. *American Ethnologist* 35 (4): 682–700.

Shiva, V. 2002. *Water wars: Privatization, pollution and profit*. New York: South End Press.

Solá Montserrat, R. 2008. *Estructura económica y su contexto centroamericano y mundial* [Economic structure and its Central American and global context]. Managua, Nicaragua: Hispamer-UCA.

TNI. 2007. Permanent People's Tribunal on European multinationals in Latin America—Unión Fenosa. Amsterdam: Transnational Institute. http://tni.org/node/61868 (last accessed 29 January 2010).

White, D. F. 2009. *Technonatures: Environments, technologies, spaces, and places in the twenty-first century*. Waterloo, ON, Canada: Wilfrid Laurier University Press.

Whittle, A., and A. Spicer. 2008. Is actor network theory critique? *Organization Studies* 29 (4): 611–29.

Wood, R. 2005. *The Nicaraguan energy sector: Characteristics and policy recommendations*. Washington, DC: Johns Hopkins SAIS.

Zamora, R. 2006. Sitios web en Nicaragua: Comunicación irreverente [Nicaraguan Web sites: Irreverent communication]. *El Nuevo Diario* 8 October: 5A.

Zimmerer, K. S. 2011. New geographies of energy: Introduction to the special issue. *Annals of the Association of American Geographers* 101 (4): 705–11.

"Because You Got to Have Heat": The Networked Assemblage of Energy Poverty in Eastern North Carolina

Conor Harrison* and Jeff Popke†

*Department of Geography, University of North Carolina, Chapel Hill
†Department of Geography, East Carolina University

Current discussions of energy policy seldom acknowledge the problem of energy poverty, a situation in which a household cannot afford to adequately heat or cool the home. In this article, we examine the concept of energy poverty and describe some of its contours in a rural part of North Carolina. Energy poverty, we suggest, is best viewed as a geographical assemblage of networked materialities and socioeconomic relations. To illustrate this approach, we focus on the geographical patterns of three key determinants of energy poverty in eastern North Carolina: the socioeconomic characteristics of rural households, the networked infrastructures of energy provision, and the material conditions of the home. Throughout, we highlight the lived effects of energy poverty, drawing on transcripts from interviews conducted with recipients of weatherization assistance in the region. The challenges of the energy poor, we suggest, deserve greater attention in public policy and as part of a broader understanding of welfare and care. *Key Words: energy poverty, infrastructure, networks, North Carolina, weatherization.*

能源政策的当前讨论很少承认能源贫困的问题，这种情况使一个家庭负担不起家庭所需充分加热或冷却的费用。在这篇文章中，我们考察了能源贫穷的概念和描述它在北卡罗莱纳州乡村地区的一些轮廓。我们建议，能源贫穷，最好被看作是网络物资和社会经济关系的地理组合。为了说明这种方法，我们着重于北卡罗莱纳州东部的能源贫困的三个关键因素的地理格局：农村家庭的社会经济特点，能源供应的网络基础设施，以及家庭的物质条件。自始至终，我们通过借鉴对该地区防寒保暖援助受惠人的采访记录，强调能源贫困的生活影响。我们建议，能源贫穷的挑战，应该在公共政策中，和作为福利和关怀的更广泛的理解的一部分，得到更多的关注。*关键词：能源贫困，基础设施，网络，北卡罗莱纳州，防寒保暖。*

Las actuales discusiones sobre política energética rara vez reconocen el problema de la pobreza de energía, una situación en la que un hogar no puede solventar el costo para calentar o enfriar adecuadamente la casa. En este artículo examinamos el concepto de pobreza de energía y describimos algunos de sus contornos en la parte rural de Carolina del Norte. Sugerimos que la pobreza de energía se puede ver mejor como un ensamble de materialidades en red y relaciones socio-económicas. Para ilustrar este enfoque, nos concentramos en los patrones geográficos de tres determinantes claves de la pobreza de energía en la parte oriental de Carolina del Norte: las características socio-económicas de los hogares rurales, las infraestructuras en red para el suministro de energía y las condiciones materiales de la casa. En todo esto, destacamos los efectos vividos en situación de pobreza de energía, a partir de las transcripciones de entrevistas administradas entre quienes reciben ayudas de climatización en la región. Sugerimos que los retos de los pobres en energía merecen mayor atención de la política pública, como parte de un más amplio entendimiento del bienestar y cuidado social. *Palabras clave: pobreza de energía, infraestructura, redes, Carolina del Norte, climatización.*

Contemporary discussions about energy policy in the United States have tended to focus on the need to transition from our current fossil fuel regime to one in which renewable energy sources play a much larger role. Amid the debates about energy security, climate change, and a new green economy, however, it is easy to overlook the fact that millions of Americans currently lack access to reliable, affordable energy regardless of source. In this contribution, we highlight some of the dimensions of this uneven landscape of energy consumption by examining the concept of energy poverty and tracing some of its contours in a rural part of North Carolina.

We use the term *energy poverty* to describe a situation in which a household cannot afford to maintain the home's indoor temperature at a level that allows for a comfortable or healthy lifestyle, a condition also known as *fuel poverty*. Our aim in what follows is twofold: First, we wish to put forward a view of energy poverty as a particular kind of techno-social assemblage, made up of an array of networked actors and materialities. A focus on the networked nature of energy poverty, we suggest, can help to highlight its historical foundations and multidimensional character. Our second aim is to offer a glimpse of the lived realities of energy poverty. To do so, we draw on results from a research project centered around in-depth interviews with households receiving weatherization assistance in eastern North Carolina. By giving voice to those coping with energy poverty, we hope to shed some light on a phenomenon that is seldom given a prominent role in contemporary policy debates and also to suggest some of the ways that the challenge of energy poverty might be viewed in the context of an expanded conceptualization of welfare and care.

Energy as an Assemblage

On the surface, energy poverty is a straightforward relationship between household income and the cost of energy. In Boardman's (1991) often-used formulation, a household spending more than 10 percent of household income on energy bills should be defined as energy poor. It should be clear, however, that this financial equation is not simply a matter of incomes and energy prices but is mediated by a variety of factors and relationships, including the provision of energy to the home, the home's energy efficiency, and the unique requirements for an individual to be comfortable in his or her home. Boardman (1991) and Healy (2004) have examined some of these aspects of energy poverty in the European context, with a particular focus on the links between energy poverty and household risk factors. Both authors have produced empirically rich studies detailing the prevalence of energy poverty using large-scale census and survey data. Building from the work of Healy and Boardman, Buzar (2007a, 2007b) examines energy poverty at a finer scale in the former socialist countries of Central and Eastern Europe. Buzar (2007b) advocated a "relational approach" to the study of energy poverty, one that combines an understanding of postsocialist energy policy and housing infrastructures with an appreciation for the lived experiences of the energy poor.

Although we are broadly supportive of this kind of approach, we wish to suggest that new avenues of inquiry can be opened up through an engagement with recent geographical discussions of relational thinking (Jones 2009) and "assemblage geographies" (Robbins and Marks 2010), which have called attention to the ways in which networks and relations are assembled and stabilized to produce what we take to be "the social." More specifically, we would say that energy poverty is best viewed as a geographical assemblage of networked relations of various kinds, including flows of energy, infrastructures of production and distribution, the properties of the built environment, and the social and economic networks that sustain communities.

In developing this perspective, we draw loosely on work in a number of theoretical traditions and areas of inquiry, including actor-network theory (Murdoch 1997; Farías and Bender 2010), Deleuzian philosophy (Marston, Jones, and Woodward 2005), the field of science and technology studies (Graham 2001; Hommels 2005), and recent work in urban political ecology (Heynen, Kaika, and Swyngedouw 2006; Swyngedouw 2006). Although not wishing to gloss over their distinctions, we find in this work a common orientation, which provides a useful frame for viewing the problem of energy poverty in new ways. For one thing, and in contrast to much work in traditional Marxist political economy, such work refuses to take social structures and relations at face value, instead asking how "the social" has come to be assembled in particular, historically contingent ways (Kirsch and Mitchell 2004; Holifield, 2009). A focus on networked assemblages also calls attention to their heterogeneous character, as well as the cultural attitudes that shape their use. Seen this way, the lived reality of energy poverty can be approached as a condition that arises from the ways in which nature, technology, cultural norms, and the individual biographies of households have been drawn together into particular networked configurations.

In what follows, then, we take a closer look at the nature of energy poverty in eastern North Carolina. Our aim is not to assess the scope of the problem but rather to suggest some of the theoretical resources and ethnographic insights that might be brought to bear in developing a richer understanding of energy poverty in the U.S. context. In so doing, we focus attention on the geographies of three key dimensions of the energy poverty assemblage: the socioeconomic characteristics of rural North Carolina, the varied landscape of energy provision in the region, and the material character of housing. Although geographical variations in any one of these factors could influence energy use and access, it is their combination into particular techno-social assemblages, we suggest, that results in energy poverty.

Networked Infrastructures

The networked infrastructures of energy and other substances have been the focus of considerable recent discussion. This research has highlighted a number of ways in which networked infrastructures and their associated technological systems work to shape landscapes and spaces (Zimmerer 2011). First, infrastructures are important sites for mediating the relationships between, and thus producing particular understandings of, nature (on the one hand) and social or domestic space (on the other). This has been described as a *metabolism*, through which the flows of natural substances such as water, energy, or waste are isolated and channeled toward, or away from, the socially produced spaces of the city and home (Gandy 2004; Swyngedouw 2006).

A second insight of recent work on networked infrastructures is that their deployment is both obdurate over time (Hommels 2005) and uneven across space, which can often lead to significant forms of spatial inequality. Graham and Marvin (2001), for example, have identified what they call a *splintering urbanism*, in which infrastructures are able to bypass certain groups and regions, resulting in an increasing gap between "premium networked spaces" and those suffering a "poverty of connections" (Graham and Marvin 2001, 249, 288). The more general point, as Monstadt (2009, 1934) put it, is that "the quality of networked infrastructures and the degree of social and geographical access to them has a huge impact on distributional justice and social well-being." Third, regulatory regimes and public policy can shape network availability in significant ways. Recent trends toward liberalization and deregulation, for example, have increased the sociospatial inequality in the cost of, and access to, many forms of networked infrastructure (Graham 2001).

For the most part, examinations of networked technologies and infrastructures have focused on their constitutive role in shaping specifically urbanized landscapes (e.g., Graham 2000; Hommels 2005; Farías and Bender 2010). Here we wish to suggest that rural spaces are in part constituted through and defined by networked materialities of various kinds. Although such networks might be less evident than those within the entangled spaces of large cities, rural settings are no less made up of what Woods (2007) called "hybrid assemblages of human and non-human entities, knitted-together intersections of networks and flows" (499). Among these networks are the rural infrastructures that gather together and distribute energy and other material substances. These infrastructures, like their urban counterparts, are characterized by fragmented regulatory regimes and significant spatial inequalities in access and cost. For these reasons, we believe, the geographies of rural energy poverty are worthy of examination.

Energy Poverty in Eastern North Carolina

Our particular investigation is focused on three counties in eastern North Carolina (Figure 1). The area is typical of lagging rural regions throughout the U.S. South (Table 1) and is characterized by relatively high rates of poverty and unemployment, a declining economic base, low levels of educational attainment, and endemic health problems (North Carolina Rural Economic Development Center n.d.). It is this kind of socioeconomic landscape that presents a high risk for energy poverty (Colton and Leviton 1991; U.S. Department of Health and Human Services 2005).

In the United States, two federal programs have been established to assist those in energy poverty. The first is the Low Income Home Energy Assistance Program (LIHEAP), which provides income subsidies to low-income households, most often in the winter when heating bills are the highest. The second is the Weatherization Assistance Program (WAP), which updates and renovates houses occupied by low-income residents in an attempt to increase their energy efficiency. Weatherization improvements typically include adding or improving insulation and repairing nonfunctioning heating and cooling equipment. These improvements usually result in lower bills and safer indoor conditions for recipients. Since the program's inception, more than 6.2 million homes in the United States have been weatherized by the WAP (National Association for Community Service Providers 2009).

In North Carolina, the WAP is administered by the state's Department of Commerce, which distributes funding to community action agencies across the state. One such agency is the Wayne Action Group for Economic Solvency (WAGES, Inc.), which administers the WAP in Greene, Lenoir, and Wayne counties. Our research was undertaken in partnership with WAGES, which provided access to its recipient database and also facilitated our interviews with weatherization recipients. Potential interview participants were identified using a typical case sampling technique (Bradshaw and Stratford 2005). An initial

Table 1. Socioeconomic data for counties in study area

	Total population	White population (%)	Black population (%)	Median age of population (Years)	Population 25 years + with less than 12th grade education (%)	Median household income ($)	Median income, Black ($)	Median income age 65 and older ($)
Greene County								
2008 ACS	20,542	49.1	39.2	35.8	29.3	38,654	28,586	22,039
WAGES	55	21.8	76.4	61.9	40.0	15,442	13,152	13,152
Lenoir County								
2008 ACS	56,840	54.5	40.1	40.8	25.7	31,475	22,875	20,901
WAGES	106	26.4	73.6	60	41.0	13,271	12,588	10,965
Wayne County								
2008 ACS	113,223	60.8	32.1	37	19.3	40,464	27,554	27,024
WAGES	193	27.5	71.0	59.4	41.7	13,260	10,596	10,128
North Carolina								
2008 ACS	9,036,449	70.3	21.2	36.8	17.1	46,107	31,580	30,175

Note: ACS = American Community Survey; WAGES = Wayne Action Group for Economic Solvency.
Source: 2006–2008 American Community Survey 3-Year Estimates and WAGES, Inc. weatherization recipient database.

group of thirty potential participants was selected from the WAGES database that loosely approximated a cross section of ages, locations, housing, and fuel types in the study area. Once identified, potential participants were contacted by WAGES to gauge their willingness to participate in the interviews, which were then scheduled and conducted by the authors in the recipients' homes. From the initial group, seventeen weatherization recipients agreed to participate in interviews conducted during the summer months of 2009. When used in conjunction with data collected from the WAGES database and U.S. Census, the interviews allow for a richer understanding of how individual people experience energy poverty, with an emphasis on the meanings people attach to their lives (Valentine 2005).

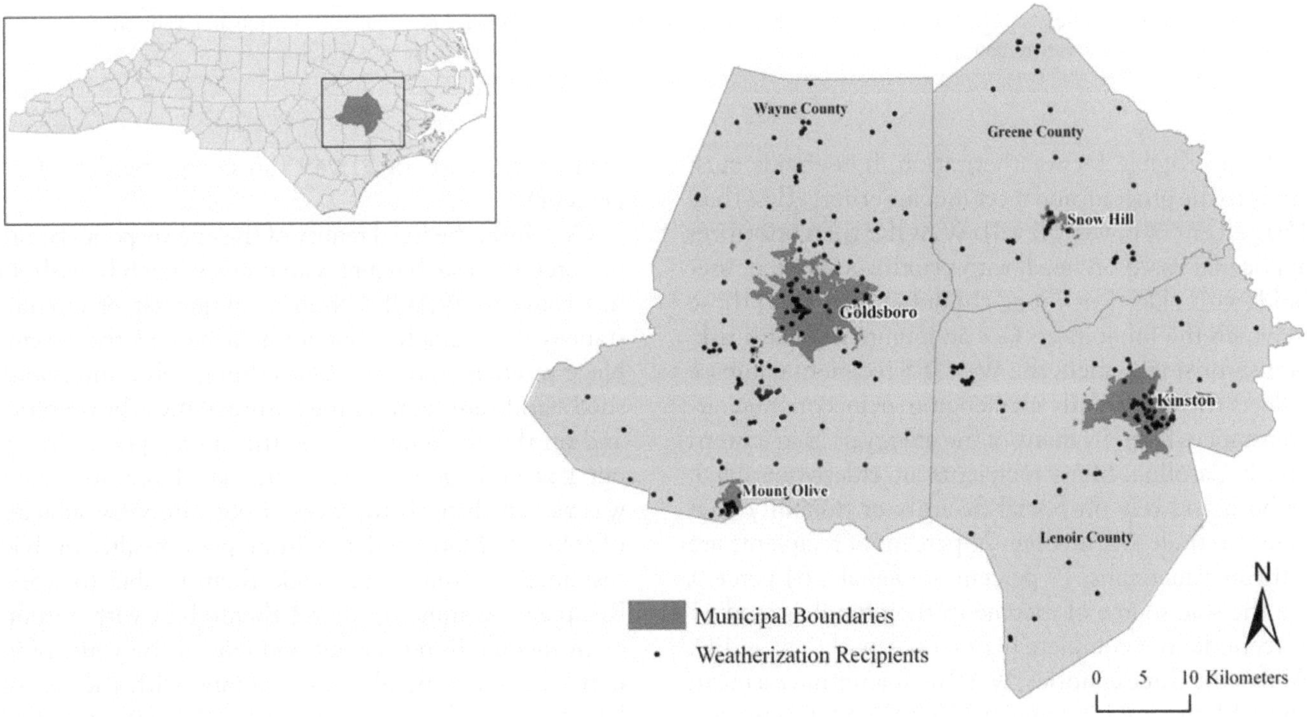

Figure 1. Study area showing municipal boundaries and location of all WAGES weatherization recipients.

Table 2. Factors contributing to energy poverty

Individual biographies
Health problems
•"There were days when I was sick, there were days when she was sicker than I was. . . . See I was out of work, back surgery . . . that put me behind on a lot of bills."
•"I was doing pretty good, but I had a leg amputated . . . and I would work right now if I could get around good."
Precarious economic circumstances
•"I about lost the house two or three times because the bills got so high."
•"Right after my dad died all my mom had was $638 (a month). And it has just been very difficult."
Isolation
•"I am here by myself, and don't nobody take care of me but me."
•"You see when you get sick . . . you don't have nobody to come help you out."
Energy situation
High electricity rates
•"When everything, the light bill . . . went up, everybody's bill went up, whether you burn a lot or not. . . . One man I know at school said he got one for $500, no way he can pay that."
Spikes in liquid propane gas prices
•"I'm scared to turn on the gas. Sometimes I have to, but last year, you know how high gas was."
Housing situation
Older energy-inefficient homes
•"Where the cold air was coming in, around the cabinets, and around the attic, that would be cold back there."
Older inefficient or poorly functioning appliances
•"It was a July 1967 refrigerator so . . . [WAGES] saw fit to replace it."
•"That unit out there . . . was an old unit. . . . It was running a whole lot of time that the Freon was out . . . and it sat there and ran and ran and ran and it brought the light bill up."
Energy-inefficient mobile homes
•"We had a garden tub . . . and the weatherization guys came out here and they pulled that off and there was a big hole in the floor, and all the cold air in the bathroom, and it was always so cold . . . and that's what it was."
Inability to perform upkeep and maintenance of housing
•"When I was walking, I would put plastic up to the windows, then that would help keep the house warmer. . . . In this chair I can't do stuff like that."
•"When I first moved here . . . I had the money to really take care of my home. . . . When I got injured on my job and went on disability . . . it was overwhelming."

To be eligible for weatherization, households must apply to the program and meet income criteria (less than 150 percent of poverty level). WAGES then prioritizes applicants based on need, with priority going to households with elderly or young children present and those in which the inhabitants face an imminent health risk. Not surprisingly, then, the WAGES recipient database reflects the particularly challenging socioeconomic circumstances faced by many of the energy poor in eastern North Carolina. Many recipients are elderly, disabled, or both and rely on Social Security or disability payments as their sole income; 74 percent of recipients are African Americans, 79 percent are female, 84 percent are the sole source of income in their families, and 41 percent did not complete high school or obtain a GED. Across the three counties, WAP recipients have a mean annual household income of $13,602 and average annual energy costs of $1,949 (an energy burden of 14 percent).

Of course, the lived reality of the energy poor is more personal than such figures can convey. Each household has come to WAGES from a unique set of circumstances, a biography that forms a part of the assemblage of energy poverty. Our ethnographic interviews shed significant light on the nature of these biographies and on the challenges facing the energy poor. One of our key findings was that a thread of precariousness was woven through the lives of our interview subjects (Table 2). Many suffered from poor health or had sustained an injury that made them unable to work. Recipients commonly found themselves with insufficient income in retirement and felt on the verge of financial ruin. A number were coping with the recent loss of a loved one. For some participants, personal circumstances meant that energy bills that were once affordable were now simply too high. In other cases, little had changed in the household itself, yet the networks to which the home is connected or the materiality of the home itself had changed. It is difficult, then, to point to a single factor underlying the descent into energy poverty. As an assemblage, it is rather the outcome of the ways in which physical health, financial exigencies, social networks, the materiality of the home, and the infrastructure of energy provision interact to produce an uneven geographical landscape of energy cost and availability.

The Variable Landscape of Energy Provision

The landscape of eastern North Carolina is crisscrossed by a dense web of networked energy infrastructures. The ability of any particular household to effectively access these networks, however, is conditioned by a number of factors that vary geographically. Although such constraints are experienced by all energy users, they pose particular challenges for low-income residents. The primary heating sources used in the study area are electricity, liquid propane gas (LPG), and natural gas. These forms of energy have distinctive properties and vary in their ability to be efficiently converted from raw fuel to heat. Natural gas and LPG are much more efficient than electricity as a source of heat and, all things being equal, would be a more cost-effective method of space heating. All things are not equal, however, as the networks of energy provision can vary significantly by both cost and availability.

Figure 2 shows the percentage of residences using natural gas, electricity, and LPG, respectively, as their primary heating fuel. The patterns are largely a result of the ways in which infrastructure networks have been laid down over time. Natural gas requires significant distributional infrastructure and, as a result, it tends to cluster in dense urban areas where utilities can expect to profit. Electricity, which is available virtually anywhere in the study area, and LPG, which is distributed by vehicles and thus less bounded geographically, are more prevalent in rural areas where natural gas is unavailable. In addition, most mobile homes leave the factory with electric heat sources, which predetermines the energy network to which they are connected.

The nature of these networked energy infrastructures also plays a significant role in the cost of energy for consumers. In North Carolina, natural gas distribution, and in some areas electricity, is regulated by the North Carolina Utilities Commission (NCUC), meaning that price increases must be approved by the commission, thus decreasing volatility. LPG is not regulated by the state, and compared to natural gas its price is more closely tied to the vagaries of the market for international crude oil (U.S. Energy Information Administration 2010). This means that LPG prices, in particular, can go from being relatively affordable one month to being a significant burden the next. In addition, the price for LPG, and to a lesser extent natural gas, fluctuates based on demand and so tends to peak during particularly cold stretches. Natural gas and LPG prices during the winter months between the years 2004 and 2009 are shown in Figure 3. Whereas natural gas prices exhibit some fluctuation from year to year, LPG prices are increasing long term and experienced a significant price spike in the winter of 2008.

The most complicated energy landscape is presented by electricity, the networks for which have their roots in the North Carolina Electric Membership Corporation Act of 1935, which encouraged the formation of nonprofit membership corporations to expand the availability of electricity to rural areas. At the same time, investor-owned utilities were focusing their efforts on cities and towns, where population densities held the promise of high returns on infrastructure investment. Smaller municipalities, which seldom received the attention of investor-owned utilities, were often left without electricity and many pursued electrification on their own. The three types of electricity providers in the study area are subject to different regulation regimes. Investor-owned utilities are regulated by the NCUC, and rate increases are subject to Commission approval. Not-for-profit rural cooperative rates are not regulated and provide at-cost service to customers. Municipalities are also exempt from NCUC regulation. Many are saddled with debt from energy generation ventures, and revenues are used not only to cover the cost of electricity service but also to support non-energy-related municipal programs. As a result, municipal providers charge the highest rates of any electricity supplier in the study area (see Table 3).

Irrespective of provider, many interview participants reported that they had a difficult time with high energy bills. Roger McDonald[1] had electricity bills as high as $400 per month, nearly 50 percent of his monthly income. During that time, he "about lost the house two or three times because the bills got so high." Carl Williams described his electricity bill this way:

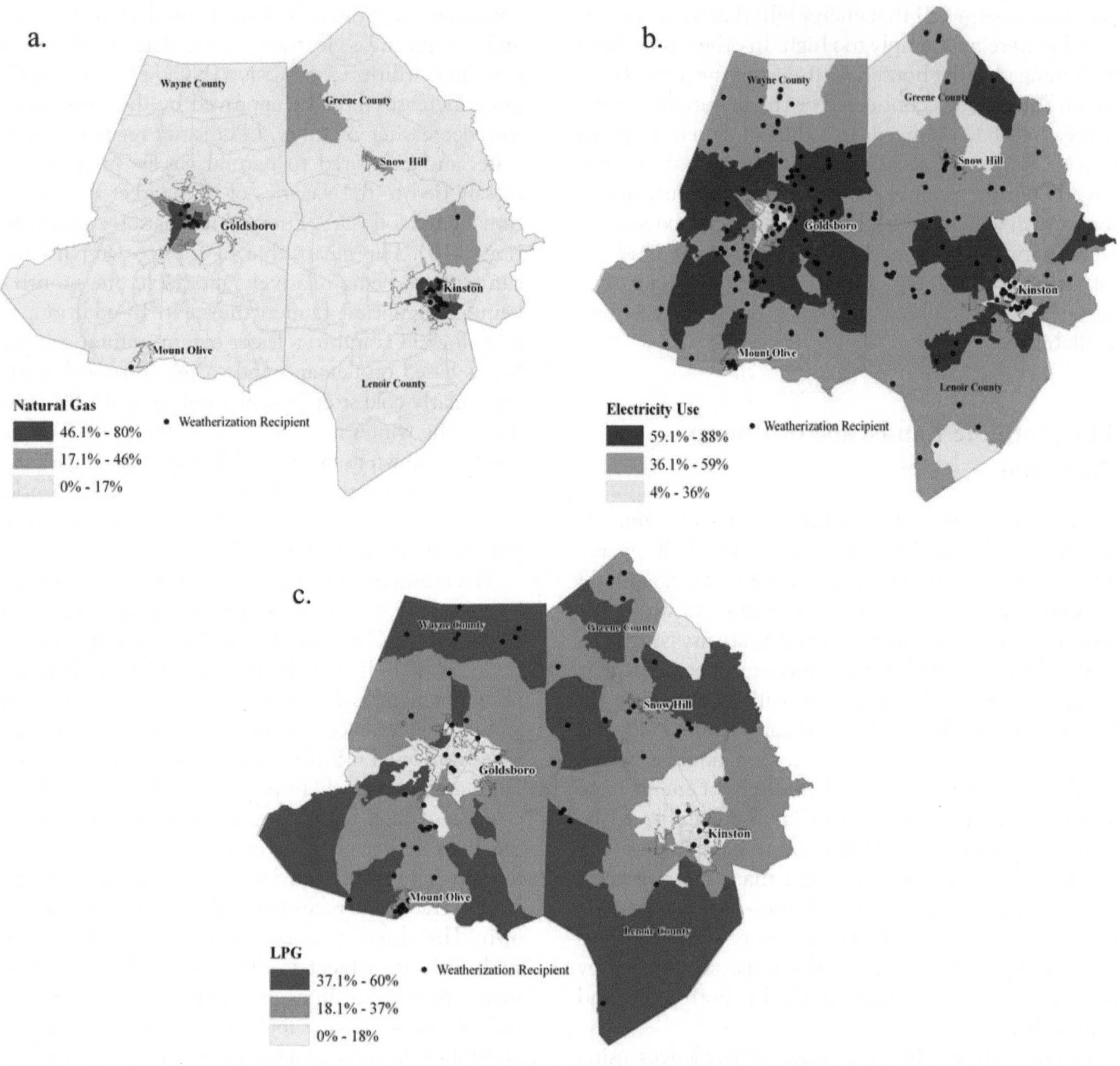

Figure 2. Spatial intensity of heating fuel types in study area by (A) natural gas, (B) electricity, and (C) liquid propane gas (LPG). Weatherization recipients using the respective fuel are identified. *Sources:* WAGES database and U.S. Census (2000) Population and Housing Summary File 3.

> The killer. I mean, you are talking about $3[00], $345, $350. That's basically what I was paying for this house. And to me it is ridiculous to pay that much, but you have to.

The Porous Materiality of the Home

The experience of energy poverty emerges not only from the geographically uneven character of energy networks but also the ways in which those networks intersect with the materiality of the home. For the energy poor, a decline in the home's thermal efficiency brings on new challenges in the form of higher energy bills and indoor conditions that can at times become dangerous to the occupant.

The connection of the home to networked infrastructure beginning in the early twentieth century remade the house into a new space, a symbol of progressive society. Life was made easier by a range of electric appliances and the home more comfortable through the climate control enabled by air conditioners and furnaces. The home thus became the site of a socio-natural metabolism, in which flows of water, fuels, electricity, and other substances were regulated

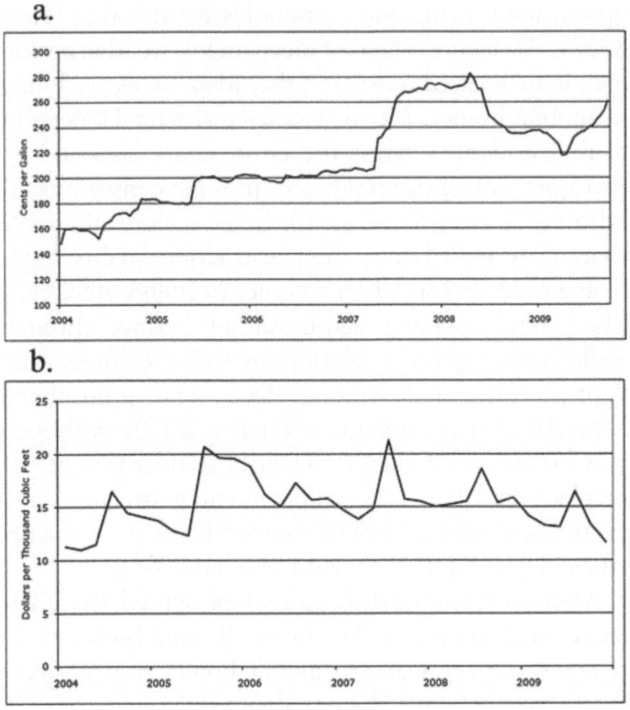

Figure 3. North Carolina estimated aggregate residential heating fuel prices for (A) liquid propane gas (LPG) and (B) natural gas. Data collected for peak heating months (October–March), 2004–2009. *Source:* U.S. Energy Information Administration (2010), independent statistics and analysis.

in and through the home (Kaika 2004). The efficiency of the modern home, however, is belied by the leaky materiality of the typical house. As Graham and Thrift (2007) remind us, the U.S. housing stock is in a perpetual state of breakdown and decay, such that commonplace activities of repair and maintenance (for example, weatherization) can be viewed as constitutive of modern life. For the energy poor, in particular, the structural features of the home, and the proficiency with which it regulates its metabolism with nature and energy, play a large role in the quality of their daily living environment (Healy 2004; Santamouris et al. 2007).

The aging of the materials and systems making up any particular house will obviously affect its energy efficiency, and houses in the study area are older than the state average (Table 4). The older homes tend to be located in the urban neighborhoods of Kinston, Goldsboro, and Mt. Olive (Figure 4). For those living in such homes, the lack of insulation, as well as the leaks and drafts that inevitably appear in the house, will eventually begin to show up in the form of high heating and cooling bills. Karen Thompson, for example, describes the variable conditions inside her home, noting that:

> The living room and the kitchen, was the coldest thing in the house. 'Cause we got . . . them old windows, and air comes up there through them, and you can hear the glass sometimes rattling in the wind.

In addition to the gradual decay of housing materials, older homes are susceptible to a failure of their heating or cooling system. According to WAGES, 76 percent of weatherization recipients had heating systems that were not functioning correctly at the time of inspection. As

Table 3. Electricity rates and estimated monthly charges for utilities serving Greene, Lenoir, and Wayne counties in November 2009

Supplier	Utility type	Rate per kWh ($)	Base charge ($)	Avg. monthly charge[a] ($)
City of Kinston	Municipal power	0.1435	13.40	140.83
City of Wilson[b]	Municipal power	0.1438	8.99	136.68
Town of Pikeville	Municipal power	0.1359	8.95	129.63
Town of Walstonburg	Municipal power	0–50 kWh 0.0758; 51–250 kWh 0.1573; 251+ kWh 0.1257	13.00	128.45
Town of Hookerton	Municipal power	0–500 kWh 0.138; next 1,000 kWh 0.1238	8.55	125.58
Town of Fremont	Municipal power	0.1265	10.01	122.34
Pitt & Greene Electric	Rural electric co-op	0.1122	20.00	119.63
Town of LaGrange	Municipal power	0–800 kWh 0.135; 800+ kWh 0.1193	0.00	118.50
Town of Stantonsburg	Municipal power	0.1103	8.99	106.94
Progress Energy	Investor-owned utility	0.10634	6.75	101.18
Tri County Electric	Rural electric co-op	0.0888	8.98	87.83

Note: Because kWh rates vary for some utilities between winter and summer, we use summer rates and standard residential base charges.
Source: Rate data collected by authors during November 2009 from published rates available by phone or Internet.
[a]Average monthly charge is calculated using average monthly residential electricity consumption data from *Residential Energy Consumption Survey 2001* (U.S. Energy Information Administration 2004).
[b]City of Wilson electric utility serves a small portion of Wayne County despite being headquartered in Wilson County.

Table 4. Housing data

	Median house age (years)	Median mobile home age (years)	Houses that are mobile homes (%)
Greene County			
2000 Census	22	N/A	36.9
WAGES	39	19	44.4
Lenoir County			
2000 Census	28	N/A	23.9
WAGES	55	18	32.1
Wayne County			
2000 Census	25	N/A	25.4
WAGES	49	21	38.9
North Carolina			
2000 Census	22	N/A	16.4

Note: Age of mobile home not available from U.S. Census. WAGES = Wayne Action Group for Economic Solvency.
Source: U.S. Census (2000) Population and Housing Summary File 3 and WAGES, Inc. weatherization recipient database.

a result, we were told, many households are forced to seek other sources of heat for their homes, most often in the form of dangerous and inefficient electric, kerosene, or propane space heaters.

For many of the energy poor in eastern North Carolina, it is not so much the age of the house that determines the energy relationship but rather the properties of the house. This is particularly true of mobile homes, which have become an important housing option for low-income households in the state (Rust 2007). As Figure 4 makes clear, many weatherization recipients located outside of the urban areas are living in mobile homes. Mobile homes make up 11 percent of the residences across the entire study area and up to 75 percent of the residences in some census blocks. Although buying a new mobile home is more affordable than a site-built home, the construction quality tends to add hidden costs when it comes to energy efficiency. Many tend to have significant air leakage through walls; little or no insulation in walls, ceilings, and floors; uninsulated heating ducts; and uninsulated doors (U.S. Department of Energy 2010). Although new building and energy efficiency standards were put in place in 1992, code enforcement is irregular, and even many newer mobile homes have poor energy efficiency (Hart, Rhodes, and Morgan 2002).

Many interview participants highlighted the challenges of living in a mobile home. Ronald Bush's experience is not atypical. "As mobile homes go it is a very pretty mobile home," he noted, but it is

> One of the most poorly built structures I have ever seen. When the wind is not even gusting, it's like someone is banging gongs up there. . . . Water comes in from the vents in the bathroom, and water pours, I have to put buckets on the oven, to accommodate all the water that comes in. . . . I don't know if there is an ounce of insulation in this thing.

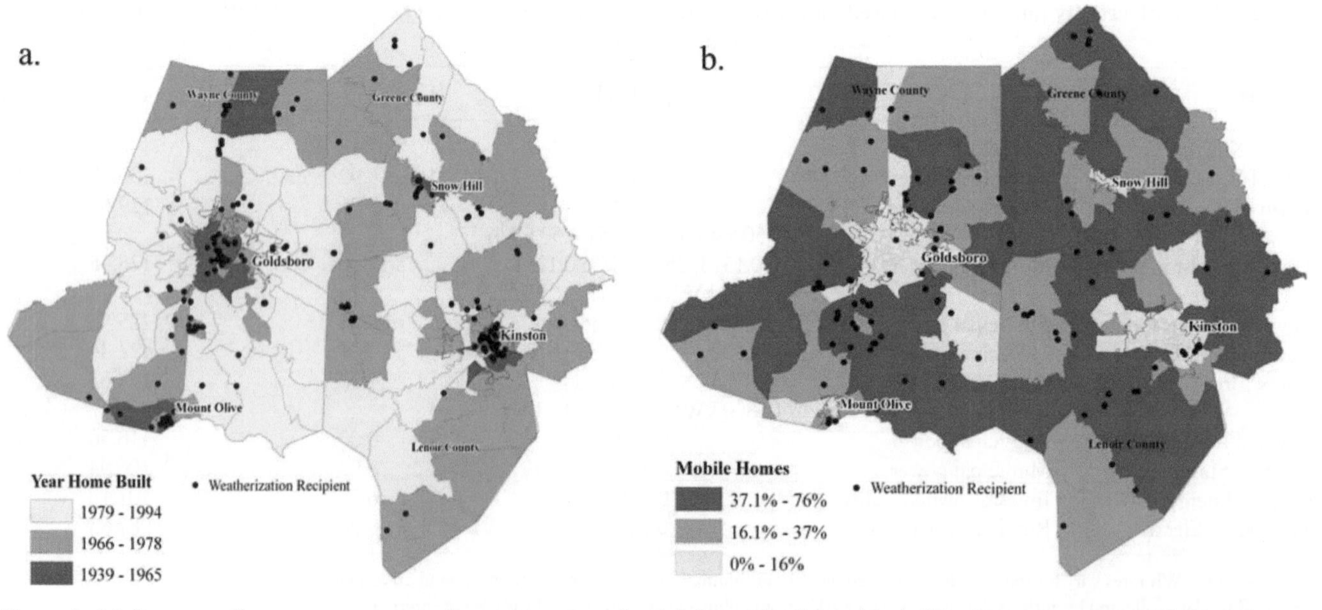

Figure 4. Median year of home construction and spatial intensity of mobile homes by Census block. Weatherization recipients living in houses are shown in (A), and recipients living in mobile homes are shown in (B). *Source:* WAGES database and U.S. Census (2000) Population and Housing Summary File 3.

Table 5. Common coping strategies of interview participants

Utility bill nonpayment
Use of oven or stove to warm kitchen and house
Closing off portions of the home to limit space heating/cooling
Use of electric, kerosene, and propane space heaters
Buying less food
Staying in bed during cold winter days
Selected quotes

- "Yeah, sometimes I used to have to use my oven to keep the house warm."
- "You know, sometimes you might have to let a bill go . . . put a bill off until you can get that paid and then you deal with that."
- "I even put padding in the windows, I blocked out all the sunlight in these two-pane windows, and the lights bills continued to be enormous."
- "I just put these comforters up, one there, one back there where the washing machine is, and we pretty much lived in one room."
- "The heat wouldn't work no more. So I ended up getting a kerosene heater."
- "It was too much . . . groceries are so expensive. I have health problems . . . I am just trying to tell you, when I go buy something, I got where I did without."
- "At Christmas, I don't buy no Christmas presents, I can't afford it. Even if they were ten dollars a piece . . . It is hard on a senior citizen, you know, it's really hard on us."
- "We tried to burn not much of the heat, but with the kids, you know?"

The poor energy efficiency of Ronald's house led to unaffordable bills, which reached as high as $400 during the warmer summer months.

The Lived Experience of Energy Poverty

We have tried to suggest some of the ways that energy poverty emerges as a kind of assemblage, in which networked infrastructures, the materiality of the house, and the biographies of individual households combine to produce a physical or financial inability to properly heat or cool the home. For residents who are already facing various social and economic challenges, this is experienced as a "domestic network crisis," to borrow a term from Kaika (2004, 277). This is particularly true for those who lack extensive social or familial networks of support, which might mitigate the impacts of energy poverty.

Interview participants described a variety of coping mechanisms they employed to deal with the discomfort that comes from living in energy poverty (Table 5). These ranged from relatively small changes in their day-to-day behavior to large changes that fundamentally alter their interactions with their home, their social networks, and their community. Some householders were aware of the role their homes played in the high bills and made crude alterations to the home itself. Others limited their movement throughout the house in an attempt to keep their bills down. Marilyn Wallace essentially limited the size of her home to two rooms. As she describes it:

> I had a kerosene heater, and an electric heater. The kerosene heater I would take to the bathroom to warm up to bathe, and the electric heater I kept in the bedroom, and I kept the door closed, to keep the room warm . . . I don't go about the house much during the day.

At the extreme, some interview participants were forced by the low temperatures to stay in bed, fully dressed and wrapped up under blankets, until it was warm enough inside to get out. Karen Thompson, when asked how she had coped with the low winter temperatures when her furnace failed, replied:

> We just dealt with it. We put on more clothes, get in the bed with some covers.

Such reactions are not only personally trying; they can also disrupt the social networks of friends and relatives who can provide assistance to those coping with energy poverty. In the case of Theodore Cole, one particularly hot Father's Day his grandchildren:

> Were complaining that it wasn't comfortable in here because it was so hot. I had a fan . . . and the ceiling fans all going, and you couldn't even tell they was on.

For Marilyn Wallace, cold winter nights often meant sleeping at her daughter's house because despite using space heaters and blankets, she "couldn't even keep the bedroom warm."

In spite of the great lengths to which the interview participants went to save money, the relational nature of energy poverty means that even individual conservation and coping behavior are not always enough in the face of networked relationships over which the energy poor might have little control: the obduracy of the technosocial infrastructure through which fuel is made available; the recalcitrant materialities of old, leaky houses or mobile homes; low incomes exacerbated by the shrinking welfare state, economic restructuring, and poor health or disability; and high energy prices resulting from past legacies and regulatory structures within the energy system, as well as distant geopolitical events. Within such assemblages, even the most mindful user of energy might be faced with unaffordable energy bills, a reality that forces many eastern North Carolina residents to make unenviable choices about which bill to pay and what they can do without. As Sally Moore put it:

> When you get that gas bill, and you pay your electric bill . . . now I do have Medicaid . . . it helps me out with my medicine. I don't know what in the world I would do. I would have to choose between . . . I don't know what I would do. Because you got to have heat, you got to have food, and you got to have medicine.

Conclusion

Our aim in this article has been to examine, both conceptually and empirically, some of the characteristics and relationships defining energy poverty in eastern North Carolina. Our investigation, we believe, has theoretical, policy, and ethical implications. Theoretically, we have explored some of the avenues opened up by characterizing energy poverty as a geographical assemblage of networked materialities. Such an approach, we believe, can provide a valuable supplement to traditional accounts of energy poverty, whether Marxist or positivist, because they focus attention on new and different questions (Holifield 2009). Specifically, the relational perspective evident in assemblage geographies can help us to be attentive to the ways in which lives, networked infrastructures, and the porous materiality of the built environment come to be assembled in very particular ways and not in others, thus opening these relations up to investigation and critique. More broadly, this kind of open and relational account has affinities with recent geographical debates in a wide range of fields, including science and technology studies (Hommels 2005; Lovell 2005; Coutard and Simon 2007), urban political ecology (Heynen, Kaika, and Swyngedouw 2006), and studies of rural networks (Murdoch 2006; Woods 2007).

From a policy perspective, our work calls attention to the lived experience of energy poverty, an experience that often results in painful choices and fundamental changes in the daily lives of already disadvantaged rural residents. Pending energy legislation represents an opportunity to address some of the relational factors that help to create energy poverty but only if those suffering or at risk become a part of the conversation. Energy efficiency standards for low-income housing and mobile homes should be improved and rigorously enforced, and funds that might be raised by carbon taxes or cap-and-trade programs should be used to assist chronically underfunded programs like the LIHEAP and the WAP, though some pending legislation proposes cutting their funding. Such programs will be at the forefront of assisting the energy poor to cope with an evolving energy system. Indeed, weatherization in particular has been shown to have a positive impact on energy poor households and communities (Schweitzer and Tonn 2003; Shortt and Rugkåsa 2007) and has made a significant difference in the lives of WAP recipients in eastern North Carolina. Says Roger McDonald:

> Since they come in and did all the weatherstripping and stuff, that brought the bills down like it should. If they hadn't have done that, man, we would have still been in a mess.

Carl Williams, now finally relieved of his "killer" electricity bill, concurs:

> I got nothing but high praises for what they did . . . they helped me to help my family and help keep up my home, so this is a blessing for me.

Such outcomes are worth applauding, but we would point out that the WAP and similar programs focus on only one strand within the constellation of relationships comprising energy poverty. Although reengineering the physical structure of the home can modify the built environment's metabolism with nature, it fails to address the many other networks and relationships through which energy poverty can emerge. It is here, then, where larger questions of ethics and responsibility come to the fore. At issue is the extent to which we can be said to bear a collective responsibility toward the various forms of connection—not just technological but social and economic—that might either mitigate or exacerbate the challenges faced by the energy poor. There are good reasons, we believe, to fold energy well-being into an expanded sense of welfare and a wider net of care. We agree with Smith (2005) when she asserts that "the aim . . . is to emphasize the values of interdependence over individualism and to . . . build an ethic of care fully into models of social policy and into the practice of welfare" (11). Our hope is that energy poverty, and its networked relations and materialities, might be included in any assessment of our progress toward such a goal.

Acknowledgements

The preparation of this article was greatly aided by the comments of Ron Mitchelson and the anonymous reviewers. The authors would also like to express their gratitude to the staff of WAGES, Inc. for their willing assistance throughout the project.

Note

1. The names of the interview participants are pseudonyms to ensure confidentiality.

References

Boardman, B. 1991. Fuel poverty: *From cold homes to affordable warmth*. New York: Belhaven.

Bradshaw, M., and E. Stratford. 2005. Qualitative research design and rigour. In *Qualitative research methodologies in human geography*, ed. I. Hay, 67–76. Oxford, UK: Oxford University Press.

Buzar, S. 2007a. The "hidden" geographies of energy poverty in post socialism: Between institutions and households. *Geoforum* 38:224–40.

———. 2007b. When homes become prisons: The relational spaces of postsocialist energy poverty. *Environment & Planning A* 39:1908–25.

Colton, R., and R. Leviton. 1991. *Poverty and energy in North Carolina: Combining public and private resources to solve a public and private problem*. Prepared for Energy Assurance Study Commission, North Carolina General Assembly. Boston: National Consumer Law Center.

Coutard, O., and G. Simon. 2007. STS and the city: Politics and practices of hope. *Science, Technology, & Human Values* 32 (6): 713–34.

Farías, I., and T. Bender, eds. 2010. *Urban assemblages: How actor-network theory changes urban studies*. London and New York: Routledge.

Gandy, M. 2004. Rethinking urban metabolism: Water, space and the modern city. *City* 8 (3): 363–79.

Graham, S. 2000. Introduction: Cities and infrastructure networks. *International Journal of Urban and Regional Research* 24 (1): 114–19.

———. 2001. The city as sociotechnical process: Networked mobilities and urban social inequalities. *City* 5 (3): 339–49.

Graham, S., and S. Marvin. 2001. *Splintering urbanism: Networked infrastructures, technological mobilities and the urban condition*. London and New York: Routledge.

Graham, S., and N. Thrift. 2007. Out of order: Understanding repair and maintenance. *Theory, Culture & Society* 24 (3): 1–25.

Hart, J. F., M. Rhodes, and J. Morgan. 2002. *The unknown world of the mobile home*. Baltimore: Johns Hopkins University Press.

Healy, D. 2004. *Housing, fuel poverty, and health: A pan-European analysis*. London: Ashgate.

Heynen, N., M. Kaika, and E. Swyngedouw, eds. 2006. *In the nature of cities: Urban political ecology and the politics of urban metabolism*. London and New York: Routledge.

Holifield, R. 2009. Actor-network theory as a critical approach to environmental justice: A case against synthesis with urban political ecology. *Antipode* 41 (4): 637–58.

Hommels, A. 2005. Studying obduracy in the city: Toward a productive fusion between technology studies and urban studies. *Science, Technology, & Human Values* 30 (3): 323–51.

Jones, M. 2009. Phase space: Geography, relational thinking, and beyond. *Progress in Human Geography* 33 (4): 487–506.

Kaika, M. 2004. Interrogating the geographies of the familiar: Domesticating nature and constructing the autonomy of the modern home. *International Journal of Urban and Regional Research* 28 (2): 265–86.

Kirsch, S., and D. Mitchell. 2004. The nature of things: Dead labor, nonhuman actors, and the persistence of Marxism. *Antipode* 36 (4): 687–705.

Lovell, H. 2005. Supply and demand for low energy housing in the UK: Insights from a science and technology studies approach. *Housing Studies* 20 (5): 815–29.

Marston, S., J. P. Jones III, and K. Woodward. 2005. Human geography without scale. *Transactions of the Institute of British Geographers* NS 30:416–32.

Monstadt, J. 2009. Conceptualizing the political ecology of urban infrastructures: Insights from technology and urban studies. *Environment and Planning A* 41: 1924–42.

Murdoch, J. 1997. Towards a geography of heterogeneous associations. *Progress in Human Geography* 21: 321–37.

———. 2006. Networking rurality: Emergent complexity in the countryside. In *Handbook of rural studies*, ed. P. Cloke, T. Marsden, and P. Mooney, 171–84. London: Sage.

National Association for Community Service Providers. 2009. Waptach home page. http://www.waptach.org (last accessed 14 February 2009).

North Carolina Rural Economic Development Center. n.d. Rural data bank. http://www.ncruralcenter.org (last accessed 15 August 2010).

Robbins, P., and B. Marks. 2010. Assemblage geographies. In *The Sage handbook of social geographies*, ed. S. Smith, R. Pain, S. Marston, and J. P. Jones III, 176–94. London: Sage.

Rust, A. 2007. *This is my home: The challenges and opportunities of manufactured housing*. Durham, NC: Carolina Academic Press.

Santamouris, M., K. Kapsis, D. Korres, I. Livada, C. Pavlou, and M. N. Assimakopoulos. 2007. On the relation between the energy and social characteristics of the residential sector. *Energy & Buildings* 39:893–905.

Schweitzer, M., and B. Tonn. 2003. Non-energy benefits of the U.S. weatherization assistance program: A summary of their scope and magnitude. *Applied Energy* 76:321–35.

Shortt, N., and J. Rugkåsa. 2007. "The walls were so damp and cold": Fuel poverty and ill health in Northern Ireland: Results from a housing intervention. *Health & Place* 13:99–110.

Smith, S. J. 2005. States, markets and an ethic of care. *Political Geography* 24:1–20.

Swyngedouw, E. 2006. Circulations and metabolisms: (Hybrid) natures and (cyborg) cities. *Science as Culture* 15 (2): 105–21.

U.S. Census. 2000. *2000 census of population and housing (SF3): North Carolina*. Washington, DC: Data User Services Division.

U.S. Department of Energy. 2010. Energy savers: Manufactured home energy-efficient retrofit measures. http://www.energysavers.gov/your_home/designing_remodeling/index.cfm/mytopic=10230 (last accessed 27 August 2010).

U.S. Department of Health and Human Services. 2005. *LIHEAP home energy notebook*. Washington, DC: U.S. Department of Health and Human Services.

U.S. Energy Information Administration. 2004. *Residential energy consumption survey 2001*. Washington, DC: U.S. Energy Information Administration.

———. 2010. Factors affecting propane prices—Energy explained: Your guide to understanding energy. http://www.eia.gov/energyexplained/index.cfm?page=propane_factors_affecting_prices (last accessed 27 August 2010).

Valentine, G. 2005. Tell me about . . . : Using interviews as a research methodology. In *Methods in human geography*, ed. R. Flowerdew and D. Martin, 110–27. Essex, UK: Pearson Education.

Woods, M. 2007. Engaging the global countryside: Globalization, hybridity and the reconstitution of rural place. *Progress in Human Geography* 31 (4): 485–507.

Zimmerer, K. S. 2011. New geographies of energy: Introduction to the special issue. *Annals of the Association of American Geographers* 101 (4): 705–11.

Powering "Progress": Regulation and the Development of Michigan's Electricity Landscape

Jordan P. Howell

Department of Geography, Michigan State University

This article examines state-level regulation as it affected Michigan's electricity infrastructure system. Whereas many studies make central the utility companies who built and operated the system, I argue that Michigan's utilities laws and the specific actions of the regulatory body, the Michigan Public Service Commission, played a crucial role in shaping the state's electricity landscape. The regulatory process, and in particular rate-of-return accounting, converged with a specific notion of progress that linked the deployment of new infrastructure to statewide socioeconomic advancement, producing a landscape dominated by huge projects and excess generating capacity. In spite of industry restructuring in the 1990s, historical attitudes toward both infrastructure and consumption continue to limit conservation and renewable fuels programs today. Beginning with a brief history of Michigan's electricity landscape, the article shifts to the Commission's role in setting rules whereby utilities could operate, profit, and make investments. I examine impacts of Commission policies as typified by hearings surrounding the Midland Nuclear Facility. Next, I consider the role of progress in utility and regulatory decisions and conclude with a look at the legacies of earlier policies, decisions, and attitudes. *Key Words: electricity, energy studies, historical geography, Michigan, regulation.*

本文探讨州级电力调节，因为它影响到密西根州的电力基础设施体系。尽管许多研究以建立的和经营该系统的事业单位为中心，我认为，密西根州的公用事业法律和管理机构，即密西根州公共服务委员会的具体行动，在塑造本州电力风景上起至关重要的作用。监管程序，尤其是会计的回报率，与连接了新的基础设施部署和全州社会经济推进进步的一个具体概念融合起来，产生了巨大项目和过剩的发电能力为主导的景观。尽管行业在 20 世纪 90 年代进行了结构调整，对基础设施和消费的历史态度如今仍继续限制保护和可再生燃料计划。从密西根的电力风景的简史开始，本文继而探讨委员会在制定使公用事业可经营的规则，利润和投资方面的作用。我以围绕米德兰核设施听证会为代表，调查委员会的政策造成的影响。接下来，我考虑进步在公用事业和管理决定的作用，并以对先前的政策，决策和态度的传统的着色而结束。*关键词：电力，能源研究，历史地理学，密西根州，调节。*

Este artículo examina la regulación a nivel de estado en la medida de su influencia sobre el sistema de la infraestructura de la electricidad en Michigan. En tanto que mucho estudios ponen en el centro de la cuestión a las compañías de servicios que construyeron y operan el sistema, yo arguyo que las leyes sobre estos servicios en Michigan y las acciones específicas del cuerpo regulador, la Comisión de Servicios Públicos de Michigan, jugaron un papel crucial en la conformación del paisaje de la electricidad en este estado. El proceso regulador y, en particular, la contabilidad de la tasa de rendimiento convergieron con una noción específica de progreso que ligó la instalación de nueva infraestructura con el avance socio-económico a nivel general del estado, produciendo un paisaje en el que dominan proyectos enormes y capacidad de generación en exceso. A pesar de la reestructuración industrial de los años 1990, las actitudes históricas hacia infraestructura y consumo siguen limitando en la actualidad los programas de combustibles renovables y de conservación. Comenzando con una breve historia del paisaje de la electricidad en Michigan, el artículo pasa a examinar el papel de la Comisión en la expedición de normas mediante las cuales las compañías de servicios públicos pudieron operar, obtener ganancias y hacer inversiones. Examino los impactos de las políticas de la Comisión, tal como fueron tipificados en las audiencias relacionadas con el Servicio Nuclear Midland. Enseguida, me refiero al papel de promoción del progreso representado por las decisiones regulatorias y de servicios, y concluyo con una mirada a los legados de políticas anteriores, sus decisiones y actitudes. *Palabras clave: electricidad, estudios sobre energía, geografía histórica, Michigan, regulación.*

Coal dominates electricity generation in Michigan. . . . Although Michigan is a major generator of electricity from wood and wood waste, has many small hydroelectric plants, and has several plants that generate electricity using methane recovered from landfills and anaerobic digesters, renewable power generation contributes minimally to the State electricity grid. Electricity generation in Michigan is high, as is overall per capita electricity consumption.

—State Energy Profile, Michigan (Energy Information Administration 2009d)

Although official statements like this epigraph offer a straightforward assessment of Michigan's electricity production and consumption patterns, they fail to capture the complexities influencing the development of the electricity system. In this article, I consider the development of Michigan's electricity landscape since its inception in the later part of the nineteenth century.[1,2] Although investor-owned utilities, federally funded rural cooperatives, and municipal bodies have all directly contributed to the physical electricity infrastructure serving the state, I argue that Michigan's utilities oversight body, the Michigan Public Service Commission (MPSC, or Commission), has played a critical role in shaping the state's electricity landscape. The regulatory process, and in particular rate-of-return accounting, converged with a specific notion of progress to produce an electricity landscape characterized by huge projects and excess generating capacity. In spite of industry restructuring in the 1990s, historical actions and attitudes toward both infrastructure deployment and electricity consumption conspire to limit the effectiveness of conservation and renewable fuels programs in the state today.

This article seeks to extend national-level analyses of the U.S. electricity landscape (e.g., Elmes 1996; Elmes and Harris 1996; Heiman and Solomon 2004; Heiman 2006; Sovacool 2008) to the state scale, while also shifting emphasis away from actors in the industry itself (cf. Miller 1971; Bush 1973; Kuhl 1998) and the processes of technological change (cf. Hughes 1983; Hirsh 1989), toward the role of the state regulatory body. This highlights the fact that most of the critical decisions surrounding electricity infrastructure have historically been made at the state level by, or in conjunction with, an appointed group of public officials. This also opens the way to greater understanding of how regulatory bodies—similar across the United States—have shaped states' electricity landscapes in response to unique geographic, socioeconomic, political, and environmental conditions.

This article uses narrative analysis to interrogate records, orders, and other documents produced by the state Commission; publications and archival materials generated by the state's investor-owned utilities and cooperatives; and state and federal utilities laws. Narrative analysis is a prominent methodology in many types of critical geographic inquiry (cf. Dalby 1996; Peet and Watts 2004; Robbins 2004) and is increasingly popular for environmental policy research as well (e.g., Sharp and Richardson 2001). Narrative analysis recognizes that the development of the electricity landscape is contingent on environmental, political, cultural, and geographic particularities that typically go unexamined. I use this approach to examine the stories that key actors in the development of Michigan's electricity system tell, not only about the infrastructure in question but also about their motives, goals, and visions for the state. Through this lens, it is clear that the development of Michigan's electricity landscape has been underpinned by a particular notion of progress equating the deployment of electricity infrastructure—regardless of necessity—with socioeconomic advancement. Crucially, this particular narrative of progress was prevalent among state commissioners and Michigan's major utilities during the period from 1915 to 1985, when most of the state's electricity infrastructure was built. As such, this study contributes to the growing body of research exploring the relationship between policy, discourse, and infrastructure deployment (cf. Kaika and Swyngedouw 2000; Vojnovic 2006; Vogel 2008).

Beginning with a brief history of Michigan's electricity landscape, the article moves to the Commission's role in shaping that landscape. I examine the roots of regulation and the impact of Commission policies as typified by hearings surrounding the Midland Nuclear Facility. Next, I consider the narrative of progress in utility and regulatory decisions, and conclude with a look at the contemporary legacy of that narrative.

Michigan's Electricity Landscape

In Michigan, electricity provision began in the early 1880s. Virtually anyone with adequate financing could sell electricity, but technology limited the scope of early generation projects and distribution networks, concentrating service in densely populated areas and spurring the proliferation of providers. By 1919, Michigan already had more than 150 utilities, most with proprietary generation and distribution infrastructure (Michigan Public Utilities Commission 1919). The

1920s saw technological advances permit the larger, more complex projects necessary to meet surging demand from the state's industries. During this period, Michigan claimed some of the largest hydroelectric projects and coal-fired power plants in the world, complete with lengthy cross-country transmission systems linking remote generating facilities to growing urban populations (Miller 1971; Bush 1973).

These advances encouraged industry consolidation. One of the state's two largest utilities, Consumers Power, exploited its expansive network to seize control of electricity markets in all of Michigan's major cities except Detroit and Lansing, which were in turn dominated by Detroit Edison (the other large utility) and a municipally owned entity, respectively. Although corporate giants controlled urban areas, by 1934 fewer than a quarter of Michigan's farms had electricity (Kuhl 1998). Farmers faced a choice between self-generation and a steep premium for the construction of new power lines, but a New Deal–era program, the Rural Electrification Administration, sought to change this. By 1937, federally funded cooperatives were building both distribution networks and generating facilities throughout the state, and in just five years more than 14,400 km (9,000 miles) of line had been electrified and over 70 percent of Michigan's farms were serviced, some two times the national average (Rural Electrification Administration 1941a, 1941b).

As in Canada (Evenden 2009), involvement in World War II prompted rapid and significant growth in Michigan's electricity infrastructure. The surge in electricity demand and corresponding construction boom that began with the war continued uninterrupted for nearly thirty years (J. R. M. Anderson 1994), producing the majority of plants still serving the state today. Between 1944 and 1978, Consumers Power added fifty generating units, including the company's four largest facilities, and Detroit Edison added 109 new generating units to its system (Energy Information Administration 2000). Each facility trumped its predecessor in size and complexity: in 1949, Detroit Edison built its St. Clair plant, which in conjunction with the company's River Rouge facility produced nearly 2.5 percent of all electricity in the United States. After expansion in 1960, it was for years the single largest power plant in the country (Miller 1971). Despite the prominence of large coal-fired facilities, nuclear technology was also attracting great interest from investor-owned utilities and cooperatives alike. Consumers Power built Michigan's first commercial nuclear plant near Charlevoix in 1965, followed by three more nuclear plants before 1988 when investor-owned utility construction of new power plants in the state all but ceased.

All of these generating facilities came to be linked by an expansive transmission network. Transmission links between Consumers Power and Detroit Edison were established for emergency purposes in the 1920s and shortly thereafter opened to bulk power transfers between utility systems by the late 1930s. Internally, by the mid-1970s all electricity networks in the state were connected, tying major utilities, cooperatives, and municipal systems together into a single network. Externally, the 1950s and 1960s saw Michigan's power supply connected to utilities in Ohio, Indiana, and the Canadian provinces of Ontario and Québec, which were themselves connected to upstate New York and New England. Westward, links were made to Indiana, Wisconsin, and Minnesota. By 1980 this broad regional network linked electricity supply in one state to demand several hundred miles away. The desire to capitalize on geographic differences in production costs, especially in light of energy crises and a broader interest in market liberalization, prompted significant changes to the electricity landscape.

Although a detailed analysis of electricity market liberalization is beyond the scope of this article (cf. Brennan, Palmer, and Martinez 2002), the primary intent of such programs was to introduce new sources of electricity generation into the marketplace and, in particular, those employing either high-efficiency cogeneration designs or renewable fuels.[3] Across the United States, uptake of these technologies was limited, and Michigan was no exception (Energy Information Administration 1993). Between the passage of the National Energy Act in 1978 and the turn of the millennium, just sixty nonutility power producers came online in the state (Energy Information Administration 2000). Representing approximately 10 percent of the state's capacity, nonutility generation would be dominated by a single facility, the Midland Cogeneration Venture, a 1,900-MW natural gas-fired plant (Energy Information Administration 2009a). Much of the remainder is also powered by natural gas, although by legislative fiat Michigan hosts several wood-, biomass-, and solid waste–powered plants as well.

Many studies reasonably argue that reluctance toward smaller and renewably fueled facilities was a function of these technologies' higher per kilowatt costs and limited ability to meet demand relative to centralized fossil fuel plants (Elmes 1996; Heiman and Solomon 2004; Heiman 2006; Sovacool 2008). Indeed, most studies characterize the tendency toward large facilities in economic terms, typically presented through the perspective of major investor-owned utilities themselves (e.g., Miller 1957, 1971; Bush 1973; Hughes 1983; Hirsh 1989; Platt 1991; Rose 1995; Hyman, Hyman, and Hyman 2000). Although I do not reject these techno-economic arguments, I assert that such studies overlook the crucial role that state regulation played in setting the rules by which utilities could operate, profit, and make investments. Understanding regulation as well as the motives behind regulators' specific policy goals is paramount to understanding the historical geography of Michigan's electricity landscape.

Regulation in Michigan

Shoddy construction practices in conjunction with frequent utility mergers and acquisitions prompted the state legislature to grant the Michigan Railroad Commission oversight of a burgeoning electric power industry in 1909.[4] The Commission was charged with enforcing safety parameters and, more important, setting companies' rates for electricity service. Ostensibly designed to protect consumers, utilities regulation was also intended to stabilize a volatile industry (Hausman and Neufeld 2002). High construction costs, cutthroat competition, and low profit margins made financing electricity projects difficult; furthermore, rapid advances in technology made new systems quickly obsolete and insufficient to meet demand. By 1910, industry leaders like Samuel Insull and Detroit Edison's Alex Dow had publicly advocated exchanging control over prices for guaranteed profits and protected service areas from a state regulator (Hausman and Neufeld 2002). New prices would reflect the costs of generating and delivering electricity to customers but also a rate of return determined by the regulatory Commission (Yakubovich, Granovetter, and McGuire 2005).

Following Wisconsin in 1907, other states in the Midwest and indeed most of the country devised similar regulatory regimes before 1930. In those states, including Michigan, a small number of investor-owned utilities came to dominate the market, due in large part to the financing and territorial protections afforded by state oversight regimes. In Michigan, Consumers Power and Detroit Edison became, de jure, the largest utilities in the state, as the de facto service areas each had carved out through mergers and system expansions were codified into state law in 1929.[5]

The Impact of Commission Policies

Establishing territorial monopolies was one way in which regulatory oversight shaped Michigan's electricity landscape. Another was through the Commission's control of utility pricing structures. "Declining block" pricing for residential accounts rewarded customers passing certain usage thresholds with lower per-unit prices (MPSC 1961), and bulk contracts for heavy industrial users subsidized electricity consumption in the name of economic growth, to offer just two concrete examples of the Commission's belief that cheap electricity and the power plants producing it were firmly "in the public interest" (MPSC 1965, 3).

Of all the ways in which the Commission shaped Michigan's electricity landscape, however, most important was the rate-making process itself: the determination of retail prices and thus the extent to which utility capital expenditures would be recovered. For decades, one could argue that the intricacies of this process scarcely mattered: between 1920 and 1970, utilities' growing capital expenditures were matched by exploding electricity consumption (J. R. M. Anderson 1994). The long lead times and significant sunk costs of a new plant made construction in anticipation of future demand the utility industry's standard practice by the 1950s (Hirsh 1989). Economies of scale achieved through plant construction pushed prices downward, satisfying any concerns the Commission had about abusive pricing. Commission oversight did not extend to individual project planning, meaning that utilities could begin construction on new facilities, assured of cost recovery later through the rate-making process. Michigan's cheap electricity prices became central to attracting new investment (Michigan Department of Economic Development 1949), but the late 1960s saw the first of many economic downturns in the state, suppressing demand for electricity. By 1974 demand actually fell for the first time since the Great Depression (J. R. M. Anderson 1994), never again returning to rapid postwar rates of growth. Nevertheless, Michigan's utilities continued plans for new construction as before, anticipating a dramatic recovery, and thereby exposing critical flaws in the state's regulatory structure.

The rate-of-return accounting system at the heart of state oversight depended on capital expenditures, not monthly sales revenues, to determine profitability. This meant that any limit to construction efforts would endanger a utility's capacity to provide electric power. Accordingly, the only means by which companies could maintain profitability was to invest in new, progressively larger, and more expensive pieces of infrastructure, even if they could not be justified to meet demand. Basing profitability on long-term capital expenditures instead of monthly revenues removed any incentive for utilities to either temper demand or improve transmission, because both would depress consumption figures and contradict claims that additional power plants were needed. Moreover, the Commission's repeated commitment to low prices precluded utilities from simply charging more per unit of electricity as a means to maintain revenue amidst falling demand (MPSC 1972). These structural faults inevitably produced a cycle wherein utilities would seek rate increases to cover short-term revenue shortfalls and use the money to service outstanding debt, all while increasing the supply of electricity (through new construction) and further depressing its price, triggering a new round of revenue deficiency.

This is comprehensively illustrated in hearings surrounding Consumers Power's Midland Nuclear Facility.[6] Intended to be among the nation's largest power plants on completion in the early 1970s, the company first sought a rate increase from the Commission in 1968, claiming that the costs of new generating facilities were no longer being matched by sales revenues. Furthermore, a previous nuclear plant, Palisades, was overbudget and weighing down the company's balance sheet. Both problems were magnified by the declining block and special contract pricing that financially penalized the company with every sale.

The state Attorney General and others argued that a rate increase to fund construction was unnecessary because new cogeneration and conservation programs, alongside expansion of existing facilities, would suffice to meet Michigan's softening electricity demand. Fearing the company's collapse, however, the Commission approved the rate hike, threatening that "if there is to be sufficient electric energy available to the citizens of Michigan, it is essential that all parties soberly address themselves to . . . finance the new construction which the public requires" (MPSC 1972, 26).

For Midland, the Commission allowed for the first time "projected-data test years," meaning that Consumers Power could present forecasts of future costs and demand instead of historical consumption data to justify an immediate rate increase. Defying observed trends, the company predicted significant increases in both. The Commission assented to another round of rate increases, again citing the hypothetical situation in which any move to withhold or deny rate increases would jeopardize Michigan's supply of electricity in the future (MPSC 1978, 18–19), justifying its decision with the claim that "Midland . . . is designed to be a major baseload plant and . . . should provide power at a price which is comparable to many other units," upon completion (18).

Financial markets were unconvinced. By 1982, Consumers Power's credit rating had been downgraded and the company faced skyrocketing capital costs, leaving the Commission and customers responsible for the mounting debts. Noting that it "does take seriously the continuous downgrading of [Consumers Power] securities" (MPSC 1982, 8), the Commission approved an additional $120.5 million annual rate increase to boost the company's short-term balance sheet and help it finance construction costs. Only when costs had ballooned to over $3 billion did Commissioners admit that their body had "failed to undertake a comprehensive review of that massive construction project" and, more generally, that:

> The . . . Michigan Public Service Commission majority does not examine these relevant issues before or during construction. Instead, it allows a utility company's investors to put millions, or in this case, billions of dollars into facilities that it claims will be evaluated for viability at the time a company requests inclusion of such facilities in its rate base. The majority [of Commissioners] says such issues should be examined in power plant siting proceedings. . . . Yet [Michigan] has no power plant siting legislation and thus no such proceedings. (E. Anderson 1982, 8–10)

Consumers Power teetered on the brink of bankruptcy, threatening to halt electricity and natural gas service in Michigan and disrupt availability in the wider Midwest. The Commission took emergency action to stabilize the company's finances, ordering another six-year, $99 million-per-year rate increase, on the condition that the company abandon all construction at Midland (MPSC 1985). After fifteen years of delays, $4 billion in cost overruns, and more than $1 billion of Commission-approved rate increases and securitizations, Midland was sold at 85 percent completion to a consortium of buyers (including a Consumers Power subsidiary), who would convert the plant into the Midland Cogeneration Venture natural gas facility under new market liberalization rules.

Midland remains unmatched in Michigan's electricity history for the scale of mismanagement that it represents, but the project typifies the structural flaws common to both utilities and regulatory organizations as well as the narratives of progress that underpin Michigan's experience with electricity infrastructure.

Consider that, according to Consumers Power's own data, had Midland actually come online the state would have possessed reserve capacity some 50 percent beyond any load ever experienced (E. Anderson 1982), at a time when actual demand was stagnant, the economy was contracting, and the power industry standard for reserves was just 10 percent. Nevertheless, on either side of Midland, Michigan's largest ever coal-fired unit, Campbell III (another Consumers Power project), and Detroit Edison's Enrico Fermi II nuclear facility both came online. These massive facilities and associated glut of generating capacity are the direct consequence of a regulatory regime that fully believed its own story, namely, that more electricity meant more economic prosperity, and accordingly rewarded new infrastructure, punishing both efficiency and conservation efforts.

Progress and Michigan's Electricity Landscape

Concerns had been raised about the sustainability of the rate-of-return model as early as the 1960s (J. R. M. Anderson 1994), and it is clear that even during the Midland hearings, the Commission itself was well aware of the model's shortcomings. Nevertheless, it remained in place, virtually unmodified, through the mid-1990s. The reasons for this spring from utilities' and regulators' shared vision of progress for the state. "Progress" is a common theme in geographies of infrastructure (cf. Berman 1982; Mitchell 1988; Brigham 1998; Wainwright 2008), particularly as infrastructure might transform everyday life. Although the scope of this article limits the consideration of transformation in the home (cf. Nye 1991; Tobey 1996; Goldstein 1997), it focuses instead on progress as directly related to electricity infrastructure itself.

In Michigan this link was forged in the late nineteenth century, as promotional material from a nascent Consumers Power promised "nights as bright as day" for towns adopting its lighting systems (Bush 1973, 48). Early commissioners defended the construction of large hydroelectric projects on account of the resulting "advancement of the general welfare of the State" (MPUC 1925, 6–8), and in the context of the post-Depression, postwar United States, Michigan's utilities touted their ability to create a "new world of order" through electricity infrastructure and planning (Miller 1971, 297).

Consumers Power and Detroit Edison's corporate histories define themselves by progress. New power plants were "milestones of progress, not only for the utility company that built the plants, but even more so for the people, for the consumption of electric power is a measure of the nation's standard of living: power means prosperity" (Bush 1973, 341). The transmission system linking power plants to customers formed a "network of usefulness" for the state (Bush 1973, 83), just as the "mystery and excitement about the power plant. . . . The cathedral-like vistas, the awesome might and majesty of the flaming furnace, and, above all, the turbine generator room, where almost unbelievable power is marshaled in dramatic orderliness" signaled a new era of prosperity (Miller 1971, 105).

Progress required that any conceivable need for electricity be met, providing a technosocial narrative to the practice of building supply ahead of demand. By 1949, Michigan boasted generating capacity already exceeding "any load ever experienced or anticipated" by 15 percent and promised that "by 1952, generation capacity will be expanded by another 15 percent" (Michigan Department of Economic Development 1949). The state's utilities claimed wide reserves were essential, "[defining] the Company's obligation—the planning, financing, and building of the entire system [was] aimed at this one 15-minute period in a late December afternoon, or in a summer heat wave" (Miller 1971, 172). The Commission robustly defended this logic, defiantly stating during the height of the Midland crisis that "overcapacity would not be a justification for excluding the plant from the rate base" (MPSC 1978, 18).

The narratives linking progress and infrastructure deployment cast conservation as the arch-villain. The Commission actually argued that it was outside its "legislative or constitutional mandate . . . to pursue a draconian and socially disruptive program of forced conservation" (MPSC 1978, 17). Such attitudes meant that by the end of the Midland hearings, the state Commission and Michigan's utilities were contributing just $1.32 per capita to conservation efforts while the rest of the nation spent on average $5.30 per person (Audubon Society of Kalamazoo 1991, D-15).

Limits to Progress

Testimony from the state Attorney General during the Midland hearings argued that "Consumers' [Power] forecasts of electricity sales were seriously flawed" and that "new capacity would not be required before the year 2000" (MPSC 1985, 15). Shortly thereafter, the state Department of Commerce organized the Michigan Electricity Options Study with the express goal of "making economically sound judgments . . . for meeting Michigan's uncertain electricity needs over the next 20 years" (Michigan Electricity Options Study 1987, 1-1). Given the state's surplus capacity, considerable attention was finally paid to electricity conservation (demand-side management, or DSM). Both Consumers Power and Detroit Edison were ordered to produce comprehensive plans for meeting future demand without new construction, and a chastened Commission had to reconsider rate-of-return accounting and the relationship among revenue, electricity sales, capital investments, and profits.

In their plans, utilities argued against conservation targets, citing their historic financing structures and the rate-of-return accounting system still at the core of the regulatory regime (MPSC 1991a, 44). The Commission had to agree but nevertheless issued sharp criticisms of utility plans, arguing that Detroit Edison "[discounted] the view that . . . demand-side resource options (including conservation) can delay investment in new capacity, improve efficiency" (MPSC 1990, i) and Consumers Power's "planned incorporation of DSM is many orders of magnitude away from achieving a meaningful integration of DSM as a utility resource" (MPSC 1991a, 48). Contradicting utility figures, Commission engineers estimated that at least 25 percent of future demand growth could be met exclusively through conservation (MPSC 1991a). Accordingly, the Commission suggested progressive modifications to regulatory oversight, offering rate increases to offset aggressive DSM implementation and a 2 percent rate of return on capital invested in conservation efforts (MPSC 1991b).

These innovations were scarcely considered, however as the market liberalization efforts of the mid-1990s turned the attention of the Commission, utilities, and general public alike toward the formation of a regional market for bulk power sales.[7] This new task further complicated efforts to implement meaningful electricity conservation programs, as decisions surrounding electricity's production in Michigan would now be taken in a multistate, and even international, electricity landscape replete with its own unique economic, political, and regulatory challenges.

Legacies of Progress

A new regional context notwithstanding, historical attitudes persist, limiting gains in efficiency and environmental protection that a new marketplace might produce. For the utilities, progress continues in a very literal sense: their territorial protections and profit models intact, Detroit Edison and Consumers Power still supply the vast majority of Michigan's electricity, retaining a central role in any and all new energy initiatives. As the two extend the life of their own inefficient and aging facilities, they oppose distributed generation, electricity conservation, and renewable fuels programs in the state on the basis of these technologies' alleged inability to meet Michigan's future demands for electricity. Similarly, the Commission continues to embrace narratives of limitless growth from years past, even as it seeks to address contemporary energy problems like pollution and climate change. Despite reforms, the Commission regularly approves deep discounts and preferential pricing contracts even as the state spends 70 cents of every energy dollar on imported fossil fuels and loses a quarter of all electricity generated to transmission inefficiencies (MPSC 2008).

Indeed, the celebration of golden years past has only increased in recent times. Consumers Power's state-sanctioned "achievement landmark" commemorates its Big Rock Point nuclear plant, incorporating decommissioned containment steel to recognize "the dream that nuclear energy could . . . reliably produce electricity" (Petrosky 2007, 12). A nearby State Historical Marker, granted in 2007, highlights a facility whose "first goal [was] to prove that nuclear power was economical" (11), despite the fact that nuclear decommissioning has cost far more than originally anticipated. More pragmatically, the Commission's 21st Century Electric Energy Plan makes a clear call for reregulation and the rate-of-return model, lamenting the floating prices and "difficult, if not impossible" project financing associated with regional electricity markets that the Commission itself helped organize (MPSC 2007, 2).

It is only in exploring the narratives of progress that the centrality of regulatory commission actions in shaping each state's response to electricity infrastructure deployment challenges becomes clear. Undoubtedly, Michigan's utilities faced tough choices regarding electricity supply and reliability, but this article has demonstrated that those utilities acted within the rules set for them by the state Commission. As such, consideration of regulators' actions is helpful in explaining differences between states' electricity landscapes. Regulatory narratives offer insight into fuel source and technology choices; for instance, illuminating Michigan's heavy reliance on imported coal from the American West, whereas nearby states (like Illinois, containing 10 percent of national coal reserves) embraced nuclear power (Energy Information Administration 2009c). Whereas Michigan's commissioners have historically considered environmental protection and conservation outside of their duties, commissioners in California have weighted environmental concerns heavily in regulatory decisions, reducing reliance on coal and nuclear facilities there (Energy Information Administration 2009b). Similarly, analyzing narratives related to infrastructure deployment offers explanations for why California and Illinois have meaningful renewable energy targets and Michigan does not.

The surplus capacity typifying Michigan's electricity landscape results directly from Commission beliefs about electricity availability, the integrated monopoly, and progress. The legacy of these historic attitudes recurs today as Michigan's commissioners are presented with different sets of solutions to power supply challenges than those available to commissioners in "reregulated" California, which recently experienced major capacity shortfalls, or in fully liberalized Illinois. Further research must explore the impact of regulatory decisions on other states' electricity landscapes, with particular attention to the relationship among regulators, utilities, and the siting and construction of individual projects. Additionally, the liberalization project has made writing historical geographies of regional electricity markets of the utmost importance. Through such efforts we could better understand the forces shaping electricity landscapes in the United States and thereby improve on a system of infrastructure that is critical to life in the twenty-first century.

Acknowledgments

Special thanks to Kyle Evered, Igor Vojnovic, and Antoinette WinklerPrins for assistance with this article; the comments of anonymous reviewers for helping sharpen the article; and the Michigan State University Graduate Office Fellowship for offsetting research costs.

Notes

1. I use *electricity landscape* to collectively reference electricity generation, transmission, and distribution infrastructure (see also Zimmerer 2011).
2. This article emphasizes Consumers Power (now operating as Consumers Energy) and Detroit Edison and their activities in the Lower Peninsula of Michigan. Although there are other private, cooperative, and municipal utilities in Michigan, the electricity landscape has been dominated by these two firms.
3. *Cogeneration* is the combined production of heat and electricity.
4. The Michigan Railroad Commission became the Michigan Public Utilities Commission in 1919 and the Michigan Public Service Commission in 1939.
5. Other utilities operate in Michigan, particularly in the Upper Peninsula, and municipal systems remain unaffected by Michigan's utilities laws.
6. See MPUC cases U-3179, U-3749, U-4174, U-4324, U-4332, U-4576, U-4840, U-5331, U-5353, U-5388, U-5438, U-5734, U-5979, U-6923, U-7263, and U-7830.
7. See MPSC Annual Reports for 1989–2009 for analysis of Commission efforts in implementing federal rule changes.

References

Anderson, E. 1982. Dissenting opinion, U-6923. Lansing: Michigan Public Service Commission.

Anderson, J. R. M. 1994. Michigan utility regulation: The perspective of the dissenters. PhD dissertation, Michigan State University, East Lansing.

Audubon Society of Kalamazoo. 1991. *Comments on Consumers Power Company's integrated resource plan*. Lansing: Michigan Department of Commerce.

Berman, M. 1982. *All that is solid melts into air: The experience of modernity*. New York: Simon and Schuster.

Brennan, T. J., K. L. Palmer, and S. Martinez. 2002. *Alternating currents: Electricity markets and public policy*. Washington, DC: Resources for the Future.

Brigham, J. L. 1998. *Empowering the west: Electrical politics before FDR*. Lawrence: University Press of Kansas.

Bush, G. 1973. *Future builders: The story of Michigan's Consumers Power Company*. New York: McGraw-Hill.

Dalby, S. 1996. Reading Rio, writing the world: The New York Times and the "Earth Summit." *Political Geography* 15:593–613.

Elmes, G. 1996. The changing geography of electric energy in the United States—Retrospect and prospect. *Geography* 81:347–60.

Elmes, G. A., and T. M. Harris. 1996. Industrial restructuring and the United States coal-energy system, 1972–1990: Regulatory change, technological fixes, and corporate control. *Annals of the Association of American Geographers* 86:507–29.

Energy Information Administration. 1993. *The changing structure of the electric power industry, 1970–1991*. Washington, DC: U.S. Department of Energy.

———. 2000. Form 860a—Existing generators, 2000. Washington, DC: U.S. Department of Energy.

———. 2009a. Existing nameplate and net summer capacity by energy source, producer type and state. Washington, DC: U.S. Department of Energy. http://www.eia.doe.gov/cneaf/electricity/epa/existing_capacity_state.xls (last accessed 23 March 2010).

———. 2009b. State energy profile: California. Washington, DC: U.S. Department of Energy. http://tonto.eia.doe.gov/state/state_energy_profiles.cfm?sid=CA (last accessed 20 February 2010).

———. 2009c. State energy profile: Illinois. Washington, DC: U.S. Department of Energy. http://tonto.eia.doe.gov/state/state_energy_profiles.cfm?sid=IL (last accessed 20 February 2010).

———. 2009d. State energy profile: Michigan. Washington, DC: U.S. Department of Energy. http://tonto.eia.doe.gov/state/state_energy_profiles.cfm?sid=MI (last accessed 20 February 2010).

Evenden, M. 2009. Mobilizing rivers: Hydro-electricity, the state, and World War II in Canada. *Annals of the Association of American Geographers* 99:845–55.

Goldstein, C. 1997. From service to sales: Home economics in light and power, 1920–1940. *Technology and Culture* 38:121–52.

Hausman, W. J., and J. L. Neufeld. 2002. The market for capital and the origins of state regulation of electric utilities in the United States. *Journal of Economic History* 62:1050–73.

Heiman, M. K. 2006. Expectations for renewable energy under market restructuring: The U.S. experience. *Energy* 31:1052–66.

Heiman, M. K., and B. D. Solomon. 2004. Power to the people: Electric utility restructuring and the commitment to renewable energy. *Annals of the Association of American Geographers* 94:94–116.

Hirsh, R. F. 1989. *Technology and transformation in the American utility industry*. Cambridge, UK: Cambridge University Press.

Hughes, T. P. 1983. *Networks of power: Electrification in Western society, 1880–1930*. Baltimore: The Johns Hopkins University Press.

Hyman, L. S., A. S. Hyman, and R. C. Hyman. 2000. *America's electric utilities: Past, present, and future*. 7th ed. Vienna, VA: Public Utilities Reports.

Kaika, M., and E. Swyngedouw. 2000. Fetishizing the modern city: The phantasmagoria of urban technological networks. *International Journal of Urban and Regional Research* 24:120–38.

Kuhl, R. G. 1998. *On their own power: A history of Michigan's electric cooperatives*. Okemos: Michigan Electric Cooperative Association.

Michigan Department of Economic Development. 1949. *Michigan power resources for industry*. Lansing: Michigan Department of Economic Development.

Michigan Electricity Options Study. 1987. *Electricity options for the State of Michigan: Results from the MEOS Project*. Lansing: Michigan Department of Commerce.

Michigan Public Service Commission (MPSC). 1961. U-787. Lansing, MI: MPSC.

———. 1965. U-2028. Lansing, MI: MPSC.

———. 1972. U-4174. Lansing, MI: MPSC.

———. 1978. U-5331. Lansing, MI: MPSC.

———. 1982. U-6923. Lansing, MI: MPSC.

———. 1985. U-7830–3A. Lansing, MI: MPSC.

———. 1990. *The Michigan Public Service Commission staff report on Detroit Edison Company's 1990 Integrated Resource Plan*. Lansing: Michigan Department of Commerce.

———. 1991a. *Commission staff report on Consumers Power Company's 1990 Integrated Resource Planning Report*. Lansing: Michigan Department of Commerce.

———. 1991b. U-9346. Lansing, MI: MPSC.

———. 2007. *Michigan's 21st century electric energy plan*. Lansing: Michigan Department of Labor and Economic Growth.

———. 2008. *Michigan energy overview*. Lansing: Michigan Department of Labor and Economic Growth.

Michigan Public Utilities Commission. 1919. *Annual report*. Lansing: Michigan Department of Commerce.

———. 1925. *Annual report*. Lansing: Michigan Department of Commerce.

Miller, R. C. 1957. *Kilowatts at work: A history of the Detroit Edison Company*. Detroit, MI: Wayne State University Press.

———. 1971. *The force of energy: A business history of the Detroit Edison Company*. East Lansing: Michigan State University Press.

Mitchell, T. 1988. *Colonising Egypt*. Cambridge, UK: Cambridge University Press.

Nye, D. 1991. *Electrifying America: Social meanings of a new technology, 1880–1940*. Cambridge, MA: The MIT Press.

Peet, R., and M. Watts. 2004. *Liberation ecologies: Environment, development, social movements*. London and New York: Routledge.

Petrosky, T. 2007. Michigan historical marker for Big Rock Point site. *Radwaste Solutions Buyers Guide* November/December:10–12.

Platt, H. L. 1991. *The electric city: Energy and growth of the Chicago area, 1880–1930*. Chicago: The University of Chicago Press.

Robbins, P. 2004. *Political ecology: A critical introduction*. Malden, MA: Blackwell.

Rose, M. H. 1995. *Cities of light and heat: Domesticating gas and electricity in urban America*. University Park: Pennsylvania State University Press.

Rural Electrification Administration. 1941a. Allotment, construction, operating, and financial statistics of REA-financed systems. Washington, DC: U.S. Department of Agriculture.

———. 1941b. *Report of the administrator*. Washington, DC: U.S. Department of Agriculture.

Sharp, L., and T. Richardson. 2001. Reflections on Foucauldian discourse analysis in planning and environmental policy research. *Journal of Environmental Policy and Planning* 3:193–209.

Sovacool, B. K. 2008. *The dirty energy dilemma: What's blocking clean power in the United States*. Westport, CT: Praeger.

Tobey, R. C. 1996. *Technology as freedom: The New Deal and the electrical modernization of the American home*. Berkeley: University of California Press.

Vogel, E. 2008. Regional power and the power of the region: Resisting dam removal in the Pacific Northwest. In *Contentious geographies: Environmental knowledge, meaning, scale*, ed. M. K. Goodman, M. T. Boykoff, and K. T. Evered, 165–86. Burlington, VT: Ashgate.

Vojnovic, I. 2006. Urban infrastructures. In *Canadian cities in transition: Local through global perspectives*, ed. T. Bunting and P. Filion, 123–37. New York: Oxford University Press.

Wainwright, J. 2008. *Decolonizing development: Colonial power and the Maya*. Malden, MA: Blackwell.

Yakubovich, V., M. Granovetter, and P. McGuire. 2005. Electric charges: The social construction of rate systems. *Theory and Society* 34:579–612.

Zimmerer, K. S. 2011. New geographies of energy: Introduction to the special issue. *Annals of the Association of American Geographers* 101 (4): 705–11.

BOOK REVIEW ESSAY

The Geography of Energy and the Wealth of the World

Martin J. Pasqualetti

School of Geographical Sciences and Urban Planning, Arizona State University

Curse of the Black Gold: 50 Years of Oil in the Niger Delta. Michael Watts, ed. New York: Power House Books, 2010. 224 pp. $39.95 (ISBN-13 978–1576875476).

Encyclopedia of Energy. Cutler Cleveland, ed. New York: Elsevier Science, 2007. 5,376 pp. $2730.00 (ISBN 978-0121764807).

Energiegeographie. Wechselwirkungen Zwischen Ressourcen, Raum und Politik [Energy geography. Interactions between resources, space and policy]. Wolfgang Brücher. Berlin and Stuttgart, Germany: Borntraeger, 2009. 280 pp. €29.80 (ISBN-13 978-3443071455).

Energy and the New Reality 1: Energy Efficiency and the Demand for Energy Services. L. D. Danny Harvey. London: Earthscan, 2010. 614 pp. $79.95 paper (ISBN-13 978-1849710725).

Energy and the New Reality 2: Carbon-Free Energy Supply. L. D. Danny Harvey. London: Earthscan, 2010. 600 pp. $79.95 paper (ISBN-13 978-1849710732).

Energy Efficiency and Climate Change: Conserving Power for a Sustainable Future. B. Sudhakara Reddy, Gaudenz B. Assenza, Dora Assenza, and Franziska Hasselmann. Thousand Oaks, CA, and New Delhi: Sage, 2009. xiv and 349 pp. $39.95 cloth (ISBN 978-8132102281).

Energy Myths and Realities. Vaclav Smil. Washington, DC: American Enterprise Institute, 2010. ix and 232 pp. $34.95 cloth (ISBN 978-0844743288).

Energy Poverty in Eastern Europe: Hidden Geographies of Deprivation. Stefan Buzar. Aldershot, UK: Ashgate, 2007. xiii and 175 pp. $124.95 cloth (ISBN 978-0754671305).

Energy Transitions. Vaclav Smil. Santa Barbara, CA: Praeger, 2010. ix and 178 pp. $34.95 cloth (ISBN 978-0313381775).

Fueling War: Natural Resources and Armed Conflicts. Phillippe Le Billon. London and New York: Routledge, 2006. 92 pp. $34.95 paper (ISBN 978-0415379700).

Géographie De L'énergie. Acteurs, Lieux et Enjeux [Geography of energy. Actors, places and issues]. Bernadette Mérenne-Schoumaker. Paris: Belin, 2008. €22.80 paper (ISBN 978-2701144658).

Landscapes of Energy. New Geographies 02. Rania Ghosn, ed. Cambridge, MA: Harvard University Press, 2009. $20.00 paper (ISBN 9781934510254).

Renewable Energy and the Public: From NIMBY to Participation. Patrick Devine-Wright, ed. London: Earthscan, 2010. xix and 336 pp. $99.95 paper (ISBN 978-1844078639).

Routledge Handbook of Energy Security. Benjamin K. Sovacool, ed. London and New York: Routledge, 2011. xviii and 436 pp. $195.00 cloth (ISBN 978-0415591171).

The New Energy Crisis: Climate, Economics and Geopolitics. Jean-Marie Chevalier. Basingstoke, UK: Macmillan, 2009. xv and 295 pp. $100.00 cloth (ISBN 978-0230577398).

The Political Economy of Sustainable Energy. Catherine Mitchell. Basingstoke, UK: Macmillan. 2009. 248 pp. $28.00 paper (ISBN 978-0230241725).

The Renewable City: A Comprehensive Guide to an Urban Revolution. Peter Droege. Hoboken, NJ: Wiley, 2006. xii and 309 pp. $60.00 paper (ISBN 978-0470019269).

Urban Energy Transition—From Fossil Fuels to Renewable Power. Peter Droege, ed. New York: Elsevier Science, 2008. 664 pp. $185.00 cloth (ISBN 978-0080453415).

Wind Power in View: Energy Landscapes in a Crowded World. Martin J. Pasqualetti, Paul Gipe, and Robert Righter. San Diego, CA: Academic Press, 2002. 248 pp. $166.10 cloth, $53.89 Kindle (ISBN 978-0125463348).

Search Google Books for the phrase "geography of energy" and you will receive about 2,150 hits. A search for "geopolitics of energy" reveals 3,360 hits. "Energy geography" returns 453. In contrast, the term "geography" returns 4.6 million, and "energy" returns 15.3 million. By such measures, the mix of geography and energy produces only a tiny subset within two areas of considerable interest. On the contrary, I believe just the opposite, that the mix of geography and energy is so common it escapes casual notice.

Let me offer a few examples. Pirates, mainly from Somalia, lurk off the Horn of Africa, ready to commandeer oil tankers that pass nearby, betting that owners will readily pay a few million dollars in ransom to regain $100 million worth of product. Men and women of the U.S. Navy routinely—and expensively—patrol the Persian Gulf and other crucial shipping lanes to ensure that oil reaches our shores regularly and without interruption. The former Soviet Republic of Georgia commonly experiences unrest because it offers the most convenient terrain for avoiding Russian territory in the movement of oil and gas from the landlocked Caspian Basin to markets in the West. In Nigeria, millions suffer from deprivation even as vast oil wealth beneath their feet is pumped to the surface and sold to foreign consumers. Unrest in the coalfields of West Virginia boils over as mountaintop removal destroys landscapes and clogs rivers. Boat captains in Louisiana worry that oil production off its shores will reduce the harvest of fish that occupy the same waters. For all these examples and countless others, the geography of energy is the common denominator.

When discussing the geography of energy, no resource attracts more attention than oil. The world, especially the First World, runs on it. Its discovery, development, and sale have for about 150 years brought wealth to a few, convenience to some, and avarice to many. Many of the problems that accompany our reliance on oil are fundamentally spatial because reserves are not evenly distributed. A few countries have more than they can use. Most do not. If we accept that the most important activity in the energy business is reliability, then we cannot argue a minor role for the geographic exercise of matching supply with demand. Rather, it is a daunting responsibility; every day, more than 3.5 billion gallons of oil must be brought to the surface and distributed for use in an interactive network so complex and laden with intrigue that its successful operation must be considered the equivalent of magic.

As useful as the development and distribution of oil is for illustrating the importance of geography to energy, many other aspects of energy display equally strong spatial dimensions. These include siting power plants, refineries, pipelines, and transmission wires; tracking the origins and distribution of contaminants from energy activities in our air and water; and recognizing the social inequities that result when energy supplies are available to people in some locations but not to people in others.

One of the clearest connections between geography and energy is through maps. Maps of gas pipelines illustrate why Romania suffered when Russia punished Ukraine for delinquent payments. Maps reveal why oil tankers from the Persian Gulf are more at risk from interdiction than tankers from Nigeria. Maps of Afghanistan help us understand why supporting each member of the U.S. military deployed there can cost $200,000 to $350,000 a year just in fuel costs (National Public Radio 2011). And maps are superbly helpful in locating energy activities, past or present, just by the place names they display: Carbondale, Illinois; Carbon County, Utah; Carbondale, Colorado; Coalville, Utah; Colstrip, Montana; Petroleum County, Montana; Petrolia, Texas; Oil City, Pennsylvania; Oildale, California; Bairoil, Wyoming; Uranium City, Saskatchewan; Atomic City; Idaho; and Nucla, Colorado.

The scholarly literature of energy geography exists mostly in the form of journal articles and book chapters, as has been discussed elsewhere (Pasqualetti 1986; Solomon and Pasqualetti 2004; Solomon, Pasqualetti, and Luchsinger 2004). Books are understandably less common, but still there are hundreds of them, produced by people in dozens of specialties and professions. This essay is limited to books written by geographers, identified here by training, employment, or membership in a professional geographic society. The books listed are my selection of some of the more notable books published during the past decade—but because they rest on the shoulders of those who came before, that is where we begin.

Setting the Stage (1950–2000)

The earliest publications on energy geography focused on the location of resources. One example is Pratt and Good's (1950) volume on petroleum published by the American Geographical Society, the same year George (1950) published *Géographie de l'Energie*. It took fifteen years for other books of a similar nature to appear, including *The Geography of Energy* (Manners 1964), *Energy in the Perspective of Geography* (Guyol 1971), *A Geography of Energy* (Wagstaff 1974), *Géographie et Économie Comparée de l'Énergie* (Sevette 1976), and *Energy: Needs and Resources* (Odell 1977). Taken as a group, they offered a convincing demonstration that mating geography and energy can produce many offspring, each with its own personality. Indeed, each book stressed something different, such as transportation, location, logistics, modeling, supply, demand, markets, and policy. Reading them all could be an enjoyable reminder of the breadth of geography, but if one wants to read only a single volume, I would recommend *Energy, Man, Society* (Cook 1976). In my view, it remains the best book for the widest range of university students, and it took a self-described geographer to do it.

For those who prefer to examine energy regionally, geographers have over the years contributed books on New Zealand (Farrell 1962), Ghana (Hart 1980), the Caspian Basin (Croissant and Aras 1999), and China (Kuby 1995); several have also appeared on the USSR and post-Soviet Russia (Hooson 1965; Dienes and Shabad 1979; Hoffman and Dienes 1985; Dienes, Dobozi, and Radetzki 1994)—but none on the United States.

For those who prefer books with more of a thematic approach, geographers were also filling that niche on topics such as oil (Odell and Rosing 1980; Gever, Kaufmann, and Skole 1991), renewables (Pryde 1983), recreation (Knapper, Gertler, and Wall 1983), modeling (Lakshmanan and Nijkamp 1980, 1983; Lakshmanan and Johansson 1985), and ecological economics (Hall, Cleveland, and Kaufmann 1992).[1]

By the 1970s, the Association of American Geographers (AAG) was playing a role in advancing the study of energy. It did this in two ways: first by supporting the publication of several monographs, including *Energy: The Ultimate Resource?* (Cook 1977), one on coal facility siting (Calzonetti and Eckert 1981), another on renewables (Sawyer 1986), and a fourth on global change (Kuby 1996). The second role played by the AAG was the inauguration of specialty groups. Martin Pasqualetti and Jerome Dobson organized the Energy Specialty Group at the 1979 annual meeting in Philadelphia, a group since renamed the Energy and Environment Specialty Group and now boasting more than 500 members (Energy and Environment Specialty Group 2011).[2] A few years later, several of the founding members contributed to a one-off collection of twenty-five articles that represented a snapshot of energy geography in the early 1980s (Calzonetti and Solomon 1985).

Meanwhile, just as energy geography was developing greater coordination in the United States, it was also maturing abroad. Several geographers in the United Kingdom, for instance, began taking up the energy theme, including three books with a Scottish flavor. These included a national overview of U.K. energy by a Scotsman (Fernie 1980) and two books on the North Sea oil developments east of Aberdeen (K. Chapman 1975; Hogg and Hutcheson 1975).[3] Elsewhere, Manners (1981) continued his interest in energy geography with a well-received book on the British coal scene. Morgan and Moss (1981) took up the study of fuel wood in the humid tropics, and on the continent Odell was continuing to update his global survey of the influence of oil on world power (Odell 1986). Keith Chapman (1991) returned with another book on oil just after John Chapman (1989) in Vancouver had tackled the tricky complexities of commercial energy systems.

For a time during the 1960s and 1970s, nuclear power was gaining momentum as a supplement and alternative to coal-burning power plants. This shift by itself attracted a share of curious geographers, but the 1979 accident at Pennsylvania's Three Mile Island nuclear power plant boosted this interest further by raising many questions ripe for geographic consumption. These included understanding risk perceptions and behavioral responses (Pasqualetti and Pijawka 1984; Blowers and Peppers 1987), safe power plant siting (Openshaw 1986; O'Riordan, Kemp and Purdue 1988), decommissioning and its social costs (Pasqualetti 1990), the transportation and disposal of nuclear waste (Openshaw, Carver, and Fernie 1989; Jacob 1990; Beaumont and Berkhout 1991; Blowers, Clark, and Smith 1991; Blowers, Lowry, and Solomon 1991; Flynn et al. 1995), lessons about democratic principles that one could draw from the Soviet Union's 1986 Chernobyl explosion (Gould 1990), and the degree to which nuclear power had spread around the world by the early 1990s (Mounfield 1991). These books underscored that a geographic perspective was just as appropriate for the study of technological hazards as Gilbert White had shown they were for natural hazards. Indeed, many geographers studying nuclear power came out of this very tradition (Pasqualetti 1986).

Although books by energy geographers continued appearing now and then through the end of the century, nowhere did they originate with more frequency than from Vaclav Smil at the University of Manitoba. Smil, one of the most prolific geographers in our midst, has kept on addressing one theme after another, consistently demonstrating a firm grasp of the technical complexities and societal reach of both energy and geography (Smil 1976, 1982, 1983, 1987, 1988, 1991, 1994, 1998; Smil and Knowland 1980).

The final years of the twentieth century saw the publication of a spate of books pertinent to energy geography. It is with a bit of disquiet, however, that I mention that many were produced by historians. One of the earliest dealt with the influence of the electrification of Western society (Hughes 1983), but several others were of interest to geographers. I am particularly attracted to the geographical leanings of Nye. Three of his stand out: *Electrifying America*, *Technologies of Landscape*, and *Consuming Power* (Nye 1990, 1998, 1999). Other historians examined the role of energy in entire states (Williams 1997), parts of states (Black 2000), and urban areas (Platt 1991). That these and more recent books of a similar nature (Melosi and Pratt 2007; Condee 2005) could have been written by geographers—but were not—is both good news and bad. It is good news because they have been written anyway, allowing geographers as well as others to reap the benefits. It is bad news because it might signal that geographers are being overtaken by those in other disciplines through insufficient attention to a natural area of study (Brücher 1997, 2004).[4] But during the past ten years, geographers have returned to the topic of energy.

The Growing Relevance of Energy Geography (2000–2011)

Many books published since 2000 could reasonably be listed under the heading "geography of energy." Because I am personally attracted to titles that combine landscape with words like *energy* or *power*, let me point readers first to *Landscapes of Energy*, a collection of short essays and evocative photographs (Ghosn 2009). Although one might assume this slim book is of minor significance, it represents a growing theme in geography.[5]

Moving from the diminutive to the massive, I next wish to recognize a series of publications from Cutler Cleveland and associates such as Robert Kaufman at Boston University. We start with the monumental *Encyclopedia of Energy*, with its 500 authors and 5,376 pages (Cleveland 2007), scale back to the *Concise Encyclopedia of the History of Energy* (Cleveland 2009), and culminate with the *Dictionary of Energy* (Cleveland and Morris 2009). A fourth book, *The Future of Energy*, is now in preparation with Cleveland's associate, Adil Najam. These valuable and impressive compilations enhance our appreciation and understanding of energy and deserve to be in every academic library. That the authors have several ties to geography is, I think, an important clue as to how appropriate geography is for the study of energy.

The remaining books recognize the shifts that have been occurring over the years as the geographical analysis of energy has evolved (Figure 1). My order of presentation follows the list in the right column of Table 1.

Energy geography has many ties to climate change, a topic Knight addressed in this space in the previous special issue of this journal (Knight 2010). One might add to Knight's list the work of the Intergovernmental

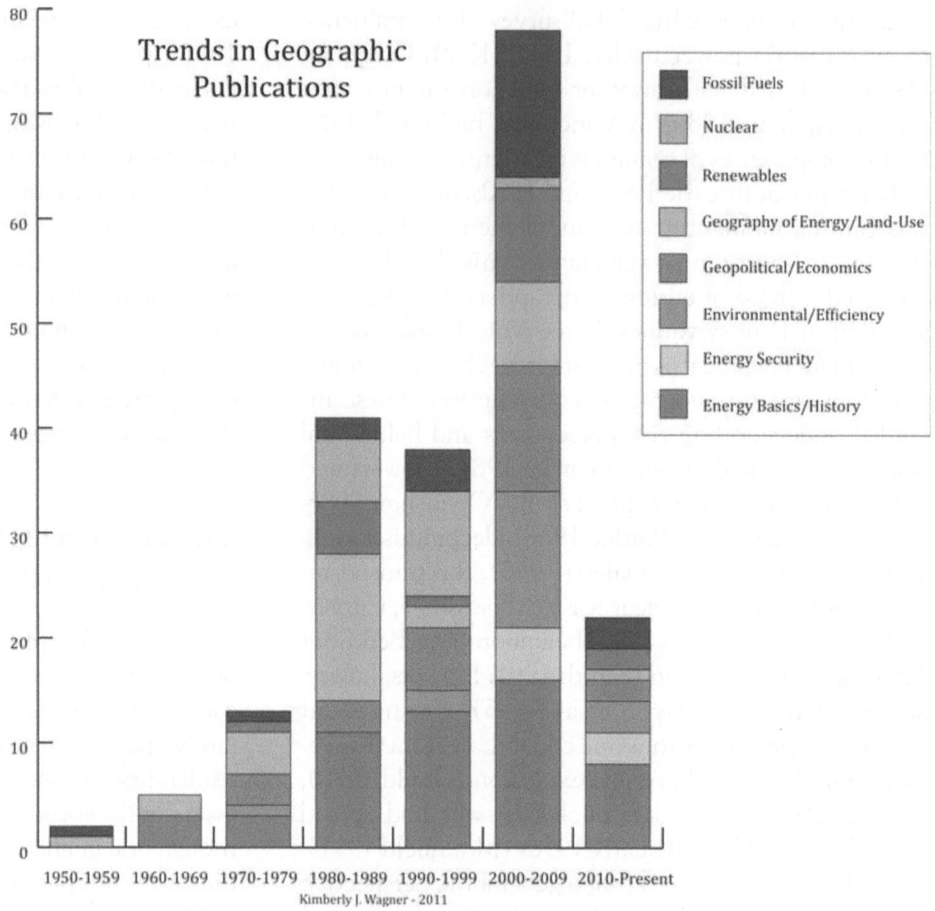

Figure 1. Trends in books on the geography of energy ($N = 203$).

Panel on Climate Change (Parry et al. 2007), for several geographers had a hand in its creation, including Thomas Wilbanks, a former president of the AAG. I provide a personal nod of thanks in his direction because he has published dozens of papers and book chapters on energy, and he is a member of a rather small club within the discipline that has maintained high-profile professional activity over several decades.[6] It is also appropriate in the context of climate change to recognize that the cheapest, fastest, and easiest way to allay some of our concerns about climate change is through energy efficiency, something Reddy and his colleagues (2009) address in a concise volume of several persuasive papers.

Table 1. Changing the geography of energy

1950–1999	2000–2011
Discoveries	Climate change
Facility siting	Energy poverty and social justice
Geography of energy	Energy security
Land use	Geography of energy
Nuclear power	Renewable energy
Risk assessment	Urban environments

Note: Sample included 203 books.

Among the more appealing recent books on the costs of our insatiable appetite for carbon-based energy is *The New Energy Crisis* (Chevalier 2009). It highlights key energy challenges by taking an appealing regional approach that includes examples from Asia, the Caspian Basin, Russia, the Middle East and North Africa, and the United States. It is not surprising that among several books that examine these same key regions, this one presents its material with a strong understanding of the links between geopolitics and energy; almost all of the authors are affiliated with the Centre de Géopolitique de l'Energie et des Matières Premières.

The themes highlighted in the last two chapters in Chevalier's book—energy poverty and energy security—have been taken up in great detail by other geographers. On the former topic, geographers have contributed two books that stand out: Watts's (2010) *The Curse of Black Gold* and Buzar's (2007) *Energy Poverty in*

Eastern Europe. Although both incorporate a regional focus, the conditions and causes they investigate are vastly different. Watts's treatment of Nigeria centers on the "resource curse," whereby impoverished people live atop energy resources of great value, a juxtaposition not uncommon in such other countries as Ecuador, Angola, and the newly formed South Sudan. Nigeria, however, presents the most egregious example of exploitation. This is a country where people are routinely subjugated and intimidated, often being made to fear for their lives; in fact, Watts himself was recently shot (and has since fully recovered) while conducting field work in the Niger Delta.

Illustrating that energy poverty can exist closer than we think, Buzar focuses not on Africa, India, or South America but on Europe, albeit the postsocialist countries of Eastern Europe. Instead of facing the conditions of indentured servitude and violence common in West Africa, Buzar's volume offers more prosaic examples, including households that cannot afford to heat their homes. He examines different spatial contexts and scales, compares them with other parts of the world, and links household-level deprivation with broader organizational and political dynamics. As Buzar argues, there is more to the shortages than the vagaries of Russian gas supplies; he finds a direct link between the energy crises experienced by the region and the social aspects of domestic energy use.

Like Buzar and Watts, Le Billon (2006) stresses that long-term political stability depends more on establishing the basic needs and security for local populations than simply on increasing energy supplies.[7] The absence of this type of thinking is one of the reasons many of the oil-rich regions Le Billon investigates—including the Persian Gulf, the Caspian Basin, and west-central Africa—do not make more progress toward the more equitable distribution of wealth. Le Billon knows that social justice is not a topic of interest to everyone, if not because of the greed that often permeates oil-rich states then because it poses a complex problem that few, especially politicians, are willing to try to solve (Le Billon 2005, 2006).

For those who have an interest in pursuing the intricacies of social theory in the context of the politics of oil, I recommend *Space, Oil and Capital* (Labban 2008). Labban explores what he calls the process of intercapitalist competition in the global economy. Unlike other studies of the geopolitical struggle for oil, Labban's book emphasizes the origin of the struggle for oil in intercapitalist competition. He looks at the instrumental role that "the production of global space, through the dialectical tensions between transnational oil corporations and resource-owning states, plays in determining the profitability of oil production and the availability of oil in the world market" (Labban 2008, 1).

Energy security serves as scaffolding for the geographic discussion of supply, demand, cost, environmental impacts, and energy transitions. Energy security can involve spatial scales from house to globe, temporal scales from the immediate to the distant future, and it can be applied in myriad ways. It is used to justify drilling in fragile environments as well as military interventions wherever there is promise of great profits. The attention of geographers to this subject has been as current as that of any other discipline. In the most recent case, a three-day workshop in Singapore was convened in late 2009 to discuss possible indicators of energy security. Several geographers were involved in this conference and the discussion was suffused with geography, including climate change, sustainable development, public policy options, environmental costs, energy poverty and social development, and energy efficiency. The greatest value of the book that resulted will be if it attracts more geographers to consider energy security (Sovacool 2010).

After a hiatus of more than two decades, our new century has again brought us a pair of books entitled *The Geography of Energy* (Brücher 2009; Mérenne-Schoumaker 2008). Brücher explores what he calls the "phenomenon" of energy, the concept of "energy geography," the preindustrial phase of the production of renewable energies from the surface, the industrial phase of nonrenewable energies for supplying large areas, the postindustrial phase of modern forms of the utilization of renewable energy supplies, energy conservation and avoidance of emissions, the rather unknown impacts of developing countries, and new renewable energy systems. Mérenne-Schoumaker's book covers approximately the same range of material on resources, shortages, environmental impacts, and energy conservation, but it has more didactic targets (with many maps, statistics, and diagrams), is less critical of the political aspects of energy, and is less interested in the lessons of history (W. Brücher, personal communication, 3 March 2011). Brücher remains skeptical, however, that progress in the development of energy resources, including renewable resources, will satisfy the growing energy hunger of the developing and transitional countries. That both authors are European suggests the growing recognition of the important interactions of geography and energy in that part of the world, something we in North America should note.

Geographers have a natural interest in renewable energy siting and development, and the reason is clear: Several resources—including geothermal, wind, hydro, and tidal—are site specific, meaning that associated land-use conflicts have inherently more in common with these resources than with those that can be moved around more readily like oil or coal. Attention to certain other renewables has been picking up in recent years, including studies of forests (Solomon and Luzadis 2009) and biomass concentrations, but not yet for solar energy development despite its potential for disturbing land-use patterns.

Of all such renewables, wind energy has been drawing the most public fire. Several factors are in play, including the large size of wind turbines, their inconsistent motion and noise, and their threats to birds, bats, and other wildlife. The dominant issue, however, is their visual presence. Drawing together the work of geographers, landscape architects, and historians from the United States and Europe, *Wind Power in View: Energy Landscapes in a Crowded World* was the first book to address land-use conflicts related to visual esthetics in detail (Pasqualetti, Gipe, and Righter 2002).

Despite the American genesis of *Wind Power in View*, renewable energy development has been attracting more attention from geographers in Europe. Two books stand out in this regard, none of them technical (Mitchell 2009). Together, these three books emphasize that energy is a social issue with a technical component, rather than the other way around. This is yet another fertile field for geographers to plow.

Any review of energy will note that the important topic of metropolitan form and function has resulted in very few books by geographers, all of them more than twenty years old (Beaumont and Keys 1982; Owens 1986, 1991). Of several others, none were organized by geographers, even though geographers were involved preparing a few of their chapters. Again, they date back at least two decades (Burchell and Listokin 1975, 1982; Harwood 1977; Burby and Bell 1978; Morris 1982; Cope, Hills, and James 1984; Cullingworth 1990). But this topic has recently resurfaced in the public eye as people once again begin to realize the importance of energy use in urban areas. We have Peter Droege to thank for calling attention to how we can make our cities and suburbs more efficient in accommodating the integration of renewable energy (Droege 2006, 2008). He is not a geographer, but he certainly thinks like one. I recommend both books in the hope that geographers will gather their talents and make their own contributions to this rich interdisciplinary subject.

For those geographers preparing for an excursion into the field of energy studies, they would do well to return to the contributions that continue to flow from the hand of Smil. His energy books include those that serve as introductions to energy and individual energy resources (Smil 1998, 2006, 2008b), one that focuses specifically on the societal dimensions of energy (Smil 2008a), another that alludes to the coming decision points in our development and use of energy (Smil 2003), and two others that I now want to highlight.

The first of these is *Energy Myths and Realities* (2010), a particularly engaging reality check for those who wish to imagine or create energy futures. It is doubly interesting because of its contrasts with a similarly titled volume published two years earlier by Sovacool and Brown (2007). Smil is intent on "debunking" myths as he perceives them, and that in itself makes for thoughtful reading. What is more interesting is that these myths are often not only different, but their treatment has frequently raised arguments that contradict each other. One should consider both, because taken together, they provide thoughtful glimpses into attitudes about energy that serve as the basis for policy decisions and public expenditures that sometimes go awry.

The second book by Smil, *Energy Transitions* (2010), digs more deeply into some of the topics he has touched on in earlier books. In *Transitions*, he outlines the difficult years ahead that face us as we are required to adjust to the limitations of energy resources of the past and the hopeful, if challenging, energy resources of the future. *Transitions* is a sobering appraisal by a geographer who understands and appreciates both the technical and societal tasks we are soon to confront.

Finally, I am pleased to recommend the massive two-volume set just published by Danny Harvey of the Department of Geography at the University of Toronto (Harvey 2010). His is an impressive achievement, one that summarizes the problems of energy supply, the costs of its usage, and the paths we might be taking into the future. His overarching theme is "energy and the new reality," and I know of no publication by anyone, let alone a geographer, that more effectively lays out what that reality will entail.

Postscript

I began this overview by suggesting that energy and geography were, by some estimates, odd bedfellows. I hope this essay has set that idea aside, because not only do geography and energy make an ideal couple—they have also conceived intelligent and insightful progeny of value to us all. But before this becomes a gushing, self-congratulatory encomium, I should also add a note of disappointment that my fellow geographers in the United States have not contributed more. If you have read this far, you have surely detected that many of the books I have noted—especially on the themes of renewables, energy landscapes, land use, and cities—have either been produced by geographers in other countries or by energy experts in other disciplines. But one must not grouse about this situation too loudly: The literature reviewed here constitutes a major contribution and a proud accomplishment, whatever its source. Moreover, it makes obvious that just as the world is the domain of the geographer, energy is the wealth of the world. The two cannot be separated.

Acknowledgments

Thanks are due to the following geographers for suggesting some of the books considered in this essay: Andy Blowers, Marilyn Brown, Wolfgang Brücher, Stefan Buzar, Michael Heiman, Matt Huber, Scott Jiusto, Greg Knight, Peter Muller, Susan Owens, Vaclav Smil, Barry Solomon, Benjamin Sovacool, Vita Valiūnaitė, Tom Wilbanks, and Karl Zimmerer. I am also grateful to Kimberly J. Wagner for assisting with the figure.

Notes

1. A somewhat more fugitive literature flowed as technical reports from the national laboratories. This was particularly true of Oak Ridge National Laboratory in the late 1970s and 1980s, when about a dozen energy geographers were active there at the same time, including Tom Wilbanks, Russ Lee, Marilyn Brown, Jerry Dobson, Ed Hillsman, David Greene, John Sorensen, and Bob Honea. Although their reports were not published commercially, they demonstrated the value of incorporating the work of geographers in energy analysis.
2. There is a comparable group in Europe organized within the German Society for Geography, one that focuses specifically on the geography of energy. See http://www.geographische-energieforschung.de/.
3. For the 1978–1979 academic year, Chapman was a Fulbright Scholar at the Center for Energy Studies at the University of Texas, Austin.
4. As if to reinforce this idea, the reader is referred to a listing of books on energy geography by Jean-Marie Chevalier found on the French Web site for Amazon books at http://www.amazon.fr/s?_encoding=UTF8&search-alias=books-fr&field-author=Jean-Marie%20Chevalier#.
5. See also a special issue on energy landscapes published in the journal *Landscape Research*, which includes an introduction by Nadaï and van der Horst (2010).
6. Other geographers with long and consistent records of publication include Andrew Blowers, Marilyn Brown, Wolfgang Brücher, John Chapman, David Greene, Bruce Hannon, Robert Kaufmann, Michael Kuby, Gerald Manners, Bernadette Mérenne-Schoumaker, Peter Odell, Martin Pasqualetti, Mathias Ruth, Vaclav Smil, Barry Solomon, and Derek Spooner. A somewhat younger generation of scholars, too numerous to list in full here, includes Rob Bailis, Patrick Devine-Wright, Michael Heiman, Scott Jiusto, Anelia Milbrandt, Alain Nadaï, Dan van der Horst, and Charles Warren.
7. The reader is also referred to the many publications on the geography of energy by Robert E. Ebel at the Center for Strategic and International Studies.

Key Words: energy, geography, resources, spatial.

References

Beaumont, J. R., and F. Berkhout. 1991. *Radioactive waste: Politics and technology*. London and New York: Routledge.
Beaumont, J. R., and P. Keys. 1982. *Future cities: Spatial analysis of energy issues*. New York: Wiley.
Black, B. 2000. *Petrolia: The landscape of America's first oil boom*. Baltimore: The Johns Hopkins University Press.
Blowers, A. T., M. Clark, and D. Smith, eds. 1991. *Waste location: Spatial aspects of waste management, hazards and disposal*. London: Routledge.
Blowers, A. T., D. Lowry, and B. D. Solomon. 1991. *The international politics of nuclear waste*. New York: Palgrave Macmillan.
Blowers, A. T., and D. Peppers. 1987. *Nuclear power in crisis*. London and New York: Routledge.
Brücher, W. 1997. Mehr Energie! Plädoyer für ein vernachlässigtes Objekt der Geographie [More energy! A plea for a neglected subject of geography]. *Geographische Rundschau* 49:330–35.
———. 2004. Energy, geography of. In *International encyclopedia of the social and behavioral sciences*, ed. N. J. Smelser and P. B. Baltes, 4520–23. Detroit, MI: Gale-Cengage.
Burby, R. J., III, and A. F. Bell. 1978. *Energy and the community*. Cambridge, MA: Ballinger Publishing Co.
Burchell, R. W., and D. Listokin, eds. 1975. *Future land use: Energy, environmental and legal constraints*. New Brunswick, NJ: Rutgers University, Center for Urban Policy Research.
———. 1982. *Energy and land use*. New Brunswick, NJ: Rutgers University, Center for Urban Policy Research.
Calzonetti, F. J., and B. Solomon, eds. 1985. *Geographical dimensions of energy*. Dordrecht, The Netherlands: D. Reidel.
Calzonetti, F. J. (with M. S. Eckert). 1981. *Finding a place for energy: Siting coal conversion facilities*. Resource Publications in Geography. Washington, DC: Association of American Geographers.
Chapman, J. D. 1989. *Geography and energy: Commercial energy systems and national policies*. New York: Longman.
Chapman, K. 1975. *North Sea oil and gas*. London: David & Charles.
———. 1991. *The international petrochemical industry*. Malden, MA: Blackwell.

Cleveland, C. J. 2009. *Concise encyclopedia of the history of energy*. Oxford, UK: Elsevier Science.

Cleveland, C., and C. Morris, eds. 2009. *Dictionary of energy: Expanded edition*. Amsterdam and New York: Elsevier Science.

Condee, W. F. 2005. *Coal and culture: Opera houses in Appalachia*. Athens: Ohio University Press.

Cook, E. F. 1976. *Man, energy, society*. San Francisco: Freeman.

———. 1977. Energy, the ultimate resource? Resource Papers For College Geography, No. 77–4, Washington, DC: Association of American Geographers.

Cope, D. R., P. Hills, and P. James, eds. 1984. *Energy policy and land-use planning: An international perspective*. Elmsford, NY: Pergamon.

Croissant, M. P., and B. Aras. 1999. *Oil and geopolitics in the Caspian Sea Region*. Westport, CT: Praeger.

Cullingworth, J. B., ed. 1990. *Energy, land, and public policy*. New Brunswick, NJ: Transaction Books.

Dienes, L., I. Dobozi, and M. Radetzki. 1994. *Energy and economic reform in the former Soviet Union: Implications for production, consumption and exports*. New York: Palgrave Macmillan.

Dienes, L., and T. Shabad. 1979. *The Soviet energy system: Resource use and policies*. New York: Wiley.

Energy and Environment Specialty Group. 2011. Welcome. http://www.eeel.centre-cired.fr/spip.php?article13 (last accessed 6 April 2011).

Farrell, B. H. 1962. *Power in New Zealand: A geography of energy resources*. Wellington, New Zealand: A.H. & A.W. Reed.

Fernie, J. 1980. *The geography of energy in the United Kingdom*. Englewood Cliffs, NJ: Prentice-Hall.

Flynn, J., J. Chalmers, D. Easterling, R. Kasperson, H. Kunreuther, D. K. Mertz, A. Mushkatel, K. D. Pijawka, and P. Slovic. 1995. *One hundred centuries of solitude: Redirecting America's high-level nuclear waste policy*. Boulder, CO: Westview.

George, P. 1950. *Géographie de l'energie* [Geography of energy]. Paris: Genin.

Gever, J., R. Kaufmann, and D. Skole. 1991. *Beyond oil: The threat to food and fuel in the coming decades*. Boulder: University of Colorado Press.

Gould, P. 1990. *Fire in the rain: The democratic consequences of Chernobyl*. Baltimore: Johns Hopkins University Press.

Guyol, N. B. 1971. *Energy in the perspective of geography*. Englewood Cliffs, NJ: Prentice-Hall.

Hall, C., C. Cleveland, and R. Kaufmann. 1992. *Energy and resource quality: The ecology of the economic process*. Boulder: University of Colorado Press.

Harwood, C. C. 1977. *Using land to save energy*. Cambridge, MA: Ballinger.

Hart, D. 1980. *The Volta River project: A case study in politics and technology*. Edinburgh, UK: Edinburgh University Press.

Hoffman, G., and L. Dienes. 1985. *European energy challenges: East and West*. Durham, NC: Duke University Press.

Hogg, A., and A. M. Hutcheson, eds. 1975. *Scotland and oil*. 2nd revised ed. London: Oliver & Boyd.

Hooson, D. J. M. 1965. *Introduction to studies in Soviet energy*. Vancouver, BC, Canada: Tantalus Research.

Hughes, T. P. 1983. *Networks of power: Electrification in western society, 1880–1930*. Baltimore: Johns Hopkins University Press.

Jacob, G. 1990. *Site unseen: The politics of siting a nuclear waste repository*. Pittsburgh, PA: University of Pittsburgh Press.

Knapper, C., L. Gertler, and G. Wall. 1983. *Energy, recreation and the urban field*. Waterloo, ON, Canada: University of Waterloo, Department of Geography.

Knight, G. 2010. Climate change: The health of a planet in peril. *Annals of the Association of American Geographers* 100:1036–45.

Kuby, M. 1995. Investment strategies for China's coal and electricity delivery system. Report No. 12687-CHA. Washington, DC: World Bank.

———. 1996. *Population growth, energy use, and pollution: Understanding the driving forces of global change*. Washington, DC: Association of American Geographers.

Labban, M. 2008. *Space, oil, capital*. London: Routledge.

Lakshmanan, T. R., and B. Johansson, eds. 1985. *Large-scale energy projects: Assessment of regional consequences*. New York: North Holland.

Lakshmanan, T. R., and P. Nijkamp, eds. 1980. *Economic, environmental and energy interactions: Modeling and policy analysis*. Hingham, MA: Martinus Nijhoff.

———. 1983. *Systems and models for energy and environmental analysis*. London: Gower.

Le Billon, P., ed. 2005. *The geopolitics of resource wars: Resource dependence, governance and violence*. London and New York: Routledge.

Manners, G. 1964. *The geography of energy*. London: Hutchinson University Library.

———. 1981. *Coal in Britain: An uncertain future*. London: Allen & Unwin.

Melosi, M. V., and J. A. Pratt. 2007. *Energy metropolis: An environmental history of Houston and the Gulf Coast*. Pittsburgh, PA: University of Pittsburgh Press.

Morgan, W. B., and R. P. Moss. 1981 *Fuelwood and rural energy, production and supply in the humid tropics*. Dublin, Ireland: Tycooling International.

Morris, D. 1982. *Self-reliant cities: Energy and the transformation of urban America*. San Francisco: Sierra Club Books.

Mounfield, P. R. 1991. *World nuclear power*. London: Routledge.

Nadaï, A., and D. van der Horst. 2010. Introduction: Landscapes of energies. *Landscape Research* 35 (2): 143–55.

National Public Radio. 2011. The cost of a soldier deployed in Afghanistan. *Marketplace* 22 February. http://marketplace.publicradio.org/display/web/2011/02/22/am-the-cost-of-a-soldier-deployed-in-afghanistan/ (last accessed 6 April 2011).

Nye, D. E. 1990. *Electrifying America: Social meanings of a new technology, 1880–1940*. Cambridge, MA: MIT Press.

———. 1998. *Consuming power: A social history of American energies*. Cambridge, MA: MIT Press.

———. 1999. *Technologies of landscape: From reaping to recycling*. Amherst: University of Massachusetts Press.

Odell, P. 1977. *Energy: Needs and resources (Aspects of geography)*. 2nd revised ed. London: Macmillan.

———. 1986. *Oil and world power*. 8th revised ed. Harmondsworth, UK: Penguin.

Odell, P. R., and K. E. Rosing. 1980. *The future of oil*. New York: Nichols.

Openshaw, S. 1986. *Nuclear power: Siting and safety*. London and New York: Routledge & Kegan Paul.

Openshaw, S., S. Carver, and J. Fernie. 1989. *Britain's nuclear waste: Safety and siting*. London: Belhaven Press.

O'Riordan, T., R. Kemp, and M. Purdue. 1988. *Sizewell B: An anatomy of the inquiry*. London: Macmillan.

Owens, S. 1986. *Energy, planning and urban form*. London: Pion.

———. 1991. *Energy-conscious planning*. London: Council for the Protection of Rural England.

Parry, M. L., O. F. Canziani, J. P. Palutikof, P. J. van der Linden, and C. E. Hanson, eds. 2007. *Contribution of Working Group II to the Fourth Assessment Report of the Intergovernmental Panel on Climate Change*. Cambridge, UK: Cambridge University Press.

Pasqualetti, M. J. 1986. The dissemination of geographic findings on nuclear energy. *Transactions of the Institute of British Geographers* 11:326–36.

———, ed. 1990. *Nuclear decommissioning and society: Public links to a new technology*. London and New York: Routledge.

Pasqualetti, M. J., and K. D. Pijawka, eds., 1984. *Nuclear power: Assessing and managing hazardous technology*. Boulder, CO: Westview.

Platt, H. L. 1991. *The electric city*. Chicago: University of Chicago Press.

Pratt, W. E., and D. Good, eds. 1950. *World geography of petroleum*. American Geographical Society Special Publication No. 31. Princeton, NJ: Princeton University Press.

Pryde, P. R. 1983. *Nonconventional energy resources*. New York: Wiley.

Sawyer, S. 1986. *Renewable energy: Progress and prospects*. Washington, DC: AAG Resource Publications.

Sevette, P. 1976. *Geographie et économie comparee de l'énergie* [Geography and comparative economics of energy]. Grenoble, France: Institut Economique et Juridique de l'Énergie, Université de Grenoble.

Smil, V. 1976. *China's energy*. Westport, CT: Praeger.

———. 1982. *Energy analysis in agriculture: An application to U.S. corn production*. Boulder, CO: Westview.

———. 1983. *Biomass energies*. New York: Plenum.

———. 1987. *Energy, food, environment: Realities, myths, options*. New York: Oxford University Press.

———. 1988. *Energy in China's modernization*. Armonk, NY: Sharpe.

———. 1991. *General energetics: Energy in the biosphere and civilization*. New York: Wiley.

———. 1994. *Energy in world history*. Boulder, CO: Westview.

———. 1998. *Energies: An illustrated guide to the biosphere and civilization* Cambridge, MA: MIT Press.

———. 2003. *Energy at the crossroads: Global perspectives and uncertainties*. Cambridge, MA: MIT Press.

———. 2006. *Energy: A beginner's guide*. London: OneWorld Publications.

———. 2008a. *Energy in nature and society: General energetics of complex systems*. Cambridge, MA: MIT Press.

———. 2008b. *Oil: A beginner's guide*. London: OneWorld Publications.

Smil, V., and W. Knowland, eds. 1980. *Energy in the developing world*. New York: Oxford University Press.

Solomon, B. D., and V. A. Luzadis, eds. 2009. *Renewable energy from forest resources in the United States*. London and New York: Routledge.

Solomon, B. D., and M. J. Pasqualetti. 2004. History of energy in geographic thought. In *Encyclopedia of energy*, ed. C. Cleveland, 831–42. San Diego, CA: Academic Press.

Solomon, B. D., M. J. Pasqualetti, and D. A. Luchsinger. 2004. Energy geography. In *Geography in America at the dawn of the 21st century*, ed. G. Gaile and C. Willmott, 302–13. New York: Oxford University Press.

Sovacool, B., ed. 2010. *Routledge handbook for energy security*. London: Routledge.

Sovacool, B. K., and M. A. Brown, eds. 2007. *Energy and American society—Thirteen myths*. Dordrecht, The Netherlands: Springer.

Wagstaff, H. R. 1974. *A geography of energy*. Dubuque, IA: Brown.

Williams, J. C. 1997. *Energy and the making of modern California*. Akron, OH: University of Akron Press.

Index

Act is a Feed-in Tariff (FIT) 197
AES-Changuinola representatives 167
Ai, B.: *et al* 19
Amazônia: agriculture 230; bioeconomy 230–1; biofuel impact 228–36; definition **235**; deforestation increment *232*; Green Redoubt 232–4; indirect land use change *232*; landscape cascades 231–2
American Society of Civil Engineers 13
anthropogenic climate change 52
Appalachia 108
Archer Daniels Midland (ADM) 185
Ashby, M.F. 36
assessments 2

Bagchi-Sen, S.: and Kedron, P. 181–93
Bailis, R.: and Baka, J. 126–34
Baka, J.: and Bailis, R. 126–34
Barthe, Y.: Callon, M.: and Lacoumes, P. 173
Benjamin, W. 117, 120
Bichler, S.: and Nitzan, J. 118
Billion-Ton Annual Supply study 42
Bilsborrow, R. 220
bioenergy 4
bioenergy policy: marginal land 152–3
biofuel-livestock linkages 133
biofuels 130–1; economy 126–34; governance 127–30; mandates *128*; marginal lands 132–3; sustainability 131–3
Biomass Crop Assistance Program (BCAP) 152
biomass fuel 32
biomass resources 198–9
biophysical definitions 153
black carbon (BC) 218
"Black Giant" 119–22
Boardman, B. 249
Bocas del Toro: dam locations **166**; hydro development 164–8
Bouzarovski, S.: and Bassin, M. 82–90
Brannstrom, C.: Jepson, W.: and Persons, N. 138–49
Brazil **233**
Bridge, G. 182
British Columbia: hydroelectricity 174–9; Peace River **175**
Buzar, S. 249

California 207–8; wind turbines **208**
Callon, M.: Lacoumes, P.: and Barthe, Y. 173
Canada 33, 263; Site C hydroelectric dam 172–9
Canadian Census of Agriculture 198
Cape Cod *212*
Cape Wind 209
carbon 32
carbon balance: spatial assessment 66
carbon emissions 52
carbon estimation 65–7
carbon footprints 37–8
carbon labels 30
carbon loss 34–5
carbon markets 4
Cass, N.: and Walker, G. 85
Castellanos, E.J.: Elias, S.: Taylor, M.I.: and Moran-Taylor, M.J. 217–25
Census of Architecture 198
Central America *241*
21st Century Electric Energy Plan 267
Cervantes, A.: and Le Billon, P. 118
Chan: 75 Dam 167; spatial control 166–8
Chapin, T.S.: Horner, M.W.: and Zhao, T. 63–9
China: carbon footprints **35**; CO2 emissions 94–102; coal 97; coke production 101; economic reforms 94; energy 94–102; energy background 95–6; energy consumption **97**; energy flow diagram 98–100, **98**, **99**; energy production and consumption trends **96**; energy supply 94–102; energy system changes 96–8; fossil fuels 97; hydropower 97; oik 97; oil imports 97; paper LCI **31**; technological change 98
Clean Air Act (1990) 187
Clean Development Mechanism (CDM) 162
Clean Energy Act 177
climate change: energy regulation 74–80; litigation 77–80
CO2 emissions: top eight countries **95**
coal 3
Cohen, S.: *et al* 85
Community Advocate 147
Complementary Index of Wind and Solar Radiation (CIWS) **23**
Consolidated Farm and Rural Development Act 185
Consumers Power 265
consumption 5–6
Cooke, P. 183
Cope, M.A.: McLafferty, S.: and Rhoads, B.L. 151–9
Copenhagen Accords 75
Corn Belt 191
Corporate Average Fuel Economy (CAFE) 12
Cupples, J. 238–45

Daily Oklahoman **121**
de Beurs, K.M.: and Henebry, G.M. 45
Department of Defence (U.S.) 12
diesel engines 13
direct land use 131–2

Disenchanted About Tax Abatements 146
Dusyk, N. 172–9

East Central Illinois: perennial energy grasses 151–9
East Texas 120
electricity 240–1
electricity generation 197
Elias, S.: Taylor, M.I.: Moran-Taylor, M.J.: and Castellanos, E.J. 217–25
Elwood, S. 159
emissions control 79–80
energy infrastructure 67
energy modeling 2
Energy Policy Act (2005) 187, 230
energy research 63–9
environmental degradation 17
Ethanol Promotion and Information Council (EPIC) 191
Europe 139

Fairhead, J. 117
Farm Bill (2008) 152
farmers decisions **153**
Favorable Toward Tax Abatements 145, 146–7
Felder, D. 19
Finley-Brook, M.: and Thomas, C. 162–8
Fitzpatrick, D.R. **122**
Florida 51–60; carbon assimilation **57**; carbon balance **57**; carbon balance study 53–9; energy and fuel consumption 56; household consumption *59*; household energy consumption **56**; map location **54**; settlement category *59*; settlement densities **58**; vegetation carbon assimilation 55
Food and Agriculture Organization (FAO) 35
forest ecologies 32
Forest Resources of Ontario 198
fossil fuel landscapes 3–4
France 12, 13
Fridly, D.G.: and Sinton, J.E. 96
Fritschi, F.B.: Stacey, G.: Yang, Z.: and Wang, C. 41–9
fuel types 34
Furlong, K. 242

Gaventa, J. 106
geographic information systems (GIS) 64
geography 1; of energy 270–7
GIScience literature 66
GIScience research 63–9
Glaser, L.S. 84
Glomsrod, S.: and Wei, T. 95
Goldblatt, D.L. 85
Gordon, J.S.: and Woods, B.R. 105–13
Graham, S.: and Marvin, S. 250, 255
green authoritarianism 165–6
Green Energy Act 201
Green Energy and Green Economy Act 197
green energy regulation 76–7
greenhouse gas (GHG) emissions 127; reduction criteria *129*
Grigor'ev, L. 89
Guatemala 217–25; biomass stoves 219; development 219–20; environment 219–20; firewood and forests 219; forest change 220–1; forest cover *224*; household survey outcomes *222*; international migration 219–20; Western Highlands 219–24

Harrison, C.: and Popke, J. 248–59
Harvey, D. 116
He, C.: Trapido-Lurie, B.: Moore, N.: and Kuby, M. 94–102
Heltberg, R. 223
Henebry, G.M.: and de Beurs, K.M. 45
Hinchcliffe, S. 242
Holifield, R. 239
Honda 12
Horner, M.W.: Sulik, J.: and Zhao, T. 51–60
Horner, M.W.: Zhao, T.: and Chapin, T.S. 63–9
household energy conservation 67–9
Howell, J.P. 261–8
Huber, M.T. 115–23
Hudson's Hope 174
human geographers 83
human rights 162–8
hybrid systems 17
hydropower 5, 163–4

Ickes, H. 121
identity 84–6
India 132
indirect land use change 131–2
Ingold, T. 243
initiatives 1
innovation 4
insolation resources 16–27
Inter-American Commission on Human Rights (IACHR) 165
Inter-American Human Rights Court (CIDH) 168
Intergovernmental Panel on Climate Control 32
International Energy Agency (IEA) 94
international financial institutions (IFIs) 239
Iowa: biofuel landscape affecting factors *186*; biorefineries characteristics *189*; biorefineries distribution *185*; ethanol industry value chain 190–2, **191**; ethanol policies *187*, *188*; ethanol production 183–93; local biorefiners 192; POET Biorefining 190; producer firms 188–90; R&D 190; renewable energy literature *184*; renewable energy sector 181–93; support networks 191–2
Isle of Lewis 209–10, **210**, *212*; Standing Stones **211**
Isthmus of Tehauantepec 210–11

Jackson, J.B. 213
Jacobsen, M.Z.: Stoutenberg, E.D.: and Jenkins, N. 19
Jacobson, M.Z. 19
Japanm 13
Jasanoff, S.: and Kim, S.-H. 84
Jenkins, N.: Jacobsen, M.Z.: and Stoutenberg, E.D. 19
Jepson, W.: Persons, N.: and Brannstrom, C. 138–49
job creation 105–13
Jones, J.P.: Marston, S.A.: and Woodward, K. 244

Kedron, P.: and Bagchi-Sen, S. 181–93
Kennedy Jr. R.F. 209
Kern, F.: and Smith, A. 85
Kim, S.-H.: and Jasanoff, S. 84
Kuby, M.: He, C.: Trapido-Lurie, B.: and Moore, N. 94–102
Kuby, M.: *et al* 3
Kyoto Protocol 52

La Venta: wind turbines **212**
Labban, M. 118

Lacoumes, P.: Barthe, Y.: and Callon, M. 173
Land-Based Wind Welcomers 145–6; factor loadings *145*
Latin America 164, 220
Latour, B. 242
Le Billon, P.: and Cervantes, A. 118
Levine, M. 95
Lewis Peatlands Special Protection Area 210
Li, W.: Stadler, S.: and Ramakumar, R. 16–27
life cycle assessment (LCA) 30
life cycle inventory (LCI) 30
Loftus, A. 240
Low Income Home Energy Assistance Program (LIHEAP) 250
Lowlands of Oaxaca *212*

Ma, H.: and Oxley, L. 95
Mabee, W.E.: and Mirck, J. 196–203
McLafferty, S.: Rhoads, B.L.: and Cope, M.A. 151–9
Macon County 154–8; farming 157; GIS-aided focus group **158**; land cover **157**; survey responses *156*
Macon County Soil and Water Conservation System 154
Malecki, E.J.: and Oinas, P. 182
marginal land: bioenergy policy 152–3
Marston, S.A.: Woodward, K.: and Jones, J.P. 244
Marvin, S.: and Graham, S. 250, 255
Massey, D. 173
Mexico **211**
Michigan: electricity 261–8; electricity regulation 264; electricity study 262–8
Michigan Electricity Options Study 266–7
Michigan Public Service Commission (MPSC) 262
mining permits 110
Moore, N. Kuby, M.: He, C.: and Trapido-Lurie, B. 94–102
Moran-Taylor, M.J.: Castellanos, E.J.: Elias, S.: and Taylor, M.I. 217–25
mountaintop removal (MTR) 3, 105–13
Murau Energy Objectives (2015) 183
Murray, B. 120

Nantucket Sound: protest banner **209**
Naso community 163
National Energy Administration 94–5
National Program 78
National Wind Watch 213
natural resources management 153–4
neocolonial carbon projects 168
neoliberal economic reforms 164
Newell, J.P.: and Vos, R.O. 29–38
Nicaragua: activism 243–4; electricity 238–45; public utility privitization 239; urban communication 243–4
Nitzan, J.: and Bichler, S. 118
Nixon, R.M. 14
Nolan County 139–48; map **140**
Nordhauser, N.E. 120
normalization 21–2
North American tall grass prairie: biomass and economic impacts 48–9; crop map 46–8; energy crops 41–9; Moderate Resolution Imaging Spectroradiometer (MODIS) **43**; normalized difference vegetation index (NDVI) **44**; phenology-based decision tree **45**; study 43–4
North Carolina: energy poverty 248–59, *252*; energy poverty experiences 257–8; energy poverty study 250–8; energy provision 253–4; heating prices **255**; porous materiality 254–7
North Carolina Electric Membership Corporation Act (1935) 253
Norway 183

oil 1
Oil Compact Commission 122
oil fields: martial law 119–20
Oinas, P. 182; and Malecki, E.J. 182
Oklahoma 16–27; complementarity 17–18, 19–27; geographic analysis 23–7; hybrid projects 20; initial eigenvectors *24*; Mesonet station data **21**; principal components *24*; weighted regression model intercepts **25**; weighted regression model landscape dimension coefficients **26**; weighted regression model local R2 **26**; weighted regression model moisture dimension **25**; weighted regression model residuals **26**; weighted regression model temperature dimension coefficients **25**; wind and solar radiation data **22**; wind and solar radiation research 18–19
Omnibus Reconciliation Act (1980) 185
Ontario: bioconversion process 200; bioenergy production 196–203; bioenergy technologies 199–200; biomass contribution *201*, *202*; biomass productivity *198*; electricity generation 199–201; residential heating 199; study statistics **202**; thermochemical process 200; transport sector 200
Orlov, D. 88
Osnos, E. 95
Osofsky, H.M. 74–80
Oxley, L.: and Ma, H. 95

Palm Springs *212*
Panama: indigenous territories 162–8
Panama National Assembly 166
paper industry 29–38
paper production 34
Pasqualetti, M.J. 206–14, 270–7
Peace River 174, 178
Peace Valley Environmental Association (PVEA) 176
Perelman, L.J. 85
Permanent People's Tribunal 244
Persons, N.: Brannstrom, C.: and Jepson, W. 138–49
Plant Science Institute 192
policymakers 154
Popke, J.: and Harrison, C. 248–59
Potter, E. 85
private-state energy sector partnerships 164
ProAlcook program 230
production 2
Putin, V. 86–7

Q-method 141

Ramakumar, R.: Li, W.: and Stadler, S. 16–27
Red de Defensa de los Consumidores 241
regulation 75, 80; climate change 74–80
remote sensing 42
renewable energy 4–5, 17
Renewable Energy Directive 131
Renewable Fuels Association (RFA) 191
renewable portfolio standard (RPS) 18
resources 76
Retort 118
Rhoads, B.L.: Cope, M.A.: and McLafferty, S. 151–9

Russia 82–90; energy 86–90, 87–90

Sahin, A.Z. 18–19
San Bernadino Country 78–9
scarcity 115–23; violence 116–17
scientific community 64
Searchinger, T.: *et al* 32
Second Circuit 79
Shaw, R.H.: and Tackle, E.S. 18
Shi, D. 96
Simonov, K.V. 87
Sinton, J.E. 96; and Fridly, D.G. 96
Smil, V. 11–14
Smith, A.: and Kern, F. 85
sociotechnical networks 173–4
solar and wind technologies 17
space restriction 83
spatial scales 66
splintering urbanism 250
Stacey, G.: Yang, Z.: Wang, C.: and Fritschi, F.B. 41–9
Stadler, S.: Ramakumar, R.: and Li, W. 16–27
Star, S.L. 84
State Energy Profile, Michigan 262
Sterzinger, G. 182
Stoutenberg, E.D.: Jenkins, N.: and Jacobsen, M.Z. 19
Sulik, J.: Zhao, T.: and Horner, M.W. 51–60
Swyngedouw, E. 127–8

Tackle, E.S.: and Shaw, R.H. 18
Taylor, M.I.: Moran-Taylor, M.J.: Castellanos, E.J.: and Elias, S. 217–25
techno-economic paradigm (TEP) 182
technological innovation 4
temporal scales 66
Texas National Guard 120
Texas Railroad Commission (TRC) 119
Thompson, E.O. 119
Thompson, K. 255
top eight countries: CO2 emissions **95**
transport 34
Trapido-Lurie, B.: Moore, N.: Kuby, M.: and He, C. 94–102
Treaty 8 Tribal Association 177
Tsing, A. 117

United Kingdom 209
United States (U.S.) 3, 65, 230; carbon emissions 53; carbon footprints **35**; domestic oil extraction 11; electricity landscape 262; energy debates 11; energy supply 182; Midwest 152; oil imports 11–14; oil overproduction 118–19; paper LCI **31**; power plants 75
U.S. Census Bureau 109
U.S. Department of Energy (D.O.E.) 42
U.S. District Court 106
U.S. Energy Information Administration (EIA) 53, 86
U.S. Environmental Protection Agency (EPA) 75
U.S. Geological Survey Population Places **110**
U.S. Geological Survey (USGS) 108
U.S. Great Plains 138–9
U.S. Phenology Network 49
U.S. Senate Environment and Public Works Committee 106
U.S. Supreme Court 75
Utilities Commission Act 175, 177

Virginia: MTR **109**
Vos, R.O.: and Newell, J.P. 29–38

Wales 183
Walker, G.: and Cass, N. 85
Walker, R. 228–36
Wall Street Journal 116
Wang, C.: Fritschi, F.B.: Stacey, G.: and Yang, Z. 41–9
Weatherization Assistance Program (WAP) 250
Wei, T.: and Glomsrod, S. 95
West Texas: windpower 138–49
West Virginia: coal mining *112*; mining boundaries **111**; mining environment **108**; MTR counties *109*; MTR study 108–12
West Virginia Coal Association 107
Williams, C. 253–4
Williams Lake 199
wind energy 5, 138–49; landscapes 206–14; resources 16–27
wind turbines 207–13
Wind Welcomers 141–5; statements *142-4*
Wolters, J. 120
Woods, B.R.: and Gordon, J.S. 105–13
Woods, M. 250
Woodward, K.: Jones, J.P.: and Marston, S.A. 244
World Trade Organization (WT0) 96
World War II 263

Yang, Z.: Wang, C.: Fritschi, F.B.: and Stacey, G. 41–9

Zhang, X.: *et al* 46
Zhao, T.: Chapin, T.S.: and Horner, M.W. 63–9
Zhao, T.: Horner, M.W.: and Sulik, J. 51–60
Zimmerer, K.S. 1–6

T - #0699 - 101024 - C0 - 276/216/16 - PB - 9781138810372 - Gloss Lamination